Targeting Protein–Protein Interactions for Drug Discovery

Targeting Protein–Protein Interactions for Drug Discovery

Edited by Jian Zhang

WILEY-VCH

Editor

Prof. Jian Zhang
Shanghai Jiao Tong University
227 South Chongqing Road
Shanghai 200025
China

Cover Design: Wiley
Cover Image: © Juan Gaertner/
Shutterstock

Typesetting: Straive, Chennai, India

Library of Congress Card No.: applied for

British Library Cataloguing-in-Publication Data A catalogue record for this book is available from the British Library.

Bibliographic information published by the Deutsche Nationalbibliothek
The Deutsche Nationalbibliothek lists this publication in the Deutsche Nationalbibliografie; detailed bibliographic data are available on the Internet at <http://dnb.d-nb.de>.

Print ISBN: 978-3-527-35360-6
ePDF ISBN: 978-3-527-84580-4
ePub ISBN: 978-3-527-84581-1
oBook ISBN: 978-3-527-84582-8

Printed and bound by
CPI Group (UK) Ltd, Croydon, CR0 4YY

C9783527353606_160925

Contents

Preface

Protein–protein interactions (PPIs) lie at the heart of cellular processes, governing diverse biological activities such as signal transduction, gene regulation, immune responses, and enzymatic catalysis. The investigation of PPIs has not only deepened our understanding of fundamental biology but has also catalyzed breakthroughs in disease treatment and drug discovery. Dysregulation of PPIs is implicated in a wide array of diseases, including neurodegenerative disorders, infectious diseases, and cancers, underscoring the importance of elucidating their molecular mechanisms.

This book, dedicated to the exploration of PPIs, provides a comprehensive and systematic overview of the field. It encompasses foundational concepts, advanced methodologies, and emerging therapeutic strategies, reflecting the multidisciplinary nature of PPI research. Spanning 15 chapters, this volume serves as a valuable resource for researchers, educators, and students who are engaged in the study of cellular mechanisms, drug discovery, and biotechnology.

Chapter 1 introduces the essential concepts, roles, and research methodologies of PPIs, setting the stage for the in-depth discussions that follow. Chapter 2 delves into the challenges and strategies of drug discovery targeting PPIs, including high-throughput screening, structure-based drug design, and computational prediction of PPI inhibitors. The subsequent chapters highlight key experimental and computational approaches to deciphering PPIs. Chapter 3 focuses on fluorescence resonance energy transfer (FRET) technology, a powerful tool for the real-time analysis of molecular interactions in live cells, while Chapter 4 discusses the use of mass spectrometry (MS) for large-scale mapping of PPI networks with high sensitivity and resolution. Chapter 5 explores proximity ligation assays (PLAs) for detecting PPIs in situ, emphasizing their spatial precision and adaptability for multiplexed detection. Chapter 6 reviews the application of surface plasmon resonance (SPR) for characterizing PPI specificity, affinity, and kinetics. Chapter 7 outlines computational methods for predicting PPI structures, with a special emphasis on protein docking and deep learning-based approaches.

The discussion of therapeutic innovations continues in Chapters 8–14. Chapters 8 and 9 introduce diverse chemical scaffolds for PPI modulation, including foldamers as inhibitors of aberrant PPIs and sulfonyl-γ-AApeptides as novel modulators, respectively. Chapter 10 explores the development and application of stapled peptides as modulators of intracellular PPIs, offering enhanced stability, binding affinity, and cellular permeability. Chapter 11 addresses cyclic peptides, focusing on their unique conformational stabilization and therapeutic potential across a range of diseases. Chapter 12 shifts the focus to small-molecule inhibitors targeting BCL-family proteins, revealing their potential in cancer therapy. In Chapter 13, molecular glues are introduced as activators for PPIs, categorized into degraders, stabilizers, and inhibitors based on their biological effects. Chapter 14 highlights the targeting of the APC–Asef interaction for drug discovery in colorectal cancer therapy, offering a case study of specificity and clinical relevance.

The final Chapter 15 reflects on the future of PPI research and therapeutic discovery, discussing the application of computational methods in drug discovery, offering insights into how artificial intelligence and machine learning are shaping the discovery of PPI modulators.

Together, the chapters in this book provide a holistic perspective on PPIs, integrating basic science with translational applications. The exploration of both classical and novel approaches to PPI characterization and modulation offers a comprehensive reference for researchers aiming to unlock the therapeutic potential of PPIs. We hope this book will inspire further advancements in the field, driving the development of new technologies, therapeutics, and innovations that will ultimately benefit human health.

February 2025 *Jian Zhang*

1

Exploring Protein–Protein Interactions: Concepts, Methods, and Implications

Mi Zhou and Renxiao Wang

Fudan University, School of Pharmacy, Department of Medicinal Chemistry, 826 Zhangheng Road, Shanghai 201203, People's Republic of China

1.1 General Concepts of Protein–Protein Interactions

Proteins are the fundamental machinery that drive the vast majority of cellular processes. These versatile biomolecules rarely perform in isolation; instead, up to 80% of proteins engage in intricate interactions with one another, forming dynamic networks that underpin the functional complexity of living organisms [1]. This chapter aims to provide a comprehensive overview of protein–protein interactions (PPIs). We will begin by elucidating the fundamental concepts of PPIs, covering their definition, structural characteristics, and pivotal roles in both physiological and pathological processes. Building upon this foundation, we delve into the diverse spectrum of current methods employed in PPI studies, showcasing both experimental and computational approaches, along with their varied applications. The chapter culminates in a discussion of the profound implications of PPI research, illuminating its potential in advancing medical understanding, revolutionizing drug discovery, and catalyzing technological innovations across various fields.

1.1.1 Definition of Protein–Protein Interactions

PPIs refer to physical contacts between two or more proteins that occur within a defined biological context. First, these interactions are intentional, not arising from random encounters, but rather precisely orchestrated by specific biomolecular forces and mechanisms. Second, they should be nongeneric, distinct from basic cellular processes like protein synthesis or degradation. Also of note, the formation and regulation of PPIs are highly organized in time and space, governed by a complex interplay of factors such as cell type, cell cycle phase, developmental stage, protein modifications, absence or presence of cofactors and binding partners, and environmental conditions [2, 3].

Targeting Protein–Protein Interactions for Drug Discovery, First Edition. Edited by Jian Zhang.

1.1.2 Structural Properties of Protein–Protein Interactions

The intricate architecture of proteins is critical in determining the specificity, strength, and functional outcomes of their interactions. By unraveling these structural characteristics, researchers can gain deeper insights into the molecular mechanisms underlying PPIs and develop strategies to manipulate them for therapeutic purposes.

Proteins are sophisticated macromolecules comprising an array of structural and functional units that work in concert to perform diverse biological tasks. Among these, domains and short linear motifs (SLiMs) serve as two main classes of functional modules that mediate PPIs. Domains, typically spanning 50–200 residues, are independently folding structural units within proteins that often harbor distinct biological activities. SLiMs, on the other hand, are compact, recurring functional peptides consisting of 3–10 residues, primarily found within intrinsically disordered regions (IDRs) [4]. Indeed, many PPIs can be categorized into domain–domain interactions (DDIs) or domain–motif interactions (DMIs). DDIs usually underpin the formation of stable and long-lasting complexes, while DMIs are associated with transient and low-affinity interactions [5].

The surface regions where the direct physical interactions between two or more proteins occur are termed interfaces. They are highly diverse in terms of size, shape, and chemical properties, determining the specific recognition and binding between proteins engaged in different biological processes. According to statistics, one-sided size of an interface typically ranges from 200 to 2800 Å^2, with the majority falling between 200 and 1200 Å^2 [6]. Although considered relatively flat, interfaces possess a complex topography characterized by cavities, grooves, and protruding regions. Complementary geometric features ensure that proteins bind to each other in the correct orientation and with high specificity. Moreover, interfaces encompass a variety of chemical interactions, including hydrogen bonds, hydrophobic interactions, electrostatic forces, salt bridges, and disulfide bonds, which collectively account for the specificity and stability of PPIs [7].

Within the interfaces, there exist clusters of residues known as "hot spots," which make disproportionately large contributions to the binding affinity. Single-point mutations of these residues to alanine may cause a substantial increase in the binding free energy ($\Delta\Delta G \geq 2$ kcal/mol), highlighting their critical roles in stabilizing the PPIs [8]. Hot spots have a distinctive amino acid composition, enriched with tryptophan, arginine, and tyrosine, due to the unique physicochemical properties of these residues like bulky side chains, the propensity to form hydrophobic surfaces, and the capacity to engage in hydrogen bonding [9]. With regard to spatial organization, hot spots cluster within tightly packed regions rather than being randomly distributed, facilitating the removal of water molecules upon binding [10]. Besides, they are surrounded by moderately conserved and energetically less important residues, which form an O-ring to further occlude bulk solvent from the hot spots [9].

Proteins are not static, rigid structures; rather, they are dynamic entities capable of adopting multiple conformations. This inherent flexibility is critical for their diverse functions, particularly in PPIs. The traditional "lock and key" model, which implies a preexisting perfect fit between interacting proteins, fails to capture the dynamic nature of PPIs [11]. Subsequently, more precise descriptions have been put forward. The "induced fit" model suggests that upon initial contact, proteins undergo conformational changes to achieve an optimal fit with their binding partners [12]. Meanwhile, the "conformational selection" model proposes that proteins exist in an equilibrium of various conformations, with binding events selecting and stabilizing the most favorable conformation [13].

1.1.3 Diverse Types of Protein–Protein Interactions

PPIs manifest in a striking diversity of forms, each tailored to perform a specific biological role. In the following sections, we will explore several prominent types of PPIs, varying in structural properties, molecular recognition mechanisms, and functional outcomes.

1.1.3.1 Enzyme–Substrate Interactions

Enzymes are biological catalysts that bind to specific substrates through their active sites and facilitate the conversion of substrates into product molecules. Key examples that illustrate the diverse roles of enzyme–substrate interactions include: (i) protein phosphorylation and dephosphorylation, where kinases and phosphatases add and remove phosphate groups to substrates, respectively, act as molecular switches in signal transduction [14]; (ii) proteolytic cleavage, an irreversible process mediated by proteases that catalyze the hydrolysis of peptide bonds in target proteins, precisely governs protein maturation, activation, stability, and localization [15]; (iii) histone modifications, such as acetylation and methylation, are carried out by enzymes like histone acetyltransferases (HATs) and histone methyltransferases (HMTs), contributing to epigenetic regulation of gene expression [16].

1.1.3.2 Receptor–Ligand Interactions

Cells possess an intricate communication network to sense and respond to environment cues and stimuli. This network is built upon cell surface receptors that bind external signaling molecules termed ligands, initiating downstream signaling cascades within the cell and manipulating various physiological processes like growth, development, immune response, and metabolism [17]. G protein-coupled receptors (GPCRs) constitute the largest family of these receptors, characterized by a unique architecture comprising seven transmembrane helices joined by intracellular and extracellular loops [18]. Chemokines, a class of small secreted proteins, exemplify protein ligands that bind to a specific subfamily of GPCRs called chemokine receptors. By sequentially binding to the N-terminal region and extracellular loops of their receptors, chemokines induce receptor conformational rearrangement and activation that direct immune cell migration and positioning during inflammation [19].

1.1.3.3 Antigen–Antibody Interactions

Antibodies are glycoproteins produced by B lymphocytes to recognize and bind to foreign substances known as antigens, which can be present on the surface of invading pathogens like viruses and bacteria. This interaction launches an immune response to neutralize and eliminate pathogens, thus protecting the body from infection and disease [20]. An antibody is composed of two identical heavy chains and two identical light chains connected by disulfide bonds. Antigen recognition is achieved by the complementarity-determining regions (CDRs) located at the N-terminus of both heavy and light chains that precisely match antigen's epitope [21]. Beyond their pivotal role in the immune system, the exquisite specificity of antigen–antibody interactions has also been exploited for diagnostic and therapeutic applications.

1.1.3.4 Chaperone–Client Interactions

Chaperones function as protein quality control machinery, engaging with client proteins to assist in their de novo folding, subcellular translocation, and recovery from stress-induced misfolding and aggregation. Many chaperones belong to the heat shock protein (HSP) families, including HSP60, HSP70, HSP90, and HSP100 [22]. Chaperones from different families exhibit structural and functional diversity, reflected in their distinct binding patterns to clients. Some utilize ATP binding to allosterically regulate their conformation, facilitating efficient client binding and release. Others assemble into oligomeric complexes, creating an enclosed environment for client encapsulation [23]. Structural biology studies have revealed that, in some cases, chaperones bind their clients in interconverting conformational ensembles that are locally highly dynamic. This binding behavior allows clients to undergo folding while bound to chaperones and enables chaperones to accommodate a broad range of clients with varying folding properties and limited shape complementarity [24, 25].

1.1.3.5 Scaffold Interactions

Scaffold proteins are molecular organizers that bring together two or more signaling molecules like kinases and receptors into higher-order complexes to modulate their activities, thereby enabling the efficient spatial–temporal coordination of cellular signaling events [26]. This remarkable organizational capability of scaffolds is attributed to their unique modular architecture. Typically, they encompass discrete interaction domains capable of recognizing and binding to SLiMs (e.g. phosphotyrosine- and proline-containing sequences) on target proteins, along with IDRs that confer flexibility and functional versatility [27]. In addition to simply tethering targets in proximity, scaffolds can exert allosteric regulation on the targets as well. In this way, they can handle both the linear input–output signal transmission and complicated feedback loops within distinct pathways and even promote signal integration and crosstalk among various pathways [28, 29].

1.2 Functional Significance of Protein–Protein Interactions

Far from being static or isolated events, PPIs are dynamically woven into complex networks that enable proteins to communicate, coordinate, and cooperate in carrying out the essential functions of life. Meanwhile, disruptions or aberrations in PPIs can lead to a cascade of dysfunction, contributing to the development of various diseases.

1.2.1 Cellular Signal Transduction

Signal transduction is the process by which extracellular signals are converted into cellular responses. It is organized into three distinct stages: reception, transduction, and response, each involving PPIs to ensure high fidelity of signal transmission (Figure 1.1).

The reception stage commences with receptor–ligand interactions. To be exact, a variety of extracellular signaling molecules, ranging from neurotransmitters and hormones to proteins like chemokines, thrombin, and growth factors, are recognized by their corresponding receptors embedded in the membrane, such as

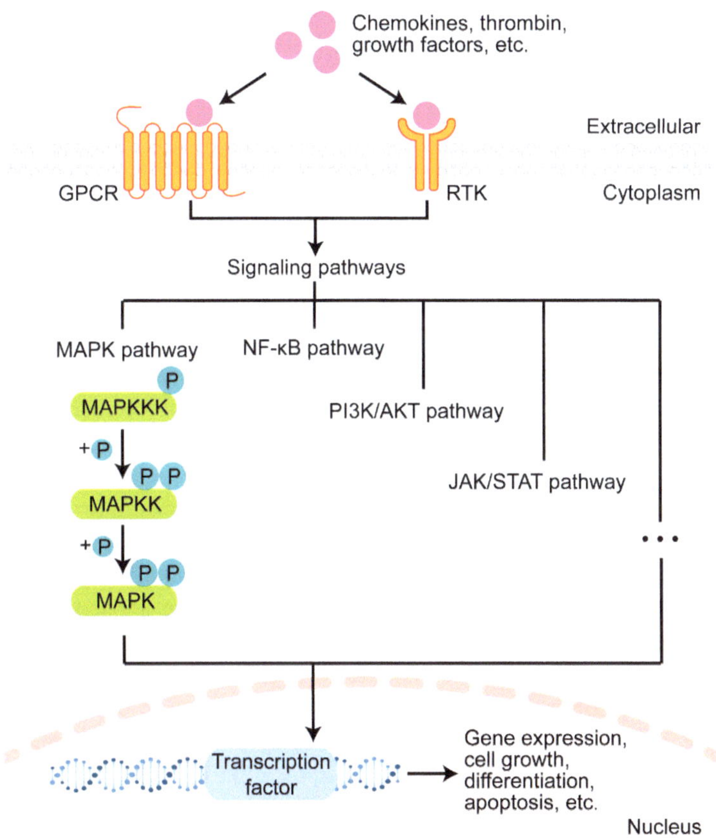

Figure 1.1 A graphic representation of the signal transduction process. Extracellular signals, upon binding to their receptors, are transmitted through intracellular signaling pathways, culminating in a series of cellular responses. (Source: Mi Zhou and Renxiao Wang.)

GPCRs and receptor tyrosine kinases (RTKs) [17]. Upon ligand binding, receptors generally undergo conformational changes, activating associated proteins or triggering homodimerization and autophosphorylation to initiate downstream signaling pathways [30, 31].

During the transduction stage, the signal is propagated from the cell surface to the intracellular targets through an elaborate network of signaling cascades. These cascades are characterized by a series of posttranslational modifications, predominantly phosphorylation events driven by kinase–substrate interactions. A classic example is the mitogen-activated protein kinase (MAPK) pathway, where a cascade of three kinases, namely MAPK kinase kinase (MAPKKK), MAPK kinase (MAPKK), and MAPK, phosphorylate and activate each other in a sequential manner [32]. A similar phosphorylation-dependent activation mechanism is also observed in other crucial pathways, including nuclear factor kappa B (NF-κB) pathway, phosphoinositide 3-kinase (PI3K)/protein kinase B (AKT) pathway [33], and Janus kinase (JAK)/signal transducer and activator of transcription (STAT) pathway [34]. While some interactions serve to relay and amplify the signal, others involving phosphatases and inhibitory proteins lead to signal decay and termination [35]. Moreover, adaptor, scaffold, and docking proteins further fine-tune signal transduction by manipulating the assembly, function, and localization of relevant proteins [36].

Finally, the signal cascades converge on the regulation of their target proteins (e.g. transcription factors, enzymes, cytoskeletal proteins, and apoptotic proteins), eliciting the desired cellular responses such as changes in gene expression, cell growth, differentiation, and apoptosis.

1.2.2 Regulation of Gene Expression

Gene expression in eukaryotes encompasses the processes of transcription, RNA processing, and translation, wherein PPIs play a critical regulatory role, contributing to the accurate transfer of genetic information from DNA to messenger RNA (mRNA) to protein.

At the onset of transcription activation, transcriptional activators bind to specific DNA sequences and then recruit coactivators to improve transcription efficiency. Some coactivators like HMTs and HATs perform histone modifications and chromatin remodeling, rendering the DNA more accessible to the transcriptional machinery [37]. Others, such as the mediator complex, bridge RNA polymerase II (Pol II) and general transcription factors onto core promoters, resulting in the assembly of a pre-initiation complex (PIC) that directs the initiation of transcription [38] (Figure 1.2a). On the other hand, the interplay between transcriptional repressors, co-repressors, and histone-modifying enzymes, like histone deacetylases (HDACs), can suppress the transcription process [39].

Once transcribed, the precursor mRNA (pre-mRNA) undergoes several processing steps to become the mature mRNA. These include 5′-end capping, splicing, and 3′-end cleavage and polyadenylation, each orchestrated by specialized molecular machinery (Figure 1.2b). Capping involves the sequential recruitment of capping

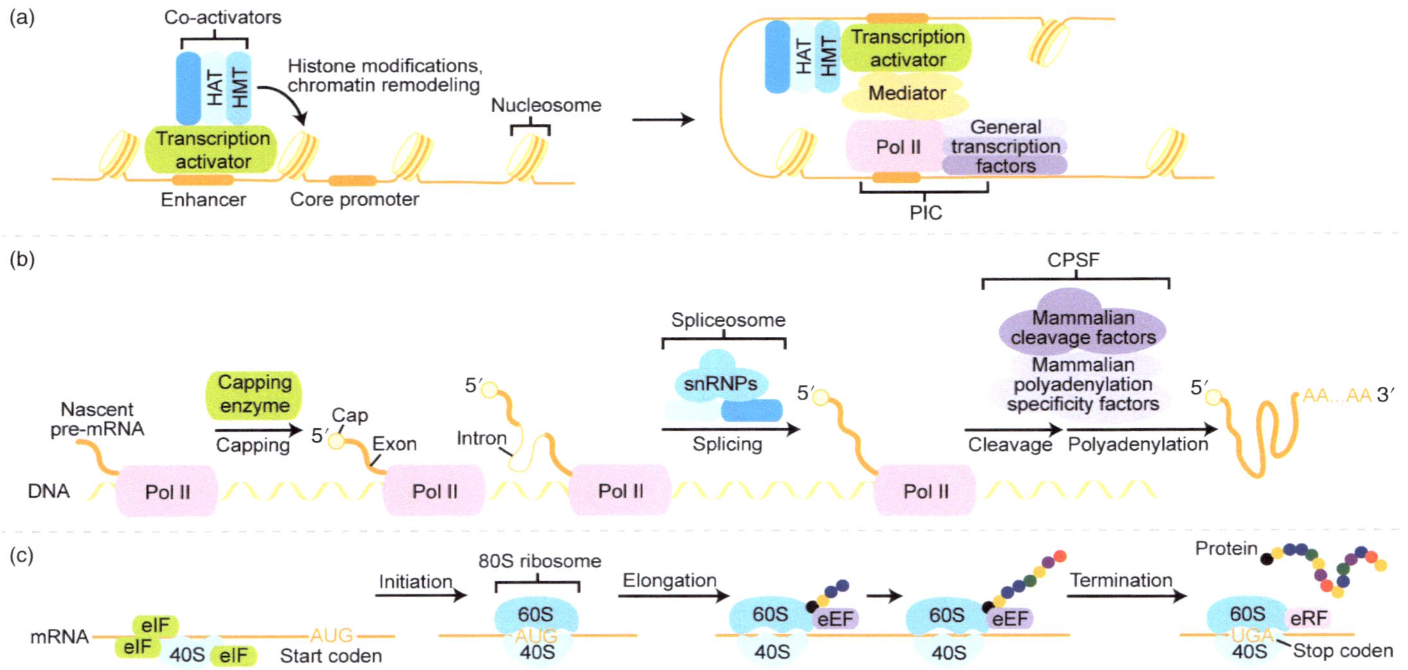

Figure 1.2 Simplified schematic of the gene expression processes in eukaryotes, including (a) transcription, (b) RNA processing, and (c) translation. (Source: Mi Zhou and Renxiao Wang.)

enzymes onto the Pol II, where they catalyze modifications to the 5′ end of the nascent pre-mRNA [40]. Next, splicing is executed by a highly dynamic and massive ribonucleoprotein complex termed the spliceosome, which is assembled by ordered binding and release of small nuclear ribonucleoproteins (snRNPs) and numerous other splicing factors [41]. Finally, 3′-end processing is performed by the cleavage and poly-adenylation specificity factor (CPSF), a large multi-subunit complex that utilizes endonuclease, poly(A) polymerase, and accessory proteins to cleave the pre-mRNA and add a poly(A) tail [42].

During translation initiation, a suite of eukaryotic translation initiation factors (eIFs) mediate the assembly of the 40S and 60S ribosomal subunits at the mRNA start codon, culminating in the formation of the 80S ribosome ready for protein synthesis [43]. Elongation ensues, with eukaryotic elongation factors (eEFs) assisting the ribosome in moving along the mRNA and adding amino acids to the growing polypeptide chain [44]. At last, upon encountering a stop codon, the ribosome triggers the recruitment of eukaryotic release factors (eRFs), promoting the release of the nascent peptide and marking the termination of translation [45] (Figure 1.2c).

1.2.3 Immune Response

The immune response is built upon a foundation of intricate cell–cell communication, with PPIs serving as the key messengers. These interactions, occurring between and within immune and nonimmune cells, are indispensable for recognizing pathogens, activating signal cascades, and mounting an effective defense against invading threats.

1.2.3.1 Immune Cell Migration

Migration is crucial for immune cells to patrol for foreign antigens and traffic to inflammation or infection sites. This process often involves transmigration across epithelial barriers, relying on adhesion factors-governed PPIs. The initial capture and rolling of immune cells along the endothelial surface are mediated by interactions between selectins (e.g. P-selectin) on endothelial cells and their ligands (e.g. P-selectin glycoprotein ligand-1 [PSGL-1]) on immune cells. Film adhesion is subsequently established through interactions between integrins (e.g. lymphocyte function-associated antigen-1 [LFA-1]) on immune cells and their ligands (e.g. intercellular adhesion molecule-1 [ICAM-1]) on endothelial cells [46]. Finally, platelet endothelial cell adhesion molecule-1 (PECAM-1) molecules on both endothelial and immune cells engage in homophilic interactions, promoting transendothelial migration [47].

1.2.3.2 T-Cell Antigen Recognition and Activation

When a foreign antigen invades the body, the T-cell receptor (TCR) on the T-cell surface specifically recognizes and binds to an antigen peptide presented by major histocompatibility complex (MHC) molecules on the surface of antigen-presenting cells (APCs). In parallel, co-receptors, such as cluster of differentiation 4 (CD4) and CD8, on the T-cell surface also engage with the MHC molecules, reinforcing

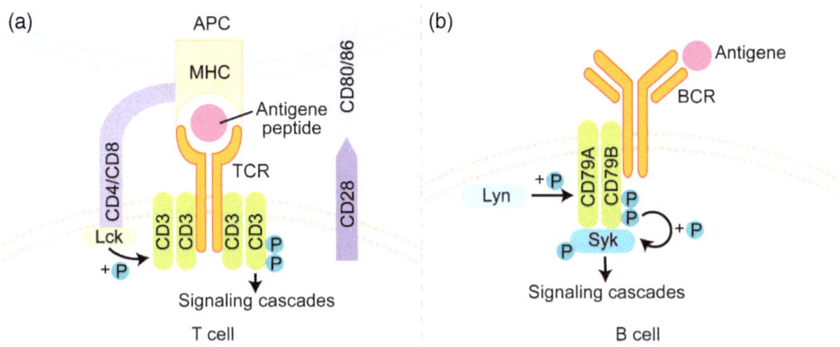

Figure 1.3 Illustration of TCR signaling (a) and BCR signaling (b) during antigen recognition. (Source: Mi Zhou and Renxiao Wang.)

the stability of the TCR-peptide-MHC complex. Co-receptor engagement brings the lymphocyte-specific protein tyrosine kinase (Lck) into proximity with the TCR-CD3 complex. Lck then induces the phosphorylation of CD3 molecules, initiating downstream signaling cascades that ultimately lead to T-cell activation, proliferation, and differentiation. Additionally, PPIs between co-stimulatory molecules, such as CD28 on T cells and CD80/86 on APCs, enhance TCR-induced signals and prevent T-cell anergy [48] (Figure 1.3a).

1.2.3.3 B-Cell Antigen Recognition and Activation

Upon engagement with its cognate antigen, the B-cell receptor (BCR) undergoes clustering on the plasma membrane. This event triggers the phosphorylation of CD79A and CD79B signaling subunits on BCR by Src family tyrosine kinases, primarily Lck/Yes-related novel protein tyrosine kinase (Lyn). Phosphorylated CD79A and CD79B then serve as docking sites for the recruitment and activation of spleen tyrosine kinase (Syk). Once activated, Syk propagates signals to downstream signaling cascades [49, 50] (Figure 1.3b). Similarly, PPIs involving co-receptors and co-stimulatory molecules can augment BCR-induced signals [51]. Ultimately, these signaling events can drive B-cell differentiation into plasma cells, specialized in secreting antibodies that target specific antigens and neutralize or clear them from the body.

1.2.4 Protein Degradation Pathway

To maintain protein homeostasis, cells employ sophisticated quality control mechanisms that promptly identify and eliminate aberrant or unwanted proteins. Two major protein degradation systems, the ubiquitin-proteasome system (UPS) and the autophagy-lysosomal pathway, are essential for this vital task.

UPS selectively targets substrate proteins, tagging them with poly-ubiquitin chains for degradation by the proteasome. The ubiquitination process is carried out by the sequential action of three key enzymes: E1 ubiquitin-activating enzymes, E2-ubiquitin-conjugating enzymes, and E3 ubiquitin ligases. E1 initiates the process by activating ubiquitin, forming a thioester bond between ubiquitin's C-terminus and its own catalytic cysteine. After that, E1 recruits E2 and transfers ubiquitin to E2's catalytic cysteine. Finally, E3 recognizes and binds to the substrate along with the E2-ubiquitin complex, facilitating the transfer of ubiquitin from E2 onto the substrate [52]. The degradation process is executed by the 26S proteasome, a large multi-subunit protease complex consisting of two main components: 20S core particle (CP) and 19S regulatory particles (RPs). Each component is made up of multiple distinct subunits, with dedicated chaperones assisting in stabilizing intermediate structures and ensuring the proper incorporation of individual subunits. Once assembled, the 20S CP and 19S RPs unite to form the functional 26S proteasome, poised for the degradation of poly-ubiquitylated substrates [53].

The autophagy-lysosomal pathway focuses on the sequestration of cargo (e.g. damaged organelles and protein aggregates) within autophagosomes, which then fuse with lysosomes for cargo degradation and recycling. This process is mainly driven by a set of autophagy-related (ATG) proteins and their interactions. The unc-51-like kinase 1 (ULK1) complex initiates phagophore nucleation, followed by the recruitment of the class III phosphatidylinositol 3-kinase (PI3K) complex that promotes autophagosome formation and elongation. Two ubiquitin-like conjugation systems, namely the ATG12 and microtubule-associated protein 1 light chain 3 (LC3)/GABA type A receptor-associated protein (GABARAP) systems, further contribute to autophagosome maturation [54]. Cargo recognition and delivery are mediated by cargo receptors like sequestosome 1 (p62/SQSTM1), which interact with LC3 on the autophagosome membrane, thereby guiding the cargo for degradation within autophagolysosomal compartment [55].

1.2.5 Disease Mechanisms

PPIs are fundamental for maintaining cellular homeostasis. However, their dysregulation, whether caused by genetic anomalies or the introduction of foreign proteins, plays a significant role in the development and progression of various diseases.

1.2.5.1 Cancer

Neoplastic progression entails the acquisition of a panel of functional capabilities that enable tumor growth and metastasis, termed "hallmarks of cancer" [56]. These hallmarks are inextricably linked to dysregulated PPIs. For instance, aberrant activation of oncoproteins, such as rat sarcoma virus (RAS), induces sustained proliferative signaling by engaging downstream effectors [57]. Negative regulation of the tumor suppressor p53 by mouse double minute 2 homolog (MDM2) is one mechanism whereby cancer cells evade growth suppression [58]. Abnormal expression of anti-apoptotic B-cell lymphoma-2 (BCL-2) family proteins, through sequestering pro-apoptotic counterparts, often contributes to cell death resistance [59]. Hypoxic tumor microenvironment stimulates the production of pro-angiogenic factors like vascular endothelial growth factor (VEGF), which activate their receptors on endothelial cells to induce angiogenesis [60]. Various cancer cells overexpress programmed cell death ligand 1 (PD-L1) and exploit the PD-L1/programmed cell death protein 1 (PD-1) pathway to escape immune destruction [61].

1.2.5.2 Neurodegenerative Diseases

Alzheimer's disease (AD) is characterized by two hallmark pathologies, amyloid-β (Aβ) plaques and neurofibrillary tangles. The former arises from the sequential proteolytic cleavage of amyloid precursor protein (APP) by secretases, generating Aβ peptides that accumulate and deposit into amyloid plaques [62]. The latter comprises insoluble aggregates of hyperphosphorylated tau protein, a consequence of an imbalanced phosphorylation and dephosphorylation [63]. Parkinson's disease (PD) is marked by the pathological aggregation of α-synuclein (α-syn) into insoluble fibrillar structures called Lewy bodies. While genetic and environmental factors contribute to α-syn aggregation, its engagements with various other proteins, including those implicated in neurodegenerative diseases (e.g. tau) and regulators of actin cytoskeleton and neurotransmission, can accelerate this process [64]. Likewise, Huntington's disease (HD) is driven by the self-aggregation of mutant huntingtin (mHTT) protein. Abnormal PPIs between mHTT and other proteins disrupt crucial cellular processes such as axonal transport, mitochondrial fission, and gene transcription, resulting in neuronal and mitochondrial dysfunction [65].

1.2.5.3 Infectious Disease

The pathogenesis of infections caused by viruses, bacteria, fungi, or parasites hinges critically on a complex interplay between pathogen proteins and host proteins. These PPIs underpin every stage of infections, including:

(i) Invasion: Viruses gain entry into the host by utilizing their surface proteins to recognize specific host cell receptors. A notable example is the binding of the severe acute respiratory syndrome coronavirus 2 (SARS-CoV-2) spike (S) protein to the human angiotensin-converting enzyme 2 (ACE2) receptor, which mediates the viral attachment and entry into the respiratory epithelial cells [66]. Bacteria and fungi, on the other hand, rely on surface-exposed adhesins that interact with corresponding host cell receptors or extracellular matrix proteins, enabling their attachment and colonization [67, 68].

(ii) Immune Evasion: Pathogens employ various strategies to escape or suppress host's immune defenses. Some, like human cytomegalovirus (HCMV), encode specific proteins that target and interfere with MHC molecules, disrupting antigen presentation [69]. Others, such as *Streptococcus pyogenes*, secrete immunoglobulin-degrading enzymes to cleave antibodies, rendering them ineffective [70].

(iii) Replication: Pathogens often hijack host cellular machinery for their replication needs. For example, expression of the human immunodeficiency virus (HIV) genome is regulated by its trans-activator of transcription (TAT) protein, which assembles with host's positive transcription elongation factor b (P-TEFb) onto the viral promoter, thus enhancing the production of viral transcripts [71].

1.3 Methods for Analyzing Protein–Protein Interactions

The field of PPI research has undergone a transformative evolution with the advent of diverse and sophisticated methodologies, ranging from traditional biochemical techniques to high-throughput technologies and advanced computational algorithms. Together, these complementary strategies form a powerful toolkit for researches to uncover insights into PPIs, from atomic-level details to intricate systemic networks.

1.3.1 Experimental Methods

A diverse array of experimental methods, grounded in either biophysical, biochemical, or genetic principles, have been established to identify and characterize PPIs across various biological contexts and scales. These methods offer multifaceted capabilities, including:

1.3.1.1 Structure Determination

Elucidating the 3D structure of PPIs may provide valuable insights into the geometry and physicochemical characteristics of the binding interfaces and even shed light on potential therapeutic interventions [72]. The primary techniques currently employed for high-resolution protein structure determination are X-ray crystallography, nuclear magnetic resonance (NMR) spectroscopy, and cryogenic electron microscopy (cryo-EM). X-ray crystallography offers detailed information on atom positions and chemical bonds by analyzing X-ray diffraction patterns generated from protein crystals [73]. As the workhorse of structural biology, it accounts for over 80% of experimental structures deposited in the Protein Data Bank (PDB), with the median resolution plateauing at ~2 Å as of 1990 [74]. Inherent limitations of crystallography are the requirement to obtain well-diffracting crystals and the occurrence of crystal packing artifacts. NMR spectroscopy, eliminating the crystallization process, enables structural studies of proteins in solution with the aid of structural restraints acquired from different NMR experiments. Given its strength in uncovering protein dynamics and flexibility over a range of timescales, this technique is well-suited for investigating intrinsically disordered proteins, proteins in transient states, and weak interactions [75, 76]. Whereas drawbacks, such as the necessity of isotope labeling and size limit (typically <50 kDa), should be taken into account. Driven by technological breakthroughs over the past decade, cryo-EM has experienced a "resolution revolution" and is gaining widespread popularity [77]. It relies on direct visualization of proteins embedded in a thin layer of vitreous ice, thus unraveling protein structures in close-to-physiological state with a routine resolution of 2.5–4.5 Å [78]. Compared with other techniques, it is more tolerant to sample quantity and purity. Although high-resolution reconstruction of small-size proteins (<100 kDa) remains challenging for cryo-EM, novel strategies have been applied to break through the barrier [79]. In conclusion, each technique has its own pros and cons, rendering them complementary tools to decipher the intricate 3D structures of proteins and protein complexes [80, 81].

1.3.1.2 Affinity, Kinetics, and Thermodynamics Measurement

Quantitative in vitro analysis of PPIs in terms of affinity, kinetics, and thermodynamics is essential for exploring the underlying mechanisms of molecular recognition. Binding affinity, defined as the strength of the interaction between biomolecules that bind reversibly, is typically quantified through the equilibrium dissociation constant (K_d) [82]. A variety of experimental methods have been established to obtain the K_d values, including fluorescence polarization (FP), surface plasmon resonance (SPR), biolayer interferometry (BLI), isothermal titration calorimetry (ITC), differential scanning calorimetry (DSC), microscale thermophoresis (MST), fluorescence and bioluminescence resonance energy transfer (FRET and BRET), AlphaScreen, protein-fragment complementation assay (PCA), enzyme-linked immunosorbent assays (ELISA), and so forth [83, 84]. Of the methods mentioned earlier, SPR and BLI can also yield kinetic parameters, specifically the association rate constant (k_{on}) and dissociation rate constant (k_{off}) [85]. Both technology platforms enable real-time and label-free monitoring of binding events and have a wide range of applications, such as investigating the assembly and activation of signaling complexes [86], antibody evaluation and selection [87], and protein engineering [88]. On the other hand, calorimetric methods like ITC and DSC can furnish extra thermodynamic insight. In the case of ITC, a comprehensive thermodynamic profile, including Gibbs free energy change (ΔG), enthalpy change (ΔH), entropy change (ΔS), and heat capacity change (ΔC_p), can be acquired from a single experiment [89]. Combining these data with structural information can further interpret the driving forces for the binding [90]. As a final note, prior to choosing the appropriate methods, factors such as the need for immobilization or labeling, affinity range, sample and time consumption, throughput performance, and the potential risks of false positives and false negatives should be carefully considered. Besides, integrating and validating results from multiple methods is crucial to ensure data accuracy and reliability.

1.3.1.3 Large-Scale PPI Network Mapping

Systematic mapping of PPI network within a cell or an organism, termed the "interactome," is an ongoing endeavor aimed at deciphering previously uncharacterized proteins, global proteome organization and function, and genotype–phenotype relationships. For nearly three decades, proteome-wide interactomes of yeast, human, and other model organisms have been delineated and progressively expanded using diverse high-throughput experimental techniques [91, 92]. One powerful method to identify binary PPIs is yeast two-hybrid (Y2H) screening. In the original Y2H, PPIs are detected inside the yeast nucleus based on the functional reconstitution of a transcription factor. Thereafter alternative versions have broadened its applicability to membrane proteins, proteins with posttranslational modifications, and PPIs in mammalian cells [93]. In 2020, an ORFeome resource covering 17,408 protein-coding genes was established, and over 150 million protein pairs were screened using Y2H, generating the largest human binary protein interactome consisting of ~53,000 high-quality PPIs [94]. In contrast, co-complex mapping is typically carried out via affinity purification coupled to mass spectrometry (AP-MS). In AP-MS, a bait protein and its binding partners are isolated from cell lysate using bait-specific antibody and then subjected to mass spectrometric analysis for identification of the purified components. AP-MS excels at detecting PPIs under near physiological conditions but falls short in capturing weak or transient interactions [95]. It was recently employed in the BioPlex project for the identification of dual human interactomes that encompass 118,162 and 70,966 PPIs in 293T and HCT116 cells, respectively [92]. Despite significant experimental efforts, current human interactomes remain incomplete and present high levels of noise [96]. Methodological advances, integration of various data sources, and incorporation of context-specific (e.g. cell-, tissue-, and disease-specific) information may enhance both interactome coverage and quality, providing researchers with a more comprehensive and dynamic perspective.

1.3.2 Computational Methods

Despite remarkable efforts made by experimental techniques, they often struggle to capture transient and weak interactions or those involving intrinsically disordered proteins. Besides, they are time-consuming and labor-intensive in general. Computational methods offer a complementary and efficient alternative, harnessing vast datasets and predictive models to infer and analyze PPIs.

1.3.2.1 Sequence-Based Methods

The abundance of protein sequence data in public databases has propelled the development of strategies for PPI prediction solely based on sequence information. Herein are some widely adopted approaches: (i) Ortholog-based methods are premised on the evolutionary conservation of PPIs across species, utilizing known PPIs in one species to identify orthologous interactions ("interologs") in another species [97]. (ii) Domain-based methods are built on the concept that PPIs are primarily mediated by domains. By integrating experimentally determined PPI datasets with domain information sourced from databases (e.g. Pfam [98]), DDIs can be inferred and serve as the basis for predicting novel PPIs [99]. (iii) Co-evolution analysis assumes that amino acid substitutions in one protein are often accompanied by compensatory substitutions in its interacting partner to maintain functional integrity. Multiple sequence alignments are employed to quantify the co-evolutionary relationships between residue pairs within proteins of interact, with highly co-evolving pairs presumed to be in physical or functional contact, suggesting a potential PPI [100]. (iv) Machine learning-based methods have gained considerable prominence, progressing from traditional algorithms like support vector machines (SVMs) and random forests (RFs) to advanced deep learning models such as deep neural networks (DNNs) and recurrent neural networks (RNNs). These models extract sequence-based features (e.g. amino acid composition, physicochemical properties, and evolutionary information) from large data sets, capture the underlying patterns, and make accurate predictions for novel protein pairs [101, 102]. To date, protein sequence data remain the primary source for computational PPI prediction. Ongoing improvements are driven by the growing availability of high-quality PPI data sets, the introduction of innovative sequence encoding schemes, as well as the integration with diverse biological data such as structure information [103].

1.3.2.2 Structure-Based Methods

Structural characterization of PPIs remains a challenge due to their inherent complexity and technological limitations. In fact, fewer than 5% of human PPIs have been structurally resolved through experimental techniques or homology modeling [104]. Molecular docking has emerged as a rapid and cost-effective alternative to bridge this gap by predicting 3D structures of protein complexes. During the docking process, unbound protein structures or structural models are used as input, generating numerous putative binding modes through sampling. The most plausible ones are then selected by means of scoring and ranking [105]. A wide variety of sophisticated docking programs have been developed, each employing different sampling algorithms and scoring functions, including ZDOCK [106], HADDOCK [107], ClusPro [108], and RossettaDock [109]. Continuous efforts have been made to improve prediction accuracy and reliability, such as accounting for backbone and side-chain flexibility, incorporating experimental data like interface details and distance restraints, and integrating deep learning techniques [110, 111]. Molecular dynamics (MD) simulation is another powerful computational technique widely employed to investigate the structural properties of proteins and their complexes. It applies Newtonian mechanics to study the motion of atoms and molecules within a system at nanosecond to microsecond timescales. Analysis of the output trajectories may provide atomic-level insights into the dynamic behavior of protein molecules [112]. In the realm of PPI studies, MD simulation is utilized to assess complex stability, analyze residue flexibility, estimate interaction strengths, and explore the impact of various conditions (e.g. environmental factors, solvents, and mutations) on PPIs [113]. Popular tools for MD simulation include AMBER [114], GROMACS [115], CHARMM [116], and NAMD [117]. Future developments are focused on extending the simulation timescales, improving the accuracy of force fields, and integrating artificial intelligence techniques for enhanced prediction and analysis capabilities.

1.3.2.3 Network-Based Methods

The identification of unmapped interactions within a PPI network can be regarded as the task of predicting missing links between nodes in a graph representation, where nodes denote proteins and links represent interactions between them. Such network-based prediction approaches can be broadly categorized as follows [118]: (i) Similarity-based methods leverage local properties of nodes to establish a similarity score function for link prediction. A classic example is the common neighbor method, which assumes that two proteins sharing a greater number of common neighbors exhibit higher similarity and are thus more likely to interact [119]. (ii) Probabilistic methods aim to identify model parameters that best explain the inherent structure of a network, such as community structure. Taking the stochastic block model as an example, each protein is assigned to multiple communities, reflecting its participation in various biological processes. This model operates on the principle that proteins within the same community are more prone to interact with each other compared to proteins from different communities [120]. (iii) Factorization-based methods constitute a class of graph embedding techniques that factorize the network's adjacency matrix to build low-dimensional representations for each node. Geometric Laplacian Eigenmap Embedding (GLEE) is one such method that exploits the simplex geometry of the Laplacian matrix to capture the most basic structural property of the graph, demonstrating strong competence in predicting unobserved interactions in PPI networks [121]. (iv) Machine learning-based methods make predictions by learning patterns from training data. As a case in point, a conditional generative adversarial network (cGAN) can be conditioned on the raw topological features of the network, enabling the generator to produce new probable links [122].

1.4 Implications of the Basic Research on Protein–Protein Interactions

Given their essential role in biological functions, deciphering the intricate PPI network not only holds paramount significance in the realm of molecular biology but also has far-reaching implications in medical advances, drug discovery, and biotechnology innovations.

1.4.1 Advancing Disease Understanding and Diagnosis

PPIs play pivotal roles in cellular functions, from normal physiological processes to the development and progression of various diseases. In cancer, genomic alterations give rise to oncogenic PPIs, driving uncontrolled cell proliferation and metastasis [123]. Neurodegenerative diseases often involve mutations, posttranslational modifications, or misfolding of aggregation-prone proteins, resulting in aberrant PPIs that exacerbate cytotoxicity [124]. Infectious diseases heavily rely on host–pathogen PPIs to facilitate pathogen entry, replication, and dissemination within the host organism [125]. Consequently, identifying and characterizing these interactions is essential for unraveling the molecular basis of diseases. The COVID-19 pandemic serves as a compelling case in point. Investigation of the interaction between the SARS-CoV-2 spike protein and the human ACE2 receptor advanced our knowledge of the virus's entry mechanism into host cells [126]. Further affinity measurement of different SARS-CoV-2 variants toward ACE2 helped to assess the virus's infectivity, transmissibility, and potential threat to public health, offering invaluable information for epidemiological studies [127].

Current molecular diagnostics mainly focus on identifying genetic alterations and individual protein biomarkers. However, emerging evidence suggests that altered PPI patterns are strongly linked to disease progression and clinical outcomes, holding significant potential as novel diagnostic and prognostic biomarkers [128]. One notable example lies in the application of proximity ligation assay (PLA) to detect protein complexes formed between epidermal growth factor receptor (EGFR) and growth factor receptor-bound protein 2 (GRB2) in non-small cell lung cancer

(NSCLC) patient specimens. This approach allows for *in situ* assessment of EGFR signaling activity and has proven to be a superior predictor of clinical response to targeted therapies compared to conventional strategies based on EGFR mutations or protein abundance [129]. As our understanding of disease-related PPIs deepens and advanced methods for PPI profiling in clinical specimens emerge, PPI-based diagnostics, alongside genomic and proteomic analyses, will become increasingly powerful tools for precision and personalized medicine.

1.4.2 Driving Target-Based Drug Discovery

Given their central role in numerous disease pathways, PPIs have emerged as attractive targets for therapeutic intervention. However, modulating these interactions presents challenges due to the complex and dynamic nature of PPI interfaces, which are often characterized by extensive, shallow surfaces with limited deep cavities. PPIs lack endogenous small-molecule ligands that act as a starting point for drug discovery [130]. Despite these obstacles, substantial efforts have been made to overcome these barriers. Structural biology techniques can pinpoint hot spots at PPI interfaces, serving as a basis for the structure-based design of PPI modulators. Computational methods, such as virtual screening and molecular docking, can predict the binding affinity and mode of molecules at binding sites, accelerating lead discovery. By integrating these approaches, researchers can efficiently identify and optimize PPI modulators, opening a new avenue for drug discovery.

Over the past decade, the field of PPI modulators has witnessed remarkable progress, with quite a few candidates being approved for medical use or entering clinical trials for the treatment of various diseases, particularly solid tumors [131]. Small molecules constitute the most abundant class of PPI modulators, offering advantages such as oral bioavailability and membrane permeability. Classic PPI targets include interactions between antiapoptotic BCL-2 family proteins and their proapoptotic counterparts, X-linked inhibitor of apoptosis (XIAP) and caspase-9, and MDM2-p53 [132]. A significant milestone was achieved in 2016 with the approval of venetoclax, a BCL-2 inhibitor, for the treatment of chronic lymphocytic leukemia (CLL), making a breakthrough in the development of PPI-modulating therapies [133]. The second category comprises antibodies, which are well-suited for covering large PPI interfaces but typically restricted to extracellular targets. Their primary focus lies in manipulating PPIs involving immune checkpoint molecules, such as PD-1/PD-L1 or CD40/CD40 ligand (CD40L) interaction, thereby exerting profound influences on immune responses [134, 135]. Peptides represent the third category of PPI modulators, exhibiting high selectivity but hampered by a short half-life. To address this, chemical modifications like cyclization are frequently employed to improve their pharmacokinetic properties [136].

1.4.3 Fostering Innovations in Biotechnology

Biotechnology often harnesses the power of protein interactions to develop innovative solutions and products. This is vividly illustrated in synthetic biology, where naturally sourced or computationally designed PPIs serve as building blocks for constructing sophisticated genetic and protein circuits, endowing living cells with novel functionalities [137]. Generally, PPIs can be engineered to function as toggle switches and logic gates within synthetic genetic circuits, offering precise control over gene expression [138, 139]. Furthermore, by incorporating PPIs that govern posttranscriptional modifications like phosphorylation, synthetic protein circuits can rapidly respond to environmental changes, enabling dynamic modulation of protein interactions, localization, and degradation [140]. These programmable circuits hold immense potential for numerous applications. For instance, researchers created a chemically disruptable heterodimer based on the interaction between BCL-X_L (B-cell lymphoma-extra large) and BH3 (BCL-2 homology 3)-mimic peptide. Incorporating this PPI into the design of chimeric antigen receptor (CAR) led to a controllable CAR-T cell therapy, whose activity could be modulated by timed administration of a small-molecule BCL-X_L inhibitor [141].

Protein assemblies found in nature, such as actin fibers, microtubules, and viral capsids, showcase the remarkable role of PPIs in organizing complex architectures. Inspired by this, scientists have employed rational design to fabricate protein assemblies with diverse superstructures like nanowires, nanotubes, 2D and 3D lattices, and cage structures [142]. This nanotechnology paves the way for the development of advanced biomaterials with tailored properties and functionalities. As an example, coiled-coils are ubiquitous protein oligomerization motifs that have the inherent ability to wrap each other into supercoiled helices. These structures can be engineered to undergo higher-order assembly into fibrous network-based hydrogels [143]. Compared to inorganic and polymeric hydrogels, such protein-based hydrogels exhibit higher biocompatibility and better flexibility, rendering them ideal candidates for tissue engineering scaffolds and drug delivery carriers. Another example comes from virus-like particles (VLPs), which are self-assembled nanostructures made up of viral structural proteins, mimicking the architecture and symmetry of native viruses but lacking viral genetic material. Demonstrating high immunogenicity alongside an excellent safety profile, VLPs are gaining popularity in vaccine development [144].

Immunosensors are analytical devices that leverage the specificity of antigen–antibody interactions to detect and quantify target molecules in complex samples, with the resulting immunoreaction converted into an electrochemical, optical, or other detectable signal [145]. Enabled by artificially produced polyclonal or monoclonal antibody, these biosensors offer detection of a wide array of analytes, encompassing tumor biomarkers (cancer antigen 15-3, carcinoembryonic antigen, and α-fetoprotein) [146], pathogens responsible for infectious diseases (e.g. SARS-CoV-2, *Streptococcus pneumoniae*, and *Legionella pneumophila*) [147], foodborne pathogenic microorganisms (e.g. *Salmonella*, *Escherichia coli*, and *Listeria monocytogenes*) [148], food allergens (e.g. ovalbumin in eggs, β-lactoglobulin in milk, and Ara h 1 in peanuts) [149], and pathogens in crops and water resources [145]. The simplicity, rapidity, and sensitivity of immunosensors have led to their widespread adoption across human health, food and agricultural safety, environmental monitoring, and other fields.

1.5 Conclusions and Perspectives

In the post-genomic era, where the focus has shifted from studying individual genes to exploring the complex interplay of biomolecules, the investigation of PPIs stands as a critical frontier. At the heart of virtually every biological function, from cellular signaling and gene expression to immune response and metabolic pathways, lies an intricate network of PPIs. By elucidating these interactions, researchers can gain valuable insights into how cells respond to environmental stimuli, communicate with one another, and maintain homeostasis. The significance of PPI research is further underscored by the increasing number of studies that have identified disease-associated PPIs. This knowledge has opened new avenues for therapeutic intervention, leading to the development of novel drugs that target specific PPIs involved in disease progression.

Technological advancements have dramatically accelerated the exploration of PPIs, with the number of human binary PPIs identified through large-scale screens increasing from approximately 2700 in 2005 to 14,000 in 2014 and reaching 53,000 in 2020 [94]. Even so, the available PPI data represent only a fraction of the estimated human interactome, which is believed to contain 650,000 PPIs [150]. While experimental techniques have made significant strides, challenges remain in terms of throughput, sensitivity, and the ability to capture transient or weak interactions. Leveraging advanced computational techniques, such as molecular docking, machine learning algorithms, and network analysis, can complement experimental findings and expand relevant data resources. Moreover, the emergence of cutting-edge technologies, like single-cell analysis, holds promise for elucidating cell-to-cell variability and the spatiotemporal heterogeneity of PPIs [151]. As the field of PPI research continues to evolve, our understanding of PPIs will be profoundly enhanced, leading to groundbreaking discoveries and advancements that benefit humanity.

References

1 Berggard, T., Linse, S., and James, P. (2007). Methods for the detection and analysis of protein-protein interactions. *Proteomics* 7: 2833–2842.
2 De Las Rivas, J. and Fontanillo, C. (2010). Protein-protein interactions essentials: key concepts to building and analyzing interactome networks. *PLoS Computational Biology* 6: e1000807.
3 Ngounou Wetie, A.G., Sokolowska, I., Woods, A.G. et al. (2014). Protein-protein interactions: switch from classical methods to proteomics and bioinformatics-based approaches. *Cellular and Molecular Life Sciences* 71: 205–228.
4 Van Roey, K., Uyar, B., Weatheritt, R.J. et al. (2014). Short linear motifs: ubiquitous and functionally diverse protein interaction modules directing cell regulation. *Chemical Reviews* 114: 6733–6778.

5 Seo, M.H. and Kim, P.M. (2018). The present and the future of motif-mediated protein-protein interactions. *Current Opinion in Structural Biology* 50: 162–170.

6 Yan, C., Wu, F., Jernigan, R.L. et al. (2008). Characterization of protein-protein interfaces. *The Protein Journal* 27: 59–70.

7 Keskin, O., Tuncbag, N., and Gursoy, A. (2016). Predicting protein-protein interactions from the molecular to the proteome level. *Chemical Reviews* 116: 4884–4909.

8 Moreira, I.S., Fernandes, P.A., and Ramos, M.J. (2007). Hot spots – a review of the protein-protein interface determinant amino-acid residues. *Proteins* 68: 803–812.

9 Bogan, A.A. and Thorn, K.S. (1998). Anatomy of hot spots in protein interfaces. *Journal of Molecular Biology* 280: 1–9.

10 Keskin, O., Ma, B., and Nussinov, R. (2005). Hot regions in protein-protein interactions: the organization and contribution of structurally conserved hot spot residues. *Journal of Molecular Biology* 345: 1281–1294.

11 Fischer, E. (1894). Influence of configuration on the action of enzymes. *Berichte der Deutschen Chemischen Gesellschaft* 27: 2985–2993.

12 Koshland, D.E. (1958). Application of a theory of enzyme specificity to protein synthesis. *Proceedings of the National Academy of Sciences of the United States of America* 44: 98–104.

13 Ma, B., Kumar, S., Tsai, C.J. et al. (1999). Folding funnels and binding mechanisms. *Protein Engineering* 12: 713–720.

14 Seok, S.H. (2021). Structural insights into protein regulation by phosphorylation and substrate recognition of protein kinases/phosphatases. *Life (Basel)* 11: 957.

15 Klein, T., Eckhard, U., Dufour, A. et al. (2018). Proteolytic cleavage-mechanisms, function, and "Omic" approaches for a near-ubiquitous post-translational modification. *Chemical Reviews* 118: 1137–1168.

16 Liu, R., Wu, J., Guo, H. et al. (2023). Post-translational modifications of histones: mechanisms, biological functions, and therapeutic targets. *MedComm (2020)* 4: e292.

17 Ullo, M.F. and Case, L.B. (2023). How cells sense and integrate information from different sources. *WIREs Mechanisms of Disease* 15: e1604.

18 Cheng, L., Xia, F., Li, Z. et al. (2023). Structure, function and drug discovery of GPCR signaling. *Molecular Biomedicine* 4: 46.

19 Bhusal, R.P., Foster, S.R., and Stone, M.J. (2020). Structural basis of chemokine and receptor interactions: key regulators of leukocyte recruitment in inflammatory responses. *Protein Science* 29: 420–432.

20 Kapingidza, A.B., Kowal, K., and Chruszcz, M. (2020). Antigen-antibody complexes. *Sub-Cellular Biochemistry* 94: 465–497.

21 Sela-Culang, I., Kunik, V., and Ofran, Y. (2013). The structural basis of antibody-antigen recognition. *Frontiers in Immunology* 4: 302.

22 Saibil, H. (2013). Chaperone machines for protein folding, unfolding and disaggregation. *Nature Reviews. Molecular Cell Biology* 14: 630–642.

23 Kim, Y.E., Hipp, M.S., Bracher, A. et al. (2013). Molecular chaperone functions in protein folding and proteostasis. *Annual Review of Biochemistry* 82: 323–355.

24 Hiller, S. and Burmann, B.M. (2018). Chaperone-client complexes: a dynamic liaison. *Journal of Magnetic Resonance* 289: 142–155.

25 Hiller, S. (2019). Chaperone-bound clients: the importance of being dynamic. *Trends in Biochemical Sciences* 44: 517–527.

26 Zeke, A., Lukacs, M., Lim, W.A. et al. (2009). Scaffolds: interaction platforms for cellular signalling circuits. *Trends in Cell Biology* 19: 364–374.

27 Cortese, M.S., Uversky, V.N., and Dunker, A.K. (2008). Intrinsic disorder in scaffold proteins: getting more from less. *Progress in Biophysics and Molecular Biology* 98: 85–106.

28 Good, M.C., Zalatan, J.G., and Lim, W.A. (2011). Scaffold proteins: hubs for controlling the flow of cellular information. *Science* 332: 680–686.

29 Pan, C.Q., Sudol, M., Sheetz, M. et al. (2012). Modularity and functional plasticity of scaffold proteins as p(l)acemakers in cell signaling. *Cellular Signalling* 24: 2143–2165.

30 Jiang, H., Galtes, D., Wang, J. et al. (2022). G protein-coupled receptor signaling: transducers and effectors. *American Journal of Physiology. Cell Physiology* 323: C731–C748.

31 Paul, M.D. and Hristova, K. (2019). The RTK interactome: overview and perspective on RTK Heterointeractions. *Chemical Reviews* 119: 5881–5921.

32 Yoon, S. and Seger, R. (2006). The extracellular signal-regulated kinase: multiple substrates regulate diverse cellular functions. *Growth Factors* 24: 21–44.

33 He, Y., Sun, M.M., Zhang, G.G. et al. (2021). Targeting PI3K/Akt signal transduction for cancer therapy. *Signal Transduction and Targeted Therapy* 6: 425.

34 Hu, X., Li, J., Fu, M. et al. (2021). The JAK/STAT signaling pathway: from bench to clinic. *Signal Transduction and Targeted Therapy* 6: 402.

35 Lemmon, M.A., Freed, D.M., Schlessinger, J. et al. (2016). The dark side of cell signaling: positive roles for negative regulators. *Cell* 164: 1172–1184.

36 Buday, L. and Tompa, P. (2010). Functional classification of scaffold proteins and related molecules. *The FEBS Journal* 277: 4348–4355.

37 Talukdar, P.D. and Chatterji, U. (2023). Transcriptional co-activators: emerging roles in signaling pathways and potential therapeutic targets for diseases. *Signal Transduction and Targeted Therapy* 8: 427.

38 Soutourina, J. (2018). Transcription regulation by the Mediator complex. *Nature Reviews. Molecular Cell Biology* 19: 262–274.

39 Payankaulam, S., Li, L.M., and Arnosti, D.N. (2010). Transcriptional repression: conserved and evolved features. *Current Biology* 20: R764–R771.

40 Garg, G., Dienemann, C., Farnung, L. et al. (2023). Structural insights into human co-transcriptional capping. *Molecular Cell* 83: 2464–2477.e2465.

41 Kastner, B., Will, C.L., Stark, H. et al. (2019). Structural insights into nuclear pre-mRNA splicing in higher eukaryotes. *Cold Spring Harbor Perspectives in Biology* 11: a032417.

42 Boreikaite, V. and Passmore, L.A. (2023). 3′-End processing of eukaryotic mRNA: machinery, regulation, and impact on gene expression. *Annual Review of Biochemistry* 92: 199–225.

43 Brito Querido, J., Diaz-Lopez, I., and Ramakrishnan, V. (2024). The molecular basis of translation initiation and its regulation in eukaryotes. *Nature Reviews. Molecular Cell Biology* 25: 168–186.
44 Xu, B., Liu, L., and Song, G. (2021). Functions and regulation of translation elongation factors. *Frontiers in Molecular Biosciences* 8: 816398.
45 Hellen, C.U.T. (2018). Translation termination and ribosome recycling in eukaryotes. *Cold Spring Harbor Perspectives in Biology* 10: a032656.
46 Cappenberg, A., Kardell, M., and Zarbock, A. (2022). Selectin-mediated signaling-shedding light on the regulation of integrin activity in neutrophils. *Cells* 11: 1310.
47 Privratsky, J.R. and Newman, P.J. (2014). PECAM-1: regulator of endothelial junctional integrity. *Cell and Tissue Research* 355: 607–619.
48 Shah, K., Al-Haidari, A., Sun, J. et al. (2021). T cell receptor (TCR) signaling in health and disease. *Signal Transduction and Targeted Therapy* 6: 412.
49 Efremov, D.G., Turkalj, S., and Laurenti, L. (2020). Mechanisms of B cell receptor activation and responses to B cell receptor inhibitors in B cell malignancies. *Cancers (Basel)* 12: 1396.
50 Tkachenko, A., Kupcova, K., and Havranek, O. (2023). B-cell receptor signaling and beyond: the role of Igalpha (CD79a)/Igbeta (CD79b) in normal and malignant B cells. *International Journal of Molecular Sciences* 25: 10.
51 Susa, K.J., Seegar, T.C., Blacklow, S.C. et al. (2020). A dynamic interaction between CD19 and the tetraspanin CD81 controls B cell co-receptor trafficking. *eLife* 9: e52337.
52 Toma-Fukai, S. and Shimizu, T. (2021). Structural diversity of ubiquitin E3 ligase. *Molecules* 26: e52337.
53 Kim, Y., Kim, E.K., Chey, Y. et al. (2023). Targeted protein degradation: principles and applications of the proteasome. *Cells* 12: e2300083.
54 Siva Sankar, D. and Dengjel, J. (2021). Protein complexes and neighborhoods driving autophagy. *Autophagy* 17: 2689–2705.
55 Johansen, T. and Lamark, T. (2020). Selective autophagy: ATG8 family proteins, LIR motifs and cargo receptors. *Journal of Molecular Biology* 432: 80–103.
56 Hanahan, D. and Weinberg, R.A. (2011). Hallmarks of cancer: the next generation. *Cell* 144: 646–674.
57 Tatli, O. and Dinler Doganay, G. (2021). Recent developments in targeting RAS downstream effectors for RAS-driven cancer therapy. *Molecules* 26: 7561.
58 Liu, Y., Su, Z., Tavana, O. et al. (2024). Understanding the complexity of p53 in a new era of tumor suppression. *Cancer Cell* 42: 946–967.
59 Qian, S., Wei, Z., Yang, W. et al. (2022). The role of BCL-2 family proteins in regulating apoptosis and cancer therapy. *Frontiers in Oncology* 12: 985363.
60 Liu, Z.L., Chen, H.H., Zheng, L.L. et al. (2023). Angiogenic signaling pathways and anti-angiogenic therapy for cancer. *Signal Transduction and Targeted Therapy* 8: 198.
61 Han, Y., Liu, D., and Li, L. (2020). PD-1/PD-L1 pathway: current researches in cancer. *American Journal of Cancer Research* 10: 727–742.

62 Zhou, R., Yang, G., and Shi, Y. (2020). Macromolecular complex in recognition and proteolysis of amyloid precursor protein in Alzheimer's disease. *Current Opinion in Structural Biology* 61: 1–8.

63 Muralidar, S., Ambi, S.V., Sekaran, S. et al. (2020). Role of tau protein in Alzheimer's disease: the prime pathological player. *International Journal of Biological Macromolecules* 163: 1599–1617.

64 Wang, J., Dai, L., Chen, S. et al. (2024). Protein-protein interactions regulating alpha-synuclein pathology. *Trends in Neurosciences* 47: 209–226.

65 Wanker, E.E., Ast, A., Schindler, F. et al. (2019). The pathobiology of perturbed mutant huntingtin protein-protein interactions in Huntington's disease. *Journal of Neurochemistry* 151: 507–519.

66 Jackson, C.B., Farzan, M., Chen, B. et al. (2022). Mechanisms of SARS-CoV-2 entry into cells. *Nature Reviews. Molecular Cell Biology* 23: 3–20.

67 Kline, K.A., Fälker, S., Dahlberg, S. et al. (2009). Bacterial adhesins in host-microbe interactions. *Cell Host & Microbe* 5: 580–592.

68 Kumari, A., Tripathi, A.H., Gautam, P. et al. (2021). Adhesins in the virulence of opportunistic fungal pathogens of human. *Mycology* 12: 296–324.

69 Noriega, V., Redmann, V., Gardner, T. et al. (2012). Diverse immune evasion strategies by human cytomegalovirus. *Immunologic Research* 54: 140–151.

70 Brouwer, S., Rivera-Hernandez, T., Curren, B.F. et al. (2023). Pathogenesis, epidemiology and control of Group A Streptococcus infection. *Nature Reviews. Microbiology* 21: 431–447.

71 Clark, E., Nava, B., and Caputi, M. (2017). Tat is a multifunctional viral protein that modulates cellular gene expression and functions. *Oncotarget* 8: 27569–27581.

72 Winter, C., Henschel, A., Tuukkanen, A. et al. (2012). Protein interactions in 3D: from interface evolution to drug discovery. *Journal of Structural Biology* 179: 347–358.

73 Brito, J.A. and Archer, M. (2020). Chapter 10 - Structural biology techniques: X-ray crystallography, cryo-electron microscopy, and small-angle X-ray scattering. In: *Practical Approaches to Biological Inorganic Chemistry*, 2e (ed. R.R. Crichton and R.O. Louro), 375–416. Elsevier.

74 Burley, S.K., Berman, H.M., Duarte, J.M. et al. (2022). Protein data bank: a comprehensive review of 3D structure holdings and worldwide utilization by researchers, educators, and students. *Biomolecules* 12: 1425.

75 Alderson, T.R. and Kay, L.E. (2021). NMR spectroscopy captures the essential role of dynamics in regulating biomolecular function. *Cell* 184: 577–595.

76 Hu, Y., Cheng, K., He, L. et al. (2021). NMR-based methods for protein analysis. *Analytical Chemistry* 93: 1866–1879.

77 Kuhlbrandt, W. (2014). Biochemistry. The resolution revolution. *Science* 343: 1443–1444.

78 Fukuda, Y., Stapleton, K., and Kato, T. (2023). Progress in spatial resolution of structural analysis by cryo-EM. *Microscopy (Oxf)* 72: 135–143.

79 Wentinck, K., Gogou, C., and Meijer, D.H. (2022). Putting on molecular weight: enabling cryo-EM structure determination of sub-100-kDa proteins. *Current Research in Structural Biology* 4: 332–337.
80 Wang, H.W. and Wang, J.W. (2017). How cryo-electron microscopy and X-ray crystallography complement each other. *Protein Science* 26: 32–39.
81 Geraets, J.A., Pothula, K.R., and Schroder, G.F. (2020). Integrating cryo-EM and NMR data. *Current Opinion in Structural Biology* 61: 173–181.
82 Kastritis, P.L. and Bonvin, A.M. (2013). On the binding affinity of macromolecular interactions: daring to ask why proteins interact. *Journal of the Royal Society, Interface* 10: 20120835.
83 Zhou, M., Li, Q., and Wang, R. (2016). Current experimental methods for characterizing protein-protein interactions. *ChemMedChem* 11: 738–756.
84 Poluri, K.M., Gulati, K., and Sarkar, S. (2021). Experimental methods for determination of protein–protein interactions. In: *Protein-Protein Interactions: Principles and Techniques: Volume I* (ed. K.M. Poluri, K. Gulati, and S. Sarkar), 197–264. Singapore: Springer.
85 Murali, S., Rustandi, R.R., Zheng, X. et al. (2022). Applications of surface plasmon resonance and biolayer interferometry for virus-ligand binding. *Viruses* 14: 717.
86 Zhou, L., Todorovic, V., Kakavas, S. et al. (2018). Quantitative ligand and receptor binding studies reveal the mechanism of interleukin-36 (IL-36) pathway activation. *The Journal of Biological Chemistry* 293: 403–411.
87 Ylera, F., Harth, S., Waldherr, D. et al. (2013). Off-rate screening for selection of high-affinity anti-drug antibodies. *Analytical Biochemistry* 441: 208–213.
88 Nastle, L., Deuschle, F.C., Morath, V. et al. (2023). FerryCalin: an engineered lipocalin protein directed against the transferrin receptor with potential for brain drug delivery. *Chembiochem* 24: e202200795.
89 Roselin, L.S., Lin, M.S., Lin, P.H. et al. (2010). Recent trends and some applications of isothermal titration calorimetry in biotechnology. *Biotechnology Journal* 5: 85–98.
90 Brookes, J.C., Gray, E.R., Loynachan, C.N. et al. (2023). Thermodynamic analysis of an entropically driven, high-affinity nanobody-HIV p24 interaction. *Biophysical Journal* 122: 279–289.
91 Bludau, I. and Aebersold, R. (2020). Proteomic and interactomic insights into the molecular basis of cell functional diversity. *Nature Reviews. Molecular Cell Biology* 21: 327–340.
92 Huttlin, E.L., Bruckner, R.J., Navarrete-Perea, J. et al. (2021). Dual proteome-scale networks reveal cell-specific remodeling of the human interactome. *Cell* 184: 3022–3040.e3028.
93 Stynen, B., Tournu, H., Tavernier, J. et al. (2012). Diversity in genetic in vivo methods for protein-protein interaction studies: from the yeast two-hybrid system to the mammalian split-luciferase system. *Microbiology and Molecular Biology Reviews* 76: 331–382.
94 Luck, K., Kim, D.K., Lambourne, L. et al. (2020). A reference map of the human binary protein interactome. *Nature* 580: 402–408.

95 Low, T.Y., Syafruddin, S.E., Mohtar, M.A. et al. (2021). Recent progress in mass spectrometry-based strategies for elucidating protein-protein interactions. *Cellular and Molecular Life Sciences* 78: 5325–5339.

96 Sharifi Tabar, M., Parsania, C., Chen, H. et al. (2022). Illuminating the dark protein-protein interactome. *Cell Reports Methods* 2: 100275.

97 Matthews, L.R., Vaglio, P., Reboul, J. et al. (2001). Identification of potential interaction networks using sequence-based searches for conserved protein-protein interactions or “interologs”. *Genome Research* 11: 2120–2126.

98 Mistry, J., Chuguransky, S., Williams, L. et al. (2021). Pfam: the protein families database in 2021. *Nucleic Acids Research* 49: D412–D419.

99 Deng, M., Mehta, S., Sun, F. et al. (2002). Inferring domain-domain interactions from protein-protein interactions. *Genome Research* 12: 1540–1548.

100 Burger, L. and van Nimwegen, E. (2008). Accurate prediction of protein-protein interactions from sequence alignments using a Bayesian method. *Molecular Systems Biology* 4: 165.

101 Tang, T., Zhang, X., Liu, Y. et al. (2023). Machine learning on protein-protein interaction prediction: models, challenges and trends. *Briefings in Bioinformatics* 24: bbad076.

102 Casadio, R., Martelli, P.L., and Savojardo, C. (2022). Machine learning solutions for predicting protein–protein interactions. *WIREs Computational Molecular Science* 12: e1618.

103 Khatun, M.S., Shoombuatong, W., Hasan, M.M. et al. (2020). Evolution of sequence-based bioinformatics tools for protein-protein interaction prediction. *Current Genomics* 21: 454–463.

104 Burke, D.F., Bryant, P., Barrio-Hernandez, I. et al. (2023). Towards a structurally resolved human protein interaction network. *Nature Structural & Molecular Biology* 30: 216–225.

105 Huang, S.Y. (2014). Search strategies and evaluation in protein-protein docking: principles, advances and challenges. *Drug Discovery Today* 19: 1081–1096.

106 Pierce, B.G., Wiehe, K., Hwang, H. et al. (2014). ZDOCK server: interactive docking prediction of protein-protein complexes and symmetric multimers. *Bioinformatics* 30: 1771–1773.

107 van Zundert, G.C.P., Rodrigues, J., Trellet, M. et al. (2016). The HADDOCK2.2 Web server: user-friendly integrative modeling of biomolecular complexes. *Journal of Molecular Biology* 428: 720–725.

108 Kozakov, D., Hall, D.R., Xia, B. et al. (2017). The ClusPro web server for protein-protein docking. *Nature Protocols* 12: 255–278.

109 Lyskov, S. and Gray, J.J. (2008). The RosettaDock server for local protein-protein docking. *Nucleic Acids Research* 36: W233–W238.

110 van Noort, C.W., Honorato, R.V., and Bonvin, A. (2021). Information-driven modeling of biomolecular complexes. *Current Opinion in Structural Biology* 70: 70–77.

111 Sunny, S. and Jayaraj, P.B. (2022). Protein-protein docking: past, present, and future. *The Protein Journal* 41: 1–26.

112 Sinha, S., Tam, B., and Wang, S.M. (2022). Applications of molecular dynamics simulation in protein study. *Membranes (Basel)* 12: 844.

113 Shukla, R. and Tripathi, T. (2020). Molecular dynamics simulation of protein and protein–ligand complexes. In: *Computer-Aided Drug Design* (ed. D.B. Singh), 133–161. Singapore: Springer.

114 Case, D.A., Aktulga, H.M., Belfon, K. et al. (2023). AmberTools. *Journal of Chemical Information and Modeling* 63: 6183–6191.

115 Pall, S., Zhmurov, A., Bauer, P. et al. (2020). Heterogeneous parallelization and acceleration of molecular dynamics simulations in GROMACS. *The Journal of Chemical Physics* 153: 134110.

116 Brooks, B.R., Brooks, C.L. 3rd, Mackerell, A.D. Jr., et al. (2009). CHARMM: the biomolecular simulation program. *Journal of Computational Chemistry* 30: 1545–1614.

117 Phillips, J.C., Hardy, D.J., Maia, J.D.C. et al. (2020). Scalable molecular dynamics on CPU and GPU architectures with NAMD. *The Journal of Chemical Physics* 153: 044130.

118 Wang, X.W., Madeddu, L., Spirohn, K. et al. (2023). Assessment of community efforts to advance network-based prediction of protein-protein interactions. *Nature Communications* 14: 1582.

119 Haque, M., Sarmah, R., and Bhattacharyya, D.K. (2018). A common neighbor based technique to detect protein complexes in PPI networks. *Journal, Genetic Engineering & Biotechnology* 16: 227–238.

120 Wang, X., Hu, P., and Hu, L. (2020). A novel stochastic block model for network-based prediction of protein-protein interactions. In: *Intelligent Computing Theories and Application*. Cham: Springer International Publishing.

121 Torres, L., Chan, K.S., Eliassi-Rad, T. et al. (2020). GLEE: Geometric Laplacian Eigenmap Embedding. *Journal of Complex Networks* 8: 1–17.

122 Balogh, O.M., Benczik, B., Horvath, A. et al. (2022). Efficient link prediction in the protein-protein interaction network using topological information in a generative adversarial network machine learning model. *BMC Bioinformatics* 23: 78.

123 Sharifi Tabar, M., Francis, H., Yeo, D. et al. (2022). Mapping oncogenic protein interactions for precision medicine. *International Journal of Cancer* 151: 7–19.

124 Calabrese, G., Molzahn, C., and Mayor, T. (2022). Protein interaction networks in neurodegenerative diseases: from physiological function to aggregation. *The Journal of Biological Chemistry* 298: 102062.

125 Poluri, K.M., Gulati, K., Tripathi, D.K. et al. (2023). Protein-protein interactions in host–pathogen interactions. In: *Protein-Protein Interactions: Pathophysiological and Therapeutic Aspects: Volume II* (ed. K.M. Poluri, K. Gulati, D.K. Tripathi, et al.), 207–264. Singapore: Springer Nature Singapore.

126 Zhou, P., Yang, X.L., Wang, X.G. et al. (2020). A pneumonia outbreak associated with a new coronavirus of probable bat origin. *Nature* 579: 270–273.

127 Nurrohman, D.T. and Chiu, N.F. (2024). Unraveling the dynamics of SARS-CoV-2 mutations: insights from surface plasmon resonance biosensor kinetics. *Biosensors (Basel)* 14: 99.

128 Yoon, T.Y. and Lee, H.W. (2019). Shedding light on complexity of protein-protein interactions in cancer. *Current Opinion in Chemical Biology* 53: 75–81.

129 Smith, M.A., Hall, R., Fisher, K. et al. (2015). Annotation of human cancers with EGFR signaling-associated protein complexes using proximity ligation assays. *Science Signaling* 8: ra4.

130 Alzyoud, L., Bryce, R.A., Al Sorkhy, M. et al. (2022). Structure-based assessment and druggability classification of protein-protein interaction sites. *Scientific Reports* 12: 7975.

131 Lu, H., Zhou, Q., He, J. et al. (2020). Recent advances in the development of protein-protein interactions modulators: mechanisms and clinical trials. *Signal Transduction and Targeted Therapy* 5: 213.

132 Wu, D., Li, Y., Zheng, L. et al. (2023). Small molecules targeting protein-protein interactions for cancer therapy. *Acta Pharmaceutica Sinica B* 13: 4060–4088.

133 Eradat, H. (2019). Venetoclax for the treatment of chronic lymphocytic leukemia. *Current Hematologic Malignancy Reports* 14: 469–476.

134 Twomey, J.D. and Zhang, B. (2021). Cancer immunotherapy update: FDA-approved checkpoint inhibitors and companion diagnostics. *The AAPS Journal* 23: 39.

135 Tang, T., Cheng, X., Truong, B. et al. (2021). Molecular basis and therapeutic implications of CD40/CD40L immune checkpoint. *Pharmacology & Therapeutics* 219: 107709.

136 Philippe, G.J.B., Craik, D.J., and Henriques, S.T. (2021). Converting peptides into drugs targeting intracellular protein-protein interactions. *Drug Discovery Today* 26: 1521–1531.

137 Rosa, S., Bertaso, C., Pesaresi, P. et al. (2021). Synthetic protein circuits and devices based on reversible protein-protein interactions: an overview. *Life (Basel)* 11: 1171.

138 Shui, S., Gainza, P., Scheller, L. et al. (2021). A rational blueprint for the design of chemically-controlled protein switches. *Nature Communications* 12: 5754.

139 Moon, T.S., Lou, C., Tamsir, A. et al. (2012). Genetic programs constructed from layered logic gates in single cells. *Nature* 491: 249–253.

140 Gordley, R.M., Williams, R.E., Bashor, C.J. et al. (2016). Engineering dynamical control of cell fate switching using synthetic phospho-regulons. *Proceedings of the National Academy of Sciences of the United States of America* 113: 13528–13533.

141 Giordano-Attianese, G., Gainza, P., Gray-Gaillard, E. et al. (2020). A computationally designed chimeric antigen receptor provides a small-molecule safety switch for T-cell therapy. *Nature Biotechnology* 38: 426–432.

142 Hamley, I.W. (2019). Protein assemblies: nature-inspired and designed nanostructures. *Biomacromolecules* 20: 1829–1848.

143 Hill, L.K., Meleties, M., Katyal, P. et al. (2019). Thermoresponsive protein-engineered coiled-coil hydrogel for sustained small molecule release. *Biomacromolecules* 20: 3340–3351.

144 Nooraei, S., Bahrulolum, H., Hoseini, Z.S. et al. (2021). Virus-like particles: preparation, immunogenicity and their roles as nanovaccines and drug nanocarriers. *Journal of Nanobiotechnology* 19: 59.

145 Kumar, H., Dhalaria, R., Guleria, S. et al. (2024). Immunosensors in food, health, environment, and agriculture: a review. *Environmental Chemistry Letters* 22: 2573–2605.

146 Palakollu, V.N., Veera Manohara Reddy, Y., Shekh, M.I. et al. (2024). Electrochemical immunosensing of tumor markers. *Clinica Chimica Acta* 557: 117882.

147 Pavia, C.S. and Plummer, M.M. (2021). The evolution of rapid antigen detection systems and their application for COVID-19 and other serious respiratory infectious diseases. *Journal of Microbiology, Immunology, and Infection* 54: 776–786.

148 Zhan, K., Chen, L., Li, S. et al. (2024). A novel metal-organic framework based electrochemical immunosensor for the rapid detection of Salmonella typhimurium detection in milk. *Food Chemistry* 444: 138672.

149 Hosu, O., Selvolini, G., and Marrazza, G. (2018). Recent advances of immunosensors for detecting food allergens. *Current Opinion in Electrochemistry* 10: 149–156.

150 Stumpf, M.P., Thorne, T., de Silva, E. et al. (2008). Estimating the size of the human interactome. *Proceedings of the National Academy of Sciences of the United States of America* 105: 6959–6964.

151 Wen, Y., Xie, D., and Liu, Z. (2022). Advances in protein analysis in single live cells: principle, instrumentation and applications. *TrAC Trends in Analytical Chemistry* 152: 116619.

2

Overview of Drug Discovery Targeting PPI Systems

Hao Ma[1] *and Jian Zhang*[1,2]

[1] *Ningxia Medical University, College of Pharmacy, Key Laboratory of Protection, Development and Utilization of Medicinal Resources in Liupanshan Area, Ministry of Education, No.1160, Shengli Street, Xingqing District, Yinchuan, Ningxia Hui Autonomous Region, China*
[2] *Shanghai Jiao Tong University School of Medicine, Medicinal Chemistry and Bioinformatics Center, 227 South Chongqing Road, Shanghai, China*

2.1 Introduction

PPIs are fundamental for the regulation of cellular processes, and their dysregulation is often implicated in the pathogenesis of various diseases. Targeting PPI systems has emerged as a promising strategy for drug discovery and the development of novel therapeutics to modulate molecular interactions and restore normal cellular function. This chapter aims to provide a comprehensive understanding of the challenges and opportunities in drug discovery targeting PPI systems, highlighting the importance of PPIs in biological processes and disease pathways, as well as the innovative strategies and technologies employed to identify and characterize PPI inhibitors.

The intricate network of PPIs plays a critical role in cellular signaling, enzymatic activity, gene regulation, and structural organization within cells. Disruption of essential PPIs can lead to aberrant cellular signaling, altered gene expression, and dysfunctional molecular machinery, contributing to the development and progression of diseases, including cancer, neurodegenerative disorders, and metabolic conditions [1]. Consequently, targeting PPIs offers a unique opportunity to intervene in disease pathways at the molecular level by modulating specific protein interactions, thereby restoring normal cellular function and potentially halt disease progression.

Despite the significant potential of targeting PPI systems for therapeutic intervention, drug discovery in this area presents several challenges. The complex and dynamic nature of PPIs demands the development of innovative approaches to identify and characterize small molecules that can selectively disrupt or stabilize specific protein interactions. Traditional drug discovery strategies, such as structure-based drug design and high-throughput screening, may not always be suitable for targeting PPIs, as these interactions often involve large and flat interfaces that are challenging to target with small molecules.

Targeting Protein–Protein Interactions for Drug Discovery, First Edition. Edited by Jian Zhang.

In recent years, advancements in computational biology, structural biology, and chemical biology have revolutionized the field of drug discovery targeting PPI systems [2, 3]. Computational modeling techniques, such as molecular docking and molecular dynamics simulations, enable the prediction of PPI interfaces and the design of small-molecule inhibitors that can disrupt critical protein interactions. Structural biology methods, including X-ray crystallography and nuclear magnetic resonance (NMR) spectroscopy, provide valuable insights into the three-dimensional structures of protein complexes, facilitating the rational design of PPI inhibitors with high binding affinity and specificity.

Furthermore, chemical biology approaches, such as fragment-based screening and covalent targeting strategies, have emerged as powerful tools for identifying novel PPI inhibitors with improved pharmacological properties and therapeutic potential. Fragment-based screening relies on the identification of small-molecule fragments that bind to specific regions of a protein interface, which can be further optimized into lead compounds through structure–activity relationship studies and medicinal chemistry refinement. Covalent targeting strategies involve the design of irreversible inhibitors that form covalent bonds with amino acid residues at the PPI interface, leading to enhanced potency and selectivity.

Drug discovery targeting PPI systems represents a dynamic and multidisciplinary field that holds great promise for the development of novel therapeutics to modulate protein interactions and combat complex diseases. By leveraging innovative technologies and approaches from computational biology, structural biology, and chemical biology, researchers can overcome the challenges associated with targeting PPIs and accelerate the identification of effective inhibitors for clinical translation. This chapter will delve into the key principles, strategies, and advancements in drug discovery targeting PPI systems, highlighting the potential impact of these novel therapies in addressing unmet medical needs and advancing precision medicine approaches.

2.2 Fundamentals of Protein–Protein Interactions

2.2.1 Basic Principles of Protein Structure and Function

Proteins are fundamental molecules in biological systems, performing diverse functions critical for cellular processes. Understanding the structure and function of proteins is essential for elucidating their roles in health and disease. Protein structure is intricately linked to its function, as the unique three-dimensional arrangement of amino acids determines its biochemical properties and interactions with other molecules [4]. Proteins are composed of linear chains of amino acids folded into specific configurations, guided by the sequence of amino acids and influenced by various forces, such as hydrogen bonding, hydrophobic interactions, electrostatic interactions, and disulfide bonds.

At the primary structure level, proteins are characterized by the sequence of amino acids, which encodes the genetic information necessary for protein synthesis. The specific sequence of amino acids in a protein chain dictates its overall structure and ultimately its function. The secondary structure of a protein refers to the local spatial arrangement of amino acids, typically in the form of alpha helices or beta sheets, stabilized by hydrogen bonds between backbone atoms. This regular folding pattern gives rise to distinct structural motifs within the protein.

The tertiary structure of a protein describes its overall three-dimensional conformation, resulting from interactions between distant amino acids in the sequence. Hydrophobic interactions play a key role in protein folding, as nonpolar amino acids tend to cluster in the protein's core to minimize exposure to the surrounding aqueous environment. Additional stabilizing forces, such as salt bridges, hydrogen bonds, and disulfide bonds, contribute to the protein's stability and native structure. The specific tertiary structure of a protein determines its functional properties, such as enzymatic activity, ligand binding, and molecular recognition.

Proteins may also exhibit quaternary structure, involving the assembly of multiple protein subunits to form a functional complex. Quaternary structure is critical for the functioning of many proteins, such as enzymes and regulatory proteins, which require interaction with other subunits for activity. Here, we use hemoglobin as an example to illustrate the process of the formation of protein structures from primary to quaternary levels, as detailed in Figure 2.1. The arrangement of subunits within a protein complex is crucial for its stability and function, often involving specific interfaces and binding sites that mediate subunit interactions.

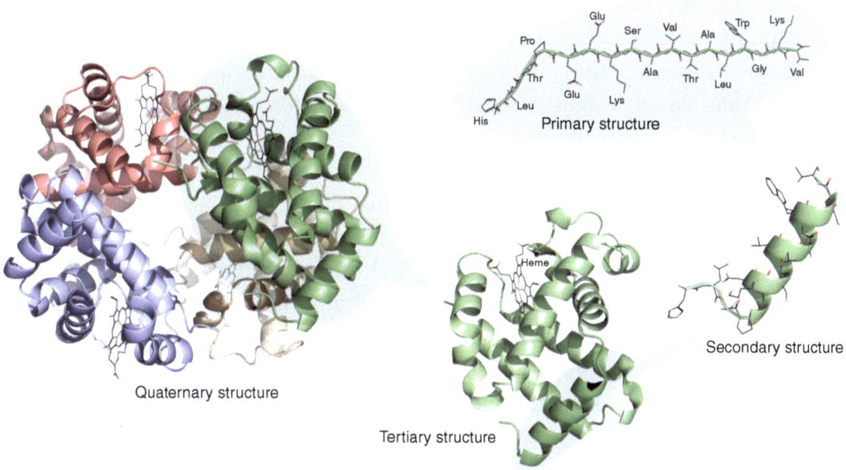

Figure 2.1 A schematic diagram of the protein's primary to quaternary structure, using hemoglobin as an example. The diagram highlights a segment of the HLTPEEKSAVT-ALWGKV sequence from one of hemoglobin's subunits. This sequence forms the first alpha helix at the N-terminus of the subunit, and subsequently, the remaining helices fold into a tertiary structure through non-covalent interactions. The individual subunits then associate through hydrophobic interactions to form the complete hemoglobin molecule, which represents the quaternary structure, enabling it to perform its biological function. (Source: Hao Ma, Jian Zhang)

In addition to structural considerations, protein function is intricately linked to its interactions with other molecules, such as substrates, cofactors, ligands, and other proteins. Proteins can undergo conformational changes upon binding to specific molecules, leading to activation or inhibition of their activity. PPIs are essential for the formation of signaling complexes, regulatory networks, and molecular machines that drive cellular processes.

Moreover, posttranslational modifications, such as phosphorylation, acetylation, and glycosylation, play a crucial role in modulating protein structure and function. These modifications can alter protein stability, activity, localization, and interactions with other molecules, adding another layer of complexity to protein regulation. The basic principles of protein structure and function are foundational to understanding the molecular mechanisms that govern biological processes. The intricate interplay between protein structure, function, and interactions underlies the diverse roles of proteins in cellular physiology and pathology [5].

2.2.2 Types of Protein–Protein Interactions

There are several types of PPIs that have been well-studied and characterized. One common type of interaction is the formation of protein complexes, where multiple proteins come together to form a stable structure [6]. These complexes can have different functions, such as catalyzing biochemical reactions, mediating signal transduction pathways, or acting as structural components of a cell.

Another type of PPI is protein binding, where two proteins interact transiently to carry out a specific function. This type of interaction is often mediated by specific binding motifs or domains on the proteins, allowing for precise and selective interactions. Protein binding interactions can also be regulated by posttranslational modifications, such as phosphorylation or methylation, which can modulate the strength and duration of the interaction.

Proteins can also interact through allosteric mechanisms, where binding of a ligand to one protein induces a conformational change that affects the activity of another protein. This type of interaction is important for the regulation of enzymatic activities, signal transduction pathways, and other cellular processes. Allosteric interactions allow for rapid and precise control of protein function in response to changes in the cellular environment.

In addition to these types of interactions, proteins can also interact through scaffolding proteins, which serve as platforms for the assembly of signaling complexes. Scaffolding proteins can bring together multiple proteins involved in a specific pathway, allowing for efficient and coordinated signaling. This type of interaction is important for the integration of multiple signals and the regulation of complex cellular processes.

2.2.3 Significance of PPIs in Cellular Processes and Disease Pathways

PPIs govern a multitude of essential biological processes, including signal transduction, enzymatic activities, gene regulation, and structural organization within cells. One key aspect of PPIs is their role in signal transduction pathways, which are responsible for transmitting extracellular signals to the nucleus to regulate gene expression and cellular responses. Signal transduction pathways often involve cascades of protein interactions that relay signals from cell surface receptors to intracellular signaling molecules, ultimately leading to changes in gene expression or cellular behavior. PPIs within these pathways are dynamic and context-dependent, allowing for precise control of signal transduction in response to various stimuli.

Enzymatic activities are also heavily reliant on PPIs, as many enzymes function as part of multi-protein complexes or require interactions with partner proteins to carry out their catalytic functions. PPIs can regulate enzymatic activity by modulating substrate recognition, catalytic efficiency, or allosteric regulation. Disruption of critical PPIs within enzymatic complexes can lead to dysregulation of metabolic pathways, aberrant cell signaling, and disease states.

Moreover, PPIs play a pivotal role in gene regulation processes by mediating interactions between transcription factors, co-regulators, and chromatin-modifying enzymes. These interactions contribute to the precise control of gene expression patterns by coordinating the recruitment of transcriptional machinery to specific genomic loci. Dysregulation of PPIs involved in gene regulation can lead to altered gene expression programs, contributing to the development of various diseases, including cancer, autoimmune disorders, and neurodegenerative conditions.

Structural organization within cells is another critical aspect influenced by PPIs, as proteins interact to form macromolecular complexes that perform specific functions within cellular compartments. These complexes provide spatial organization and functional specialization within cells, facilitating processes such as protein trafficking, cell cycle regulation, and organelle biogenesis. Disruption of PPIs that maintain cellular structure and organization can lead to cellular dysfunction and disease pathology.

In the context of disease pathways, aberrant PPIs have been implicated in the pathogenesis of numerous disorders, including cancer, neurodegenerative diseases, infectious diseases, and metabolic disorders. Dysregulated PPIs can result in the hyperactivation of oncogenic signaling pathways, the misfolding and aggregation of proteins implicated in neurodegeneration, or the perturbation of immune signaling pathways in autoimmune conditions. Identifying and targeting specific PPIs associated with disease states represents a promising strategy for the development of novel therapeutic interventions aimed at correcting molecular defects and restoring normal cellular function.

2.3 Challenges in Targeting PPI Systems

2.3.1 Structural Complexities of PPI Interfaces

One of the key features of PPI interfaces is their unique structural organization, which often involves large and flat interaction surfaces that mediate the binding between two or more protein partners. These interfaces are characterized by a combination of hydrophobic, electrostatic, and hydrogen bonding interactions that contribute to the stability and specificity of the protein complex. The complementarity of surface complementarity between interacting proteins is critical for achieving high-affinity binding and selectivity, as even subtle deviations in the shape, size, or chemical composition of the interface can significantly impact the strength of the interaction.

The structural complexities of PPI interfaces are further compounded by the dynamic nature of PPIs, which can involve conformational changes, posttranslational modifications, and allosteric regulation. Proteins are inherently flexible molecules that can adopt multiple conformations and undergo structural rearrangements in response to environmental cues or binding events. As a result, PPI interfaces may exhibit induced fit or conformational plasticity, where the interacting proteins undergo structural changes to optimize their interaction geometry and binding affinity. These structural dynamics add an additional layer of complexity to the design of inhibitors targeting PPI interfaces, as the binding site may not be rigid and static but rather dynamic and adaptable.

The heterogeneity of PPI interfaces poses challenges for the identification of small-molecule inhibitors that can specifically target and disrupt protein interactions. Unlike enzyme active sites or receptor–ligand binding sites, PPI interfaces typically lack well-defined pockets or cavities that can accommodate small molecules with high binding affinity. Instead, PPI interfaces are characterized by extensive protein–protein contacts and solvent-exposed surfaces that present a larger and more challenging target for drug discovery. The flat and featureless nature of PPI interfaces complicates the design of small-molecule inhibitors that can effectively bind to and disrupt protein complexes without interfering with other cellular processes.

2.3.2 Dynamics and Flexibility of Protein Complexes

Proteins are inherently flexible molecules that can adopt multiple conformational states and undergo structural rearrangements in response to environmental cues or binding events. Protein complexes often exhibit conformational flexibility, where individual subunits or domains within the complex can move relative to each other to accommodate different binding partners or functional states. This dynamic behavior is mediated by a combination of intra-molecular motions, inter-molecular interactions, and allosteric regulation, which collectively contribute to the structural plasticity of protein complexes.

The dynamics of protein complexes are influenced by various factors, including protein flexibility, binding kinetics, posttranslational modifications, and environmental conditions. Conformational changes in protein complexes can be induced by ligand binding, allosteric regulation, PPIs, or changes in pH, temperature, or ionic strength. These structural dynamics enable protein complexes to respond to stimuli, transmit signals, and perform specific functions within the cell. For example, the assembly of multi-protein complexes involved in signal transduction pathways or gene regulation often requires conformational changes that facilitate the formation of active signaling complexes or enhance the specificity and affinity of PPIs.

Moreover, the dynamics of protein complexes play a critical role in determining their stability, specificity, and biological activity. Protein complexes with high flexibility and dynamic properties may exhibit enhanced binding affinity, allosteric regulation, or functional adaptability, enabling them to interact with multiple partners or perform diverse functions. Conversely, rigid or static protein complexes may have limited binding specificity, restricted conformational plasticity, or reduced catalytic efficiency. The conformational dynamics of protein complexes can also influence their susceptibility to degradation, proteolysis, or misfolding, which can impact their cellular localization, half-life, and biological function.

2.3.3 Druggability Issues Associated with PPI Targets

Identifying druggable sites within PPI interfaces is a formidable challenge due to the lack of traditional targetable pockets or grooves typically found in enzymes or traditional protein–ligand interactions. The flat and extended nature of PPI interfaces often necessitates the development of alternative strategies for targeting PIs.

One of the primary challenges is the structural complexity of PPI interfaces. Unlike traditional drug targets such as enzymes or receptors that often have well-defined binding pockets, PPI interfaces are typically large, flat, and lack distinct binding sites [7–9]. This structural feature makes it difficult to design small molecules that can selectively disrupt specific PPIs without affecting other essential cellular functions. Additionally, the dynamic and flexible nature of PPI interfaces further complicates the identification of druggable sites [10], as protein complexes can adopt multiple conformations and binding modes, leading to changes in the interaction interface over time.

The size and complexity of PPI interfaces pose challenges in identifying specific hotspots or critical residues that can be targeted for disruption. The lack of well-defined pockets or grooves in PPI interfaces hinders the rational design of small-molecule inhibitors that can effectively interfere with PPIs.

Furthermore, the transient and context-dependent nature of PPIs adds another layer of complexity, as the binding interfaces may vary in different functional states or cellular environments, necessitating a deep understanding of the dynamic behavior of PPI complexes.

2.4 Approaches in Drug Discovery Targeting PPI Systems

2.4.1 High-Throughput Drug Design (HTS) Methods

High-throughput drug design methods encompass computational techniques and automated experimental procedures aimed at expediently identifying and optimizing potential drug candidates. This paradigm shift has revolutionized drug discovery by markedly reducing the time and cost traditionally associated with developing new pharmaceuticals [11]. Historically, drug discovery relied on laborious trial and error processes, involving years of screening thousands of compounds to uncover a handful of promising candidates, followed by iterative rounds of testing and refinement. This approach was not only time-intensive but also prohibitively expensive, with costs for bringing a new drug to market often exceeding billions of dollars.

Recent advancements in computational biology, bioinformatics, and high-throughput screening (HTS) technologies have dramatically accelerated the drug discovery timeline [12]. Computational methods such as molecular modeling, virtual screening, and machine learning algorithms predict the binding affinity of drug candidates to target proteins, evaluate their pharmacokinetic profiles, and optimize molecular structures to enhance efficacy. These tools enable researchers to swiftly identify potential leads from vast compound libraries and prioritize them for further investigation. Concurrently, HTS technologies, including robotic synthesis and automated cell-based assays, have transformed how drug candidates are evaluated. These methods enable rapid screening of thousands of compounds to assess biological activity, toxicity, and specificity, pinpointing the most promising candidates for subsequent development. Automation in these processes enhances speed and efficiency while minimizing the risks of false positives and negatives.

The impact of high-throughput drug design methods on drug discovery has been transformative, facilitating the rapid and efficient identification of novel treatments across diverse disease areas. By integrating sophisticated computational modeling with automated experimental methodologies, researchers can now design and refine drug candidates in a fraction of the time and cost compared to traditional approaches. Continued advancements in these technologies promise even greater strides in pharmaceutical development, fostering the creation of new and improved therapies at an accelerated pace.

2.4.2 Structure-Based Drug Design (SBDD) Techniques

Structure-based drug design (SBDD) methodologies encompass the utilization of structural insights regarding a target protein to devise and enhance drug candidates. By comprehending the three-dimensional configuration of the target protein and its interactions with small molecules, researchers can methodically craft compounds that bind with heightened efficacy and selectivity, culminating in the creation of safer and more potent drugs. A pivotal advantage of SBDD lies in its capacity to illuminate the molecular underpinnings of drug–target interactions. Through the visualization of the binding site of a target protein and the analyzing of its structure at the atomic level, researchers can pinpoint crucial residues and regions essential for ligand binding. This knowledge empowers the design of molecules tailored to specifically target these sites, thereby enhancing efficacy and mitigating off-target effects.

SBDD facilitates the refinement of drug candidates via structure-based virtual screening and molecular docking simulations. These computational methodologies enable researchers to sift through extensive compound libraries and forecast their binding affinity to the target protein, aiding in the prioritization of the most promising candidates for subsequent experimental evaluation. By harnessing these computational tools, researchers can significantly streamline the time

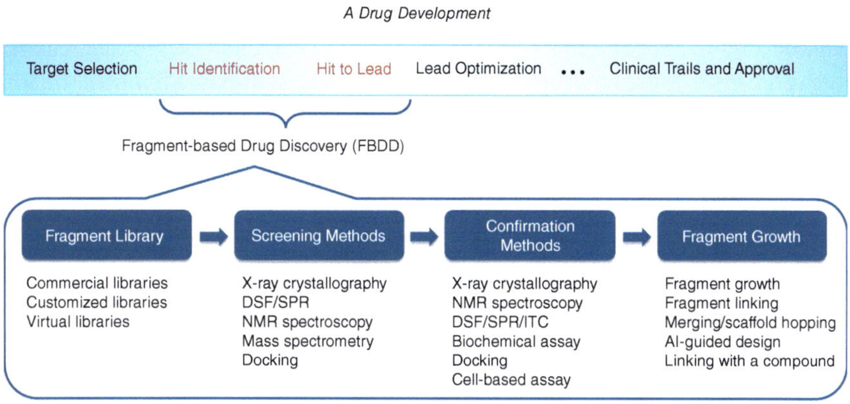

Figure 2.2 A flowchart of FBDD. The upper panel outlines the steps in target-based drug discovery, while the lower panel presents the FBDD process. X-ray crystallography is essential in FBDD and should always be considered. NMR plays a critical role in fragment screening. (Source: Hao Ma, Jian Zhang)

and resources required for drug discovery while augmenting the likelihood of identifying successful drug candidates.

SBDD has witnessed notable progress driven by advancements in computational capabilities, structural biology techniques, and the accessibility of high-resolution protein structures. The integration of artificial intelligence and machine learning algorithms has further bolstered the predictive prowess of SBDD, empowering researchers to more accurately model protein–ligand interactions and devise novel compounds possessing desired properties.

2.4.3 Fragment-Based Drug Discovery (FBDD) Strategies

Fragment-based drug discovery (FBDD, Figure 2.2) also known as fragment-based lead discovery (FBLD) strategies entail the screening of small, low molecular weight fragments to identify primary hits that can be subsequently optimized into lead compounds for drug development. This methodology has garnered significant attention in the pharmaceutical industry owing to its capacity to effectively explore chemical space, target protein–ligand interactions, and generate high-quality drug candidates. FBDD strategies specifically center on small, fragment-like molecules characterized by their reduced molecular complexity, which results in a heightened likelihood of binding to target proteins. Through the screening of a diverse repertoire of fragments, researchers can pinpoint initial hits with low micromolar affinity that can then undergo systematic optimization via structure-based design or fragment-linking approaches to enhance potency, selectivity, and drug-like attributes. This gradual optimization process facilitates the creation of therapeutics with enhanced efficacy and minimized off-target effects.

Furthermore, FBDD offers the advantage of delving into a broader chemical space in contrast to conventional HTS methods, thereby leading to the discovery of novel chemical scaffolds and pharmacophores. Leveraging fragments enables researchers to explore an array of chemical functionalities and interactions, facilitating the identification of unique binding motifs and the development of groundbreaking drug candidates. Moreover, the modular nature of fragment-based libraries streamlines the swift assembly of diverse compound collections for screening, imparting flexibility and efficiency in hit identification and optimization.

In recent years, FBDD has witnessed substantial growth and success in drug discovery initiatives within academia and the pharmaceutical sector. Technological advances in fragment screening methodologies, such as X-ray crystallography, NMR spectroscopy, and surface plasmon resonance (SPR), have empowered researchers to proficiently screen extensive fragment libraries against target proteins and unravel their binding mechanisms at the atomic level. These structural insights have played a pivotal role in guiding the optimization of fragment hits into lead compounds with enhanced potency and drug-like characteristics. FBDD continues to serve as a valuable and adaptable strategy in drug discovery, furnishing a rational and methodical approach to lead generation and refinement.

2.4.4 Computational Methods for Predicting PPI Inhibitors

Predicting PPIs and developing inhibitors constitutes a critical challenge in drug discovery, particularly for therapeutic strategies targeting complex molecular pathways implicated in diseases such as cancer and neurodegenerative disorders. Computational methods have emerged as indispensable tools in this endeavor, providing efficient and cost-effective avenues for screening extensive chemical libraries to identify potential inhibitors.

Among the primary computational approaches utilized, molecular docking stands out by simulating the interaction between small-molecule ligands and target proteins. By predicting the binding affinity and orientation of ligands within protein binding sites, docking algorithms facilitate the design of novel inhibitors. These algorithms assess the complementarity between ligand structures and protein surfaces to prioritize compounds capable of disrupting PPIs effectively.

Molecular dynamics (MD) simulations play a pivotal role in refining initial docked poses by exploring the dynamic behavior of protein–ligand complexes over time. They offer insights into the stability and flexibility of interactions, crucial for identifying compounds that form stable complexes under physiological conditions. This dynamic perspective is essential for understanding the temporal requirements of PPI inhibition in therapeutic contexts.

Quantitative structure–activity relationship (QSAR) models represent another valuable computational tool for predicting the biological activity of small molecules based on their chemical structure. By correlating structural features with experimental activity data, QSAR models prioritize compounds likely to exhibit desired inhibitory effects on specific PPI targets. Machine learning approaches further enhance QSAR models by uncovering intricate relationships between molecular descriptors and biological activities, thereby accelerating the discovery of potent PPI inhibitors.

Additionally, virtual screening techniques such as pharmacophore modeling and ligand-based similarity searching facilitate the identification of structurally diverse compounds with potential PPI inhibitory activity. Pharmacophore models define essential spatial arrangements of functional groups crucial for binding to target proteins, guiding the selection and optimization of lead compounds. Likewise, ligand-based similarity searches identify molecules sharing structural similarities with known inhibitors, thereby expanding the chemical space explored in inhibitor discovery efforts.

Recent advancements in computational chemistry, particularly the integration of artificial intelligence techniques like AlphaFold (AF) and RFdiffusion, have significantly influenced the prediction of PPI inhibitors. AF, developed by DeepMind, has transformed structural biology by accurately predicting protein structures using deep learning. By enhancing our understanding of the 3D structures of proteins involved in PPIs, AF provides a solid foundation for rational drug design aimed at protein interfaces. The latest iteration, AF3, represents a substantial milestone, offering improved accuracy and reliability in predicting protein structures, including those crucial to intricate PPI networks. This advancement holds great promise for designing novel PPI inhibitors with enhanced specificity and efficacy.

Many protein structures predicted by AF have reached or even surpassed experimental accuracy. However, AF is still far from fully replacing experimental methods for the precise prediction of all protein structures. Bonin et al. reported significant differences between the experimentally determined structure of IL-18 and the structure predicted by AF, as shown in Figure 2.3a. The detailed structural differences include several key aspects. The pro-domain (1–36) at the N-terminus is positioned far from the core protein region in the AF structure, whereas in the NMR structure, it is part of the core. Additionally, the β-sheet length varies, with the experimental structure having a longer β-sheet. Significant differences are also observed in other protein folding regions, such as β_{AF} and α_{AF}, where both the position and length of the secondary structures are inconsistent. Furthermore, the positions and lengths of $\beta 1$ and β^* differ substantially, with a backbone RMSD of 13.3 Å (shown as Figure 2.3b). Notably, this region is one of the binding sites for the substrate proteins caspase. The structure of the IL-18 precursor protein predicted by AF differs significantly from the experimental structure. However, the structure predicted by AF closely matches the conformation of the mature IL-18 (the protein after the caspase cleavage of 1–36 amino acid residues), with an RMSD of only 0.2 Å (shown as Figure 2.3c).

One significant limitation is that some intrinsically disordered proteins (IDPs) do not adopt a single, stable conformation in solution. In these cases, AF often predicts a structure that resembles the bound state, typically that of the substrate, leading to false positives, as observed with α-synuclein. Furthermore, AF is unable to predict the structural consequences of point mutations, such as those associated with diseases, nor can it capture the structural changes induced by posttranslational modifications. AF also faces challenges in predicting the effects of environmental factors, such as membrane environments, which are crucial for the structure of membrane proteins. Additionally, AF is not capable of accurately modeling the

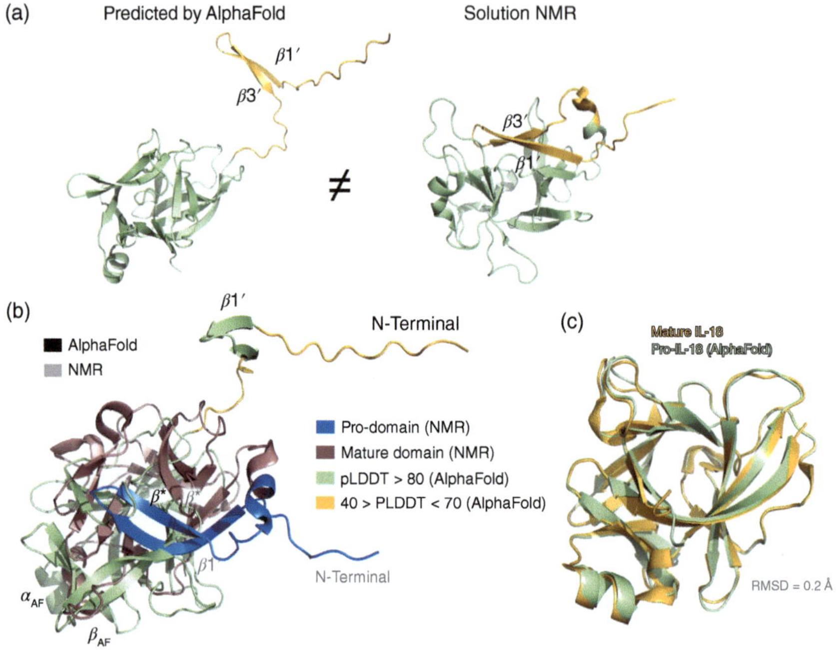

Figure 2.3 The protein structure predicted by AF exhibits some differences compared to the experimental measurements. (a) The structure of pro-IL-18 predicted by AlphaFold differs from the experimental results. (b) Solution NMR structure of pro-IL-18 differs significantly from a high confidence predicted AF2 model. (c) AF prediction of pro-IL-18 is virtually identical to the structure of mature IL-18. Source: Bonin et al. [13]/with permission of Elsevier.

conformational changes that occur upon protein–substrate binding, nor can it predict the multiple conformations a protein may adopt in solution. The accuracy of AF predictions is typically lower for loop regions, and even when the pLDDT confidence score is high, discrepancies remain between AF-predicted structures and experimental crystal structures. Lastly, there may be additional yet unidentified factors that could further influence the accuracy of AF predictions.

Combining the computational methods described above into drug discovery in a systematic way accelerates the identification and optimization of PPI inhibitors, providing notable advantages over traditional experimental approaches, particularly in terms of efficiency and resource utilization.

2.5 Case Studies and Success Stories

2.5.1 BCL Family

2.5.1.1 Bcl-2

Bcl-2 and Bcl-xL are anti-apoptotic proteins that play a significant role in cancer by inhibiting programmed cell death, making them highly attractive targets for therapeutic intervention. Despite this, the development of effective small-molecule inhibitors has proven challenging due to the necessity of targeting protein–protein interactions. Over the past few decades, numerous strategies have been employed to identify inhibitors of these proteins, yet only a handful have successfully advanced to clinical trials, with a few more in preclinical development.

In 2005, Oltersdorf et al. [14] reported the discovery of ABT-737, a small-molecule inhibitor of the anti-apoptotic proteins Bcl-2, Bcl-xL, and Bcl-w. This molecule was identified using a combination of NMR-based screening, parallel synthesis, and structure-based design. ABT-737 was shown to enhance pro-apoptotic signaling and induce cancer cell death both in vitro and in vivo, demonstrating high affinity (<1 nM) for Bcl-xL, Bcl-2, and Bcl-w. It exhibited single-agent activity against small-cell lung cancer and lymphoid malignancies and also potentiated the effects of several therapeutic agents in broader contexts. Notably, ABT-737 induced apoptosis in primary patient-derived lymphoma and chronic lymphocytic leukemia cells in vitro. However, its lack of oral bioavailability limited its clinical potential, particularly for chronic monotherapy and combination regimens. To address this limitation, Tse et al. [15] introduced the second-generation Bcl-2/Bcl-xL inhibitor, navitoclax (ABT-263), in 2008. Navitoclax not only displayed robust single-agent activity but also significantly enhanced the antitumor efficacy of conventional chemotherapeutics, leading to the development of practical combination regimens with parenteral chemotherapy drugs. These findings laid a strong theoretical foundation for the clinical use of navitoclax in treating small-cell lung cancer and hematologic malignancies.

However, the clinical application of navitoclax has been hampered by thrombocytopenia, a side effect resulting from Bcl-xL inhibition. In response, Souers et al. [16, 17] developed ABT-199 (venetoclax), a potent, orally bioavailable, and Bcl-2-selective inhibitor that circumvents this issue. ABT-199 has demonstrated efficacy in inhibiting the growth of Bcl-2-dependent tumors in vivo while sparing human platelets, and it is currently employed in the treatment of chronic lymphocytic leukemia. Meanwhile, in 2005, the team at Gemin X [18] discovered GX15-070 (Obatoclax), which exhibited significant antitumor activity across various hematologic malignancies, including leukemia, lymphoma, myelofibrosis, and mastocytosis (NCT00438178, NCT00600964).

In 2020, Mukherjee et al. [19] investigated the combined efficacy of two BH3 mimetics, ABT-199 and MCL1 inhibitors S63845 or S64315/MIK665 (NCT04629443), in treating melanoma, both in vitro and in vivo. They concluded that the combination of MCL1 and Bcl-2 inhibitors holds promise for treating advanced melanoma. Additionally, in 2020, Luo et al. [20] demonstrated that APG-2575, a novel Bcl-2 selective inhibitor, effectively targets recurrent or refractory diffuse large B-cell lymphoma (DLBCL) with high Bcl-2 expression by activating the mitochondrial apoptosis pathway. They also showed that combining Bruton's tyrosine kinase (BTK) inhibitor ibrutinib with a Bcl-2 inhibitor produced synergistic antitumor effects in DLBCL characterized by high Bcl-2 and MCL-1 expression. For DLBCL with wild-type p53 and high Bcl-2 expression, APG-2575, in combination with the MDM2-p53 inhibitor APG-115, demonstrated strong synergistic effects, significantly extending survival in animal models. This suggests that combining Bcl-2 and BTK or MDM2-p53 inhibitors may be an effective treatment strategy for DLBCL, warranting further clinical investigation.

In 2022, Wang and colleagues [21] discovered AT-101, a Bcl-2 family inhibitor that targets all anti-apoptotic Bcl-2 family members except Bfl-1. AT-101 is among the first small-molecule Bcl-2 inhibitors to enter clinical trials and is currently in phase II trials for hormone-refractory prostate cancer and other malignancies (NCT00286780, NCT05338931).

Efforts to develop peptide ligands targeting the Bcl-2 family have also made significant progress. In 2004, Walensky et al. [22] developed stapled peptides, known as "stabilized alpha-helix of Bcl-2 domains (SAHBs)," using a hydrocarbon-stapling strategy. These peptides specifically activate apoptotic pathways, effectively killing leukemia cells and inhibiting the growth of human leukemia xenografts in vivo. This all-hydrocarbon stapling approach may provide a viable strategy for experimentally and therapeutically modulating PPIs in various signaling pathways. In 2016, David P. Fairlie's team [23] successfully shortened the BAD BH3 domain-based long-stapled peptides to 8–14 residue peptides, retaining significant affinity for disrupting Bcl-xL PPI.

2.5.1.2 Bcl-xL

In 2013, Lessene et al. [24] identified WEHI-539, a novel small molecule that selectively targets Bcl-xL. This compound exhibits high affinity and selectively kills cells by antagonizing Bcl-xL's pro-survival activity. In 2014, Tao et al. [25, 26] reported on A-1155463, which demonstrated superior potency against Bcl-xL-dependent cell lines compared to WEHI-539, without the latter's inherent drug defects. By 2021, Tao's team had developed A-1293102, a new small-molecule inhibitor with picomolar binding affinity for Bcl-xL, which effectively and selectively eradicates Bcl-xL-dependent tumor cells.

In 2019, Gorostizaa and Ernest Giralt [27] conceptualized a modular design strategy using a universal template to develop nanoscale switchable peptides applicable to PPIs mediated by α-helices. By combining this universal template peptide approach with structural information from PPI hotspots, they developed a series of light-regulated peptide inhibitors targeting both Bcl-xL and Bak.

2.5.1.3 MCL-1

MCL-1, another anti-apoptotic protein in the Bcl-2 family, prevents cell apoptosis by binding to pro-apoptotic Bcl-2 proteins. Overexpression of MCL-1 is common in various tumor types and is closely associated with tumor progression, poor prognosis, and resistance to therapy. Despite its potential as a therapeutic target, clinical trials involving MCL-1 inhibitors face challenges due to resistance. Most reported MCL-1 inhibitors suffer from poor pharmacokinetics, low membrane permeability, and high serum protein binding, which limit their antitumor efficacy. Additionally, some MCL-1 inhibitors, such as AZD5991 and AMG-176, have exhibited off-target effects in cells, leading to the termination of clinical trials (NCT03218683, NCT03797261).

2.5.2 p53–MDM2

In 2004, Roche [28] conducted a screening of various synthetic chemicals and identified a series of cis-imidazoline analogs that inhibit the p53–MDM2 interaction, which they named Nutlins. These compounds were subsequently advanced to clinical trials. In 2013, Vu et al. [29] further optimized the structure of Nutlin-3, leading to the development of RG7112 (NCT01143740), the first experimental MDM2 inhibitor to enter clinical research for the treatment of advanced solid tumors. RG7112 effectively activates wild-type p53 and inhibits tumor growth in models characterized by MDM2 gene amplification and overexpression of MDM2 protein [30]. Later, the team reported the discovery and characterization of RG7388, a second-generation MDM2 inhibitor with enhanced potency and selectivity, which is also in clinical development (NCT02545283) [31].

In 2014, Rew et al. [32] discovered the piperidine-based MDM2 inhibitor AMG232 (NCT05027867) through a combination of de novo design, X-ray crystallography, molecular modeling, and iterative medicinal chemistry. By optimizing the conformational control, conducting extensive structure-activity relationship (SAR) studies, and capturing additional interactions with the "Gly58 scaffold" region, they identified AMG232 as a potent and selective inhibitor of the MDM2-p53 PPI, which is currently undergoing clinical trials for cancer treatment.

Simultaneously, Weisberg et al. [33] identified the HDM2 inhibitor CGM097 (NCT01760525), which suppresses the proliferation of acute myeloid leukemia (AML) by inducing apoptosis and inhibiting cell cycle progression. CGM097 selectively enhances the activity of wild-type p53 in AML cases expressing wild-type p53. Additionally, when combined with FLT3 inhibitors and MEK inhibitors, CGM097 effectively reduces leukemia burden in vivo, highlighting its potential as both a monotherapy and in combination with other targeted therapies for AML.

In 2016, Furet and colleagues [34], considering the MDM2 binding conformation of dihydropyridone inhibitors such as NVP-CGM097, designed a series of highly efficient inhibitors with a pyrazolopyrrolone core structure, which laid the foundation for the discovery of HDM201 (NCT05447663). In 2017, Aguilar and colleagues [35], building on their laboratory-designed spiroindole MDM2 inhibitors, addressed the challenge of slow isomerization in solution. They reported the design of a novel spiroindole containing two identical substituents at the C-2 position of the pyrrole ring, enabling rapid and irreversible conversion to a single enantiomer. Extensive SAR studies on these MDM2 inhibitors led to the discovery of AA-115/APG-115 (NCT02935907), an effective and chemically stable MDM2 inhibitor.

In 2024, Liu and colleagues [36] synthesized a novel compound, pvv-rg7388, which integrates the advantages of MDM2 inhibitors with platinum-based drugs. Through the synergistic action of RG7388 and cisplatin, this compound induces cell cycle arrest at the S and G2/M phases, effectively promoting apoptosis. Complexes 5a and 5b demonstrated significantly enhanced antiproliferative activity against both wild-type and mutant p53 cancer cells, with complex 5a showing effective in vivo tumor growth inhibition in the A549 xenograft model without notable toxicity.

This study suggests that the combination of these compounds exhibits stronger antitumor activity both in vitro and in vivo compared to either drug alone.

Peptides also play a crucial role in regulating the p53–MDM2 interaction. In 2011, Niu et al. [37] pioneered the synthesis of a series of γ-amino acid (γ-AA) peptides with distinct side-chain functional groups. These peptides demonstrated superior efficacy in inhibiting the p53–MDM2 interaction compared to traditional p53 mimetic peptides. The design was based on critical p53 residues , phenylalanine 19 (Phe19), tryptophan 23 (Trp23), and leucine 26 (Leu26) , which are essential for the MDM2-p53 interaction. The study highlighted γ-AA peptides as promising lead compounds, expanding the scope of peptide mimetics in disrupting the p53–MDM2 interaction.

By 2018, researchers recognized that previous bioactive peptides struggled to form stable helical structures, limiting their ability to fully mimic the p53 interface. In response, Teng et al. [38] introduced sulfonamide groups into existing γ-AA peptides to induce backbone bending and enhance folding tendencies. In 2020, Sang et al. [39] developed a subclass of sulfonated-γ-AA peptides designed to better simulate the α-helical conformation of p53. These peptides maintained stability through intramolecular hydrogen bonding, promoting α-helical formation in solution and exhibiting enhanced efficacy in blocking PPIs. The design of the sulfonated-γ-AA peptide PS1 led to the creation of the more effective PS10, with PS11 being cross-linked to improve cellular permeability. Additionally, the left-handed sequence of sulfonated-γ-AA peptides was used to develop the D-sulfonated-γ-AA4 peptide, a right-handed helical foldamer, which substituted leucine 26 with tryptophan, demonstrating comparable binding affinity to its lead compounds. The γ-AA4 peptide showed low proteolytic sensitivity, enhancing the potential of this novel D-sulfonated-γ-AA peptide for future biomedical applications [40].

To further diversify peptide mimetic inhibitors, researchers selected 1 : 1 right-handed α/sulfonated-γ-AA peptides as building block monomers for additional modification, resulting in the most effective helical heterodimer, 1 : 1 α/Sulfono-γ-AA1. The cross-linking strategy was also applied to optimize the best cross-linked peptide, α/Sulfono-γ-AA14, which exhibited an IC50 value of 4.9 μM. This peptide effectively disrupts the p53–MDM2 interaction and resists enzymatic hydrolysis, increasing its potential for future biomedical applications [41].

2.5.3 XIAP/c-IAP1

XIAP and c-IAP1 are members of the inhibitor of apoptosis proteins (IAPs) family, which play crucial roles in regulating anti-apoptotic and pro-survival signaling pathways. The overexpression of IAPs is frequently observed in various cancers and is associated with tumor progression and resistance to therapy.

In 2012, Flygare et al. [42] developed GDC-0512, a compound that effectively inhibits the proliferation of breast cancer cells. By substituting one of the nitrogen atoms with a carbon atom and incorporating an oxazole ring into the GDC-0512 structure, they optimized the inhibitor to create CUDC-427 (formerly known as GDC-0917). Studies have demonstrated that CUDC-427 exhibits dose-dependent anti-tumor activity in breast cancer xenograft models and holds potential for treating refractory solid tumors or lymphoma patients (NCT01226277) [43]. LCL-161, another IAP inhibitor, has been used in the treatment of advanced solid tumors and is structurally similar to GDC-0512 and CUDC-427. While the parent nucleus of LCL-161 has shown anticancer effects across various cancer cell lines, its clinical development has been hindered by poor cellular permeability [44]. In 2022, it was discovered that TNF-related apoptosis-inducing ligand (TRAIL) could potentially be combined with the Smac mimetic LCL-161 to induce cancer cell death (NCT01617668) [45].

In 2004, Sun et al. [46] identified an efficient conformationally constrained Smac mimetic, SM-122, through structure-based drug design. Subsequent modifications to SM-122 led to the development of SM-337, a potent oral diazabicyclic small molecule [47]. In 2017, Tamanini et al. [48] discovered AT-IAP, a potent dual antagonist of XIAP and c-IAP1, through structure-based drug design. To enhance metabolic stability and cardiac safety, a hydroxymethyl group was added to the pyridine ring, resulting in ASTX660. ASTX660 is a powerful inhibitor of c-IAP1 and XIAP, capable of inducing TNF-dependent apoptosis, inhibiting tumor growth in cancer cells, and improving metabolic stability and cardiac safety. ASTX660 is currently undergoing phase I/II clinical trials (NCT02503423).

2.5.4 CD40-CD40L

The overexpression of the CD40-CD40L signaling pathway is a key pathogenic mechanism underlying various autoimmune diseases [49]. In clinical practice, inhibitors targeting CD40 or CD40L, particularly monoclonal antibodies, are being explored for the treatment of multiple autoimmune conditions, including systemic lupus erythematosus, rheumatoid arthritis, and psoriasis. One such CD40 inhibitor is a fully human IgG1 antagonist antibody that binds specifically to CD40 [50], thereby blocking the activation of the endogenous CD40/CD40L pathway and inhibiting adaptive immune responses.

Frexalimab (SAR441344) is a novel monoclonal antibody that targets CD40L, aiming to block the co-stimulatory CD40-CD40L cell signaling pathway. This pathway is critical for the activation and function of adaptive immune cells (T and B cells) and innate immune cells (macrophages and dendritic cells) without requiring lymphocyte depletion [51]. In 2021, Sanofi [52] reported near one-year treatment results from a Phase II trial of frexalimab in patients with multiple sclerosis (MS) (NCT04879628). At 48 weeks, data analysis showed a sustained decrease in disease activity in patients treated with frexalimab, along with a favorable safety profile. On April 17 of this year, the National Medical Products Administration of China announced that Sanofi's Class 1 new drug frexalimab injection had received two silent approvals for clinical trials, aimed at developing treatments for adult relapsing multiple sclerosis and adult non-relapsing secondary progressive multiple sclerosis. Previously, the 12-week treatment trial results of frexalimab for MS were published in *The New England Journal of Medicine*. Concurrently, Sanofi has initiated Phase III clinical trials for frexalimab in the treatment of relapsing MS and non-relapsing secondary progressive MS. Beyond MS, frexalimab is also undergoing clinical trials for systemic lupus erythematosus (NCT05039840).

Historically, the development of CD40 inhibitor antibodies has received less attention compared to CD40 agonists, with pharmaceutical companies showing limited interest in this area. However, in recent years, there has been a resurgence of interest, with several major pharmaceutical companies, including Biogen, Sanofi, and AstraZeneca, actively pursuing the development of CD40 inhibitors for various autoimmune diseases. This renewed focus has led to accelerated research and development efforts. Biogen's Dapirolizumab Pegol, a pegylated Fab fragment of a monoclonal antibody targeting CD40L [53], is currently undergoing three Phase III clinical trials for systemic lupus erythematosus (NCT04976322), having shown promising efficacy in early trials. This molecule has also received approval for clinical trials in China (CTR20222715). Additionally, ASKP1240, another monoclonal antibody targeting CD40, is currently in a Phase IIa clinical trial for kidney transplantation (NCT01780844), with results expected to provide valuable insights into the potential of this pathway to improve transplant outcomes and treat autoimmune diseases [54].

Despite the therapeutic potential of targeting the CD40L pathway in autoimmune diseases, early research in this area has been marred by safety concerns, leading to the discontinuation of several clinical trials [55]. For example, Ruplizumab (hu5C8, BG9588), developed by Biogen, had its clinical development halted due to safety issues, including an increased incidence of thrombotic events, during its Phase II lupus nephritis trial [56]. Similarly, Toralizumab (IDEC-131) failed to meet efficacy targets in systemic lupus erythematosus patients and was associated with an elevated risk of thrombotic events, resulting in the termination of its clinical development [57]. In 2021, the Phase III trial for Iscalimab (CFZ533), a CD40L monoclonal antibody developed by Novartis, was terminated due to its failure to demonstrate superiority over cyclosporine in preventing organ rejection in kidney transplant recipients [58].

2.5.5 Cyclin-Dependent Kinase (CDK)

The synthesis of Roscovitine was first reported by Havlicek et al. To address concerns related to the structure and stereochemistry of Roscovitine, Wang et al. [59] undertook a study to determine its absolute configuration. They found that the R isomer of Roscovitine (CYC202) is a more potent and selective inhibitor of CDK2 cyclin E. CYC202 is a highly effective CDK inhibitor and is currently undergoing Phase II clinical trials for lung and B-cell malignancies.

CYC065, an advanced derivative of Roscovitine, is available for both intravenous and oral administration and has entered early clinical trials in patients with refractory acute myeloid leukemia (AML), myelodysplastic syndrome (NCT04017546), and chronic lymphocytic leukemia (NCT03739554). CYC065 primarily targets CDK9/2 and offers significantly improved metabolic stability, efficacy, and potency both in vitro and in vivo compared to its first-generation parent compound, Roscovitine. Previous studies have demonstrated that CYC065 can cross the blood–brain barrier and reduce tumor growth in an orthotopic mouse model of glioblastoma. Recent clinical trials have further highlighted the good tolerance of CYC065 in patients with advanced lymphoma or solid tumors (NCT02552953) [60].

2.5.6 PD-1/PD-L1

PD-1/PD-L1 signaling pathway inhibitors include monoclonal antibodies, peptides, and small-molecule inhibitors. In China, there are currently 16 approved PD-1/PD-L1 inhibitors available for treating conditions such as melanoma and non–small-cell lung cancer, offering a broad spectrum of therapeutic options for various cancers. Among these, Pembrolizumab is the first PD-1 inhibitor approved by the FDA, primarily used for treating advanced melanoma, while Nivolumab holds the distinction of being the first commercially available PD-1 inhibitor globally, suitable for a range of malignant tumors [61]. Toripalimab and Xindilimab, both independently developed in China, have demonstrated strong efficacy across multiple cancer types and have been approved for several indications.

Despite the widespread clinical application of PD-1/PD-L1 immunotherapies, the use of monoclonal antibodies can interfere with T-cell proliferation and activation, potentially leading to severe immune-related adverse reactions [62]. Unlike monoclonal antibodies, peptides and small-molecule drugs avoid these limitations. In 2015, Chang et al. developed the first anti-hydrolysis D-peptide antagonist, with the structurally optimized compound DPPA-1 exhibiting an in vitro binding affinity to PD-L1 of 0.51 μM [63]. Both cellular and animal studies demonstrated that DPPA-1 effectively disrupted the PD-1/PD-L1 interaction in vivo. That same year, Aurigene developed AUNP-12, a small peptide that targets PD-1 as an immunotherapy for cancer (though the structural details remain undisclosed) [64]. AUNP-12 inhibited PD-1/PD-L1 binding in vitro with an IC50 value of 0.72 nM, though it suffers from a short metabolic half-life [65]. Animal trials indicated that AUNP-12 displayed strong anti-PD-L1 activity, effectively inhibiting tumor growth and metastasis.

The development of small-molecule PD-1/PD-L1 inhibitors has lagged behind that of antibody-based therapies, with no small-molecule drugs currently approved for clinical use. However, with advancements in understanding the structure of PD-1 and PD-L1 proteins, Bristol Myers Squibb (BMS) disclosed its first patent for immunomodulators in 2015 [66]. These compounds were designed to disrupt the PD-1/PD-L1 interaction, with some showing nanomolar-range activity. Another BMS patent from that year introduced structural modifications, such as replacing the benzene ring in one segment with a 1,4-benzoquinone and incorporating an isocyanophenyl group through an ether bond into another segment [66]. These optimizations significantly enhanced the inhibitory activity of the compounds against PD-1/PD-L1, with IC50 values ranging from 0.6 to 10 nM. In 2018, BMS revealed new "centrally symmetrical" compounds, which featured identical or similar structural groups on both sides, resulting in activities typically below 1 nM, with a representative compound achieving an IC50 value of 0.04 nM [67].

CA-170, a small-molecule inhibitor inspired by the interface amino acids of PD-L1, represents a prototypical PPI inhibitor in this context [68]. CA-170 demonstrates strong inhibitory activity against the PD-1/PD-L1 interaction, with an EC50 value of 17 nM. It is characterized by a novel heterocyclic pyrazole scaffold, a urea linker, and connections to L-Ser and L-Thr, making it the first small-molecule PD-1/PD-L1 inhibitor to advance to clinical trials (NCT02812875).

2.5.7 Hsp90-Cdc37

Heat shock protein 90 (Hsp90) plays a critical role in tumor growth and progression. Cell division cycle 37 (Cdc37) selectively mediates the interaction between Hsp90 and specific client proteins, such as protein kinases, facilitating Hsp90's role in recognizing and stabilizing proteins essential for the cell cycle. Consequently, disrupting the interaction between Hsp90 and Cdc37 can destabilize key proteins in tumor cells, thereby inhibiting tumor growth and metastasis [69].

Since the introduction of 17-allylamino-17-demethoxygeldanamycin (17-AAG, tanespimycin) [70] as the first Hsp90 inhibitor in clinical trials in 1999, over 100 oncology trials involving Hsp90 inhibitors across various cancer types have been conducted. Over the past few decades, numerous Hsp90 inhibitors have entered clinical trials. However, as of 2021, only pimitespib (TAS-116) has recently received approval for clinical research in China and the United States, targeting gastrointestinal stromal tumors (NCT05245968) and breast cancer (NCT05655598). Unfortunately, most candidates have been deemed clinical failures due to low efficacy, toxicity, or resistance [71]. Despite the decline in the number of Hsp90 inhibitors in clinical trials, the proportion of drugs targeting Hsp90 and its associated interacting proteins has increased.

In 2019, Wang et al. [72] screened key residue sites at the Hsp90-Cdc37 interface, identifying potential binding sites on Cdc37 (Arg167, Gln208, Arg246) and Hsp90 (Arg46, Glu47, Tyr61, Gln133), and subsequently identified the lead compound 47 from a pool of 33 candidates. After structural optimization, the PPI inhibitor DDO-5936 was discovered. In 2020, this research team further optimized DDO-5936 by replacing the tetrahydrofuran ring with a piperazine ring, resulting in a compound with the highest in vitro binding affinity. Compared to DDO-5936, this optimized compound demonstrated enhanced antiproliferative activity, improved stability in plasma and microsomes, and increased in vivo oral efficacy, offering a promising approach for Hsp90-targeted cancer treatment.

In 2021, Siddiqui et al. [73] developed a series of derivatives with micromolar affinity capable of disrupting the Hsp90-Cdc37 interaction without binding to the ATP pocket. The inhibitors X1540 and X1742 have been shown to inhibit cancer cell proliferation and microtumor growth in both in vitro and in vivo experiments, presenting a novel alternative strategy for targeting this PPI.

2.5.8 Menin-MLL

Recent research on small-molecule Menin-MLL inhibitors has been relatively limited, with only two inhibitors, SNDX-5613 (NCT04065399) and KO-539 (NCT04067336), advancing to clinical trials. In 2022, it was discovered that both

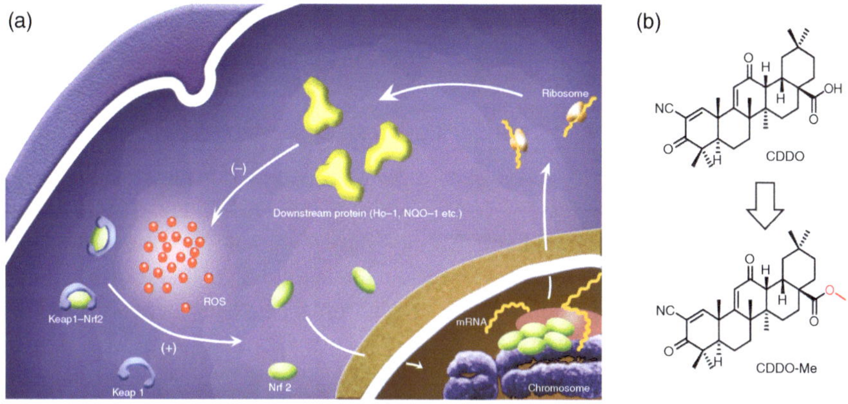

Figure 2.4 Schematic of the Keap1-Nrf2 Pathway and Modulators in Clinical Trials. (a) A brief schematic of the mechanism by which the Keap1-Nrf2 PPI influences reactive oxygen species within cells. (b) Chemical structure of CDDO and CDDO-Me.

inhibitors demonstrate efficacy when used in combination therapy against AML characterized by MLL rearrangements or NPM1 mutations (NPM1c) resulting from Menin inhibition [74, 75].

2.5.9 Kras-SOS1

In 2021, Ramharter et al. [76] identified an inhibitor named BI-68BS, which exhibited an IC50 in the picomolar range. Subsequent modifications led to the development of BI-3406, with an improved IC50 of 4 nM. When combined with the upstream EGFR inhibitor erlotinib, BI-3406 demonstrated significant efficacy against Kras-mutated colorectal cancer cells in vivo.

In 2022, Ketcham et al. [77] began their research with a simplified 6,7-dimethoxyphthalazine-based compound. Through a series of modifications, they developed the inhibitor MRTX0902 – a selective, brain-penetrant, and orally bioavailable SOS1-Kras inhibitor with an IC50 of 29 nM. This inhibitor features a novel scaffold distinct from those of previously discovered inhibitors, offering valuable insights for the design of new inhibitors targeting the Kras-SOS1 interaction.

2.5.10 Keap1-Nrf2 PPI

As shown in Figure 2.4a, the Keap1-Nrf2 system has been recognized as a crucial defense mechanism against oxidative stress, and modulating this pathway may offer an effective therapeutic strategy for a wide range of diseases. Activating Nrf2 has shown preliminary efficacy in treating cardiovascular, respiratory, and neurodegenerative diseases [78, 79].

Among the numerous anti-inflammatory drugs targeting the Keap1-Nrf2 pathway, three have been the most widely applied in recent years: dimethyl fumarate (DMF), silymarin, and bardoxolone methyl (CDDO-Me) [80, 81]. Researchers continue to advance the clinical development of these drugs, with several others also entering clinical trials.

Dimethyl fumarate (DMF), which gained significant attention, was approved by Biogen in 2014 for the treatment of relapsing-remitting multiple sclerosis [82]. DMF is also undergoing further clinical trials for the treatment of this condition (NCT05798520). Recent years have seen a surge in patents for new formulations, compositions, and derivatives of DMF, sparking renewed interest in the neuroprotective effects of sulforaphane. Studies have demonstrated the substantial potential of sulforaphane in the prevention and adjunctive therapy of various neurological disorders, including acute and chronic neurodegenerative diseases and neurodevelopmental disorders.

Bardoxolone methyl (CDDO-Me, Figure 2.4b), developed by Reata Pharmaceuticals, is another prominent Nrf2 activator. This semi-synthetic triterpenoid, derived from the natural product silymarin, robustly activates the Keap1-Nrf2 pathway [83]. Research suggests that CDDO-Me holds promise as a novel therapeutic approach for various kidney diseases [84]. Currently, a Phase II clinical trial is assessing the safety, tolerability, and efficacy of CDDO-Me in patients with chronic kidney disease at risk of rapid progression (NCT04702997).

However, the covalent binding of Nrf2 activators to cysteine residues on other proteins raises concerns about off-target effects. Consequently, preclinical research has focused on developing PPI inhibitors and PROTACs to enhance specificity [85]. Researchers have identified numerous non-covalent molecules that directly target the Keap1-Nrf2 PPI interface, including peptide inhibitors and small-molecule inhibitors.

Several linear and cyclic peptides have been reported, some designed to enhance cellular permeability. However, these peptides often exhibit insufficient or significantly diminished activity in cellular assays, limiting their effectiveness as Keap1-Nrf2 inhibitors. Iegre et al. [86] introduced a dual-component peptide docking strategy to rapidly access constrained and functional peptides targeting the Keap1-Nrf2 interaction. Among these, the P8-H peptide, which features a fatty acid tag, demonstrated nanomolar affinity for Keap1 and effectively induced ARE gene transcription, making it the most potent Keap1-Nrf2 peptide inhibitor to date. To further improve peptide activity and cellular permeability, Zou Jihua et al. [87] designed a series of 9-mer linear peptides based on Nrf2, including Ac-LDEETGEFL-NH2. By incorporating this linear peptide with the crystal structure of Keap1 binding, they developed a cyclic 9-mer, named ZC9, which exhibited a Kd value of 51 nM for Keap1, demonstrating high selectivity and affinity. In vitro, ZC9 showed significant dose-dependent reversal activity in models of LPS-induced oxidative damage and acute lung injury, with minimal toxicity.

Research has highlighted the immense potential of natural product-derived Keap1-Nrf2 pathway inhibitors in treating inflammatory diseases. Specifically, terpenoids, flavonoids, alkaloids, and quinones, which have shown remarkable efficacy in pharmacological studies, offer new directions for the clinical treatment of inflammatory conditions [88–91]. Moreover, inspired by the success of covalent Nrf2 activators, researchers have designed and screened a series of non-covalent molecules that directly target the Keap1-Nrf2 PPI interface. These novel molecules, including naphthylsulfonamides, tetrahydroisoquinolines, triazoles, and indoles, aim to enhance therapeutic selectivity, stability, and safety [92–95]. As research into these non-covalent molecules progresses, further refinements in their structures and activities are anticipated, paving the way for the development of more effective and safer therapeutic agents for inflammatory diseases.

We have compiled a summary of modulators designed to target PPIs that have entered clinical trials in recent years in the table below for the reader's review (Table 2.1).

2.6 Conclusion

Understanding the fundamental principles of protein structure and function is crucial for elucidating the roles of PPI systems in health and disease. Proteins perform a wide range of functions essential to cellular processes, as their structure intricately linked to their function. This chapter explores various types of PPIs, including the formation of protein complexes, protein binding, allosteric mechanisms, and interactions mediated by scaffolding proteins. PPIs are vital for processes such as signal transduction, enzymatic activities, gene regulation, and the structural organization within cells. Aberrant PPIs are implicated in the pathogenesis of numerous diseases, including cancer, autoimmune disorders, and neurodegenerative conditions, making them promising targets for therapeutic intervention.

The structural complexities and dynamics of PPI interfaces present significant challenges for drug discovery. Traditional drug discovery strategies may be inadequate for targeting PPIs due to their large, flat interfaces, which are challenging to target with small molecules. This chapter introduces various strategies for drug discovery targeting PPI systems, such as HTS methods that combine computational techniques with automated experimental workflows to accelerate the identification and optimization of drug candidates. SBDD utilizes insights into the target protein's structure to design and enhance drug candidates, while FBDD screens small, low-molecular-weight fragments to identify initial hits that can be optimized into lead compounds. Computational methods, including molecular docking, molecular dynamics simulations, and QSAR models, are also employed to predict PPI inhibitors and their interactions with protein binding sites.

The chapter highlights specific examples of drug development targeting PPIs, including frexalimab for multiple sclerosis, dapirolizumab pegol for systemic lupus erythematosus, and various CDK inhibitors for cancer treatment. It also discusses the development of inhibitors targeting PD-1/PD-L1, Hsp90-Cdc37, Menin-MLL, Kras-SOS1, and Keap1-Nrf2 interactions, highlighting both advancements and challenges in each area. The field of PPI systems is dynamic and interdisciplinary, offering immense potential for the development of novel therapeutics aimed at modulating protein interactions and combating complex diseases. By leveraging innovative technologies and methodologies from computational biology, structural biology, and chemical biology, researchers are well-equipped to overcome the

Table 2.1 Application of PPI drugs in clinical trials.

Target(s)	Trial drug	Indication(s)	Clinical trial number	Approval
Bcl-2/Bcl-xL	ABT-263	Chronic lymphocytic leukemia	NCT01087151	2008
Bcl-2	GX15-070	Acute myeloid leukemia, chronic lymphocytic leukemia	NCT00438178	2005
	APG-2575	Diffuse large B-cell lymphoma	NCT05147467	2020
	AT-101	B-cell lymphoma	NCT05338931	2022
MCL1	S64315	Acute myeloid leukemia	NCT04629443	2020
p53–MDM2	RG7112	Liposarcomas	NCT01143740	2010
	RG7388	Relapsed or refractory acute myeloid leukemia	NCT02545283	2015
	AMG232	Relapsed or refractory small-cell lung cancer	NCT05027867	2021
	CGM097	Acute myeloid leukemia	NCT01760525	2014
	HDM201	Acute myeloid leukemia	NCT05447663	2016
	APG-115	Advanced malignant solid neoplasm, lymphoma	NCT02935907	2017
XIAP/c-IAP1	GDC-0917	Solid tumors, lymphoma	NCT01226277	2010
	LCL-161	Triple negative breast cancer	NCT01617668	2012
	ASTX660	Advanced solid tumors, lymphomas	NCT02503423	2015
CDK	CYC065	Advanced cancers	NCT02552953	2015
PD-1/PD-L1	CA-170	Advanced solid tumors, lymphomas	NCT02812875	2016
Hsp90-Cdc37	TAS-116	Gastrointestinal stromal tumors	NCT05245968	2021
		Advanced breast cancer	NCT05655598	2022
Menin-MLL	NDX-5613	Acute myeloid leukemia, acute lymphoblastic leukemia, mixed lineage acute leukemia	NCT04065399	2019
	KO-539	Acute myeloid leukemia, mixed lineage leukemia	NCT04067336	2019
Keap1-Nrf2	DMF	Multiple sclerosis relapse	NCT05798520	2014
	CDDO-Me	Chronic kidney diseases	NCT04702997	2021

challenges associated with PPIs, facilitating the expedited identification of effective inhibitors and accelerating their translation into clinical applications.

Looking ahead, the discovery of PPI-targeted drugs faces several challenges. A deeper understanding of the dynamic behavior of PPI complexes is necessary to identify transient binding sites and hotspots for drug targeting. The development of novel inhibitors that effectively target the unique structural features of PPI interfaces is crucial. Advances in computational biology and bioinformatics are essential for predicting PPI interfaces and designing drugs capable of modulating these interactions. Additionally, designing drugs that selectively target PPIs without disrupting other cellular functions remains a significant challenge. Overcoming physical and chemical barriers to targeting PPIs with conventional small molecules is also imperative. Finally, translating discoveries in PPI-targeted drug discovery into clinically effective treatments requires ensuring safety, efficacy, and minimal side effects, while integrating PPI-targeted therapies into precision medicine approaches to tailor treatment plans to the unique disease characteristics of individual patients.

References

1 Ran, X. and Gestwicki, J.E. (2018). Inhibitors of protein-protein interactions (PPIs): an analysis of scaffold choices and buried surface area. *Current Opinion in Chemical Biology* 44: 75–86.

2 Wells, J.A. and McClendon, C.L. (2007). Reaching for high-hanging fruit in drug discovery at protein-protein interfaces. *Nature* 450 (7172): 1001–1009.

3 Huttlin, E.L., Ting, L., Bruckner, R.J. et al. (2015). The BioPlex Network: a systematic exploration of the human interactome. *Cell* 162 (2): 425–440.

4 Rual, J.F., Venkatesan, K., Hao, T. et al. (2005). Towards a proteome-scale map of the human protein-protein interaction network. *Nature* 437 (7062): 1173–1178.

5 van der Lee, R., Buljan, M., Lang, B. et al. (2014). Classification of intrinsically disordered regions and proteins. *Chemical Reviews* 114 (13): 6589–6631.

6 Roberts, C.J. (2014). Therapeutic protein aggregation: mechanisms, design, and control. *Trends in Biotechnology* 32 (7): 372–380.

7 Cheng, A.C., Coleman, R.G., Smyth, K.T. et al. (2007). Structure-based maximal affinity model predicts small-molecule druggability. *Nature Biotechnology* 25 (1): 71–75.

8 Smith, M.C. and Gestwicki, J.E. (2012). Features of protein-protein interactions that translate into potent inhibitors: topology, surface area and affinity. *Expert Reviews in Molecular Medicine* 14: e16.

9 Buchwald, P. (2010). Small-molecule protein-protein interaction inhibitors: therapeutic potential in light of molecular size, chemical space, and ligand binding efficiency considerations. *IUBMB Life* 62 (10): 724–731.

10 Ivanov, A.A., Khuri, F.R., and Fu, H. (2013). Targeting protein-protein interactions as an anticancer strategy. *Trends in Pharmacological Sciences* 34 (7): 393–400.

11 Carnero, A. (2006). High throughput screening in drug discovery. *Clinical & Translational Oncology : Official Publication of the Federation of Spanish Oncology Societies and of the National Cancer Institute of Mexico* 8 (7): 482–490.

12 Inglese, J., Johnson, R.L., Simeonov, A. et al. (2007). High-throughput screening assays for the identification of chemical probes. *Nature Chemical Biology* 3 (8): 466–479.

13 Bonin, J.P., Aramini, J.M., Dong, Y. et al. (2024). AlphaFold2 as a replacement for solution NMR structure determination of small proteins: Not so fast! *Journal of Magnetic Resonance* 364: 107725.

14 Oltersdorf, T., Elmore, S.W., Shoemaker, A.R. et al. (2005). An inhibitor of Bcl-2 family proteins induces regression of solid tumours. *Nature* 435 (7042): 677–681.

15 Tse, C., Shoemaker, A.R., Adickes, J. et al. (2008). ABT-263: a potent and orally bioavailable Bcl-2 family inhibitor. *Cancer Research* 68 (9): 3421–3428.

16 Souers, A.J., Leverson, J.D., Boghaert, E.R. et al. (2013). ABT-199, a potent and selective BCL-2 inhibitor, achieves antitumor activity while sparing platelets. *Nature Medicine* 19 (2): 202–208.

17 Pan, R., Hogdal, L.J., Benito, J.M. et al. (2014). Selective BCL-2 inhibition by ABT-199 causes on-target cell death in acute myeloid leukemia. *Cancer Discovery* 4 (3): 362–375.

18 Nguyen, M., Marcellus, R.C., Roulston, A. et al. (2007). Small molecule obatoclax (GX15-070) antagonizes MCL-1 and overcomes MCL-1-mediated resistance to apoptosis. *Proceedings of the National Academy of Sciences of the United States of America* 104 (49): 19512–19517.

19 Mukherjee, N., Amato, C.M., Skees, J. et al. (2020). Simultaneously inhibiting BCL2 and MCL1 is a therapeutic option for patients with advanced melanoma. *Cancers* 12 (8).

20 Luo, Q., Pan, W., Zhou, S. et al. (2020). A novel BCL-2 inhibitor APG-2575 exerts synthetic lethality with BTK or MDM2-p53 inhibitor in diffuse large B-cell lymphoma. *Oncology Research* 28 (4): 331–344.

21 Renner, O., Mayer, M., Leischner, C. et al. (2022). Systematic review of gossypol/AT-101 in cancer clinical trials. *Pharmaceuticals (Basel, Switzerland)* 15 (2): 144.

22 Walensky, L.D., Kung, A.L., Escher, I. et al. (2004). Activation of apoptosis in vivo by a hydrocarbon-stapled BH3 helix. *Science (New York, NY)* 305 (5689): 1466–1470.

23 Shepherd, N.E., Harrison, R.S., Ruiz-Gomez, G. et al. (2016). Downsizing the BAD BH3 peptide to small constrained α-helices with improved ligand efficiency. *Organic & Biomolecular Chemistry* 14 (46): 10939–10945.

24 Lessene, G., Czabotar, P.E., Sleebs, B.E. et al. (2013). Structure-guided design of a selective BCL-X(L) inhibitor. *Nature Chemical Biology* 9 (6): 390–397.

25 Tao, Z.F., Hasvold, L., Wang, L. et al. (2014). Discovery of a potent and selective BCL-XL inhibitor with in vivo activity. *ACS Medicinal Chemistry Letters* 5 (10): 1088–1093.

26 Tao, Z.F., Wang, X., Chen, J. et al. (2021). Structure-based design of A-1293102, a potent and selective BCL-X(L) inhibitor. *ACS Medicinal Chemistry Letters* 12 (6): 1011–1016.

27 Nevola, L., Varese, M., Martín-Quirós, A. et al. (2019). Targeted nanoswitchable inhibitors of protein-protein interactions involved in apoptosis. *ChemMedChem* 14 (1): 100–106.

28 Vassilev, L.T., Vu, B.T., Graves, B. et al. (2004). In vivo activation of the p53 pathway by small-molecule antagonists of MDM2. *Science (New York, NY)* 303 (5659): 844–848.

29 Vu, B., Wovkulich, P., Pizzolato, G. et al. (2013). Discovery of RG7112: a small-molecule MDM2 inhibitor in clinical development. *ACS Medicinal Chemistry Letters* 4 (5): 466–469.

30 Wang, S., Zhao, Y., Aguilar, A. et al. (2017). Targeting the MDM2-p53 protein-protein interaction for new cancer therapy: progress and challenges. *Cold Spring Harbor Perspectives in Medicine* 7 (5): a026245.

31 Ding, Q., Zhang, Z., Liu, J.J. et al. (2013). Discovery of RG7388, a potent and selective p53-MDM2 inhibitor in clinical development. *Journal of Medicinal Chemistry* 56 (14): 5979–5983.

32 Rew, Y. and Sun, D. (2014). Discovery of a small molecule MDM2 inhibitor (AMG 232) for treating cancer. *Journal of Medicinal Chemistry* 57 (15): 6332–6341.

33 Weisberg, E., Halilovic, E., Cooke, V.G. et al. (2015). Inhibition of wild-type p53-expressing AML by the novel small molecule HDM2 inhibitor CGM097. *Molecular Cancer Therapeutics* 14 (10): 2249–2259.

34 Furet, P., Masuya, K., Kallen, J. et al. (2016). Discovery of a novel class of highly potent inhibitors of the p53-MDM2 interaction by structure-based design starting from a conformational argument. *Bioorganic & Medicinal Chemistry Letters* 26 (19): 4837–4841.

35 Aguilar, A., Lu, J., Liu, L. et al. (2017). Discovery of 4-((3'R,4'S,5'R)-6″-Chloro-4′-(3-chloro-2-fluorophenyl)-1′-ethyl-2″-oxodispiro[cyclohexane-1,2′-pyrrolidine-3′,3″-indoline]-5′-carboxamido)bicyclo[2.2.2]octane-1-carboxylic acid (AA-115/APG-115): a potent and orally active murine double minute 2 (MDM2) inhibitor in clinical development. *Journal of Medicinal Chemistry* 60 (7): 2819–2839.

36 Liu, W., Ma, Y., He, Y. et al. (2024). Discovery of novel p53-MDM2 inhibitor (RG7388)-conjugated platinum(IV) complexes as potent antitumor agents. *Journal of Medicinal Chemistry* 67 (11): 9645–9661.

37 Niu, Y., Hu, Y., Li, X. et al. (2011). γ-AApeptides: design, synthesis and evaluation. *New Journal of Chemistry* 35: 542–545.

38 Teng, P., Niu, Z., She, F. et al. (2018). Hydrogen-bonding-driven 3D supramolecular assembly of peptidomimetic zipper. *Journal of the American Chemical Society* 140 (17): 5661–5665.

39 Sang, P., Shi, Y., Lu, J. et al. (2020). α-Helix-Mimicking Sulfono-γ-AApeptide Inhibitors for p53-MDM2/MDMX Protein-Protein Interactions. *Journal of Medicinal Chemistry* 63 (3): 975–986.

40 Sang, P., Shi, Y., Higbee, P. et al. (2020). Rational design and synthesis of right-handed d-sulfono-γ-AApeptide helical foldamers as potent inhibitors of protein-protein interactions. *The Journal of Organic Chemistry* 85 (16): 10552–10560.

41 Shi, Y., Sang, P., Lu, J. et al. (2020). Rational design of right-handed heterogeneous peptidomimetics as inhibitors of protein-protein interactions. *Journal of Medicinal Chemistry* 63 (21): 13187–13196.

42 Flygare, J.A., Beresini, M., Budha, N. et al. (2012). Discovery of a potent small-molecule antagonist of inhibitor of apoptosis (IAP) proteins and clinical candidate for the treatment of cancer (GDC-0152). *Journal of Medicinal Chemistry* 55 (9): 4101–4113.

43 Tolcher, A.W., Bendell, J.C., Papadopoulos, K.P. et al. (2016). A phase I dose-escalation study evaluating the safety tolerability and pharmacokinetics of CUDC-427, a potent, oral, monovalent IAP antagonist, in patients with refractory solid tumors. *Clinical Cancer Research : An Official Journal of the American Association for Cancer Research* 22 (18): 4567–4573.

44 Bai, L., Smith, D.C., and Wang, S. (2014). Small-molecule SMAC mimetics as new cancer therapeutics. *Pharmacology & Therapeutics* 144 (1): 82–95.

45 Granqvist, V., Holmgren, C., and Larsson, C. (2022). The combination of TRAIL and the Smac mimetic LCL-161 induces an irreversible phenotypic change of MCF-7 breast cancer cells. *Experimental and Molecular Pathology* 125: 104739.

46 Sun, H., Nikolovska-Coleska, Z., Yang, C.Y. et al. (2004). Structure-based design of potent, conformationally constrained Smac mimetics. *Journal of the American Chemical Society* 126 (51): 16686–16687.

47 Peng, Y., Sun, H., Nikolovska-Coleska, Z. et al. (2008). Potent, orally bioavailable diazabicyclic small-molecule mimetics of second mitochondria-derived activator of caspases. *Journal of Medicinal Chemistry* 51 (24): 8158–8162.

48 Tamanini, E., Buck, I.M., Chessari, G. et al. (2017). Discovery of a potent nonpeptidomimetic, small-molecule antagonist of cellular inhibitor of apoptosis protein 1 (cIAP1) and X-linked inhibitor of apoptosis protein (XIAP). *Journal of Medicinal Chemistry* 60 (11): 4611–4625.

49 Pucino, V., Gardner, D.H., and Fisher, B.A. (2020). Rationale for CD40 pathway blockade in autoimmune rheumatic disorders. *The Lancet Rheumatology* 2 (5): e292–e301.

50 Angelou, A., Antoniou, E., Garmpis, N. et al. (2018). The role of soluble CD40L ligand in human carcinogenesis. *Anticancer Research* 38 (5): 3199–3201.

51 Vermersch, P., Granziera, C., Mao-Draayer, Y. et al. (2024). Inhibition of CD40L with frexalimab in multiple sclerosis. *The New England Journal of Medicine* 390 (7): 589–600.

52 Cree, B.A.C., Hartung, H.P., and Barnett, M. (2022). New drugs for multiple sclerosis: new treatment algorithms. *Current Opinion in Neurology* 35 (3): 262–270.

53 Berman, D., Kenyon, N., Willman, M. et al. (2021). AT-1501, a novel and clinically applicable CD40L specific monoclonal antibody, promotes islet allograft survival in nonhuman primates [abstract]. *American Journal of Transplantation* 21 (suppl 3).

54 Oura, T., Yamashita, K., Suzuki, T. et al. (2012). Long-term hepatic allograft acceptance based on CD40 blockade by ASKP1240 in nonhuman primates. *American Journal of Transplantation : Official Journal of the American Society of Transplantation and the American Society of Transplant Surgeons* 12 (7): 1740–1754.

55 Kawai, T., Andrews, D., Colvin, R.B. et al. (2000). Thromboembolic complications after treatment with monoclonal antibody against CD40 ligand. *Nature Medicine* 6 (2): 114.

56 Elster, E.A., Xu, H., Tadaki, D.K. et al. (2001). Treatment with the humanized CD154-specific monoclonal antibody, hu5C8, prevents acute rejection of primary skin allografts in nonhuman primates. *Transplantation* 72 (9): 1473–1478.

57 Davis, J.C. Jr., Totoritis, M.C., Rosenberg, J. et al. (2001). Phase I clinical trial of a monoclonal antibody against CD40-ligand (IDEC-131) in patients with systemic lupus erythematosus. *The Journal of Rheumatology* 28 (1): 95–101.

58 Novartis announces discontinuation of CIRRUS-1 study of CFZ533 (iscalimab) in kidney transplant patients. Novartis.com2021.

59 Wang, S., McClue, S.J., Ferguson, J.R. et al. (2001). Synthesis and configuration of the cyclin-dependent kinase inhibitor roscovitine and its enantiomer. *Tetrahedron Asymmetry.* 12: 2891–2894.

60 Juric, V., Hudson, L., Fay, J. et al. (2021). Transcriptional CDK inhibitors, CYC065 and THZ1 promote Bim-dependent apoptosis in primary and recurrent GBM through cell cycle arrest and Mcl-1 downregulation. *Cell Death & Disease* 12 (8): 763.

61 Mullard, A. (2015). 2014 FDA drug approvals. *Nature Reviews Drug Discovery* 14 (2): 77–81.

62 Naidoo, J., Page, D.B., Li, B.T. et al. (2015). Toxicities of the anti-PD-1 and anti-PD-L1 immune checkpoint antibodies. *Annals of Oncology : Official Journal of the European Society for Medical Oncology* 26 (12): 2375–2391.

63 Vlieghe, P., Lisowski, V., Martinez, J. et al. (2010). Synthetic therapeutic peptides: science and market. *Drug Discovery Today* 15 (1–2): 40–56.

64 Chang, H.N., Liu, B.Y., Qi, Y. et al. (2015). Blocking of the PD-1/PD-L1 interaction by a d-peptide antagonist for cancer immunotherapy. *Angewandte Chemie (International ed. in English)* 54 (40): 11760–11764.

65 Sasikumar, P. G. N., Ramachandra, M., Vadlamani, S. K. et al. Immunosuppression modulating compounds. W02011161699 (2011).

66 Chupak, L. S. and Zheng, X. Compounds useful as immunomodulators. WO2015034820A1 (2015).

67 Yeung, K.-S., Grant-Young, K. A., Zhu J. et al. Compounds useful as immunomodulators. US patent WO2018044963A1 (2018).

68 Musielak, B., Kocik, J., Skalniak, L. et al. (2019). CA-170 - a potent small-molecule PD-L1 inhibitor or not? *Molecules (Basel, Switzerland)* 24 (15): 2804.

69 Wang, L., Zhang, L., Li, L. et al. (2019). Small-molecule inhibitor targeting the Hsp90-Cdc37 protein-protein interaction in colorectal cancer. *Science Advances* 5 (9): eaax2277.

70 Kamal, A., Boehm, M.F., and Burrows, F.J. (2004). Therapeutic and diagnostic implications of Hsp90 activation. *Trends in Molecular Medicine* 10 (6): 283–290.

71 Xie, X., Zhang, N., Li, X. et al. (2023). Small-molecule dual inhibitors targeting heat shock protein 90 for cancer targeted therapy. *Bioorganic Chemistry* 139: 106721.

72 Wang, L., Jiang, J., Zhang, L. et al. (2020). Discovery and optimization of small molecules targeting the protein-protein interaction of heat shock protein 90 (Hsp90) and cell division cycle 37 as orally active inhibitors for the treatment of colorectal cancer. *Journal of Medicinal Chemistry* 63 (3): 1281–1297.

73 Siddiqui, F.A., Parkkola, H., Vukic, V. et al. (2021). Novel small molecule Hsp90/Cdc37 interface inhibitors indirectly target K-ras-signaling. *Cancers* 13 (4): 927.

74 Fiskus, W., Boettcher, S., Daver, N. et al. (2022). Effective Menin inhibitor-based combinations against AML with MLL rearrangement or NPM1 mutation (NPM1c). *Blood Cancer Journal* 12 (1): 5.

75 Fiskus, W., Daver, N., Boettcher, S. et al. (2022). Activity of menin inhibitor ziftomenib (KO-539) as monotherapy or in combinations against AML cells with MLL1 rearrangement or mutant NPM1. *Leukemia* 36 (11): 2729–2733.

76 Ramharter, J., Kessler, D., Ettmayer, P. et al. (2021). One atom makes all the difference: getting a foot in the door between SOS1 and KRAS. *Journal of Medicinal Chemistry* 64 (10): 6569–6580.

77 Ketcham, J.M., Haling, J., Khare, S. et al. (2022). Design and discovery of MRTX0902, a potent, selective, brain-penetrant, and orally bioavailable inhibitor of the SOS1:KRAS protein-protein interaction. *Journal of Medicinal Chemistry* 65 (14): 9678–9690.

78 Madden, S.K. and Itzhaki, L.S. (2020). Structural and mechanistic insights into the Keap1-Nrf2 system as a route to drug discovery. *Biochimica et Biophysica Acta. Proteins and Proteomics* 1868 (7): 140405.

79 Lin, L., Wu, Q., Lu, F. et al. (2023). Nrf2 signaling pathway: current status and potential therapeutic targetable role in human cancers. *Frontiers in Oncology* 13: 1184079.

80 Cuadrado, A., Rojo, A.I., Wells, G. et al. (2019). Therapeutic targeting of the NRF2 and KEAP1 partnership in chronic diseases. *Nature Reviews Drug Discovery* 18 (4): 295–317.

81 Yore, M.M., Kettenbach, A.N., Sporn, M.B. et al. (2011). Proteomic analysis shows synthetic oleanane triterpenoid binds to mTOR. *PLoS One* 6 (7): e22862.

82 Zhou, H., Wang, Y., You, Q. et al. (2020). Recent progress in the development of small molecule Nrf2 activators: a patent review (2017-present). *Expert Opinion on Therapeutic Patents* 30 (3): 209–225.

83 Torres, V.E., Chapman, A.B., Devuyst, O. et al. (2017). Tolvaptan in later-stage autosomal dominant polycystic kidney disease. *The New England Journal of Medicine* 377 (20): 1930–1942.

84 Kanda, H. and Yamawaki, K. (2020). Bardoxolone methyl: drug development for diabetic kidney disease. *Clinical and Experimental Nephrology* 24 (10): 857–864.

85 Adinolfi, S., Patinen, T., Jawahar Deen, A. et al. (2023). The KEAP1-NRF2 pathway: targets for therapy and role in cancer. *Redox Biology* 63: 102726.

86 Iegre, J., Krajcovicova, S., Gunnarsson, A. et al. (2023). A cell-active cyclic peptide targeting the Nrf2/Keap1 protein-protein interaction. *Chemical Science* 14 (39): 10800–10805.

87 Zou, J., Yan, J., Lu, Y. et al. (2024). Cyclic peptide Keap1-Nrf2 protein-protein interaction inhibitors: design, synthesis, and in vivo treatment of acute lung injury. *Journal of Medicinal Chemistry* 67 (6): 4889–4903.

88 Cheng, C., Zhang, J., Liu, K. et al. (2023). Ginsenoside CK targeting KEAP1-DGR/Kelch domain disrupts the binding between KEAP1 and NRF2-DLG motif to ameliorate oxidative stress damage. *Phytomedicine : International Journal of Phytotherapy and Phytopharmacology* 119: 154992.

89 Huang, J., Zhu, Y., Li, S. et al. (2023). Licochalcone B confers protective effects against LPS-Induced acute lung injury in cells and mice through the Keap1/Nrf2 pathway. *Redox Report : Communications in Free Radical Research* 28 (1): 2243423.

90 Zhang, Y., Yan, T., Sun, D. et al. (2020). Rutaecarpine inhibits KEAP1-NRF2 interaction to activate NRF2 and ameliorate dextran sulfate sodium-induced colitis. *Free Radical Biology & Medicine* 148: 33–41.

91 Lee, S. and Hu, L. (2020). Nrf2 activation through the inhibition of Keap1-Nrf2 protein-protein interaction. *Medicinal Chemistry Research : An International Journal for Rapid Communications on Design and Mechanisms of Action of Biologically Active Agents* 29 (5): 846–867.

92 Jiang, Z.Y., Lu, M.C., Xu, L.L. et al. (2014). Discovery of potent Keap1-Nrf2 protein-protein interaction inhibitor based on molecular binding determinants analysis. *Journal of Medicinal Chemistry* 57 (6): 2736–2745.

93 Hu, L., Magesh, S., Chen, L. et al. (2013). Discovery of a small-molecule inhibitor and cellular probe of Keap1-Nrf2 protein-protein interaction. *Bioorganic & Medicinal Chemistry Letters* 23 (10): 3039–3043.

94 Davies, T.G., Wixted, W.E., Coyle, J.E. et al. (2016). Monoacidic inhibitors of the Kelch-like ECH-associated protein 1: nuclear factor erythroid 2-related factor 2 (KEAP1:NRF2) protein-protein interaction with high cell potency identified by fragment-based discovery. *Journal of Medicinal Chemistry* 59 (8): 3991–4006.

95 Cosimelli, B., Greco, G., Laneri, S. et al. (2019). Identification of novel indole derivatives acting as inhibitors of the Keap1-Nrf2 interaction. *Journal of Enzyme Inhibition and Medicinal Chemistry* 34 (1): 1152–1157.

3

Fluorescence Resonance Energy Transfer Technology and its Applications

Jing-Yu Lang

Kunming Medical University, School of Pharmaceutical Science & Yunnan Key Laboratory of Pharmacology for Natural Products, 1168 West Chunrong Road, Chenggong District, Kunming 650500, P.R. China

3.1 Introduction

Fluorescence resonance energy transfer (FRET) is a photophysical process in which energy is transferred nonradiatively from an excited donor fluorophore to a nearby acceptor fluorophore through dipole–dipole interactions [1]. This process is highly dependent on the distance between the donor and acceptor, making it a useful tool for studying molecular proximity and interactions at the nanometer scale. For efficient FRET to occur, several conditions must be met: (i) There must be a significant overlap between the emission spectrum of the donor and the absorption spectrum of the acceptor; (ii) The distance between the donor and acceptor must be within 1–10 nm of each other, as FRET efficiency decreases rapidly with increasing distance; (iii) The dipoles should be optimally aligned for maximum energy transfer, as the relative orientation of the donor and acceptor transition dipoles can affect FRET efficiency.

3.2 Mechanism of FRET

The FRET process begins when the donor fluorophore absorbs a photon and transitions to an excited electronic state. If an acceptor fluorophore is within a close range (typically 1–10 nm), the excited state energy can be transferred to the acceptor without the emission of a photon. This energy transfer occurs through dipole–dipole coupling, where the efficiency of the transfer is inversely proportional to the sixth power of the distance between the donor and acceptor:

$$E = 1/\left(1 + (r/R_0)^6\right),$$

where E is the FRET efficiency, r is the distance between the donor and acceptor, and R_0 is the Förster distance. The Förster distance is the distance between donor and acceptor at which energy transfer efficiency is 50%, and it depends on the

Targeting Protein–Protein Interactions for Drug Discovery, First Edition. Edited by Jian Zhang.

spectral properties of the donor and acceptor, their quantum yields, and the relative orientation of their transition dipoles.

The selection of appropriate donor and acceptor fluorophores is crucial for the success of FRET experiments. There are several factors to consider when choosing a FRET pair, including spectral overlap, photostability, brightness, and the biological context. Cyan and yellow fluorescent proteins that are genetically encodable and well-characterized are ideal for live-cell imaging, while photobleaching and maturation time can be issues. Green fluorescent protein and mCherry (GFP/mCherry) pair is another choice for cell-living imaging due to the excellent photostability and brightness of red fluorescent protein mCherry. Synthetized Alexa Fluor dyes have been used, which have a wide wavelength range and higher brightness compared to fluorescent proteins, although they require chemical conjugation to target molecules. Quantum dots and lanthanide chelates have also been used for measuring protein–protein interactions, which showed exceptional photostability and reduced background noise from autofluorescence.

Labeling biomolecules with donor and acceptor fluorophores is a critical step in FRET experiments. Unlike genetically encoded fluorescent proteins, systems such as SNAP-tag, CLIP-tag, and HaloTag can covalently bind to synthetic fluorescent ligands, offering flexibility in choosing different fluorescent probes. Labeling with synthetic dyes often involves reactive groups like NHS esters (which react with amines), maleimides (which react with thiols), or click chemistry reagents. The Cu(I)-catalyzed azide-alkyne cycloaddition (CuAAC) and strain-promoted azide-alkyne cycloaddition (SPAAC) are highly specific reactions that can occur in living cells without interfering with natural biochemical processes [2, 3]. The inverse-electron-demand Diels–Alder reaction between tetrazines and strained alkenes is fast, bioorthogonal, and suitable for live-cell labeling. Techniques such as amber suppression and genetic code expansion allow for the incorporation of unnatural amino acids with reactive handles, enabling site-specific labeling. These methods allow precise, efficient, and stable labeling of proteins, nucleic acids, or other biomolecules while preserving the biological activity.

Proper calibration and the inclusion of appropriate controls are essential for ensuring the reliability and reproducibility of FRET measurements. Negative controls (systems in which the donor and acceptor are present but are not expected to interact) and positive controls (systems where donor and acceptor are known to interact or are linked by a short peptide) are very helpful for determining whether any observed FRET signal is specific to the interaction being tested. Dynamic controls such as inducible systems in which FRET interactions can be turned on or off by external stimuli are ideal for validating whether the FRET system can detect dynamic changes in interactions or conformational states.

FRET efficiency can be measured using several methods: (i) intensity-based methods, which compare the fluorescence intensities of the donor and acceptor in the presence and absence of energy transfer; (ii) fluorescence lifetime imaging (FLIM), which measures the decrease in donor fluorescence lifetime in the presence of the acceptor using widefield microscopy, confocal microscopy, or two-photon microscopy; and (iii) acceptor photobleaching, which measures the increase in donor fluorescence after selectively photobleaching the acceptor. These methods provide quantitative data on the efficiency of energy transfer, which can be correlated with the distance between the donor and acceptor and used to infer molecular interactions and conformational changes.

Data analysis is also crucial in FRET experiments. Spectral unmixing can separate overlapping fluorescence signals from donor and acceptor fluorophores, which requires precise knowledge of the spectral properties of each fluorophore and the use of specific software. Quantitative FRET calculations are based on intensity ratios or fluorescence lifetimes to determine FRET efficiency using software such as Image J. Proper calibration and correction of background fluorescence, direct excitation of the acceptor, and bleed-through are essential for accurate calculations. Correction for artifacts such as bleed-through correction, direct excitation, and photobleaching are required to ensure that the detected signal is due to FRET and not due to donor emission overlapping with the acceptor detection channel, avoiding for the possibility that the excitation light might directly excite the acceptor fluorophore and adjusting for the loss of the fluorescence signal over time due to photobleaching.

3.3 Applications of FRET

FRET is frequently used to study protein–protein interactions, nucleic acid interactions, and other biomolecular interactions, as well as to monitor conformational changes in biomolecules and dynamic processes in live-cell imaging. FRET is also used for high-throughput screening (HTS) in drug discovery.

3.3.1 Molecular Interactions

One of the primary applications of FRET is the study of molecular interactions, which include protein–protein, protein–DNA, and protein–RNA interactions. FRET allows researchers to observe these interactions in real time and within the context of living cells.

For protein–protein interaction, FRET is used to study the mechanisms by which proteins interact, providing insights into binding affinities, interaction kinetics, and conformational changes [4]. FRET-based biosensors have been developed to monitor signaling pathways. For instance, FRET sensors can track the activity of kinases and phosphatases by detecting changes in the interaction between sensor components. FRET can also be employed to investigate the binding of ligands to their receptors, offering insights into receptor activation, dimerization, and the dynamics of ligand binding.

For protein–nucleic acid interaction, FRET is used to study how transcription factors bind to DNA, revealing the dynamics of transcriptional regulation. For example, a study by Park et al. employed FRET to monitor the protein–DNA binding affinity [5]. The binding of proteins to RNA and the formation of ribonucleoprotein complexes can also be monitored using FRET, providing insights into post-transcriptional regulation.

3.3.2 Conformational Changes

FRET is particularly useful for studying conformational changes in biomolecules, which are crucial for understanding their function and regulation. Research by Agyemang et al. employed FRET to study the conformational dynamics of GPCRs at a single-molecule level, providing insights into its structural transitions [6]. FRET has been used to study allosteric sites and conformational changes in allosteric regulation, providing insights into the allosteric regulation of enzymes.

FRET is used to investigate the folding pathways of proteins, identify intermediate states, and explore the kinetics of folding. FRET is also beneficial for studies on protein misfolding and aggregation, which are implicated in diseases such as Alzheimer's and Parkinson's by monitoring the formation of oligomers and aggregates.

3.3.3 Cellular Imaging

FRET-based imaging techniques enable the visualization of dynamic processes in live cells, providing spatial and temporal resolution that is essential for understanding complex cellular behaviors.

For intracellular signaling, FRET-based calcium sensors, such as cameleon, allow real-time monitoring of calcium dynamics within cells, providing insights into signaling cascades. Research by Miranda et al. utilized FRET sensors to monitor changes in calcium ion concentrations and study calcium signaling dynamics in neurons [7]. FRET sensors for cyclic nucleotides like cAMP and cGMP enable the study of their spatiotemporal distribution and the regulation of downstream signaling pathways. For instance, FRET biosensors were used to measure cyclic AMP levels and investigate kinase signaling dynamics [8]. FRET was also used to monitor the dynamics of phosphoinositides, such as PIP2 and PIP3, and shed light on their roles in cellular processes like membrane trafficking and interactions [9].

For membrane dynamics, FRET is used to study the organization and dynamics of lipid rafts, which are microdomains within cell membranes that play key roles in signaling and trafficking. Research by Hernandez-Adame et al. utilized FRET to investigate the oligomerization and localization dynamics of lipid rafts at a single-molecule level [10]. FRET is also helpful for monitoring the processes of endocytosis and exocytosis, providing insights into vesicle formation, trafficking, and fusion [11].

For subcellular localization, FRET-based probes help investigate interactions between organelles, such as the endoplasmic reticulum and mitochondria, revealing the coordination of cellular functions [12]. The trafficking of proteins between different cellular compartments can be monitored using FRET, providing insights into the regulation of protein localization and function.

3.3.4 Drug Discovery

FRET has been widely used in drug discovery, including target identification and validation, drug screening and optimization, mechanism studies, *etc.* [13]. For example, FRET assays have been employed to study protein–protein interactions in cancer cells to identify potential targets for anticancer therapies [4, 14]. FRET can reveal how allosteric modulators affect the conformation and activity of target proteins and provide critical information in the optimization of drug candidates with improved specificity and reduced side effects [15]. FRET can be used to study the binding kinetics of a compound to its target, providing insights into its mechanisms of action and thereby offering important information for the optimization of drug candidates. For example, He et al. utilized FRET to characterize the activity and substrate specificity of kinases, providing insights into their enzymatic mechanisms, and facilitate compound optimization. FRET was used to monitor the binding affinity between monocarboxylate transporter 1 (MCT1) and its antibody [16]. FRET-based assays have been used to discover protease inhibitors by monitoring the kinetics of substrate cleavage. For instance, FRET has been employed in screening libraries of compounds to identify inhibitors of HIV protease [17]. FRET-based sensors could also be utilized to study ion channel function and screen for modulators. These sensors monitor changes in intracellular ion concentrations upon channel activation, facilitating drug discovery efforts targeting ion channels [18]. FRET-based assays are employed to study membrane transporter function and screen for transporter modulators. Research has used FRET-based assay to assess substrate binding and transport kinetics of membrane transporters involved in drug absorption and efflux [19].

FRET-based assays are compatible with HTS platforms, allowing for rapid screening of large compound libraries. This capability accelerates the discovery of compounds targeting specific molecular interactions or disease pathways [13]. FRET-based biosensors can monitor the efficacy of drugs in real time within living cells [20]. The use of FRET in patient-derived cells allows for the assessment of individual cellular responses to therapies, supporting the development of personalized treatment strategies [21]. FRET-based imaging techniques enable *in vivo* monitoring of protein activity, the pharmacokinetic of drugs, as well as measuring efficacy. FRET assays have been employed to investigate RhoA kinase activity using a RhoA-FRET mouse [21]. Studies have utilized FRET biosensors to visualize drug distribution and metabolism and assess efficacy in animal models, providing insights into drug delivery and therapeutic outcomes [22]. The design and optimization of targeted drug delivery systems has benefitted from FRET by monitoring the release and interaction of therapeutic agents with their targets.

3.3.5 Clinical Diagnosis

FRET-based assays for the detection of disease biomarkers will enable early diagnosis and monitoring of diseases such as cancer, cardiovascular diseases, and neurodegenerative disorders. Research by Fu et al. utilized FRET assays to detect circulating tumor cells and monitor drug resistance *in vivo*, demonstrating clinical diagnostic potential [18]. Development of portable and user-friendly FRET-based diagnostic devices will facilitate point-of-care testing, improving access to healthcare and enabling rapid decision-making.

FRET will be used to monitor the efficacy of therapeutic interventions in real time, allowing for the optimization of treatment strategies and personalized medicine. Longitudinal FRET studies will provide insights into disease progression and the impact of treatments over time, aiding in the development of more effective therapies.

3.3.6 Structural Biology

FRET provides detailed information about the structure and dynamics of biomolecules, complementing traditional structural biology techniques like X-ray crystallography and NMR spectroscopy [19]. FRET can determine the relative positions of subunits within protein complexes, providing insights into their quaternary structure. The dynamics of protein complex assembly and disassembly can be monitored using FRET, revealing the mechanisms that regulate these processes.

FRET is also used to study the folding of DNA and RNA structures, such as G-quadruplexes and ribozymes, providing insights into their stability and function [23]. The hybridization and melting dynamics of nucleic acid duplexes can be monitored using FRET, aiding in the study of processes like replication and transcription.

3.3.7 Materials Science

FRET is also applied in materials science to study the properties and interactions of various materials, including polymers, nanoparticles, and biomaterials [20]. FRET can monitor the conformational changes of polymers in response to environmental stimuli, providing insights into their mechanical properties and functionality. The process of polymer self-assembly can be studied using FRET, aiding in the design of materials with specific properties and applications.

FRET is used to investigate the interactions of nanoparticles with biological molecules, revealing the effects of surface chemistry on biocompatibility and functionality. The interaction of drug-loaded nanoparticles with target cells and tissues can be monitored using FRET, providing insights into the efficiency and targeting of drug delivery systems [20].

FRET is employed to study the structural dynamics and interactions within hydrogels, aiding in the design of materials for tissue engineering and drug delivery. The interactions of cells with scaffold materials used in tissue engineering can be monitored using FRET, providing insights into cell adhesion, proliferation, and differentiation.

3.3.8 Environmental and Agricultural Sciences

FRET-based sensors and assays are increasingly used in environmental and agricultural sciences to monitor and analyze various biological and chemical processes. FRET-based sensors can detect pollutants such as heavy metals, pesticides, and toxins in water and soil, providing real-time monitoring for environmental protection. The activity and interactions of microbes in environmental samples can be studied using FRET, aiding in the understanding of microbial ecology and biogeochemical cycles.

Synthetic biosensors using FRET can detect environmental changes, such as pH shifts or the presence of specific ions, aiding in environmental monitoring and bioremediation [24]. FRET is used to study the interactions between plant proteins, hormones, and signaling molecules, providing insights into plant physiology and stress responses. FRET-based assays help monitor the effectiveness of biopesticides. The development of FRET assays for monitoring soil health and nutrient levels will support sustainable agriculture and soil conservation efforts.

3.4 Advantages and Limitations

FRET is a powerful and versatile tool that has found extensive applications across various scientific disciplines. FRET is highly sensitive to changes in the distance between donor and acceptor molecules, typically in the range of 1–10 nm. This allows researchers to study interactions and conformational changes with great precision. Parameters such as FRET efficiency, binding constants, and interaction kinetics can be accurately determined. FRET-based biosensors can be used *in vivo* to monitor physiological processes, providing valuable information for biomedical research and clinical applications. Genetically encoded FRET biosensors allow for noninvasive imaging and monitoring of cellular processes in live organisms, facilitating *in vivo* studies [22].

FRET can be adapted to study a wide range of biomolecular features, including protein–protein interactions, nucleic acid dynamics, and enzyme activities. FRET can be combined with other imaging and spectroscopic techniques, such as confocal microscopy, two-photon microscopy, and Förster resonance energy transfer using anisotropy (FRET-FLIM), to enhance its capabilities. FRET can be multiplexed with other imaging modalities or fluorescence techniques, enabling simultaneous detection of multiple biomolecular interactions within the same sample [25]. FRET exhibits a broad dynamic range, allowing for the detection of both small and large changes in molecular interactions or distances between fluorophores.

However, like any technique, it comes with its own set of limitations. FRET efficiency is highly dependent on the distance between donor and acceptor fluorophores, limiting its application to biomolecules within the distance range of ~1–10 nm. The efficiency of energy transfer decreases sharply with increasing distance between donor and acceptor, which can complicate the interpretation of FRET data if the exact distance between molecules is not well-known. Continuous exposure to excitation light can lead to photobleaching of fluorophores, affecting the stability and longevity of FRET measurements, which can limit the duration of experiments and affect the accuracy of measurements. Effective FRET requires significant spectral overlap between the emission spectrum of the donor and the absorption spectrum of the acceptor. The spectral overlap between donor emission and acceptor absorption spectra can result in spectral bleed-through, leading to false-positive FRET signals. Finding suitable fluorophore pairs with optimal spectral overlap can be challenging.

Background fluorescence from cellular components or autofluorescence can interfere with FRET signals, reducing signal-to-noise ratios and accuracy. Fluorescence bleed-through (donor emission detected in the acceptor channel) and direct excitation of the acceptor can lead to artifacts. Autofluorescence from the sample and background fluorescence can interfere with FRET measurements, particularly in complex biological samples or tissues. Thus, careful design of control experiments is essential to differentiate specific FRET signals from nonspecific interactions and artifacts. FRET efficiency is also influenced by the orientation factor (κ^2), which describes the relative orientation of the transition dipole moments of donor and acceptor fluorophores. FRET efficiency can also be influenced by environmental factors such as pH, temperature, and ion concentrations, requiring careful control and calibration.

FRET assays often require specific labeling of biomolecules with suitable fluorophores, which can be technically challenging and may change native biomolecular properties. The attachment of fluorophores to proteins or other biomolecules can potentially perturb their native structure, dynamics, and function, affecting the accuracy of FRET measurements and biological relevance. FRET experiments performed *in vitro* may not always accurately reflect interactions *in vivo*. The cellular context, including molecular crowding and the presence of other interacting partners, can influence FRET measurements. Using FRET in model organisms for *in vivo* studies can be challenging due to the complexity of the biological system and the need for suitable fluorescent proteins or dyes that are compatible with the organism's physiology.

Advanced instrumentation is often required for FRET experiments, including specialized microscopes, spectrometers, and lifetime imaging systems. This can be a barrier for some laboratories. Analyzing FRET data can be complex and requires expertise in various analytical techniques and software tools. Accurate interpretation of FRET efficiency and other parameters often involves sophisticated mathematical modeling and correction for artifacts. FRET measurements can be performed at the single-molecule level or in bulk. Single-molecule FRET provides detailed insights into heterogeneity and dynamics, but it is technically challenging and requires sophisticated instrumentation. Ensemble measurements average out these variations, which may obscure important details.

3.5 Recent Advances

Despite these limitations, various strategies can be employed to mitigate their impact and enhance the utility of FRET. In recent years, the development of FRET technology has been promoted by the technological advancements, novel fluorophores, improved data analysis techniques, and innovative experimental approaches.

For novel fluorophores, the development of new organic dyes with high quantum yields, photostability, and minimal photobleaching, such as ATTO and Alexa Fluor series [26], has significantly improved the reliability and duration of FRET experiments. The introduction of far-red and near-infrared (NIR) fluorophores has expanded the spectral range of FRET, enabling deeper tissue penetration and reduced background fluorescence in biological samples. These fluorophores, such as Cy7 and IRDye [27], are particularly useful for *in vivo* imaging. The engineering of fluorescent proteins, such as GFP variants, has led to improved brightness, photostability, and a broader range of colors. These enhanced fluorescent proteins are widely used in FRET biosensors. The development of photoactivatable and photoconvertible fluorescent proteins, such as PA-GFP and mEos, allows for precise spatiotemporal control of fluorescence, enabling more detailed studies of dynamic processes. Quantum dots (QDs) have unique optical properties, such as high brightness, photostability, and tunable emission wavelengths [28]. They have been integrated into FRET technology to improve sensitivity and enable multiplexing. The use of nanoparticles, such as gold nanoparticles and upconversion nanoparticles, in FRET assays offers advantages in terms of enhanced signal strength and stability, as well as the ability to perform FRET in different biological environments.

For technological advancements, techniques like STED, PALM, and STORM have pushed the resolution limits of fluorescence microscopy beyond the diffraction limit, allowing for the visualization of molecular interactions and dynamics at nanometer scales. Integrating FRET with complementary techniques such as cryo-electron microscopy, NMR spectroscopy, and mass spectrometry provides a more comprehensive understanding of molecular systems by combining structural, dynamic, and functional information. Super-resolution microscopy combined with FRET has provided unprecedented insights into molecular organization and interactions [29]. Two-photon excitation microscopy offers deeper tissue penetration and reduced photodamage, making it ideal for *in vivo* FRET imaging. This technique has been used to study complex biological systems in living organisms with high spatial and temporal resolution. FLIM-FRET measures changes in the fluorescence lifetime of the donor molecule, providing a robust method to quantify FRET efficiency that is less affected by concentration variations and direct excitation of the acceptor. This technique has enhanced the accuracy and reliability of FRET measurements. Time-resolved FRET uses time-resolved detection to distinguish between specific FRET signals and background fluorescence, improving sensitivity and allowing for the detection of low-abundance targets [30]. Single-molecule FRET provides detailed insights into the heterogeneity and dynamics of molecular interactions at the single-molecule level [15, 23, 31, 32]. This technique has been used to study conformational changes, folding pathways, and complex assembly processes with high temporal resolution. Innovations in single-molecule detection methods, such as total internal reflection fluorescence microscopy and confocal microscopy, have improved the sensitivity and resolution of single-molecule FRET experiments.

For innovative experimental approaches, the design of genetically encoded FRET biosensors for various biological targets, such as ions, metabolites, and signaling molecules, has enabled real-time monitoring of cellular processes [22]. The development of multicolor FRET technology using multiple donor–acceptor pairs allows for the simultaneous monitoring of different interactions and pathways within the same sample [25]. This multiplexing capability increases the throughput and information content of FRET experiments. Combinatorial FRET approaches use multiple FRET pairs to study complex molecular networks and interactions, providing a more comprehensive understanding of cellular processes. The integration of FRET assays with automated liquid handling and imaging systems has streamlined HTS processes, increasing efficiency and reducing costs. Advances in FRET imaging techniques have enabled the study of molecular interactions and dynamics in live organisms, providing valuable insights into physiological processes and disease mechanisms.

For computational advances, the development of advanced algorithms for FRET data analysis has improved the accuracy and reliability of FRET efficiency calculations, corrected artifacts, and extracted more detailed information from FRET experiments. Machine learning techniques have been applied to analyze complex FRET datasets, identify patterns, and predict molecular interactions, enhancing the interpretative power of FRET studies. Combining FRET with molecular dynamics simulations provides a powerful approach for studying the structural dynamics of biomolecules, offering insights into their conformational changes and interactions at atomic resolution. Computational models have been developed to predict FRET efficiencies and optimize experimental designs, thereby facilitating the interpretation of FRET data and guiding experimental strategies.

3.6 Conclusion

The future of FRET technology is bright with numerous exciting advancements underway. Continued innovations in fluorophore design, imaging technologies, data analysis, and integrative approaches will enhance the sensitivity, accuracy, and applicability of FRET, making it an even more powerful tool for studying molecular interactions and dynamics. As FRET technology evolves, it will undoubtedly continue to provide valuable insights into the complexities of biological systems and contribute to advancements in medicine, environmental science, agriculture, and education.

One of the key development directions of FRET technology is to measure it at the single-molecule level to study molecular conformational changes, protein folding/unfolding, and DNA/RNA interactions, providing insights into heterogeneity and dynamics that are not apparent in bulk measurements. Innovations in single-molecule detection technologies will allow for the simultaneous analysis of multiple molecules and interactions. It requires highly sensitive instrumentation and rigorous data analysis to account for noise and photobleaching. Advances in imaging technologies and fluorophore design will facilitate the application of single-molecule FRET in living organisms, providing insights into molecular interactions and dynamics in their native environments. Improved temporal resolution in single-molecule FRET will enable the study of rapid conformational changes and interactions in real time, revealing detailed kinetic information.

Another key development direction of FRET technology is to combine FRET with fluorescence anisotropy and lifetime imaging to provide detailed information on molecular orientation and dynamics by investigating molecular orientations and structural changes in complex environments. Combining FRET with other techniques, such as Förster resonance energy transfer using anisotropy (FRET-FLIM) and Förster resonance energy transfer using spectral imaging (FRET-SIM), will enhance the spatial and temporal resolution of FRET measurements. It will require specialized equipment and expertise in interpreting anisotropy data.

Finally, advances in multicolor FRET technology will allow for the simultaneous monitoring of multiple molecular interactions within the same sample, providing a comprehensive view of complex cellular processes. It uses multiple donor–acceptor pairs to study several interactions simultaneously, allowing HTS and complex pathway analysis. It needs to manage spectral overlap and ensure accurate separation of signals from different FRET pairs. Development of biosensors suitable for *in vivo* applications will enable real-time monitoring of physiological processes and disease progression in live organisms. The design of highly specific and sensitive FRET-based biosensors for a wide range of targets, including small molecules, ions, and proteins, will enhance their utility in biological and clinical research.

Acknowledgements

This work is financially supported by grants from Yunling Scholar Project and Advanced Talents Recruitment Scientific Program of Kunming Medical University.

References

1 Forster, T. (2012). Energy migration and fluorescence. *Journal of Biomedical Optics* 17 (1): 011002.

2 Ouadahi, K., Sbargoud, K., Allard, E., and Larpent, C. (2012). FRET-mediated pH-responsive dual fluorescent nanoparticles prepared via click chemistry. *Nanoscale* 4 (3): 727–732.

3 Jumeaux, C., Spicer, C.D., Charchar, P. et al. (2024). Strain-promoted cycloadditions in lipid bilayers triggered by liposome fusion. *Angewandte Chemie (International Ed. in English)* 63 (14): e202314786.

4 Liao, J., Madahar, V., Dang, R., and Jiang, L. (2021). Quantitative FRET (qFRET) technology for the determination of protein-protein interaction affinity in solution. *Molecules* 26 (21): 6339.

5 Park, H., Kang, H., Ko, W. et al. (2015). FRET-based analysis of protein-nucleic acid interactions by genetically incorporating a fluorescent amino acid. *Amino Acids* 47 (4): 729–734.

6 Agyemang, E., Gonneville, A.N., Tiruvadi-Krishnan, S., and Lamichhane, R. (2024). Exploring GPCR conformational dynamics using single-molecule fluorescence. *Methods* 226: 35–48.

7 Miranda, P., Contreras, J.E., Plested, A.J. et al. (2013). State-dependent FRET reports calcium- and voltage-dependent gating-ring motions in BK channels.

Proceedings of the National Academy of Sciences of the United States of America 110 (13): 5217–5222.
8 Jiang, J.Y., Falcone, J.L., Curci, S., and Hofer, A.M. (2017). Interrogating cyclic AMP signaling using optical approaches. *Cell Calcium* 64: 47–56.
9 Gao, Y.G., Zhai, X., Boldyrev, I.A. et al. (2021). Ceramide-1-phosphate transfer protein (CPTP) regulation by phosphoinositides. *The Journal of Biological Chemistry* 296: 100600.
10 Hernandez-Adame, P.L., Meza, U., Rodriguez-Menchaca, A.A. et al. (2021). Determination of the size of lipid rafts studied through single-molecule FRET simulations. *Biophysical Journal* 120 (11): 2287–2295.
11 Abella, M., Andruck, L., Malengo, G., and Skruzny, M. (2021). Actin-generated force applied during endocytosis measured by Sla2-based FRET tension sensors. *Developmental Cell* 56 (17): 2419–26 e4.
12 Wang, Y. and Wang, N. (2009). FRET and mechanobiology. *Integrative Biology* 1 (10): 565–573.
13 Blay, V., Tolani, B., Ho, S.P., and Arkin, M.R. (2020). High-throughput screening: today's biochemical and cell-based approaches. *Drug Discovery Today* 25 (10): 1807–1821.
14 Guala, D., Bernhem, K., Blal, H.A. et al. (2018). Experimental validation of predicted cancer genes using FRET. *Methods and Applications in Fluorescence* 6 (3): 035007.
15 Asher, W.B., Geggier, P., Holsey, M.D. et al. (2021). Single-molecule FRET imaging of GPCR dimers in living cells. *Nature Methods* 18 (4): 397–405.
16 Martinez, W.E., Arenas, J.E., Mok, L. et al. (2021). Bioelectronic measurement of target engagement to a membrane-bound transporter. *SLAS Discovery* 26 (8): 1004–1013.
17 Cella, L.N., Biswas, P., Yates, M.V. et al. (2014). Quantitative assessment of in vivo HIV protease activity using genetically engineered QD-based FRET probes. *Biotechnology and Bioengineering* 111 (6): 1082–1087.
18 Fu, A., Ma, S., Wei, N. et al. (2016). High expression of MnSOD promotes survival of circulating breast cancer cells and increases their resistance to doxorubicin. *Oncotarget* 7 (31): 50239–50257.
19 Lerner, E., Barth, A., Hendrix, J. et al. (2021). FRET-based dynamic structural biology: challenges, perspectives and an appeal for open-science practices. *eLife* 10: e60416.
20 Pini, F., Frances-Soriano, L., Andrigo, V. et al. (2023). Optimizing upconversion nanoparticles for FRET biosensing. *ACS Nano* 17 (5): 4971–4984.
21 Wilbertz, J.H., Frappier, J., Muller, S. et al. (2022). Time-resolved FRET screening identifies small molecular modifiers of mutant Huntingtin conformational inflexibility in patient-derived cells. *SLAS Discovery* 27 (4): 219–228.
22 Nobis, M., Herrmann, D., Warren, S.C. et al. (2020). Shedding new light on RhoA signalling as a drug target in vivo using a novel RhoA-FRET biosensor mouse. *Small GTPases* 11 (4): 240–247.
23 Filius, M., van Wee, R., and Joo, C. (2024). Single-molecule FRET X. *Methods in Molecular Biology* 2694: 203–213.

24 Bagot, C.C., Rappeport, E., Das, A. et al. (2022). True FRET-based sensing of pH via separation of FRET and photon reabsorption. *Advanced Optical Materials* 10 (15): 2200242.

25 Grant, D.M., Zhang, W., McGhee, E.J. et al. (2008). Multiplexed FRET to image multiple signaling events in live cells. *Biophysical Journal* 95 (10): L69–L71.

26 Schneider, U.V., Severinsen, J.K., Geci, I. et al. (2010). A novel FRET pair for detection of parallel DNA triplexes by the LightCycler. *BMC Biotechnology* 10: 4.

27 Zhang, S., Metelev, V., Tabatadze, D. et al. (2008). Fluorescence resonance energy transfer in near-infrared fluorescent oligonucleotide probes for detecting protein-DNA interactions. *Proceedings of the National Academy of Sciences of the United States of America* 105 (11): 4156–4161.

28 Nobori, T., Kishimura, A., Mori, T., and Katayama, Y. (2021). A FRET-based protein kinase assay using phos-tag-modified quantum dots and fluorophore-labeled peptides. *Analytical Sciences* 37 (10): 1361–1366.

29 Szalai, A.M., Zaza, C., and Stefani, F.D. (2021). Super-resolution FRET measurements. *Nanoscale* 13 (44): 18421–18433.

30 Ergin, E., Dogan, A., Parmaksiz, M. et al. (2016). Time-resolved fluorescence resonance energy transfer [TR-FRET] assays for biochemical processes. *Current Pharmaceutical Biotechnology* 17 (14): 1222–1230.

31 Bartels, K., Lasitza-Male, T., Hofmann, H., and Low, C. (2021). Single-molecule FRET of membrane transport proteins. *ChemBioChem* 22 (17): 2657–2671.

32 Martinac, B. (2017). Single-molecule FRET studies of ion channels. *Progress in Biophysics and Molecular Biology* 130 (Pt B): 192–197.

4

Dissect Protein Interactions Using Mass Spectrometry

Bin Liao[1,2,3] *and Liang Zhang*[1,2,3]

[1] *City University of Hong Kong, Department of Biomedical Sciences, and Tung Biomedical Sciences Center, 83 Tat Chee Avenue, Kowloon, Hong Kong SAR, China*
[2] *City University of Hong Kong Futian Research Institute, Department of Precision Diagnostic and Therapeutic Technology, 3 Penang Road, Futian District, Shenzhen, Guangdong, China*
[3] *Shenzhen Research Institute of City University of Hong Kong, Key Laboratory of Biochip Technology, Biotech and Health Centre, 8 Yuexing Yidao Road, Nanshan district, Shenzhen, Guangdong, China*

4.1 Introduction

The complicated three-dimensional structures and post-translational modifications of proteins result in their functional diversity. Besides, protein–protein interaction (PPI) network is also one vital factor that cannot be ignored in protein functional profiling since cellular processes are essentially regulated by PPIs. Over 20,000 human proteins have been well annotated by the Human Genome Project and computationally simulated and predicted for their spatial structures with high accuracy by artificial intelligence [1, 2]. On the other hand, mathematically achieving prediction and discovery of PPIs and protein-assembled complexes on a wide scale still remains a huge challenge and lacks robust accuracy. Even though the recently released AlphaFold 3, a breakthrough deep learning model, has claimed to enable the prediction of protein structures/interactions, problems such as protein structure hallucination, prediction of only static complexes, and incorrect conformation for specific proteins have restricted its application on the accurate discovery of PPIs [3]. Moreover, several public PPI platforms like BioGrid [4] and IntAct [5] mainly focus on experimentally validated protein interactions mined and integrated from thousands of literatures. Hence, comprehensive characterization of PPIs primarily relies on advanced experimental techniques and approaches, of which mass spectrometry (MS)-based proteomics methods perform superior efficiency and sensitivity.

Liquid chromatography–tandem mass spectrometry (LC–MS/MS) is a powerful and versatile technique, which has achieved efficient separation, rapid qualification, and accurate quantification of proteins from complicated samples [6]. MS-based proteomics has extended numerous techniques to accelerate the validation of PPIs of interest and the revelation of novel PPIs. For example, as a classic strategy, affinity

Targeting Protein–Protein Interactions for Drug Discovery, First Edition. Edited by Jian Zhang.

purification coupled with mass spectrometry (AP-MS) specifically captures protein of interest (POI) and its putative interacting partners by antibody of POI or fused epitope tag [7]. MS is then implemented to infer the identity of these partner proteins prior to interaction profiling. High-throughput identification of thousands of proteins within short time by MS also provides convenient objective conditions for profiling large-scale PPI networks.

For either POI interactions or unbiased PPI network profiling, emerging MS-based techniques have been developed to fulfill requirement for a variety of complex systems and conformations. Clarifying the principles, properties, and applications of these methodologies will offer valuable insights for developing new approaches and enable the selection of appropriate methods to advance research according to specific needs in drug discovery. Hence, this chapter will focus on recent and advanced developments in MS-based PPI approaches, introducing their underlying principles, existing challenges and solutions, and relevant applications (also see Table 4.1).

4.2 Affinity Purification Coupled with Mass Spectrometry (AP-MS)

AP-MS is a powerful and widely used method for investigating PPIs. It leverages the specificity of affinity purification to isolate protein complexes and the sensitivity of mass spectrometry to identify and quantify the components of these separated complexes. In other words, protein of interest (POI, bait protein) with its potential interacting partners (prey proteins) can be specifically identified, captured, and enriched by specific antibodies or affinity epitope tags prior to protein identification by MS. Specifically, the workflow of AP-MS normally consists of (i) incubating bait protein with antibody or fusing it to epitope tag followed by overexpression in cell lysates, (ii) adding beads (such as agarose beads with protein A/G and strep beads) or anti-tag and incubating with cell lysate to capture bait–prey complexes, (iii) washing away unbound proteins or non-specific binding proteins to reduce background proteins and enrich target protein complexes, (iv) eluting bait/prey proteins isolated from beads or anti-tag, and (v) PPI complexes will be processed with routine bottom-up proteomics prior to MS analysis (Figure 4.1a).

In addition to co-immunoprecipitation (Co-IP), which leverages the interaction between POI and specific antibody to end up discerning antibody–POI-interacting protein complexes, AP-MS circumvents the limitation of the antibody repertoire by fusing POI with an epitope tag (Figure 4.1a). Such flexibility of AP-MS can build up to thousands of tagged bait proteins to assemble large-scale PPI networks at high throughput and low cost. For example, Guruharsha et al. assembled *Drosophila* protein interaction network DPiM by processing about 5000 proteins fused with FLAG-HA tags in AP-MS. DPiM uncovered underlying members for some protein complexes and provided functional annotations for 586 proteins [8].

As a key part of AP-MS, epitope tags are typically short peptide sequences attached to the N/C terminal of POI to facilitate its purification. The commonly used affinity epitope tags in AP-MS mainly include peptide tags, protein tags, and multimodal

Table 4.1 Principles, advantages and limitations of advanced MS-based methods for PPI study.

Methods	Principles	Advantages	Limitations
AP-MS	Utilizing specific antigen-antibody immune responses or affinity epitope tags to isolate interacting proteins, followed by MS analysis	(1) Theoretically applicable for any POI. (2) Detecting endogenous protein interactions within a physiological environment	(1) Requiring mild lysis condition, limiting to extract membrane proteins (2) Missing low-abundance, weak and transient interactions (3) May introduce false positive pairs due to non-specific binding with tags
Proximity labeling	Employing proximity-tagging enzymes or photoactive probes to biotinylate proteins in the vicinity of POI, followed by capturing with affinity beads	(1) Capturing weak and transient PPIs in situ (2) Decreasing the risk of false positive as the short labeling radius of the biotinylating enzymes	(1) Being inactive within cellular environment for peroxidase (2) Inducing cell damage under H_2O_2 condition for ascorbate peroxidase (3) Impeding the selectivity to POI because of the large molecular weight of promiscuous biotin ligase (4) Leading to false positives from pre-existing biotin in the body and incidental positional proximity to the POI
XL-MS	Covalently linking the residues of interacting proteins via cross-linkers	(1) Unbiasedly dissecting direct protein interactions, including low-abundance and weak interactions (2) Providing protein structural information according to the length of cross-linkers (3) Specifically targeting to given residues using different cross-linkers, expanding the cover regions	(1) Difficulty in identifying cross-linked peptides because of the low proportion (2) Matching cross-linked peptides within search space can be associated with high computational cost

(Continued)

Table 4.1 (Continued)

Methods	Principles	Advantages	Limitations
CF-MS	Deciphering the co-elution signals across fractions to predict protein complexes	(1) Profiling thousands of protein complexes with high throughput and efficiency (2) Inferring endogenous and physiologically relevant protein interactions since proteins remain non-denatured and unmodified	(1) Unsuccessfully fitting elution profiles due to missing quantitative protein values, potentially overlooking true positives (2) Introducing false positives from accidental co-elution of proteins, necessitating advanced computational tools for accurate distinction
TPCA	Leveraging the co-aggregation and co-precipitation of interacting proteins at elevated temperatures	(1) Facilitating unbiased and systematic mapping of dynamic protein complex networks in situ and within living cells, devoid of exogenous tags or antibodies (2) Deciphering dynamics of protein complexes at a proteome-wide scale with high throughput and cost-effectiveness	(1) Restricting the monitoring of known PPIs without the capabilities to predict novel PPIs, necessitating complementary methodologies and algorithms (2) Being inappropriate for thermostable proteins within a defined temperature range
LiP-MS	Using limited proteolysis accessibility to generate differential peptides under varying conditions	(1) Monitoring protein conformation changes during interactions (2) Pinpointing interacting surface and structural change regions (3) Streamlining experiment protocol without modification on POI	(1) Requiring the addition of purified POI, decreasing throughput (2) Limiting to clearly elucidate mechanisms of protein structural change since LiP-MS is versatile to detect various perturbation events (3) Requiring high-resolution MS to detect peptides located within regions of protein conformation change (4) Being incapable of tracking protein structural changes in intact cells

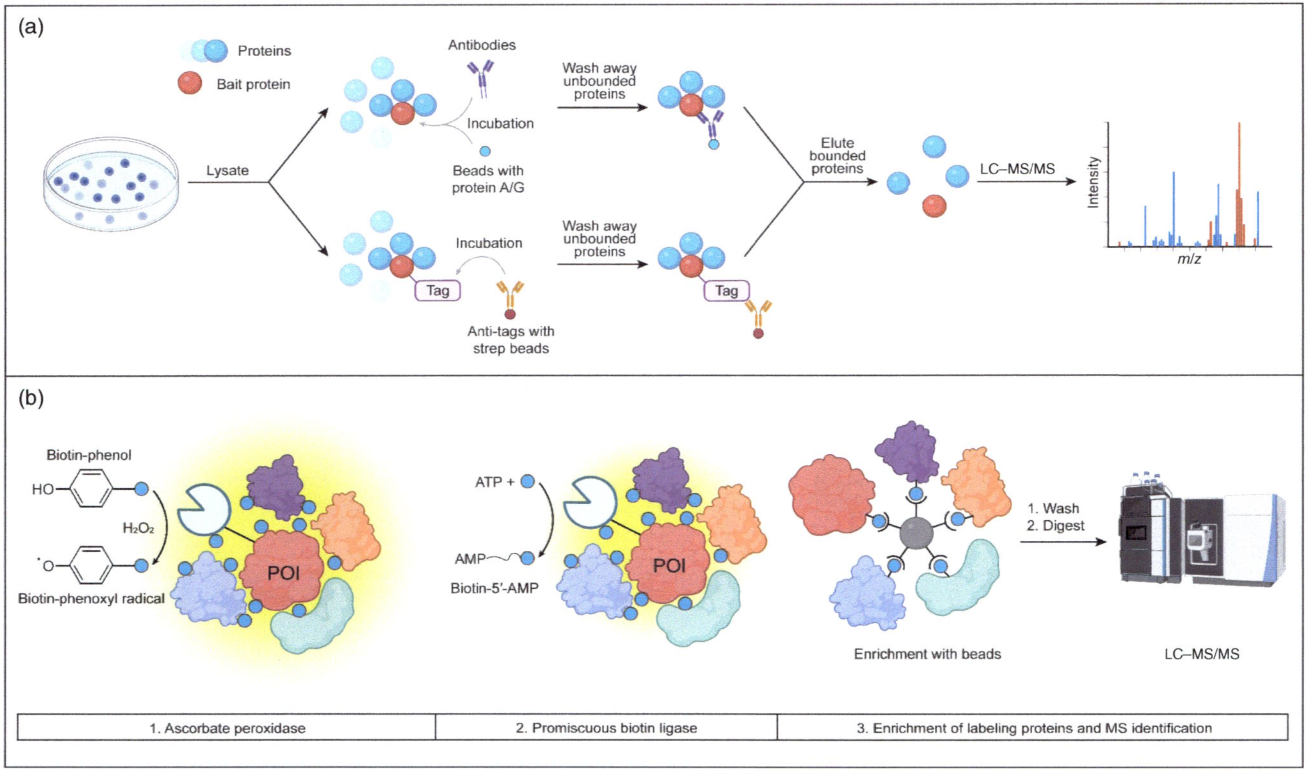

Figure 4.1 Workflow of AP-MS and proximity labeling. (a) A schematic illustration of general AP-MS. Protein extracts are incubated with bait protein antibodies (top) or anti-tag antibodies (bottom) for bait proteins without (top) or with (bottom) affinity tags, respectively. Unbound proteins are washed away, leaving only the bait and prey proteins for subsequent LC–MS/MS analysis. (b) General workflow of enzyme-mediated proximity labeling methods. Ascorbate peroxidase facilitates the biotinylation of proteins in the vicinity of POI via biotin-phenols under H_2O_2 conditions (left chart). By converting ATP and biotin into biotin-5′-AMP, promiscuous biotin ligases biotinylate neighboring proteins of POI (median chart). The biotinylated proteins are enriched by streptavidin beads and digested into peptides, followed by LC–MS/MS analysis (right chart). Source: Bin Liao and Liang Zhang.

tags. The appropriate affinity tags should be considered in several aspects, including their binding efficiency, availability of compatible antibodies or binding proteins, and their influence on the bait protein's function and localization.

For details, peptide tags, such as FLAG tag, HA tag, His-tag, and c-Myc tag, can be recognized by corresponding monoclonal antibodies, which allow for efficient protein purification [9]. These tags are advantageous due to their small size, which minimize the likelihood of interfering with the protein's function or localization. In addition, GST tag, which is a protein tag of approximately 26 kDa, facilitates protein purification through binding to glutathione resin and can enhance the solubility of fused protein [10]. To address the setbacks of traditional protein tags (e.g. GST tag and GFP tag) that may interfere with protein functions, some novel tags have been developed, like LlamaTag and Inntags [11].

Multimodal tags generally consist of two or more affinity tags so that they can improve protein purification under both native and denaturing conditions. As an original multimodal tag, tandem affinity purification (TAP) tag configuration includes a calmodulin-binding peptide (CBP) and a protein A tag, separated by a tobacco etch virus (TEV) protease cleavage site. Bait protein fused with TAP can identify specific interacting partners by two-step affinity purification. The first step of affinity purification is performed with an IgG resin to exclude non-specific binding proteins, as protein A can tightly bind with resin, followed by digestion with TEV protease to obtain a CBP-fusion protein complex. The second step involves the mixture of CBP-fusion protein complex and calmodulin beads, which is solidly bound to CBP and the native elution with EGTA [12]. The proposal of TAP has also accelerated the development of other multimodal tags, such as MAC-tag and MAP tag, which all enable protein localization and interactions [13, 14].

There are also notable efforts to simplify and accelerate AP-MS workflows. Ke et al. introduced an approach namely AP-SISPROT which combined AP-MS and the spintip-based proteomics technology (SISPROT) [15]. AP-SISPROT enabled simultaneous processing of more than 10 AP samples within 2 hours while ensuring highly efficient digestion. Further optimization was proposed by Mao et al. with FISAP, which incorporated a pull-down step to reduce sample loss and processing time [16].

Despite numerous improvements aimed at optimizing the experimental process of AP-MS, some limitations remain. First, mild lysis condition is required for AP-MS to preserve the intact status of protein complexes, leading to the challenges for extracting membrane proteins [17]. Second, transient or weak interactions involving low-abundance proteins tend to be lost during affinity purification due to the multiple washing steps. While a one-step purification process can relatively preserve transient and weak PPIs, more background contaminants may interfere. Third, AP-MS can recover both direct and indirect interactions of POI, but it is hard to distinguish them [7]. Finally, non-specific binding to the affinity matrix can introduce contaminants and inevitably cause some false-positive interactions in spite of the elution by tag purification and enrichment. Therefore, appropriate negative control needs to be introduced to differentiate real PPI from false-positive interactions.

For data analysis of AP-MS, several computational tools have enabled distinguishing positive and background proteins in terms of protein intensity and MS quantitative information across large batches, including SAINT, MIST, CompPASS, etc. [18]. Improved algorithms have been proposed recently for better profiling PPI network based on AP-MS. On the basis of CompPASS, Huttlin et al. developed CompPASS-Plus which expanded features from CompPASS and integrated additional features derived from statistics of MS data and general rank of PPIs. Importantly, CompPASS-Plus employed the Naive Bayes classifier to predict high-confidence interactors [19, 20]. By combining data-independent acquisition mode with deep neural network, DIP-MS offers a robust strategy for deciphering protein complexes sharing the same target and for deconvoluting protein complexes based on AP-MS data [21].

AP-MS has been applied to disclose novel PPIs and to dissect protein interactome. An illustrative example is that AP-MS revealed the interactions of nucleolar zinc-finger protein NOA36 with HSPA1 and HSPA8, promoting the recovery of the regular cell cycle in response to thermal stress [22]. At the large scale, Huttlin et al. built up a human protein interactome database named BioPlex, which records the expanding human protein interactome derived from thousands of AP-MS experiments in bulk [20]. In the latest BioPlex, 10,128 and 5522 bait proteins from two kinds of cell lines were employed to ultimately detect 118,162 and 10,531 high-confidence PPIs, respectively, revealing cell line- and phenotype-based PPI network in protein complex, signal pathway, and protein family [19]. Another interesting application of AP-MS is the demonstration of interactions between canonical proteins and microproteins, which are usually encoded by traditional non-coding regions with less than 100 amino acids in sequence. Koh et al. optimized the AP-MS protocol by incorporating the atypical amino acid AbK into tag and covalently cross-linking AbK with prey proteins using UV irradiation, enabling the identification of microprotein SEHBP and its interactions with canonical proteins, associating with transcription [23]. Despite these advances, AP-MS needs to integrate with other technologies introduced below to deconvolve the structural details of PPIs to further aid drug discovery.

4.3 Proximity Labeling

To complement AP-MS, proximity labeling (PL) technology has been introduced to recognize weak and transient PPIs in situ and to overcome constraints of cell lysis and protein solubility in profiling membrane protein interactions. The typical workflow of PL involves (i) fusion of POI with proximity-tagging enzymes and expression in specific cell lines or subcellular regions, (ii) generation of reactive intermediates by adding small molecule substrates to enzymes, enabling the labeling of proteins with biotin or biotin-phenol within nanometer range around POI, and (iii) enrichment of biotin-labeled proteins by affinity magnetic beads, followed by MS identification for labeled proteins (Figure 4.1b). As an important factor, the coenzyme biotin is a water-soluble B-vitamin that can tightly bind to streptavidin beads, enabling purification and identification of biotin-labelled proteins. Proximity labeling enables unbiased biotinylation of neighboring proteins, capturing both direct and indirect interactions. However, this property would introduce false-positive proteins, which "accidentally" distribute around POI. Hence, negative control and filtering algorithms, like SAINT, are important PL experiment and data processing [24].

Typically, the nearby proteins of POI are reasonably considered as the interacting partners of POI with high probability. Therefore, the choice of enzymes could be the key step of PL since they directly determine the label efficiency and radius. An ideal PL enzyme would not affect the original function, expression, and localization of POI. It would react with substrates to generate active intermediates capable of covalently labeling adjacent proteins [25]. The mainstream tool enzymes can be classified as peroxidase (such as horseradish peroxidase [HRP], ascorbate peroxidase [APEX], and APEX2) and promiscuous biotin ligase (including BioID, TurboID, BASU, etc.) [26].

In the presence of hydrogen peroxide (H_2O_2), peroxidase can catalyze substrates with free radical chemical reactions to covalently bind to surrounding proteins. Among various peroxidases, HRP was the first to be developed for application in PL-based PPI study. HRP can activate aryl azide-biotin substrates, producing free radical intermediates that enable biotinylation of proteins in the vicinity of POI. Zhang et al. proposed immunoproximity labeling for axon initial segment proteome (IPL-AIS), which applied HRP to label AIS proteins [27]. Anatomizing dynamic changes in AIS proteins during neuronal development revealed novel AIS enriched protein SCRIB, which depended on the interaction with AnkG. However, four disulfide bonds and two Ca^{2+} binding sites involved in HRP render it inactive within intracellular, limiting its application to oxidative environment or cell surfaces [28, 29].

To overcome such limitations, Martell et al. designed and developed an engineered ascorbate peroxidase namely APEX which is a monomeric peroxidase with 28 kDa, derived from dimeric pea or soybean [28]. The advantage of APEX over HRP is that it lacks disulfide bonds and calcium ions, allowing it to label cellular compartments with greater activity. In PL experiment, APEX can catalyze the conversion of biotin-phenols into biotin-phenoxyl radicals by H_2O_2 treatment (Figure 4.1b). Due to the short lifetime of these radicals, they can only diffuse into neighboring proteins, covalently labeling the sites like Tyr, Trp, and Cys, within a radius of 20 nm centered on APEX and 1 minute. Rhee et al. employed APEX to specifically identify 495 mitochondrial matrix proteins in which PPOX was localized in the matrix for the first time, enhancing the understanding of its interaction with FECH [30].

Nevertheless, the low sensitivity of APEX often makes it challenging to detect biotin-phenol signals in MS-based proteomics. To this end, Lam et al. screened for more active APEX using a directed evolution strategy in the yeast-display platform and fluorescence-activated cell sorting (FACS)-based selection, then successfully unearthed an APEX mutant, termed as APEX2 [31]. Both APEX and APEX2 have been demonstrated to be powerful in identifying PPIs and localizing proteins at subcellular resolution. Combining APEX and APEX2, Lobingier et al. deciphered a dynamic interaction network of G-protein–coupled receptors (GPCRs) induced by ligand treatment and subcellular localizations at spatial and temporal scales [32]. With such a robust approach, the interacting proteins for delta opioid receptor

(DOR) have been uncovered at spatial–temporal resolution, including some known and novel interacting proteins.

To address the problems in off-target and excessive labeling, Han et al. developed split APEX (sAPEX) using directed evolution with FACS screening, which consisted of two inactive fragments, AP and EX, performing peroxidase activity only when a molecular interaction occurs between AP and EX [33]. Therefore, sAPEX empowers labeling proteins at specific sites or regions, which has been illustrated in many studies, improving the capabilities of APEX. While APEX and its derivatives have achieved highly efficient proximity labeling with high activity, it has a caveat that the generation of phenoxy radicals depends on the reaction with H_2O_2, which is toxic to cells, inducing cell damage and interfering with the redox process [25, 29].

As an alternative PL enzyme, BirA is an *E. coli* biotin ligase that has been demonstrated to achieve efficient biotinylation in various species for localizing proteins and profiling functional protein complexes [34]. BirA with mutation in R118G (termed as BirA* with 35 kDa) was utilized in a PL proteome application (namely BioID) by Roux et al. for identifying interacting proteins in intrinsic cellular environment [35]. With biotin and ATP, BioID can catalyze the generation of biotin-5′-AMP, a highly active and unstable compound (Figure 4.1b). After the reactive center of BioID releases biotin-5′-AMP, this kind of biotin can quickly release to neighboring proteins and biotinylate them at lysine residues. With a labeling radius at ~10 nm, BioID enables labeling of most proteins nearby while reducing background noise. These biotinylated proteins can be captured by affinity beads and identified by MS [36]. Go et al. presented a BioID-based human protein subcellular localization map by fusing 192 bait proteins with BirA* to discover 4424 reliable proximate interactors scored by SAINT and define subcellular compartments of 4145 proteins [37].

The relatively large molecular weight of BioID is an obstacle to the efficient targeting of POI. Hence, Kim et al. designed a smaller tagging enzyme namely BioID2 with 233 amino acids, derived from *Aquifex aeolicus* [38]. BioID2 requires less supplement of biotin, performs better labeling of adjoining proteins, and flexibly adjusts label radius. Moreover, similar to the principle of sAPEX, Schopp et al. established split-BioID, which splits BirA into two inactive segments and fuses with two putative interacting proteins, respectively [39]. The biotinylation of split-BioID would be active when these two proteins interact, labeling proximate proteins to detect the other potential protein complex components. By utilizing BioID2, Aprosoff et al. anatomized the interactome of SLX4, which is essential for DNA repair [40]. On the basis of 221 high-confidence interactors, SLX4 was reported to associate with RNA metabolism and chromatin remodeling as well.

Considering the long labeling time and inactivity under 37 °C of BioID, BioID2, and splitBioID (~18 hours), Branon et al. implemented directed evolution and developed two engineered promiscuous biotin ligases, TurboID (35 kDa) and miniTurbo (28 kDa), from wild-type BirA [41]. Each of these ligases enables biotinylating proteins within 10 minutes while keeping high reactivity at 30 °C. A practical application of TurboD is the investigation of interactome of transcription factor FOXO1. Su et al. constructed the FOXO1-TurboID fusion gene and conducted proximity labeling of proteins at various time points, ultimately validating that hnRNPK and RBM14 interact with FOXO1 [42]. Furthermore, reconstituted on the interactions of two proteins or organelles fused with inactive fragments of TurboID, respectively, split-TurboID enables the establishment of protein interactome and subcellular proteome maps in mammalian cells [43]. Lee et al. proposed a light-mediated TurboID termed as LOV-Turbo, which can initiate biotinylation upon blue light irradiation or bioluminescence resonance energy transfer [44]. Therefore, LOV-Turbo enables the dissection of the PPI network and subcellular compartments in high temporal and spatial precision with high resolution.

The other BirA*-based promiscuous biotin ligases have also been rapidly developed. Ramanathan et al. extracted and engineered a BirA* variant from *Bacillus subtilis*, namely BASU, which performed over 1000-fold faster kinetics and a more than 30-fold higher signal-to-noise ratio compared to BirA*, leading to shorter labeling time (1 minute) [45]. Kido et al. de novo designed a BirA variant enzyme termed as AirID which performed better biotinylation activity than BioID, even though they have 82% sequence similarity [46]. AirID has been demonstrated to be applied to investigate PPIs induced by proteolysis-targeting chimeras (PROTACs) [46, 47].

Recently, photo-mediated dependent proximity labeling has also inspired advancements. Such methods differ from enzyme-catalyzed production of intermediates in that highly reactive radicals or intermediates are produced through photoactivation. Singlet oxygen generators (SOGs) can photocatalytically generate singlet oxygen (SO) which enables oxidation of proximate proteins within nanometer range. Based on SOG, Muller et al. proposed an approach named as LUX-MS for specifically profiling protein interactions on or in between living cells [48]. Coupling SOG with ligase of interest (i.e. antibodies and drugs), LUX-MS would guide them to specific cell surfaceome domain for light-induced labeling. Amino acids like histidine would be transformed into 2-oxo-histidine that stabilize biotin hydrazide so that it enables biotinylation of neighboring proteins. Also leveraging the properties of SOG, Zhai et al. fused miniSOG with POI and added an exogenous probe containing a click-chemistry handle to enrich and map organelle-specific proteomes [49]. Irradiated by blue light and modulated by irradiation time, miniSOG-generated SO biotinylate proteins in the vicinity. Applying this approach to BRD4 and E3 ligase Parkin, researchers revealed novel interactors of BRD4, Ssu72, and SNW1, which are degraded by the Parkin-mediated ubiquitination-proteosome pathway. In a separate case, Zheng et al. developed a chemoproteomics approach in 2023, namely interaction-guided crosslinking (IGC), that utilized trifunctional probes to specifically interact with receptors on the surface of living cells [50]. Triggered by UV light and click chemistry, the probes enable the biotinylation of proteins around receptors, systematically identifying ligand–receptor interactions.

4.4 Cross-linking Mass Spectrometry (XL-MS)

XL-MS is a crucial tool for characterizing PPIs and serves as a complementary technique in structural biology. In XL-MS, intra-protein or inter-protein linkages are established using cross-linking reagents (Figure 4.2a). After proteolytic digestion

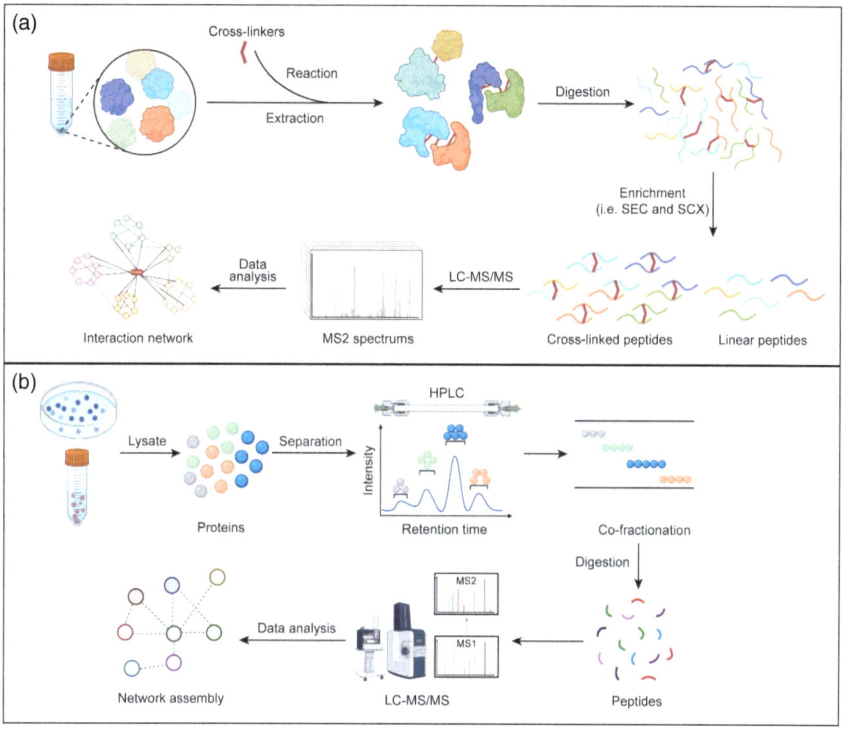

Figure 4.2 Workflow of XL-MS and CF-MS. (a) An overview of XL-MS workflow. Cross-linkers are introduced to cell samples to react with proteins. After digestion, the resulting peptide mixtures are separated by SEC or SCX to enrich cross-linked peptides. These peptides are then analyzed by LC–MS/MS, providing data from which the interaction network is assembled. (b) An illustration of CF-MS workflow. In a native environment, proteins are extracted from cell or tissue lysates and separated via chromatography or electrophoresis to obtain gradient fractions. For each fraction, proteins are digested into peptides and analyzed by LC–MS/MS before utilizing bioinformatics tools to assemble protein networks. Source: Bin Liao and Liang Zhang.

of the samples, separation procedures are generally taken to enrich cross-linked peptides from the other unmodified linear peptides. Quantitative analysis with LC–MS/MS will identify the linked peptides and match them to the original proteins for uncovering high-confidence interactions.

Generally, cross-linking reagents consist of a spacer arm and two reactive groups connected via the spacer arm. The reactive groups can covalently target the same/different amino acid residues located within a protein or different proteins [51, 52]. The spacer arm is typically an adjustable-length alkyl chain that determines the maximum distance limit of the cross-linking reagent, leading to a defined range within which cross-linkers can only react with amino acid residues. Therefore, cross-linked peptides provide distance constraints for modeling and validation of protein conformation and 3-dimensional structures [53]. Depending on the nature of the cross-linking reaction or the side chains of amino acid residues, cross-linkers can be categorized into seven types: (i) amine-reactive,

(ii) thiol-reactive, (iii) photoreactive, (iv) carboxyl-reactive, (v) formaldehyde, (vi) cleavable, and (vii) enrichable trifunctional [54].

Amine-reactive cross-linkers, such as *N*-hydroxysuccinimide (NHS), can target lysine residue in proteins with high activity. In addition, NHS has been reported to react with some amino acids containing hydroxyl side chains like serines, tyrosines, and threonines [55]. NHS is an important component of various cross-linkers. For example, Alkyne-A-DSBSO is a cross-linker processing two NHS ester groups, a spacer with approximately 14 Å in length and the other components [56]. Jiao et al. utilized an Alkyne-A-DSBSO-based XL-MS to depict the human PPI network involved in breast cancer patient-derived xenograft (PDX) models [57]. On the other hand, both thiol-reactive cross-linkers targeting cysteine and carboxyl-reactive cross-linkers targeting aspartic acid or glutamic acid exhibit some limitations and require careful control of reaction conditions [54, 58].

In order to target more residues for deciphering different kinds of proteins, photoreactive reagents have been applied to XL-MS. An ideal photoreactive reagent should be highly reactive but stable in the dark. Currently, diazirine-based photo cross-linkers, which can be active by UV light and react with many residues nearby, are widely used for this purpose. However, unbiased cross-linking reactions often result in cross-linked peptides of low abundance and complex patterns, posing a challenge for their identification by MS [59]. To decrease search cost, an alternative approach is to use a group that specifically reacts with residues at one end of the cross-linker while using a non-specific group at the other end. Gutierrez et al. developed a photoreactive cross-linker, SDASO, which consists of NHS, diazirine, and MS cleavable part [60]. SDASO has been demonstrated to elucidate protein interactome in complex samples like yeast 26S proteasome.

In XL-MS, proteolytic cleavage typically produces four types of peptides (Figure 4.2a), which are (i) inter-linked peptides (cross-linkers connect residues from different peptides), (ii) intra-linked peptides (connecting residues are within same peptide), (iii) mono-linked peptides (only one end of cross-linkers connects to a peptide), and (vi) linear peptides (non-modified peptides) [52]. Among these peptides, only inter-linked peptides are useful for analyzing protein interactions, but they are sub-stoichiometric compared with the other peptides, especially linear peptides, leading to only a small fraction of cross-linked peptides can be identified using typical bottom-up proteomics [58, 59]. With the aim of making cross-linked peptides easier to detect, a series of optimized approaches have been developed for each part of XL-MS.

First, XL-MS can benefit from the improved design of cross-linkers. During database searching, XL-MS tends to incur higher computational costs and to take much more time, compared with typical proteomics, since for each cross-linked peptide spectrum, the search engine requires pairwise proteins from the reference protein database to calculate the protein pair that best matches that spectrum, resulting in a n^2 search space, where n is the number of proteins [61]. To resolve such restriction, MS-cleavable cross-linkers have been developed, which contain an unstable bond in the spacer chain, easily cleaved into two fragments during collision-induced dissociation (CID)-MS/MS, decreasing search space to $2n$.

Specifically, a cross-linked peptide would release two characteristic peptides in MS2 using CID, and each individual peptide would be further fragmented in MS3 to identify the protein to which it belongs [56, 62, 63].

On the other hand, trifunctional probes with enrichment capability can distinguish cross-linked peptides from linear peptides by means such as biotin enrichment. Separating and enriching cross-linked peptides not only increases the likelihood of MS detection but also reduces search time. The trifunctional probes commonly use biotin as the third functional group to biotinylate and enrich cross-linked peptides using magnetic streptavidin beads. A representative example is a range of protein interaction reporters (PIR), which consists of different combinations of reaction groups, cleavable tags, and affinity tags, but they all contain mass-specific central components as reporting groups [64]. Nevertheless, natural biotinylated proteins could increase background noise. For this season, an alternative strategy employs click chemistry-based trifunctional probes to explicitly biotinylate cross-linked peptides after cross-linking reactions. As an example, Huang group developed two cross-linkers, Alkyne-A-DSBSO and Azide-A-DSBSO, which applied alkynes or azide as enrichable handle that enables cross-linked peptides to take on a biotin tag through click conjugation, followed by affinity purification with strep beads [56, 63]. Besides biotin enrichment, phosphonic acid-based cross-linkers (such as PhoX) empower cross-linked peptides to be enriched by immobilized metal ion affinity chromatography (IMAC) [65]. For allowing PhoX to penetrate intact cells, Jiang et al. developed tBu-PhoX that masks the negative charge on the hydroxyl group of phosphate with a tert-butyl group, expanding the application area *in vivo* [66].

Second, refinement of sample preparation can substantially improve the outcome of XL-MS. Subsequent to the cross-linking reaction, proteins are digested to produce peptides by proteases, the most commonly used of which is trypsin, which specifically cleaves lysine and arginine sites. However, numerous cross-linkers (i.e. NHS) targeting lysine occupy the enzymatic sites, causing cross-linked peptides that are often too long to be detected. Research works have revealed that the implementation of multiple enzymes with distinct cleavage sites can unexpectedly improve the identification of cross-linked peptides [67, 68]. Moreover, an enrichment step, utilizing size exclusion chromatography (SEC) or strong cation exchange (SCX), is usually deployed prior to LC–MS/MS analysis [57, 62, 63]. Leveraging the larger Stoke's radius and more positive charge of cross-linked peptides, SEC and SCX have, respectively, achieved the differentiation between cross-linked peptides and the other peptides [58].

Third, the enhancement of XL-MS can be achieved by improving the MS analysis pipeline. As mentioned above, while MS-cleavable cross-linkers reduce computational costs, the computational complexity of XL-MS remains high compared to standard peptide identification. To date, a wide array of computational tools tailored for XL-MS have been developed to meet diverse needs, significantly enhancing the depth and accuracy of cross-linked peptide identification. For instance, XlinkX (v2) enabled precise and efficient characterization of the conformations and interactions of endogenous proteins [69]. In addition, OpenPepXL utilizes the entire search space instead of heuristics like other algorithms, allowing it to identify unique residue pairs more quickly due to efficient data structures and built-in parallelization [70]. More tools are comprehensively reviewed by Yilmaz et al. [71].

XL-MS has proven to be a powerful technique for studying PPIs and even characterizing protein structures. By covalently binding interacting proteins prior to cell lysis, XL-MS can capture weak and transient interactions in situ. Despite some limitations of XL-MS in cross-linked peptide enrichment and data processing, ongoing advancements in both cross-linker development and MS performance have expanded its capability to elucidate protein interactions and functions in diseases and biological processes. A compelling case is the application of XL-MS in the investigation of protein interactions and conformation change during Alzheimer's disease (AD) progression, uncovering the interactions between A1AT and CO3 that may be associated with AD pathogenesis [72]. In another example, Bogdanow et al. depicted a cytomegalovirus protein interactome map using quantitative XL-MS in spatial resolution [73]. Four designated viral layers encompassed 32 viral proteins, of which UL32 participated in all layers and antagonized binding to 14–3-3 proteins via recruiting PP-1 phosphatase.

4.5 Co-fractionation Coupled with Mass Spectrometry (CF-MS)

Co-fractionation is a biochemical technique used to study the physical association of proteins in the cellular environment. Biological samples are resolved into various fractions using co-fractionation, with each fraction containing different proteins. Generally, proteins with similar biophysical or biochemical properties are more likely to occur in the same fractions, indicating their functional or structural relation [74]. Originally, this kind of protein association was applied to illustrate organelles proteome map, aiding in the dissection of protein compositions within organelles or the separation of subcellular fractions via differential centrifugation or density gradient centrifugation [75]. When such physicochemical associations were extended to PPI studies, it was discovered that the proteins sharing similar elution signals across fractions prompt underlying interactions and complex components. Therefore, CF-MS has emerged as an attractive strategy for profiling thousands of protein interactions and complexes on a proteome-wide scale.

Beginning with cell or tissue lysates under nondenatured conditions, proteins in CF-MS experiment are subsequently separated into a serial of fractions using chromatography or electrophoresis such as size-exclusion chromatography (SEC), ion-exchange chromatography (IEX) (Figure 4.2b). For enhanced separation performance, high-performance liquid chromatography (HPLC) is frequently employed during elution [74, 76]. Each fraction will be independently forwarded to tryptic digestion into peptide mixtures prior to accurate quantification by LC–MS/MS. By integrating protein intensities across various fractions and analyzing distinctive features of their elution profiles, advanced algorithms and toolkits can evaluate the correlation between proteins and achieve the scoring and prediction of protein interactions and complexes, culminating in the assembly of the comprehensive protein complex network.

In the pivotal elution step, SEC differentiates protein complexes according to their hydrodynamic radius, while IEX accomplishes this separation based on the charges in proteins [77]. By utilizing porous packing material as stationary phase, SEC enables small-size or low-molecular-weight proteins to penetrate the pores of packing when biological samples traverse the chromatographic columns, leading to longer elution time compared to proteins with larger sizes. Alternatively, IEX employs resin with charges as a stationary phase, including cation exchange resin (with negative charges) or anion exchange resin (with positive charges). Proteins bind to the stationary phase based on their charge properties, with the extent of binding modulated by the pH or salt concentration of the mobile phase, ultimately leading to a gradient elution of protein complexes.

To enhance identification depth, CF-MS can involve dozens to thousands of fractions, posing significant challenges for data analysis. Differentiating true positive interactions from redundant negative data within the elution signals of thousands of proteins necessitates the use of robust bioinformatics tools. Stacey et al. combined CF-MS and stable isotope labeling in amino acids in cell culture (SILAC) to assess protein interactions under different conditions and developed PrlnCE, a modular pipeline applying to CF-MS data, which leverages Gaussian mixture models to fit elution profiles and predict PPIs based on Naive Bayes [78]. Moreover, Hu et al. evaluated elution profiles of protein complexes using eight distinct correlation metrics which were subsequently integrated as input information for scoring complexes via support vector machine and random forest classifiers and for clustering using ClusterONE [79]. Hu et al. termed such a method as EPIC and applied it to the mapping of *Caenorhabditis elegans* interactome.

However, both PrlnCE and EPIC are constrained to inferring PPIs from elution profiles instead of directly deciphering protein complexes, which still rely on additional clustering algorithms, amplifying the network noise. To this end, Heusel et al. proposed a complex-centric proteome pipeline, which widely separates complexes by SEC, accurately quantifies proteins via SWATH-MS, and robustly dissects protein complexes using the CCProfiler algorithm [80]. CCProfiler leverages quantitative peptide information to generate protein intensity matrices which are then queried against target and decoy complex sets for ultimately assembling complex co-elution features. Another computational toolkit used to predict the novel protein complex from CF-MS is PCprophet, which is compatible with data generated from various CF-MS methods [81]. PCprophet extracts a series of features from protein elution profiles and abundance matrices for PPI prediction by ensemble learning, performs postprediction processing using gene ontology to reduce the false discovery rate and leverage a Bayesian inference-based method to evaluate protein complexes across different conditions.

The notable advantages of CF-MS are (i) high throughput: CF-MS can efficiently profile thousands of protein complexes under mild conditions and (ii) preservation of native states: CF-MS can infer endogenous and physiologically relevant protein interactions since proteins remain non-denatured and unmodified. Nevertheless, standard label-free quantification may result in large missing values, influencing the fitting of elution profiles, decreasing the predictive performance of models, and overlooking true positive complexes. As a solution, Havugimana et al. established a multiplex CF-MS platform named mCF/MS, which reproducibly quantifies proteins using isobaric tandem mass tag (TMT) and enables complexes comparison among different biological samples via comparative proteomics [76]. In addition, sophisticated data imposes more rigorous demands on bioinformatics algorithms. However, most current algorithms manually extract features from elution profiles, introducing bias, leading to overfitting in imbalanced datasets, and complicating the differentiation of false positives resulting from coincidental co-elution. An alternative strategy is the implementation of end-to-end learning, which automatically learn features from elution profiles [82]. Skinnider et al. integrated data from 206 CF-MS experiments and conducted meta-analysis for a benchmarking design and data processing of CF-MS experiments [83]. Several intriguing conclusions were drawn by Skinnider et al. such as the finding that increasing biological replicates is more beneficial than more fractionations and that using over 40 fractions does not significantly enhance protein coverage or complex identification.

CF-MS can be applied to map natural protein complexes and their changes in conditions ranging from evolution to drug treatments. Shrestha et al. employed IEX-HPLC to separate proteins into 192 fractions and analyzed the data with PrInCE, leading to the discovery of 486 complexes and 10,054 interactions in the human brain [84]. Additionally, comparative proteome allows CF-MS to reveal the evolutionary conservation of protein complexes across species. McWhite et al. conducted an in-depth analysis of 13 plant species via combining SEC, IEX, and isoelectric focusing (IEF) to extensively separate plant proteins into 2111 individual fractions [85]. This approach measured the abundance of over 2 million proteins, uncovering stable protein complexes shared among plant cells and providing a framework for interpreting plant genetics and mutant phenotypes. In another interesting example, Lazear et al. investigated chemical probes that can alter protein complexes by combining activity-based protein profiling (ABPP) strategy and CF-MS, unveiling the disruption of PA28 proteasomal regulatory complex and stabilization of a dynamic state of the spliceosome caused by stereoselective participation of covalent ligands [86].

4.6 Thermal Proximity Co-aggregation (TPCA)

In high-temperature environments, most proteins struggle to maintain their natural structures, leading to unfolding, denaturation, and precipitation. The proportion of protein precipitation varied at different temperatures, exhibiting an S-shaped decreasing trend. Consequently, the soluble protein fraction at various temperatures can be modeled with a sigmoid curve, producing a protein thermal melting curve. In 2013, Molina et al. made an innovative discovery that the thermal stability of proteins is affected by ligand binding, causing a shift in the protein melting curves [87]. This phenomenon was leveraged by Molina et al. to monitor the ligand-binding events within living cells and tissue, coining this approach the cellular thermal shift assay (CETSA). Subsequently, Savitski et al. expanded CETSA with MS, developing thermal proteome profiling (TPP), which significantly increased experiment throughput by simultaneously profiling melting curves of thousands of proteins, thereby effectively identifying potential off-target drugs [88]. Thermal proteomics was further improved in 2018 when Tan et al. proposed thermal proximity co-aggregation (TPCA), which enables the monitoring of dynamic protein complexes under various conditions on a proteome-wide scale [89].

By mining information from CETSA and TPP data, Tan et al. unearthed that proteins and their interacting partners are subjected to heat, and they exhibit a tendency to co-aggregate and co-precipitate, while maintaining correlated melting points across different temperature ranges [89]. It appears that the melting curves of protein complexes are also thermally shifted when subjected to external stimuli. By leveraging such mechanisms and phenomena, TPCA can monitor the dynamics of PPIs and protein complexes without exogenous modification. An exemplary TPCA experiment builds on CETSA and TPP, where protein samples are divided into 10 equal aliquots, each heated to different temperatures ranging from 37 °C to 67 °C, leading to the co-precipitation of interacting proteins at elevated temperatures (Figure 4.3a) [87–90]. The supernatants from these heated samples are subsequently labeled with isobaric TMT and analyzed using LC–MS/MS. In data processing, the distance between melting curves is used to assess the similarity of protein-soluble fractions and TPCA z-score, derived from distance, serves as a metric for evaluating protein interactions within complexes [89]. At the complex level, the extent of association or dissociation of the protein complexes is measured by TPCA signatures, which reflect the statistical significance of the target complex within a distribution generated by random sampling.

TPCA enables unbiased and systematic mapping of protein complex dynamic networks in situ or in living cells without exogenous tags or antibodies. By heating living cells prior to lysing them, TPCA can dissect the activities of protein complexes within intact cells. Studies have shown that certain protein complexes differ between living cells and cell lysates, likely due to the disruptive effects of the lysing process on complex structures [89]. TPCA has been demonstrated to be available across various cell lines, tissues, and species under different physiological conditions, providing an effective means for assessing the chemical kinetic profile of the same complex in multiple dimensions [89, 91–93].

Although TPCA offers time-efficient and convenient in both experimental workflows and computational pipelines compared to other methodologies, recent studies have highlighted areas in both aspects that still require optimization. Typically, detergents are employed to lyse cells and extract proteins from them. The selection of detergents is particularly critical in thermal proteomics as proteins exhibit different solubilization properties in various detergents, possibly leading to distorted results. It has been discussed that both non-ionic detergents (i.e. NP40) and zwitterionic detergents (i.e. CHAPS) diminish protein thermal stabilities [94]. Compared to the milder detergent like DDM, a more stringent detergent is necessary for TPCA to accurately profile the thermal melting curves of insoluble proteins,

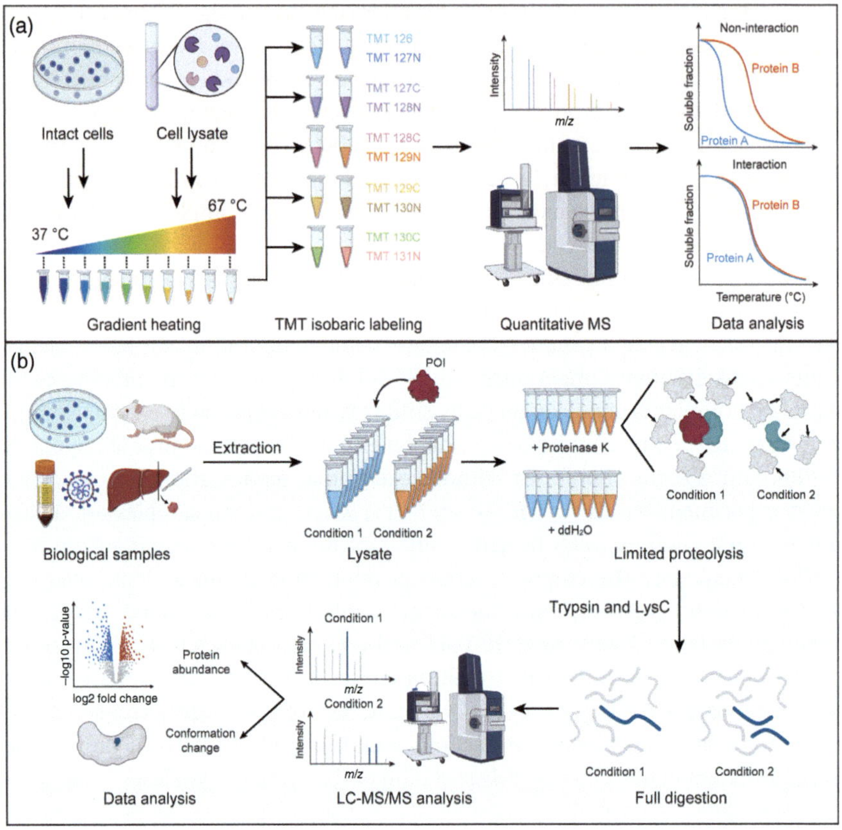

Figure 4.3 Workflow of TPCA and LiP-MS. (a) A general schematic of TPCA workflow. For either intact cells or cell lysate, samples are divided into aliquots and subjected to gradient heating at temperatures ranging from 37 to 67 °C. With the increasing temperature, the soluble fraction within samples tends to decrease. The supernatants are subsequently collected for TMT isobaric labeling and quantitative LC–MS/MS analysis. A protein melting curve is generated by calculating the soluble fraction at each temperature and protein interactions are quantified by assessing the distance between protein melting curves. (b) The workflow of LiP-MS for profiling PPIs. Biological samples from various sources are processed into two batches under different conditions, with each batch containing eight replicates. The purified protein of interest (POI) is added to Condition 1. Four replicates from each condition are subjected to limited proteolysis, while the remaining four are left untreated. The peptide mixtures undergo secondary digestion with trypsin and LysC, followed by LC–MS/MS analysis. Proteins interacting with the POI exhibit distinct peptide signal spectra, enabling protein differential analysis and assessment of conformational changes. Source: Bin Liao and Liang Zhang.

such as membrane proteins [93, 95]. However, it is also important to consider the compatibility of different detergents with mass spectrometry. Furthermore, it is vital to select appropriate temperature points in thermal proteomics since each point carries varying significance in assessing protein thermal stability. It has been found that, at an optimal temperature, even a single temperature point can provide sufficient evidence of drug-target deconvolution [96]. Selecting the applicable temperature range can not only increase the probability of target deconvolution but also enhance the prediction of protein complexes [95, 97]. Emerging evidence supports that fewer temperature points can also achieve comparable predictive accuracy in TPCA [93, 95]. Sun et al. optimized the TPCA temperature design and introduced Slim-TPCA by evaluating the contribution of different distance algorithms and combinations of temperature points [93]. Slim-TPCA demonstrated that using as low as three optimal temperature points (37, 49, and 58 °C) with the Manhattan distance algorithm can maintain relatively robust TPCA signatures compared to using ten temperatures while largely increasing experimental throughput.

Advancements in statistical and computational methods of TPCA can also enhance the quality of dynamic PPI prediction. For example, Slim-TPCA supported the sensitivity of capturing dynamic TPCA signatures of protein complexes by switching from absolute distance to relative distance. In addition, it reduced computational costs by calculating TPCA signatures using a beta distribution rather than relying on distributions generated through random sampling [93]. Moreover, current TPCA algorithms are limited to tracking known protein complexes and struggle to accurately predict novel complexes since irrelevant proteins may coincidently display similar melting curves, which TPCA algorithms find difficult to differentiate, resulting in a substantial number of false-positive predictions.

Another strategy is the combination of TPCA and the other approaches that provide multi-dimensional information. Reed et al. developed an ensemble machine learning framework named Tapioca by leveraging MS data from TPCA, CF-MS, ion-based proteome-integrated solubility alteration (I-PISA), physical properties from bioinformatics prediction based on protein structures, functional domain data from Pfam database, and prior PPI knowledge from HumanBase [95]. The integrated tapioca method can efficiently predict novel protein complexes during dynamic processes and has been successfully applied to study the dynamics of complexes during the reactivation of an oncogenic gamma herpes virus. For more convenient analysis and visualization of TPCA data, advanced toolkits such as rTPCA [98] and ProSAP [99] have been developed.

An interesting application area of TPCA is the dynamics of protein complexes in response to virus infection and immunomodulation. Hashimoto et al. employed TPCA to systematically profile the temporal changes of protein complexes during human cytomegalovirus (HCMV) infection, uncovering host–host and virus–host protein interactions during infection [92]. In another work, Justice et al. discovered that the co-regulation of DNA-dependent protein kinase and IFI16 during herpes simplex virus 1 (HSV-1) infection impacts the immune response to viral infection, as elucidated through TPCA [100]. Moreover, in a comparative proteome using TPCA, Lum et al. demonstrated that type I and type II interferons differentially stabilize interferon-stimulated gene interactions in macrophage-like monocytic cells [101].

4.7 Limited Proteolysis–Mass Spectrometry (LiP–MS)

LiP-MS employed a two-step limited proteolysis method to generate a higher abundance of distinctive digested peptides compared with a one-step full digestion condition, tracking protein conformation change on a proteome-wide scale and localizing the change sites of proteins. In 2014, LiP-MS was originally developed to probe protein structural change in response to perturbations caused by various molecular events in complex biological systems [102]. A general LiP-MS workflow involves the following steps: (i) extracting proteins from biological samples, such as purified proteins, intact cells, tissue lysates, and body fluids; (ii) dividing protein samples into two equal parts designated as Condition 1 (CON 1) and Condition 2 (CON 2), adding purified protein of interest (POI) to CON1; (iii) treating CON 1 and CON 2 samples with a non-specific protease while keeping the rest untreated and the interacting proteins of POI are protected from digestion in CON1; (iv) fully digesting proteins in both CON 1 and CON 2 with trypsin and LysC; (v) analyzing peptide mixtures with LC–MS/MS; and (vi) processing and comparing MS data (Figure 4.3b).

Limited proteolysis is a pivotal step of LiP-MS, crucial for determining the digestion patterns and effects on structure-specific proteins. It is typically performed using a controlled and non-specific protease, such as proteinase K (PK), which preferentially cleaves flexible and readily accessible protein regions [102–104]. In contrast, regions involved in folding or binding exhibit reduced PK accessibility. Consequently, the impact of molecular perturbations on protein structure is reflected by variations in PK accessibility under different conditions, which are quantified by measuring peptide abundance through MS [103, 104]. In addition to PK, there are other proteinases that are available in LiP-MS, which have been comprehensively discussed by Schopper et al. [104]. To ensure experimental reproducibility and high peptide quality in LiP-MS, it is crucial to meticulously optimize several parameters, including the enzyme-to-substrate ratio, the incubation time with PK, and the thoroughness of PK inactivation [103].

As an innovative and high-throughput tool for monitoring protein structural change within systematic biology, LiP-MS has been effectively applied to assess protein structural alterations induced by thermal stabilities and ligand-binding events [105–107]. Recently, Holfeld et al. expanded LiP-MS to the study of PPI through incubating purified POI with CON 1 sample prior to limited proteolysis of both CON 1 and CON 2 samples [108]. The peptides located on the interacting surface can be less accessible to and less readily digested by PK in CON 1 compared to the untreated CON 2, leading to the differential abundance of the same peptides between conditions, which are then utilized to determine the interacting partners of POI (Figure 4.3b). As a first proof of concept, Holfeld et al. benchmarked the interactions between respiratory syncytial virus F (RSVF) glycoprotein and its several known site-specific antibodies, effectively identifying established antigenic sites and demonstrating that this approach is able to characterize interaction surfaces and detect structure-specific interactions within complex cell lysate [108]. Further proof-of-principle test was performed in integral membrane proteins by investigating the interactions between calmodulin and membrane-integral adenylyl cyclase type 8 (AC8) [108].

The presence of a protein in different structural forms during disease progression may indicate a pathogenic mechanism. Investigating the interaction partners of these structurally diverse proteins can yield novel molecular insights into disease mechanisms and suggest potential novel therapeutic targets. Holfeld et al. first applied LiP-MS to the study of structure-specific interactions of aSyn, a protein that performs fibrillar structural states in Parkinson's disease and identified unique interacting partners between monomeric and amyloid fibril forms of aSyn [108]. They also demonstrated that despite having similar conformations, some proteins exhibit different interactomes, such as GDP- and GTP-bound Rab GTPases.

Given the robust research potential of LiP-MS across various aspects, several LiP-MS–specific bioinformatics pipelines have been developed to facilitate the rapid, sensitive, and accurate identification of differential peptides such as FLiPPR [109] and LiPAnalyzeR [110]. However, since LiP-MS can capture the structural signals induced by various perturbations, it can provide the observations of protein conformation changes but elucidations of undergoing mechanisms, requiring supplementary data from orthogonal methods [103, 104]. The sensitivity of LiP-MS hinges on the successful MS identification of peptides located in regions of protein conformation change. Although data-independent acquisition (DIA) MS has been employed for increasing protein coverage of LiP-MS, identifying low-abundance proteins remains challenging [108]. Current LiP-MS is only available to lysate samples and is unable to track protein structural change in intact cells. Overall, LiP-MS is a robust technology to dissect both protein physical interactions and conformation changes with clear binding sites on a proteome-wide scale.

4.8 Conclusion and Outlook

PPIs serve as pivotal orchestrators of diverse biological processes in living organisms, and dissecting these interactions is essential to mechanistic comprehension and drug design. Various MS-based approaches have demonstrated robust performances in PPI studies. It is important to understand the advantages and limitations of each approach to determine the optimal choice. AP-MS is a conventional methodology for PPI profiling but has limitations in throughput and weak/transient interactions. As a complement to AP-MS, proximity labeling tags proteins in the vicinity of POI, capturing interactions of various strengths/durations. However, both methods are limited in providing structural information. On the other hand, decoding cross-linking peptides in XL-MS provides ample structural information whilst facing challenges in data analysis. These PPI-deciphering approaches typically rely on exogenous additives (i.e. tags, probes, or cross-linkers), which inevitably have impacts on the cellular physiological environment, potentially altering protein conformations and functions. In comparison, CF-MS, TPCA, and LiP-MS, which leverage biophysical synergistic behavior, are less dependent on additives.

A common challenge of these approaches lies in MS data processing. Combining high-resolution mass spectrometry and state-of-the-art machine learning algorithms can enhance discovery performance and reduce the false-positive rate. In addition, refined analyses can offer multidimensional information, such as interacting surfaces and protein structures, guiding drug design. For instance, both XL-MS and LiP-MS can pinpoint interaction regions via cross-linking peptides and differential peptides, respectively. XL-MS can provide further spatial data based on the varying lengths of cross-linking peptides, while LiP-MS enables the deciphering of protein conformation changes under different conditions.

Although these MS methods are different in principle, their integrative application can be complementary in improving the number and quality of true PPIs, such as the combination of AP-MS and PL, AP-MS and XL-MS, and CF-MS and TPCA [40, 95, 111]. Future studies will inspire more innovations to advance the dissection of protein interactions and to promote drug design.

Acknowledgements

This work was supported by grants from the Shenzhen Science and Technology Innovation Commission (JCYJ20210324133812034), Hetao Shenzhen-Hong Kong Science and Technology Innovation Cooperation Zone Shenzhen Park Project (HZQB-KCZYZ-2021017), the Research Grants Council of Hong Kong (11102322 and T12-101/23-N), and grants from the Tung Foundation Biomedical Sciences Centre (9609314 and 9609315).

References

1 International Human Genome Sequencing Consortium (2004). Finishing the euchromatic sequence of the human genome. *Nature* 431 (7011): 931–945.

2 Jumper, J., Evans, R., Pritzel, A. et al. (2021). Highly accurate protein structure prediction with AlphaFold. *Nature* 596 (7873): 583–589.

3 Abramson, J., Adler, J., Dunger, J. et al. (2024). Accurate structure prediction of biomolecular interactions with AlphaFold 3. *Nature* 630 (8016): 493–500.

4 Oughtred, R., Rust, J., Chang, C. et al. (2021). The BioGRID database: a comprehensive biomedical resource of curated protein, genetic, and chemical interactions. *Protein Science : A Publication of the Protein Society* 30 (1): 187–200.

5 Del Toro, N., Shrivastava, A., Ragueneau, E. et al. (2022). The IntAct database: efficient access to fine-grained molecular interaction data. *Nucleic Acids Research* 50 (D1): D648–D653.

6 Thomas, S.N., French, D., Jannetto, P.J. et al. (2022). Liquid chromatography-tandem mass spectrometry for clinical diagnostics. *Nature Reviews. Methods Primers* 2 (1): 96.

7 Richards, A.L., Eckhardt, M., and Krogan, N.J. (2021). Mass spectrometry-based protein–protein interaction networks for the study of human diseases. *Molecular Systems Biology* 17 (1): e8792.

8 Guruharsha, K.G., Rual, J.F., Zhai, B. et al. (2011). A protein complex network of Drosophila melanogaster. *Cell* 147 (3): 690–703.

9 Gingras, A.C., Gstaiger, M., Raught, B., and Aebersold, R. (2007). Analysis of protein complexes using mass spectrometry. *Nature Reviews. Molecular Cell Biology* 8 (8): 645–654.

10 Young, C.L., Britton, Z.T., and Robinson, A.S. (2012). Recombinant protein expression and purification: a comprehensive review of affinity tags and microbial applications. *Biotechnology Journal* 7 (5): 620–634.

11 Vandemoortele, G., Eyckerman, S., and Gevaert, K. (2019). Pick a tag and explore the functions of your pet protein. *Trends Biotechnol* 37 (10): 1078–1090.

12 Rigaut, G., Shevchenko, A., Rutz, B. et al. (1999). A generic protein purification method for protein complex characterization and proteome exploration. *Nature Biotechnology* 17 (10): 1030–1032.

13 Liu, X., Salokas, K., Tamene, F. et al. (2018). An AP-MS- and BioID-compatible MAC-tag enables comprehensive mapping of protein interactions and subcellular localizations. *Nature Communications* 9 (1): 1188.

14 Ma, H., McLean, J.R., Chao, L.F. et al. (2012). A highly efficient multifunctional tandem affinity purification approach applicable to diverse organisms. *Molecular & Cellular Proteomics : MCP* 11 (8): 501–511.

15 Ke, M., Liu, J., Chen, W. et al. (2018). Integrated and quantitative proteomic approach for charting temporal and endogenous protein complexes. *Analytical Chemistry* 90 (21): 12574–12583.

16 Mao, Y., Chen, P., Ke, M. et al. (2021). Fully integrated and multiplexed sample preparation technology for sensitive interactome profiling. *Analytical chemistry* 93 (5): 3026–3034.

17 Pankow, S., Bamberger, C., Calzolari, D. et al. (2015). ΔF508 CFTR interactome remodelling promotes rescue of cystic fibrosis. *Nature* 528 (7583): 510–516.

18 Teng, B., Zhao, C., Liu, X., and He, Z. (2015). Network inference from AP-MS data: computational challenges and solutions. *Briefings in Bioinformatics* 16 (4): 658–674.

19 Huttlin, E.L., Bruckner, R.J., Navarrete-Perea, J. et al. (2021). Dual proteome-scale networks reveal cell-specific remodeling of the human interactome. *Cell* 184 (11): 3022–3040.e28.

20 Huttlin, E.L., Ting, L., Bruckner, R.J. et al. (2015). The BioPlex network: a systematic exploration of the human interactome. *Cell* 162 (2): 425–440.

21 Frommelt, F., Fossati, A., Uliana, F. et al. (2024). DIP-MS: ultra-deep interaction proteomics for the deconvolution of protein complexes. *Nature Methods* 21 (4): 635–647.

22 Sanchez-Brinas, A., Duran-Ruiz, C., Astola, A. et al. (2023). ZNF330/NOA36 interacts with HSPA1 and HSPA8 and modulates cell cycle and proliferation in response to heat shock in HEK293 cells. *Biology Direct* 18 (1): 26.

23 Koh, M., Ahmad, I., Ko, Y. et al. (2021). A short ORF-encoded transcriptional regulator. *Proceedings of the National Academy of Sciences of the United States of America* 118 (4): e2021943118.

24 Abbasi, S. and Schild-Poulter, C. (2019). Mapping the Ku interactome using proximity-dependent biotin identification in human cells. *Journal of Proteome Research* 18 (3): 1064–1077.

25 Guo, J., Guo, S., Lu, S. et al. (2023). The development of proximity labeling technology and its applications in mammals, plants, and microorganisms. *Cell Communication and Signaling : CCS* 21 (1): 269.

26 Wang, Y. and Qin, W. (2024). Revealing protein trafficking by proximity labeling-based proteomics. *Bioorganic Chemistry* 143: 107041.

27 Zhang, W., Fu, Y., Peng, L. et al. (2023). Immunoproximity biotinylation reveals the axon initial segment proteome. *Nature Communications* 14 (1): 8201.

28 Martell, J.D., Deerinck, T.J., Sancak, Y. et al. (2012). Engineered ascorbate peroxidase as a genetically encoded reporter for electron microscopy. *Nature Biotechnology* 30 (11): 1143–1148.

29 Wang, Y.Y., Li, W., Ye, B.C., and Bi, X.B. (2023). Chemical and biological strategies for profiling protein-protein interactions in living cells. *Chemistry, an Asian Journal* 18 (14): e202300226.

30 Rhee, H.W., Zou, P., Udeshi, N.D. et al. (2013). Proteomic mapping of mitochondria in living cells via spatially restricted enzymatic tagging. *Science* 339 (6125): 1328–1331.

31 Lam, S.S., Martell, J.D., Kamer, K.J. et al. (2015). Directed evolution of APEX2 for electron microscopy and proximity labeling. *Nature Methods* 12 (1): 51–54.

32 Lobingier, B.T., Hüttenhain, R., Eichel, K. et al. (2017). An approach to spatiotemporally resolve protein interaction networks in living cells. *Cell* 169 (2): 350–360.e12.

33 Han, Y., Branon, T.C., Martell, J.D. et al. (2019). Directed evolution of split APEX2 peroxidase. *ACS Chemical Biology* 14 (4): 619–635.

34 de Boer, E., Rodriguez, P., Bonte, E. et al. (2003). Efficient biotinylation and single-step purification of tagged transcription factors in mammalian cells and transgenic mice. *Proceedings of the National Academy of Sciences of the United States of America* 100 (13): 7480–7485.

35 Roux, K.J., Kim, D.I., Raida, M., and Burke, B. (2012). A promiscuous biotin ligase fusion protein identifies proximal and interacting proteins in mammalian cells. *Journal of Cell Biology* 196 (6): 801–810.

36 Kim, D.I. and Roux, K.J. (2016). Filling the void: proximity-based labeling of proteins in living cells. *Trends in Cell Biology* 26 (11): 804–817.

37 Go, C.D., Knight, J.D.R., Rajasekharan, A. et al. (2021). A proximity-dependent biotinylation map of a human cell. *Nature* 595 (7865): 120–124.

38 Kim, D.I., Jensen, S.C., Noble, K.A. et al. (2016). An improved smaller biotin ligase for BioID proximity labeling. *Molecular Biology of the Cell* 27 (8): 1188–1196.

39 Schopp, I.M., Amaya Ramirez, C.C., Debeljak, J. et al. (2017). Split-BioID a conditional proteomics approach to monitor the composition of spatiotemporally defined protein complexes. *Nature Communications* 8 (1): 15690.

40 Aprosoff, C.M., Dyakov, B.J.A., Cheung, V.H.W. et al. (2023). Comprehensive interactome mapping of the DNA repair scaffold SLX4 using proximity labeling and affinity purification. *Journal of Proteome Research* 22 (6): 1660–1681.

41 Branon, T.C., Bosch, J.A., Sanchez, A.D. et al. (2018). Efficient proximity labeling in living cells and organisms with TurboID. *Nature Biotechnology* 36 (9): 880–887.

42 Su, Y., Guo, Y., Guo, J. et al. (2023). Study of FOXO1-interacting proteins using TurboID-based proximity labeling technology. *BMC Genomics* 24 (1): 146.

43 Cho, K.F., Branon, T.C., Udeshi, N.D. et al. (2020). Proximity labeling in mammalian cells with TurboID and split-TurboID. *Nature Protocols* 15 (12): 3971–3999.

44 Lee, S.-Y., Cheah, J.S., Zhao, B. et al. (2023). Engineered allostery in light-regulated LOV-Turbo enables precise spatiotemporal control of proximity labeling in living cells. *Nature Methods* 20 (6): 908–917.

45 Ramanathan, M., Majzoub, K., Rao, D.S. et al. (2018). RNA-protein interaction detection in living cells. *Nature Methods* 15 (3): 207–212.

46 Kido, K., Yamanaka, S., Nakano, S. et al. (2020). AirID, a novel proximity biotinylation enzyme, for analysis of protein–protein interactions. *eLife* 9: e54983.

47 Yamanaka, S., Horiuchi, Y., Matsuoka, S. et al. (2022). A proximity biotinylation-based approach to identify protein-E3 ligase interactions induced by PROTACs and molecular glues. *Nature Communications* 13 (1): 183.

48 Müller, M., Gräbnitz, F., Barandun, N. et al. (2021). Light-mediated discovery of surfaceome nanoscale organization and intercellular receptor interaction networks. *Nature Communications* 12 (1): 7036.

49 Zhai, Y., Huang, X., Zhang, K. et al. (2022). Spatiotemporal-resolved protein networks profiling with photoactivation dependent proximity labeling. *Nature Communications* 13 (1): 4906.

50 Zheng, J., Zheng, Z., Fu, C. et al. (2023). Deciphering intercellular signaling complexes by interaction-guided chemical proteomics. *Nature Communications* 14 (1): 4138.

51 Ruwolt, M., Piazza, I., and Liu, F. (2023). The potential of cross-linking mass spectrometry in the development of protein–protein interaction modulators. *Current Opinion in Structural Biology* 82: 102648.

52 Yu, C. and Huang, L. (2023). New advances in cross-linking mass spectrometry toward structural systems biology. *Current Opinion in Chemical Biology* 76: 102357.

53 Stahl, K., Graziadei, A., Dau, T. et al. (2023). Protein structure prediction with in-cell photo-crosslinking mass spectrometry and deep learning. *Nature Biotechnology* 41 (12): 1810–1819.

54 Piersimoni, L., Kastritis, P.L., Arlt, C., and Sinz, A. (2022). Cross-linking mass spectrometry for investigating protein conformations and protein-protein interactions horizontal line a method for all seasons. *Chemical Reviews* 122 (8): 7500–7531.

55 Kalkhof, S. and Sinz, A. (2008). Chances and pitfalls of chemical cross-linking with amine-reactive N-hydroxysuccinimide esters. *Analytical and Bioanalytical Chemistry* 392 (1–2): 305–312.

56 Kaake, R.M., Wang, X., Burke, A. et al. (2014). A new in vivo cross-linking mass spectrometry platform to define protein-protein interactions in living cells. *Molecular & Cellular Proteomics : MCP* 13 (12): 3533–3543.

57 Jiao, F., Yu, C., Wheat, A. et al. (2024). DSBSO-based XL-MS analysis of breast cancer PDX tissues to delineate protein interaction network in clinical samples. *Journal of Proteome Research* 23: 3269–3279.

58 Britt, H., Beveridge, R., Calabrese, A. et al. (2023). Cross-linking mass spectrometry for mapping protein complex topologies in situ. *Essays in Biochemistry* 67 (2): 215–228.

59 O'Reilly, F.J. and Rappsilber, J. (2018). Cross-linking mass spectrometry: methods and applications in structural, molecular and systems biology. *Nature Structural & Molecular Biology* 25 (11): 1000–1008.

60 Gutierrez, C., Salituro, L.J., Yu, C. et al. (2021). Enabling photoactivated cross-linking mass spectrometric analysis of protein complexes by novel MS-cleavable cross-linkers. *Molecular & Cellular Proteomics : MCP* 20: 100084.

61 Matzinger, M. and Mechtler, K. (2021). Cleavable cross-linkers and mass spectrometry for the ultimate task of profiling protein-protein interaction networks in vivo. *Journal of Proteome Research* 20 (1): 78–93.

62 Schweppe, D.K., Chavez, J.D., Lee, C.F. et al. (2017). Mitochondrial protein interactome elucidated by chemical cross-linking mass spectrometry. *Proceedings of the National Academy of Sciences of the United States of America* 114 (7): 1732–1737.

63 Wheat, A., Yu, C., Wang, X. et al. (2021). Protein interaction landscapes revealed by advanced in vivo cross-linking-mass spectrometry. *Proceedings of the National Academy of Sciences of the United States of America* 118 (32): e2023360118.

64 Tang, X., Munske, G.R., Siems, W.F., and Bruce, J.E. (2005). Mass spectrometry identifiable cross-linking strategy for studying protein-protein interactions. *Analytical Chemistry* 77 (1): 311–318.

65 Steigenberger, B., Pieters, R.J., Heck, A.J.R., and Scheltema, R.A. (2019). PhoX: an IMAC-enrichable cross-linking reagent. *ACS Central Science* 5 (9): 1514–1522.

66 Jiang, P.L., Wang, C., Diehl, A. et al. (2022). A membrane-permeable and immobilized metal affinity chromatography (IMAC) enrichable cross-linking reagent to advance in vivo cross-linking mass spectrometry. *Angewandte Chemie (International ed. in English)* 61 (12): e202113937.

67 Mendes, M.L., Fischer, L., Chen, Z.A. et al. (2019). An integrated workflow for crosslinking mass spectrometry. *Molecular Systems Biology* 15 (9): e8994.

68 Zhao, L., Zhao, Q., Shan, Y. et al. (2020). Smart cutter: an efficient strategy for increasing the coverage of chemical cross-linking analysis. *Analytical Chemistry* 92 (1): 1097–1105.

69 Liu, F., Lossl, P., Scheltema, R. et al. (2017). Optimized fragmentation schemes and data analysis strategies for proteome-wide cross-link identification. *Nature Communications* 8: 15473.

70 Netz, E., Dijkstra, T.M.H., Sachsenberg, T. et al. (2020). OpenPepXL: an open-source tool for sensitive identification of cross-linked peptides in XL-MS. *Molecular & Cellular Proteomics* 19 (12): 2157–2168.

71 Yılmaz, Ş., Shiferaw, G.A., Rayo, J. et al. (2018). Cross-linked peptide identification: a computational forest of algorithms. *Mass Spectrometry Reviews* 37 (6): 738–749.

72 Zhu, Z., Zhong, X., Wang, B. et al. (2024). Probing protein structural changes in Alzheimer's disease via quantitative cross-linking mass spectrometry. *Analytical Chemistry* 96 (19): 7506–7515.

73 Bogdanow, B., Gruska, I., Mühlberg, L. et al. (2023). Spatially resolved protein map of intact human cytomegalovirus virions. *Nature Microbiology* 8 (9): 1732–1747.

74 Havugimana, P.C., Hart, G.T., Nepusz, T. et al. (2012). A census of human soluble protein complexes. *Cell* 150 (5): 1068–1081.

75 Udi, Y., Zhang, W., Stein, M.E. et al. (2023). A general method for quantitative fractionation of mammalian cells. *Journal of Cell Biology* 222 (6): e202209062.

76 Havugimana, P.C., Goel, R.K., Phanse, S. et al. (2022). Scalable multiplex co-fractionation/mass spectrometry platform for accelerated protein interactome discovery. *Nature Communications* 13 (1): 4043.

77 Farsang, E., Guillarme, D., Veuthey, J.-L. et al. (2020). Coupling non-denaturing chromatography to mass spectrometry for the characterization of monoclonal antibodies and related products. *Journal of Pharmaceutical and Biomedical Analysis* 185: 113207.

78 Stacey, R.G., Skinnider, M.A., Scott, N.E., and Foster, L.J. (2017). A rapid and accurate approach for prediction of interactomes from co-elution data (PrInCE). *BMC Bioinformatics* 18 (1): 457.

79 Hu, L.Z., Goebels, F., Tan, J.H. et al. (2019). EPIC: software toolkit for elution profile-based inference of protein complexes. *Nature Methods* 16 (8): 737–742.

80 Heusel, M., Bludau, I., Rosenberger, G. et al. (2019). Complex-centric proteome profiling by SEC-SWATH-MS. *Molecular Systems Biology* 15 (1): e8438.

81 Fossati, A., Li, C., Uliana, F. et al. (2021). PCprophet: a framework for protein complex prediction and differential analysis using proteomic data. *Nature Methods* 18 (5): 520–527.

82 Chen, Y.H., Chao, K.H., Wong, J.Y. et al. (2023). A feature extraction free approach for protein interactome inference from co-elution data. *Briefings in Bioinformatics* 24 (4): bbad229.

83 Skinnider, M.A. and Foster, L.J. (2021). Meta-analysis defines principles for the design and analysis of co-fractionation mass spectrometry experiments. *Nature Methods* 18 (7): 806–815.

84 Shrestha, H.K., Lee, D., Wu, Z. et al. (2024). Profiling protein-protein interactions in the human brain by refined cofractionation mass spectrometry. *Journal of Proteome Research* 23 (4): 1221–1231.

85 McWhite, C.D., Papoulas, O., Drew, K. et al. (2020). A pan-plant protein complex map reveals deep conservation and novel assemblies. *Cell* 181 (2): 460–474 e14.

86 Lazear, M.R., Remsberg, J.R., Jaeger, M.G. et al. (2023). Proteomic discovery of chemical probes that perturb protein complexes in human cells. *Molecular Cell* 83 (10): 1725–1742.e12.

87 Martinez Molina, D., Jafari, R., Ignatushchenko, M. et al. (2013). Monitoring drug target engagement in cells and tissues using the cellular thermal shift assay. *Science* 341 (6141): 84–87.

88 Savitski, M.M., Reinhard, F.B.M., Franken, H. et al. (2014). Tracking cancer drugs in living cells by thermal profiling of the proteome. *Science* 346 (6205): 1255784.

89 Tan, C.S.H., Go, K.D., Bisteau, X. et al. (2018). Thermal proximity coaggregation for system-wide profiling of protein complex dynamics in cells. *Science* 359 (6380): 1170–1177.

90 Franken, H., Mathieson, T., Childs, D. et al. (2015). Thermal proteome profiling for unbiased identification of direct and indirect drug targets using multiplexed quantitative mass spectrometry. *Nature Protocols* 10 (10): 1567–1593.

91 Perrin, J., Werner, T., Kurzawa, N. et al. (2020). Identifying drug targets in tissues and whole blood with thermal-shift profiling. *Nature Biotechnology* 38 (3): 303–308.

92 Hashimoto, Y., Sheng, X., Murray-Nerger, L.A., and Cristea, I.M. (2020). Temporal dynamics of protein complex formation and dissociation during human cytomegalovirus infection. *Nature Communications* 11 (1): 806.

93 Sun, S., Zheng, Z., Wang, J. et al. (2023). Improved in situ characterization of protein complex dynamics at scale with thermal proximity co-aggregation. *Nature Communications* 14 (1): 7697.

94 Ye, Y., Li, K., Ma, Y. et al. (2023). The introduction of detergents in thermal proteome profiling requires lowering the applied temperatures for efficient target protein identification. *Molecules* 28 (12): 4859.

95 Reed, T.J., Tyl, M.D., Tadych, A. et al. (2024). Tapioca: a platform for predicting de novo protein-protein interactions in dynamic contexts. *Nature Methods* 21 (3): 488–500.

96 Ball, K.A., Webb, K.J., Coleman, S.J. et al. (2020). An isothermal shift assay for proteome scale drug-target identification. *Communications Biology* 3 (1): 75.

97 Van Vranken, J.G., Li, J., Mintseris, J. et al. (2024). Large-scale characterization of drug mechanism of action using proteome-wide thermal shift assays. Cold Spring Harbor Laboratory.

98 Kurzawa, N., Mateus, A., and Savitski, M.M. (2021). Rtpca: an R package for differential thermal proximity coaggregation analysis. *Bioinformatics* 37 (3): 431–433.

99 Ji, H., Lu, X., Zheng, Z. et al. (2022). ProSAP: a GUI software tool for statistical analysis and assessment of thermal stability data. *Briefings in Bioinformatics* 23 (3): bbac057.

100 Justice, J.L., Kennedy, M.A., Hutton, J.E. et al. (2021). Systematic profiling of protein complex dynamics reveals DNA-PK phosphorylation of IFI16 en route to herpesvirus immunity. *Science Advances* 7 (25): eabg6680.

101 Lum, K.K., Reed, T.J., Yang, J., and Cristea, I.M. (2024). Differential contributions of interferon classes to host inflammatory responses and restricting virus progeny production. *Journal of Proteome Research* 23: 3249–3268.

102 Feng, Y., De Franceschi, G., Kahraman, A. et al. (2014). Global analysis of protein structural changes in complex proteomes. *Nature Biotechnology* 32 (10): 1036–1044.

103 Malinovska, L., Cappelletti, V., Kohler, D. et al. (2023). Proteome-wide structural changes measured with limited proteolysis-mass spectrometry: an advanced protocol for high-throughput applications. *Nature Protocols* 18 (3): 659–682.

104 Schopper, S., Kahraman, A., Leuenberger, P. et al. (2017). Measuring protein structural changes on a proteome-wide scale using limited proteolysis-coupled mass spectrometry. *Nature Protocols* 12 (11): 2391–2410.

105 Piazza, I., Beaton, N., Bruderer, R. et al. (2020). A machine learning-based chemoproteomic approach to identify drug targets and binding sites in complex proteomes. *Nature Communications* 11 (1): 4200.

106 Pepelnjak, M., Velten, B., Näpflin, N. et al. (2024). In situ analysis of osmolyte mechanisms of proteome thermal stabilization. *Nature Chemical Biology* 20: 1053–1065.

107 Leuenberger, P., Ganscha, S., Kahraman, A. et al. (2017). Cell-wide analysis of protein thermal unfolding reveals determinants of thermostability. *Science* 355 (6327): eaai7825.

108 Holfeld, A., Schuster, D., Sesterhenn, F. et al. (2024). Systematic identification of structure-specific protein-protein interactions. *Molecular Systems Biology* 20 (6): 651–675.

109 Manriquez-Sandoval, E., Brewer, J., and Lule, G. et al. (2023). FLiPPR: A Processor for Limited Proteolysis (LiP) Mass Spectrometry Datasets Built on FragPipe. *bioRxiv*.

110 Nagel, L., Grossbach, J., and Cappelletti, V. et al. (2024). Enhancing the biological insight from limited proteolysis-coupled mass spectrometry. *bioRxiv*.

111 Zhong, B., An, Y., Gao, H. et al. (2023). In vivo cross-linking-based affinity purification and mass spectrometry for targeting intracellular protein-protein interactions. *Analytica Chimica Acta* 1265: 341273.

5

Detection of Protein–Protein Interactions In Situ via Proximity Ligation Assay

Xinyue Zhou[1] *and Peng Zou*[1,2,3]

[1] *Peking University, College of Chemistry and Molecular Engineering, Synthetic and Functional Biomolecules Center, Key Laboratory of Bioorganic Chemistry and Molecular Engineering of Ministry of Education, Beijing National Laboratory for Molecular Sciences, No. 5 Yiheyuan Road, Haidian District, Beijing 100871, China*
[2] *Peking University, Peking-Tsinghua Center for Life Sciences, PKU-IDG/McGovern Institute for Brain Research, No. 5 Yiheyuan Road, Haidian District, Beijing 100871, China*
[3] *Chinese Institute for Brain Research, Beijing (CIBR), No. 3 Building, No. 9, Science Park Road, Zhongguancun Life Science Park, Changping District, Beijing 102206, China*

5.1 Introduction

Proteins, central to life processes, mediate most cellular activities through interactions with biomolecules. Protein–protein interactions (PPIs) are vital for an array of biological processes, ranging from signal transduction and gene expression to host–pathogen interactions and adaptive immunity. Decoding the complex network of these interactions could offer valuable insights into the mechanisms driving cellular functions, aiding in identifying novel drug targets and developing therapeutic strategies [1, 2]. To achieve these goals, various chemical biology tools have been created to monitor and quantify biomolecular interactions in situ. These methods effectively act as coincidence detectors for biomolecular pairs (Figure 5.1a).

A classic method for detecting protein interactions in situ is Förster resonance energy transfer (FRET), which measures energy transfer from a fluorescent donor molecule to a neighboring acceptor molecule through dipole–dipole coupling [3] (Figure 5.1b). FRET is most efficient when the donor–acceptor pair is within the Förster radius, typically less than 10 nm. Beyond this distance, FRET efficiency declines sharply, allowing for highly precise spatial coincidence detection. Bioluminescent resonance energy transfer (BRET), a variant of FRET, uses a bioluminescent luciferase enzyme as the energy donor instead of a fluorescent tag, emitting light upon oxidizing a substrate such as luciferin [4]. This modification improves detection sensitivity by eliminating the need for excitation light, thereby reducing autofluorescence and light scattering. However, FRET and BRET typically require the genetic modification of recombinant proteins, limiting their suitability for studying endogenous PPIs in native contexts.

An alternative approach, the proximity ligation assay (PLA), achieves coincidence detection via binder-based targeting coupled with an amplifiable split-reporter

Targeting Protein–Protein Interactions for Drug Discovery, First Edition. Edited by Jian Zhang.

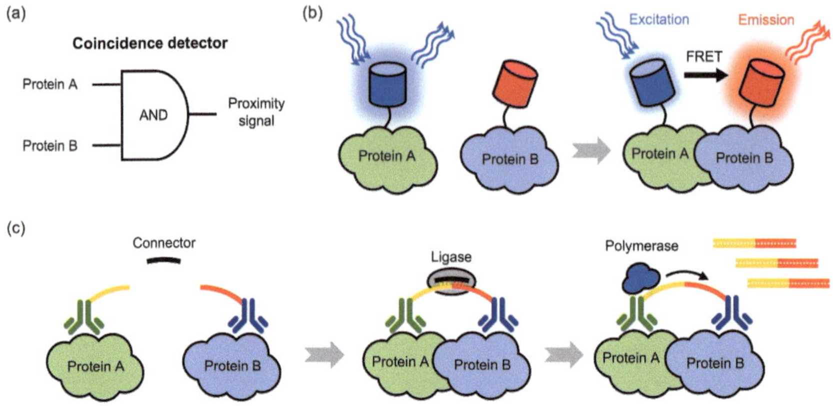

Figure 5.1 Principal of proximity-induced reactions. (a) Schematic of proximity-induced reactions. A reporting event is triggered exclusively when proteins A and B come into direct contact. (b) Förster resonance energy transfer (FRET) represents a proximity-dependent detection method involving two light-sensitive molecules, such as genetically encoded fluorescent tags. When a donor fluorophore is excited, it can transfer energy non-radiatively to a neighboring acceptor fluorophore within a close range (typically 1–10 nm), resulting in light emission from the acceptor. (c) Proximity ligation assay (PLA) is a technique used to detect and quantify protein interactions. It employs affinity binders, usually antibodies, conjugated with single-stranded oligonucleotides as proximity probes targeting protein complexes. When these proximity probes bind to adjacent epitopes (typically <40 nm apart), they generate a full-length single-stranded DNA that serves as a template for PCR. Source: Xinyue Zhou, Peng Zou.

molecule [5, 6]. In PLA, two proximity probes, each consisting of target-binding molecules conjugated with single-stranded oligonucleotides, bind to epitopes within 40 nm of each other (Figure 5.1c). A connector oligonucleotide, complementary to both probes, hybridizes to the bound probes. DNA ligase then links the oligonucleotides on the two probes, forming a full-length single-stranded DNA (ssDNA) that records the proximity of the probes. This ssDNA can subsequently be amplified via polymerase chain reaction (PCR). PLA's multirecognition design helps reduce background from nonspecific adsorption, enhances specificity by eliminating binder cross-reactivity, and achieves high sensitivity suitable for detecting low-abundance proteins and single-molecule-level interactions.

5.2 Implementations of Proximity Ligation Assay

PLA probes could be targeted with aptamers and antibodies. Aptamer-based probes offer lower costs and can be readily conjugated with oligonucleotides via solid-phase synthesis. However, the limited availability of aptamers specific to individual proteins restricts the broader application of aptamer-based PLA [7]. Alternatively, PLA can be achieved with oligo-conjugated antibodies, which offer far more versatile targeting capability [6, 8, 9]. Söderberg et al. utilized antibody-directed proximity probes to investigate the dimerization of endogenous c-Myc and Max in fixed cell samples.

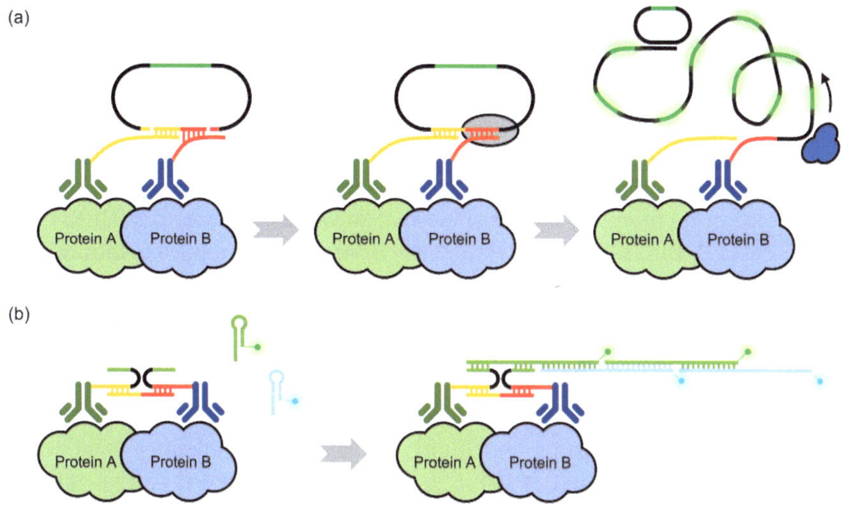

Figure 5.2 Schematic of in situ PLA. (a) RCA-based in situ PLA utilizes proximity probes to link a pair of connector oligonucleotides to a circular DNA molecule, which is subsequently amplified through rolling circle amplification (RCA). (b) HCR-based in situ PLA employs proximity probes to form a trigger that initiates the assembly of hairpin-shaped detection probes around the target protein complex. Source: Xinyue Zhou, Peng Zou.

To enable direct visualization of PPIs in cell and tissue samples, in situ PLA was developed with high spatial resolution (Figure 5.2a). Here, proximity probes bound to adjacent epitopes enable the formation of a circular DNA molecule via DNA ligase and custom-designed connector oligonucleotides. This circular DNA, marking the proximal binding event, is amplified through rolling circle amplification (RCA) using phi29 DNA polymerase. RCA products with repetitive regions are covalently linked to the proximity probes, allowing hybridization with fluorophore-conjugated detection probes. The resulting submicrometer bright spots are observable under a microscope, facilitating the high-resolution detection of PPIs in complex biological samples [8, 10, 11]. Alternatively, a detection probe conjugated with horseradish peroxidase (HRP) can be hybridized to the RCA product. In the presence of hydrogen peroxide, HRP catalyzes the conversion of the NovaRED substrate into a red chromogenic precipitate, enabling the visualization of the RCA product under bright-field microscopy [12, 13].

While RCA-based in situ PLA demonstrates robust performance and wide applicability, its heavy reliance on enzymes presents challenges for efficiency and stability. Nonenzymatic signal amplification has been achieved using hybridization chain reaction (HCR), initiated by proximity probes bound to neighboring epitopes. In HCR-based in situ PLA, a connector bridges the oligonucleotides on proximity probes, forming an initiator. This initiator hybridizes with a kinetically trapped hairpin, exposing a segment that subsequently binds to and invades an additional hairpin structure. The HCR process generates an extended, nicked double-stranded DNA that grows linearly until the available hairpin species are depleted [14, 15] (Figure 5.2b).

5.3 Applications of PLA for Detecting Protein–Protein Interactions

Over the past two decades, PLA-based methods have become increasingly popular for detecting protein interactions due to their exceptional selectivity and sensitivity. These methods have successfully elucidated diverse PPIs, including those of membrane-bound receptors such as epidermal growth factor receptors (HERs) [16], VEGFR2/3 [17], and IL-7 receptors [18]. They have also been applied to intracellular transcription factors, capturing heterodimerizations such as those between c-Myc and Max [8], Aire and Sirt1 [19], HOXA5 and galectin-1 [20], and Smad proteins [12].

PLA-based methods have facilitated studies on membrane contact sites (MCSs), which reveal signal transduction and material exchange between organelles [21]. For instance, in situ PLA has been used to explore protein interactions at mitochondria–endoplasmic reticulum contact sites, including interactions between VDAC1/Grp75 and IP3R1 [22, 23], VAPB and PTPIP51 [23–26], and Calnexin/CLIMP63 and TOM20 [27]. PLA methods have also proven valuable for monitoring dynamic protein interactions between cells. Dore et al., for example, targeted presynaptic and postsynaptic proteins with proximity probes, employing in situ PLA to track GluA1-containing AMPA receptor transfer to synapses under chemically induced long-term potentiation [28]. In another example, PLA has enabled the detection of interactions such as CD2-CD58 on T cells and antigen-presenting cells (APCs) [29].

To detect larger protein complexes, additional proximity probes are introduced to detect the coincidence of multiple proteins, enabling the construction of more complex reporter DNA molecules (Figure 5.3a). In 2006, Söderberg et al. used three proximity probes to report the assembly of the c-Myc:Max:RNA polymerase II ternary complex [8]. This method has also been applied to detect trimeric complexes, such as Smad2/3/4 [30]. Schallmeiner et al. further expanded on this strategy, developing the triple-binder proximity ligation assay (3PLA) to detect PSA–ACT interactions in solution [31]. Currently, the maximum reported application for PLA-based methods involves five affinity binders – one immobilized capture antibody and four proximity probes with oligonucleotides – to generate amplifiable DNA molecules [32]. While introducing more proximity probe–connector pairs could enhance complex detection, this also places higher demands on enzymatic reaction efficiency.

To detect multiple PPI events in a highly parallelized manner, a modified in situ PLA procedure was developed, where proximity probes are barcoded with unique tag sequences. These tags are enzymatically linked into the circular DNA template, followed by RCA (Figure 5.3b). Using tag-specific detection oligonucleotides labeled with distinct fluorophores, it is possible to achieve multiplexed detection of protein interactions [33]. A similar design supports HCR-based multiplex imaging. Schulte et al., for instance, demonstrated the simultaneous detection of the α/β-tubulin complex, β-catenin/E-cadherin complex, and SC35/SON complex through a set of orthogonal HCR amplifiers [15].

Beyond PPIs, PLA technology can detect protein associations with other biomolecules, such as nucleic acids and lipids. For protein–nucleic acid interaction studies, DNA-based proximity probes selectively hybridize with specific DNA

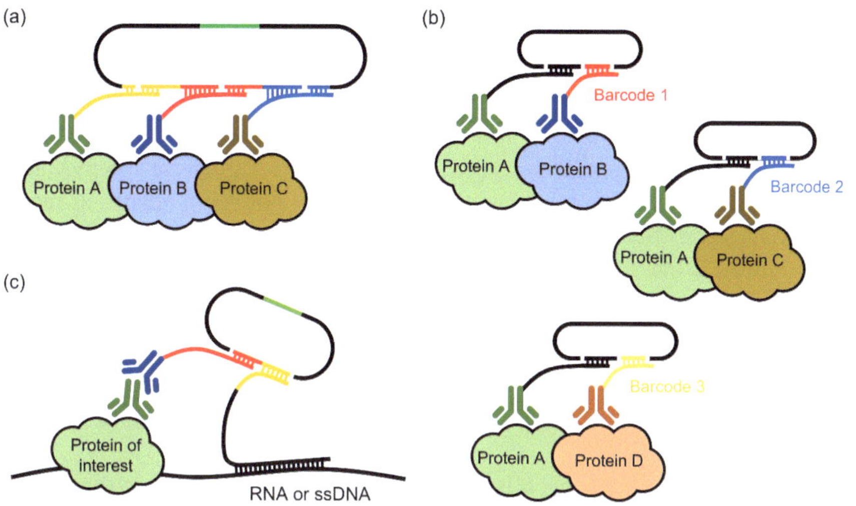

Figure 5.3 Emerging trends in in situ PLA. (a) By strategically designing connector oligonucleotides, in situ PLA can detect three or more adjacent epitopes. (b) Simultaneous monitoring of multiple sets of protein–protein interactions. (c) Oligonucleotides can act as recognition elements for proximity probes, targeting specific regions of RNA or single-stranded genomic DNA through hybridization to detect protein interactions. Source: Xinyue Zhou, Peng Zou.

or RNA through complementary sequences [34–37] (Figure 5.3c). Additionally, commercial lipid-specific antibodies can serve as affinity binders. For instance, Choi et al. used in situ PLA to examine the interaction between phosphatidylinositol 4,5-bisphosphate and p53 in specific subnuclear regions [38].

Since the core principle of in situ PLA is converting two adjacent epitopes into an amplifiable DNA molecule, a wide array of labeling strategies and recognition elements can be adapted to detect biomolecules or biological processes in close proximity to the protein of interest (POI). For example, biotin labeling of DNA double-strand breaks or nascent DNA enables PLA detection of protein interactions at DNA damage foci [39] or replication forks [40]. Similar strategies have also been employed to detect protein–nascent RNA interactions in fixed cell samples [41]. In 2015, Dieck et al. developed FUNCAT-PLA and Puro-PLA to detect newly synthesized proteins [42]. FUNCAT-PLA involves the incorporation of azidohomoalanine (AHA) into nascent proteins, followed by a click reaction with a biotin tag, while Puro-PLA uses puromycin for labeling. When antibodies against the "nascent" tags (biotin or puromycin) and the protein of interest are in close proximity, the resulting RCA product can reveal the cellular location of the newly synthesized POI.

5.4 Conclusions and Outlooks

To summarize, this chapter outlines the key principles of the PLA method, focusing on critical factors that influence detection selectivity and sensitivity. These include the recognition elements of proximity probes, signal generation and amplification, and the choice of readout methods. We have also reviewed the rapid advancement and the widespread adoption of PLA technology over the past two decades, highlighting notable technological breakthroughs and applications.

Future developments should prioritize the following areas:

i) expanding the range of target molecules by incorporating diverse affinity binder scaffolds – such as nanobodies, affibodies, or recombinant binders – as recognition elements in proximity probes;
ii) optimizing the PLA process by enhancing the stability and efficiency of enzymatic reactions to improve the method's applicability and sensitivity – alternatively, HCR-based strategies have been designed to achieve enzyme-free amplification of molecular proximity [15];
iii) increasing PLA detection throughput to enable the simultaneous readout of multiple interactions.

Current in situ PLA detection is limited by the number of available fluorescent channels in microscopy, restricting the number of fluorescent labels for encoding RCA products. To overcome this limitation, the direct sequencing of RCA products with different barcodes can substantially enhance identification throughput. For instance, integrating in situ PLA with spatial transcriptomics would allow for the simultaneous detection of multiple protein interactions while preserving the spatial information of RCA products.

With the above advancements, we envision a future of PLA technology that is more stable, precise, sensitive, and high throughput for studying protein interactions. Such improvements will open new vistas into biomolecular interaction networks that regulate various cellular functions under both physiological and pathological conditions.

Acknowledgments

This work was supported by the Ministry of Science and Technology (2022YFA1304700), the National Natural Science Foundation of China (32088101), and Beijing National Laboratory for Molecular Sciences (BNLMS-CXXM-202403). PZ is sponsored by the Bayer Investigator Award.

References

1 Braun, P. and Gingras, A.C. (2012). History of protein–protein interactions: from egg-white to complex networks. *Proteomics* 12: 1478–1498.
2 Vidal, M., Cusick, M.E., and Barabási, A.-L. (2011). Interactome networks and human disease. *Cell* 144: 986–998.
3 Truong, K. and Ikura, M. (2001). The use of FRET imaging microscopy to detect protein-protein interactions and protein conformational changes in vivo. *Current Opinion in Structural Biology* 11: 573–578.

4 Xu, Y., Piston, D.W., and Johnson, C.H. (1999). A bioluminescence resonance energy transfer (BRET) system: application to interacting circadian clock proteins. *Proceedings of the National Academy of Sciences of the United States of America* 96: 151–156.
5 Fredriksson, S., Gullberg, M., Jarvius, J. et al. (2002). Protein detection using proximity-dependent DNA ligation assays. *Nature Biotechnology* 20: 473–477.
6 Gullberg, M., Gústafsdóttir, S.M., Schallmeiner, E. et al. (2004). Cytokine detection by antibody-based proximity ligation. *Proceedings of the National Academy of Sciences of the United States of America* 101: 8420–8424.
7 Zhong, Y., Zhao, J., Li, J. et al. (2020). Advances of aptamers screened by cell-SELEX in selection procedure, cancer diagnostics and therapeutics. *Analytical Biochemistry* 598: 113620.
8 Soderberg, O., Gullberg, M., Jarvius, M. et al. (2006). Direct observation of individual endogenous protein complexes in situ by proximity ligation. *Nature Methods* 3: 995–1000.
9 Jarvius, M., Paulsson, J., Weibrecht, I. et al. (2007). In situ detection of phosphorylated platelet-derived growth factor receptor beta using a generalized proximity ligation method. *Molecular & Cellular Proteomics* 6: 1500–1509.
10 Leuchowius, K.J., Jarvius, M., Wickström, M. et al. (2010). High content screening for inhibitors of protein interactions and post-translational modifications in primary cells by proximity ligation. *Molecular & Cellular Proteomics* 9: 178–183.
11 Sartori, M., Mendes, T., Desai, S. et al. (2019). BIN1 recovers tauopathy-induced long-term memory deficits in mice and interacts with Tau through Thr(348) phosphorylation. *Acta Neuropathologica* 138: 631–652.
12 Flanders, K.C., Heger, C.D., Conway, C. et al. (2014). Brightfield proximity ligation assay reveals both canonical and mixed transforming growth factor-β/bone morphogenetic protein smad signaling complexes in tissue sections. *The Journal of Histochemistry and Cytochemistry* 62: 846–863.
13 Zieba, A., Wahlby, C., Hjelm, F. et al. (2010). Bright-field microscopy visualization of proteins and protein complexes by in situ proximity ligation with peroxidase detection. *Clinical Chemistry* 56: 99–110.
14 Koos, B., Cane, G., Grannas, K. et al. (2015). Proximity-dependent initiation of hybridization chain reaction. *Nature Communications* 6: 7294.
15 Schulte, S.J., Shin, B., Rothenberg, E.V., and Pierce, N.A. (2024). Multiplex, quantitative, high-resolution imaging of protein:protein complexes via hybridization chain reaction. *ACS Chemical Biology* 19: 280–288.
16 Barros, F.F., Abdel-Fatah, T.M., Moseley, P. et al. (2014). Characterisation of HER heterodimers in breast cancer using in situ proximity ligation assay. *Breast Cancer Research and Treatment* 144: 273–285.
17 Nilsson, I., Bahram, F., Li, X. et al. (2010). VEGF receptor 2/−3 heterodimers detected in situ by proximity ligation on angiogenic sprouts. *The EMBO Journal* 29: 1377–1388.
18 Andersen, S.S., Hvid, M., Pedersen, F.S., and Deleuran, B. (2013). Proximity ligation assay combined with flow cytometry is a powerful tool for the detection of cytokine receptor dimerization. *Cytokine* 64: 54–57.

19 Avin, A., Levy, M., Porat, Z., and Abramson, J. (2017). Quantitative analysis of protein-protein interactions and post-translational modifications in rare immune populations. *Nature Communications* 8: 1524.

20 Sharanek, A., Burban, A., Hernandez-Corchado, A. et al. (2021). Transcriptional control of brain tumor stem cells by a carbohydrate binding protein. *Cell Reports* 36: 109647.

21 Benhammouda, S., Vishwakarma, A., Gatti, P., and Germain, M. (2021). Mitochondria endoplasmic reticulum contact sites (MERCs): proximity ligation assay as a tool to study organelle interaction. *Frontiers in Cell and Development Biology* 9: 789959.

22 Tubbs, E., Theurey, P., Vial, G. et al. (2014). Mitochondria-associated endoplasmic reticulum membrane (MAM) integrity is required for insulin signalling and is implicated in hepatic insulin resistance. *Diabetes* 63: A76–A76.

23 Gomez-Suaga, P., Paillusson, S., Stoica, R. et al. (2017). The ER-mitochondria tethering complex VAPB-PTPIP51 regulates autophagy. *Current Biology* 27: 371–385.

24 De Vos, K.J., Morotz, G.M., Stoica, R. et al. (2012). VAPB interacts with the mitochondrial protein PTPIP51 to regulate calcium homeostasis. *Human Molecular Genetics* 21: 1299–1311.

25 Stoica, R., De Vos, K.J., Paillusson, S. et al. (2014). ER-mitochondria associations are regulated by the VAPB-PTPIP51 interaction and are disrupted by ALS/FTD-associated TDP-43. *Nature Communications* 5: 3996.

26 Hedskog, L., Pinho, C.M., Filadi, R. et al. (2013). Modulation of the endoplasmic reticulum-mitochondria interface in Alzheimer's disease and related models. *Proceedings of the National Academy of Sciences of the United States of America* 110: 7916–7921.

27 Ilamathi, H.S., Benhammouda, S., Lounas, A. et al. (2023). Contact sites between endoplasmic reticulum sheets and mitochondria regulate mitochondrial DNA replication and segregation. *iScience* 26: 107180.

28 Dore, K., Pao, Y., Soria Lopez, J. et al. (2020). SYNPLA, a method to identify synapses displaying plasticity after learning. *Proceedings of the National Academy of Sciences of the United States of America* 117: 3214–3219.

29 Sable, R., Jambunathan, N., Singh, S. et al. (2018). Proximity ligation assay to study protein-protein interactions of proteins on two different cells. *BioTechniques* 65: 149–157.

30 Zieba, A., Pardali, K., Söderberg, O. et al. (2012). Intercellular variation in signaling through the TGF-β pathway and its relation to cell density and cell cycle phase. *Molecular & Cellular Proteomics* 11: M111-013482.

31 Schallmeiner, E., Oksanen, E., Ericsson, O. et al. (2007). Sensitive protein detection via triple-binder proximity ligation assays. *Nature Methods* 4: 135–137.

32 Tavoosidana, G., Ronquist, G., Darmanis, S. et al. (2011). Multiple recognition assay reveals prostasomes as promising plasma biomarkers for prostate cancer. *Proceedings of the National Academy of Sciences of the United States of America* 108: 8809–8814.

33 Leuchowius, K.J., Clausson, C.M., Grannas, K. et al. (2013). Parallel visualization of multiple protein complexes in individual cells in tumor tissue. *Molecular & Cellular Proteomics* 12: 1563–1571.

34 Weibrecht, I., Gavrilovic, M., Lindbom, L. et al. (2012). Visualising individual sequence-specific protein-DNA interactions in situ. *New Biotechnology* 29: 589–598.

35 Jung, J., Lifland, A.W., Zurla, C. et al. (2013). Quantifying RNA-protein interactions in situ using modified-MTRIPs and proximity ligation. *Nucleic Acids Research* 41: e12.

36 Blanchard, E.L., Argyropoulou, D., Zurla, C. et al. (2019). Quantification and localization of protein-RNA interactions in patient-derived archival tumor tissue. *Cancer Research* 79: 5418–5431.

37 Burke, K.S., Antilla, K.A., and Tirrell, D.A. (2017). A fluorescence in situ hybridization method to quantify mRNA translation by visualizing ribosome-mRNA interactions in single cells. *ACS Central Science* 3: 425–433.

38 Choi, S., Chen, M., Cryns, V.L., and Anderson, R.A. (2019). A nuclear phosphoinositide kinase complex regulates p53. *Nature Cell Biology* 21: 462–475.

39 Galbiati, A., Beausejour, C., and d'Adda di Fagagna, F. (2017). A novel single-cell method provides direct evidence of persistent DNA damage in senescent cells and aged mammalian tissues. *Aging Cell* 16: 422–427.

40 Roy, S., Luzwick, J.W., and Schlacher, K. (2018). SIRF: quantitative in situ analysis of protein interactions at DNA replication forks. *The Journal of Cell Biology* 217: 1521–1536.

41 Das, R., Dey, A., and Uppal, S. (2023). A method for in situ visualization of protein-nascent RNA interactions in single cell using proximity ligation assay (IPNR-PLA) in mammalian cells. *Transcription* 14: 146–157.

42 tom Dieck, S., Kochen, L., Hanus, C. et al. (2015). Direct visualization of newly synthesized target proteins in situ. *Nature Methods* 12: 411–414.

6

Application of Surface Plasmon Resonance in the Characterization of Protein–Protein Interactions

*Yuanyuan Xie and Jianrong Xu**

Shanghai University of Traditional Chinese Medicine, School of Integrative Medicine, 1200 Cailun Road, Shanghai 201203, China

6.1 Introduction

6.1.1 Protein–Protein Interactions

The interactions between proteins are achieved through their specific structural domains. These domains can be part of a protein or the entire protein itself. When the structural domains of two proteins match with each other, they can bind together through noncovalent or covalent bonds. These interactions not only affect the structure and function of proteins but also regulate many biological processes, such as cell signaling, regulation of gene expression, and metabolic modulation [1]. The disruption of protein–protein interactions (PPIs), often caused by mutations at interface residues, alterations in conformation, or lack of binding partners, results in dysfunctional cellular pathways, potentially leading to various diseases, including cancer and neurodegenerative disorders [2–4]. Therefore, identifying and manipulating PPIs is pivotal to gaining insights into the biological function of proteins and the treatment of related diseases.

6.1.2 Principle of Surface Plasmon Resonance

Surface plasmon resonance (SPR) is an optical technique used to detect molecular interactions, including protein–protein and protein–ligand interactions. An SPR assay exploits the phenomenon where evanescent waves, generated when light undergoes total internal reflection, resonate with the surface plasmon waves present on a metal surface. This resonance results in a decrease in the intensity of the reflected light. When the wavelength of the incident light is fixed, the intensity of the reflected light varies with the angle of incidence. The angle at which the intensity of the reflected light is minimized is known as the SPR angle. In Figure 6.1, the SPR angle is highly sensitive to the refractive index of the dielectric medium

*Lead author.

Targeting Protein–Protein Interactions for Drug Discovery, First Edition. Edited by Jian Zhang.

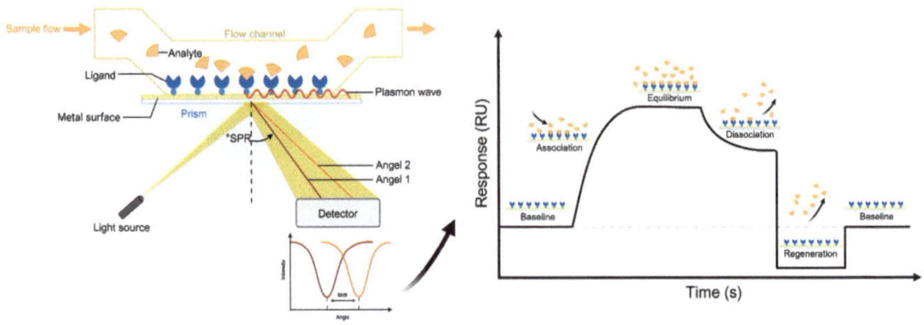

Figure 6.1 The principle of SPR technology and the sensorgram. Source: Yuanyuan Xie and Jianrong Xu. In the left panel, SPR operates on the resonance condition established between evanescent waves and surface plasmon on a metal film. This causes a marked decrease in the reflected light intensity. The SPR angle, where the reflected light intensity is minimal, depends on the incident angle of light and is very sensitive to the dielectric refractive index of medium on the metal surface. Protein–protein interactions alter the surface refractive index, leading to changes in the SPR intensity angle. These angle changes, which are proportional to the mass of the bound material, result in shifts of the resonance angle that indicate properties or quantity changes in the surface medium. In the right panel, it shows a sensorgram depicting the angle change over time.

on the metal surface. As the molecules bind, the surface refractive index changes, resulting in a change in this SPR angle. The change in the angle is proportional to the mass of the bonded material. When the properties or the amount of the surface medium changes, the resonance angle shifts accordingly. The results obtained can be represented by a sensorgram that records the angle change over time [5, 6].

6.1.3 Advantage of SPR

In PPI studies, methods such as yeast two-hybrid, enzyme-linked immunosorbent assay (ELISA), fluorescence resonance energy transfer (FRET), co-immunoprecipitation (Co-IP), western blotting (WB), and mass spectrometry are commonly used [7–9]. However, most of these techniques often yield limited information, necessitate labeling, and are restricted to analyzing inactive or semi-active interactions, only detecting those with strong affinities. In contrast, SPR assay offer significant advantages for the validation and study of PPIs. A key benefit of SPR is its ability to measure the binding affinities and association/dissociation kinetics of complexes in real time. Meanwhile, SPR assay enables the acquisition of quantitative binding constants for bimolecular interactions, rather than merely providing qualitative positive or negative binding results. Additionally, SPR requires relatively small quantities of material; often milligram or sub-milligram amounts are sufficient, and no labeling is required [10, 11]. Moreover, SPR can detect interactions between active molecules ranging from weak to strong, covering concentrations from the millimolar to the nanomolar range with high accuracy [12, 13].

The SPR assay, widely regarded as the gold standard technique for investigating molecular interactions, has been included in the US and Japanese Pharmacopoeias since 2016 and in the Chinese Pharmacopoeia since 2020. Currently, SPR is recognized as one of the most powerful techniques for determining specificity, affinity, and kinetic parameters in the binding processes of diverse biomolecules. Examples include protein–protein, protein–DNA, enzyme–substrate or inhibitor, receptor–drug, lipid membrane–protein, protein–polysaccharide, and cell or virus–protein interactions [14]. This assay can be used for PPI studies, drug discovery, antibody screening, and characterization.

6.2 Applications of SPR Assays in PPIs Characterization

In the application of SPR assays, the analysis of binary PPI systems represents the most fundamental category of SPR experiments. These experiments provide preliminary information about whether two proteins bind and how strong that binding might be. Researchers utilize SPR not only to verify interactions between proteins and to screen for compounds with high affinity to target proteins but also to accurately identify key binding sites by comparing the binding of different protein fragments or mutants to the target protein. This allows for an understanding of the impact of key amino acid residues or domains on PPIs, which is crucial for revealing the complex functional and structural relationships of proteins. In the field of drug development, SPR assay similarly demonstrates their value by enabling the screening of compounds with high affinity to target proteins and optimizing the structure of compounds through interaction analysis, thereby enhancing their efficacy and selectivity.

Furthermore, in ternary PPI systems, SPR binning assay can assess the competitive relationship of different antibodies to the same protein-binding site, offering an effective means to deeply understand antibody-binding characteristics. Similarly, SPR assays can be utilized to characterize and optimize regulatory molecules in the development of PPI modulators.

In summary, SPR assay plays an essential role in biomolecular recognition, drug design and development, and in-depth exploration of complex protein interactions. They not only improve the efficiency and quality of scientific research but also bring revolutionary changes to drug development, making it a valuable technique in life sciences and pharmaceutical research.

6.2.1 SPR Application in Binary PPI Systems

The binary PPI system refers to the specific binding event between two proteins, which forms the basis for many key biological processes within cells. When employing SPR to study binary PPI systems, it can be generally divided into three categories: verifying potential interacting proteins, validating key binding sites, and screening or optimizing a series of candidate proteins.

6.2.1.1 SPR Assay in Verifying and Measuring PPIs

In the study of PPIs, researchers typically employ a series of preliminary methods to identify and screen potential interacting proteins. These methods include an in-depth literature review to obtain detailed information and experimental evidence of known protein interactions; bioinformatics analysis, utilizing tools such as homology modeling and molecular docking to predict protein binding modes; and experimental approaches like GST pull-down and Co-IP for initial screening and confirmation of interacting proteins. Although these preliminary research methods can provide numerous hypotheses and data about protein interactions, the results often require further validation through more rigorous and precise experimental techniques. SPR assay can serve as a verification method for the findings from preliminary studies, confirming whether an interaction exists between two proteins and accurately determining their binding strength and kinetic parameters, such as association and dissociation constants, significantly enhancing the reliability and depth of the research.

Typically, when using SPR to verify the existence of PPIs, researchers immobilize the target protein as the "ligand" on the chip surface and flow potential interacting proteins as "analyte" over the sensor chip at varying concentration gradients. The resulting association and dissociation curves allow for the characterization of PPIs.

Cell-Surface PPIs Cell–surface PPIs mediate cell–cell communication, recognition, and responses. Elucidation of the map of extracellular PPIs is essential for understanding the role of cell-surface PPIs in both healthy and disease tissues. Cell surface receptors and their interactions play a central role in physiological and pathological signaling.

Wojtowicz and colleagues executed an interactome screen of 564 human cell-surface and secreted proteins, most of which are part of the immunoglobulin superfamily proteins. Screen results, augmented by phylogenetic homology analysis, revealed approximately 380 previously unreported PPIs. To enhance avidity and, consequently, sensitivity, Fc-dimerized ECD analytes and ligands were employed in the majority of experiments. Researchers utilized human IgSF cell–surface interactions to uncover the complex network of PPIs. In this study, researchers used SPR to detect the interactions between molecules, such as WFIKKN2-Fc, ICAM5-Fc, IGFBPL1-Fc and DCC-Fc, PUNC-Fc, PUNC e11-Fc, and conducted corresponding screening work. They validated 59 newly identified PPIs, including three additional PPIs observed between homologous proteins in mouse and cross-species (human–mouse) [15]. Overall, this article significantly expands the known range of human IgSF cell–surface interactions using advanced biotechnology and computational analyses. It provides new resources and methodologies for investigating complex protein interaction networks.

By analyzing large-scale genomic data, Erik Verschueren and colleagues reveal expression patterns of IgSF members across various cancer types and find significant correlations between these patterns and clinical outcomes, such as patient survival rates. The constructed IgSF interaction network further identifies potential cancer-relevant key proteins and their interactions, offering new perspectives on the role of IgSF proteins in cancer progression. Utilizing SPR, they evaluated affinity and kinetic parameters for multiple molecular interactions, such as PD1-Fc, PDL1-Fc, PDL2-Fc with EPHA3, CEACAM4, LILRA3, among others. This validates the functional roles of certain IgSF members in cancer development and provides potential biomarkers and therapeutic targets for future cancer therapeutic strategies [16].

PPIs in Immune System Within the immune system, gamma delta (γδ) T cells are unique, differing significantly from their alpha-beta (αβ) T-cell counterparts. Unlike αβ T cells, which primarily recognize peptide antigens presented by major histocompatibility complex (MHC) molecules, γδ T cells possess a distinct T-cell receptor structure made up of gamma and delta chains. This unique architecture allows γδ T cells to identify a broader range of antigens independent of the MHC, including both exogenous and endogenous antigens, encompassing both foreign and self-antigens [17]. After two decades of preclinical research and preliminary clinical studies involving very small numbers of patients, γδ T cells are now being explored as a viable and promising approach for cancer immunotherapy [18, 19].

Willcox et al. used SPR to detect the direct binding interactions between recombinant butyrophilin-like (BTNL) 3 and 8 IgV proteins and various soluble Vγ T-cell receptors (TCRs), including Vγ2, Vγ3, and Vγ4. The results showed that BTNL3 IgV injection over immobilized Vγ4 TCR produced significantly stronger signals compared to Vγ2 or Vγ3 TCRs, suggesting a specific Vγ4-TCR binding affinity. In contrast, BTNL8 IgV did not exhibit this specificity, as its binding signals on surfaces with Vγ2, Vγ3, and Vγ4 TCRs were comparable to those on control surfaces [20]. Willcox and colleagues have shown that the B7-like molecule BTNL3 physically interacts with Vγ4-bearing TCRs on γδ T cells.

In humans, most γδ T cells express the $V\gamma 9V\delta 2^+$ TCR, which responds to small molecules known as phosphoantigens (pAg) produced by pathogens and cancer cells. This response activates the γδ T cells and enables the elimination of diseased cells [21, 22]. Studies by Marc Rigau et al. propose that BTN2A1 might bind phosphoantigens and present them to the $V\gamma 9V\delta 2^+$ TCR. Through SPR assays, researchers verified the direct interaction between BTN2A1 and the $V\gamma 9V\delta 2^+$ TCR. They found that soluble BTN2A1 bound to the $V\gamma 9V\delta 2^+$ TCR with an affinity (KD) of 40 μM, mirroring the behavior of typical αβ T cells, shedding light on new pathways by which γδ T cells detect cancer and infections [23].

PPIs in Liver Fibrosis Glial cell line-derived neurotrophic factor (GDNF), a protein in the transforming growth factor-β (TGF-β) superfamily, was initially identified as a crucial survival factor for brain dopaminergic neurons. Subsequent studies revealed its broader functions across various cell types. Research by Le Tao et al. explored GDNF's influence on activating hepatic stellate cells and fostering liver fibrosis, spanning both human and mouse subjects. Employing the SPR approach, the team examined GDNF's binding affinity to the ALK5 receptor, revealing a dose-dependent, rapid association–dissociation interaction. By plotting the equilibrium response units against varying ALK5 concentrations and employing nonlinear regression analysis, they determined a dissociation constant (KD) of 85.47 nM, indicating a robust binding affinity between GDNF and ALK5 [24].

PPIs in Virus–Host Recognition Coronavirus disease 2019 (COVID-19) is caused by SARS-CoV-2, and SARS-CoV-2 enters host cells through the binding of its surface protein, specifically the spike protein (S protein), to the human angiotensin-converting enzyme 2 (ACE2) receptor on host cells. SPR has been widely used to study the binding kinetics and affinity between SARS-CoV-2 variants and host cell receptors.

Shang and colleagues detected the affinity of the receptor-binding domains (RBDs) for ACE2, as well as the kinetic parameters of the interaction. By comparing the differences in ACE2 binding between the RBD of SARS-CoV-2 and other coronaviruses (such as SARS-CoV), researchers explained why SARS-CoV-2 has a higher transmission efficiency. These differences may be attributed to changes in certain key amino acids within the RBD, which likely enhance the binding capability of the viral spike protein to ACE2, thereby facilitating interpersonal transmission of the virus [25].

Toelzer and colleagues revealed the 2.85 Å cryo-EM structure of the SARS-CoV-2 spike glycoprotein, showing that the RBDs are tightly bound to the essential free fatty acid linoleic acid (LA) within three composite binding pockets. Subsequently, researchers measured the kinetic parameters for the LA-bound spike trimer and the apo spike trimer with ACE2 by SPR. They consistently observed a markedly reduced signal for the LA-bound S compared to the apo S at the same concentrations, suggesting that targeting the LA binding pockets with small molecule inhibitors is a promising approach [26].

Han and colleagues characterize the interactions of six different SARS-CoV-2 RBD variants including alpha, beta, and gamma and present crystal structures of these ACE2–RBD complexes. To better understand the interactions, researchers measured their binding affinity using SPR, finding that SARS-CoV-2 WT RBD protein bound to ACE2 with a KD of 26.34 nM. Alpha RBD, beta RBD, gamma RBD, mink-Y453F RBD, and mink-N501T RBD displayed higher affinities to ACE2 than WT RBD, with ~7, ~3, ~5, ~8, and ~4-fold increase in binding affinity, respectively. These results were consistent with the FACS assays [27].

In the study of the Ebola virus invasion mechanism, the NPC1 molecule is a multipass transmembrane protein responsible for cholesterol transport and serves as an intracellular receptor for the Ebola virus. Gong and colleagues found that the purified recombinant hNPC1 bound to GPcl in a pH-dependent manner as measured using SPR. At pH 7.5, the measured KD value is 95 μM, whereas the affinity increases to 2.4 μM at pH 6.0, almost 40-fold higher. This ultimately led to the discovery of the mechanism by which the Ebola virus spreads more easily in an acidic environment [28].

6.2.1.2 SPR-Guided Screening and Optimization in Drug Discovery

Traditional drug screening methods often rely on cellular or animal models, which, despite providing a physiological environment, have limitations such as high costs, long time consumption, and significant data variability. Furthermore, these traditional approaches offer limited information about the interaction features between target receptors and candidate drugs, constraining the efficiency and precision of drug design. In recent years, SPR assay have been widely employed in drug discovery, particularly in the initial screening and optimization stages.

Researchers typically begin by attaching the target protein to a sensor chip and then pass potential protein candidates over it to observe binding interactions, enabling rapid identification of potential drug candidates. Comparing the association (ka) and dissociation (kd) rate constants provides understanding of both drug–target interaction mechanism and its stability.

Neutralizing Antibodies There is a corresponding increase in the demand for validation of antibody characteristics and functions. The SPR assay offers an efficient and precise method for analyzing and confirming the interactions between antibodies and antigens. Throughout the development process of antibody-based pharmaceuticals, the SPR assay not only facilitates the screening of antibodies with potential neutralizing capabilities but also assesses the affinity and specificity of these antibodies, which is crucial for the rapid identification of effective therapeutic antibodies. The SPR assay enables real-time monitoring of the entire process of neutralizing antibodies binding to and dissociating from viruses, thereby determining kinetic parameters, including ka, kd, and KD. This provides a reliable evaluation of the neutralizing effects of these antibodies. In the field of vaccine development, understanding whether antibodies possess the capability to neutralize viruses is vital for designing effective vaccines. The SPR assay assists researchers in assessing whether the antibodies produced by an immune response can effectively recognize and neutralize specific pathogens. Moreover, in response to rapidly spreading infectious diseases, the SPR assay allows for the swift identification of neutralizing antibodies, offering key information for the development of vaccines and therapeutic antibodies.

Antibody–Drug Conjugates Antibody–drug conjugates (ADCs) comprise antibodies, linkers, and cytotoxic payloads. Over the last decade, ADCs have progressively matured in the treatment of both solid and hematological tumors, such as HER2-positive breast cancer and acute myeloid leukemia, emerging as a revolutionary domain in antitumor drug research [29, 30].

In the development process of ADCs, selecting an appropriate linker to effectively conjugate the antibody with a cytotoxic molecule is crucial. SPR assays are primarily utilized to examine the binding characteristics of antibodies to antigens, assess the impact of linkers and cytotoxic molecules on antigen–antibody binding [31, 32], and determine the activity of ADCs at different drug-to-antibody ratio (DAR) values, including their affinity for targets and cytotoxicity [33]. Additionally, it serves as a powerful tool for understanding the mechanism of action of ADCs and ensuring quality control and assurance.

Healey and his colleagues embarked on the development of novel ADCs targeting ovarian cancer. The research team prepared two antibodies (RBGO1 and RBGO3) that specifically bind to the ovarian cancer target RAGE, and conjugated them with specific cytotoxic molecules using two different types of linkers (cleavable and non-cleavable). Through single-cycle kinetic analysis by an SPR assay, the researchers measured the affinity and kinetic parameters of these conjugates, comparing their binding characteristics to the antigen. The results showed that neither the cleavable nor the noncleavable linker significantly affected the binding kinetics and affinity of RBGO1-ADC with the antigen RAGE (aa198–217). Further research found that ADCs with DARs of 1.5 and 3.5, when compared to the bare antibody, did not show significant changes in binding affinity to the antigen. These findings emphasize the importance of carefully considering linker type and DAR value in the design and optimization of ADC drugs [33].

Milazzo and colleagues conjugated Trastuzumab with HDAC inhibitors to obtain the ADC candidate molecules ST8176AA1 and ST8178AA1 and utilized SPR to complete the kinetic characterization of different ADCs, Trastuzumab, and the human HER2 receptor (target protein). The results indicated that compared to the standalone antibody (Trastuzumab), the affinity of ST8176AA1 for the target protein did not significantly weaken. Subsequent related experimental results also confirmed from multiple aspects that ST8176AA1 is more conducive to the drug efficiently recognizing its site of action [31].

Anti-amyloid-beta Antibodies Immunotherapy against amyloid-beta (Aβ) is a promising option for the treatment of Alzheimer's disease (AD). Aβ exists in various species, including monomers, oligomers, protofibrils, and insoluble fibrils in plaques. Oligomers and protofibrils have been shown to be toxic, and removal of these aggregates might represent an effective treatment for AD. Soderberg et al. used SPR to compare the kinetic characteristics of lecanemab, aducanumab, and gantenerumab binding to different species of Aβ. The results showed that all three antibodies were able to bind to Aβ monomers but with relatively weak affinity. Specifically, the affinity of lecanemab was 2300 ± 910 nM, aducanumab had an affinity of 7300 ± 990 nM, and gantenerumab exhibited an affinity of 1300 ± 480 nM [34].

Anti-viral Vaccines and Antibodies The RBD of the coronavirus spike protein is a vaccine target of great interest but has been limited by its weak immunogenicity. Lianpan Dai and colleagues have provided a universal design for a β-coronavirus vaccine targeting COVID-19, MERS, and SARS. Their study describes an RBD dimer of MERS-CoV that overcomes this limitation. Compared to traditional monomers, the RBD-dimer significantly enhances the titer of neutralizing antibodies (Nab) and protects mice from MERS-CoV infection. Based on the structural design, the research team further produced a stable tandem repeat single chain (RBD-sc-dimer), successfully retaining the vaccine's efficacy. SPR assays were used to measure the binding affinity and kinetic parameters of RBD-sc-dimer and its monomeric form, RBD-monomer, with the ACE2 receptor. These data confirmed that the vaccine efficacy was successfully retained through structure-guided design. This strategy was also extended to vaccine designs for COVID-19 and SARS, resulting in a 10–100-fold increase in NAb titers [35].

Shi and colleagues isolated two specific human monoclonal antibodies, CA1 and CB6, from a patient in the recovery phase of COVID-19, which exhibit strong *in vitro* neutralizing activity specific to SARS-CoV-2. Researchers assessed the binding kinetics of CA1 and CB6 using an SPR assay. The measured KD of CA1 and CB6 with the SARS-CoV-2 RBD were 4.68 ± 1.64 nM and 2.49 ± 1.65 nM, respectively. Subsequent structural studies revealed that CB6 recognizes an epitope in the SARS-CoV-2 RBD that overlaps with the ACE2 binding site, thereby interfering with the virus–receptor interaction through steric hindrance and direct competition for interface residues [36].

Wrapp et al. isolated two single-domain camelid antibodies named MERS VHH-55 and SARS VHH-72 that can potently neutralize SARS-CoV RBD and MERS-CoV RBD, respectively. These VHHs were triggered by immunizing llamas with the β-coronavirus spike proteins of SARS-CoV and MERS-CoV, respectively. They resolved the crystal structures of these two VHHs in complex with their respective viral epitopes, indicating that the mechanism of neutralization may involve the blocking of the receptor-binding interface and maintaining the RBD in an upward conformation. Using SPR assays, researchers complete the measurement of affinity and kinetic parameters for the interactions between MERS-CoV RBD, SARS-CoV RBD, and VHHs [37].

6.2.1.3 SPR in the Validation of PPI Interface

When utilizing SPR to verify the interface of PPIs, one protein is typically immobilized as the "ligand" on the sensor chip. Different protein fragments or mutants are then flowed over the chip surface at varying concentrations as "analyte." By comparing the binding responses between different protein fragments or mutants and the ligand, it is possible to identify which fragments or mutants exhibit higher affinity or distinct binding features. For instance, if a mutant shows a significantly reduced binding response, this may indicate that the mutation affects a crucial binding site, as illustrated in Figure 6.2. Furthermore, the interface can be further confirmed through other biochemical or biological experiments, such as X-ray crystallography, nuclear magnetic resonance (NMR) spectroscopy, mass spectrometry, and so on.

PPIs in Neurodegenerative Disease The triggering receptor expressed on myeloid cells 2 (TREM2) is implicated in the risk of neurodegenerative diseases. Xie and colleagues found that TREM2 deficiency impaired phagocytic clearance of pathological TDP-43 by microglia and enhanced neuronal damage and motor impairments. To investigate the direct interactions between the extracellular domain of human TREM2 (hTREM2 ECD) and four low-complexity domain fragments or full-length hTDP-43, they employed SPR analyses. Their findings indicated that hTREM2 indeed interacts with GFN, SWG, and full-length hTDP-43 in a dose-dependent manner with KD values of 1.86×10^{-4} M, 1.33×10^{-5} M, and 1.18×10^{-7} M, respectively, supporting the direct interaction of hTREM2 with hTDP-43 that was inferred from their *in vivo*, *in vitro*, and *in silico* studies. Their data highlight that TDP-43 is a possible ligand for microglial TREM2 and that this interaction mediates neuroprotection of microglia in TDP-43-related neurodegeneration [38].

Huntingtin yeast interacting protein E-like kinase (HYPK) plays a significant role in autophagy, particularly in the cellular processing of protein aggregates. In neurodegenerative diseases such as Huntington's disease, the function of HYPK

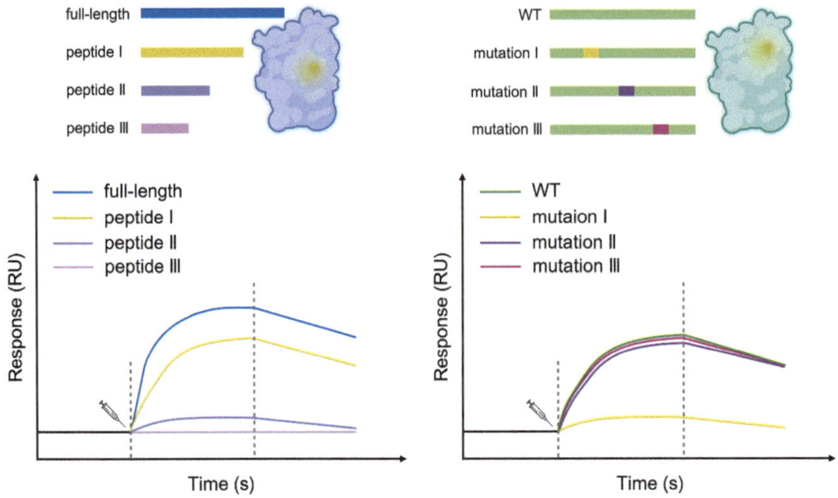

Figure 6.2 Segments and mutations employed in SPR assays for interface validation. Source: Yuanyuan Xie and Jianrong Xu. Various peptide segments of the protein as analytes (left) were applied in SPR assays to determine the difference of affinity to immobilized bait protein as ligand from the full-length format. In the left panel, peptide I caused small loss of affinity, indicating its significant role in the interface; instead, peptides II and II were less important. In the right panel, with different mutations of the bait protein as ligands immobilized on chip surfaces, the full-length protein as analyte was employed to measure the affinities and compare the difference of affinity with WT bait protein. The sensorgram showed that the affinity decreased a lot for mutation I, rather than mutations II and III, suggesting that mutation I impaired the interface of PPIs.

is especially prominent. The scaffolding function of HYPK is facilitated by its C-terminal ubiquitin-associated (UBA) domain and N-terminal tyrosine-type LC3-interacting region, which binds to NEDD8 and LC3, respectively. In SPR assays, both mutants displayed lower but not complete loss of binding affinity for NEDD8 compared to the affinities of wild-type HYPK and HYPK-UBA for NEDD8, indicating that four conserved residues within the HYPK-UBA domain are essential for its interaction with NEDD8 [39].

PPIs in Osteoporosis The receptor activator of nuclear factor-κB ligand (RANKL) is a multifunctional cytokine that plays a crucial role in osteoporosis [40]. In previous studies, RANK was considered the sole receptor for RANKL [41, 42]. In 2016, Jian Luo and colleagues discovered a new receptor for RANKL, termed LGR4, in osteoclasts. LGR4 competes with RANK for binding to RANKL and inhibits the canonical RANK signals during osteoclast differentiation. Researchers found through SPR analysis that the full-length extracellular domain of LGR4 can bind to RANKL with an affinity KD of 52.2 nM. A truncated form of the LGR4 extracellular domain exhibited significantly weakened binding to RANKL, with a KD of 1.527 μM. Further truncation of the LGR4 extracellular domain resulted in no binding to RANKL. These findings indicate that LGR4 directly binds to RANKL and that the intact extracellular domain of LGR4 is essential for its binding to RANKL, providing a concrete structural basis for future drug design [43].

Targeting the excessive activation of osteoclasts leading to bone loss, inhibiting the activity of Src kinase, a key gene in osteoclasts, is considered an effective treatment strategy [44, 45]. Using SPR, Zhenxi Li and colleagues confirmed that the C-terminus of GPR54 directly binds to Src kinase with high affinity (KD = 2.595 nM). Genetic analysis showed that the activation of GPR54 leads to a reduction in the phosphorylation level of Src kinase. The research team speculated that the activation of Gpr54 by Kp-10 might recruit a phosphatase, e.g. DUSP18, to reduce the phosphorylation of Src kinase. SPR assays further confirmed that the C-terminus of GPR54 could bind to human and mouse DUSP18 phosphatase with high affinity. These results collectively indicate that the C-terminus of GPR54 recruits both DUSP18 phosphatase and Src kinase upon stimulation by the ligand Kp-10, offering important scientific evidence for further drug development and treatment strategy design [46].

PPIs in Neuronal Migration Teneurins are ancient metazoan cell adhesion receptors that control brain development and neuronal wiring in higher animals. The extracellular C terminus binds the adhesion GPCR Latrophilin, forming a trans-cellular complex with synaptogenic functions. Daniel Del Toro and colleagues present crystal structures of Teneurin–Latrophilin complexes that reveal how the lectin and olfactomedin domains of Latrophilin bind across a spiraling beta-barrel domain of Teneurin, the YD shell. Then, chose the minimal complex of Lphn2 Lec-Ten2 for a set of molecular dynamics (MD) simulations to assess the stability of the complex and to better define the positions of side chains in the interface, molecular dynamics simulations reveal two "hotspots" on the Lphn2 Lec domain. To validate their structural and simulation results, researchers utilized SPR to complete the measurement of affinity and kinetic parameters for multiple groups, including both Wild-type and Mutant forms of murine Lphn1 (Lymphocyte Penetration 1) with Ten2 (Tenascin 2), confirming that the Latrophilin extracellular domain can bind both Teneurin and FLRT simultaneously *in vitro* [47].

PPIs in Infectious Diseases Wang and colleagues identified key regions in SARS-CoV-2 that interacts with the ACE2 receptor, speculating that SARS-CoV-2-S1 and SARS-CoV-2-CTD should exhibit stronger affinity for the receptor than RBD. To test this hypothesis, researchers conducted SPR assays. The KD values for SARS-CoV-2-S1 and SARS-CoV-2-CTD binding to ACE2 were calculated to be 94.6 ± 6.5 nM and 133.3 ± 5.6 nM, respectively. These values represent approximately a fourfold higher binding affinity compared to the SARS-RBD engaging the same receptor, which was determined to be 408.7 ± 11.1 nM [48]. Yurkovetskiy and colleagues used SPR to determine if the kinetics of SARS-CoV-2 S protein binding to human ACE2 is changed by D614G, the rate of association with ACE2 was little different between D614G and D614, but D614G dissociated from ACE2 at a rate fourfold faster than D614, resulting in a 5.7-fold reduction in binding affinity. These data demonstrate that the increased infectivity of D614G is not explained by greater ACE2 binding strength [49].

Mxra8 is a receptor for multiple arthritogenic alphaviruses that cause debilitating acute and chronic musculoskeletal disease in humans. Song and colleagues analyzed the crystal structures of the complex between hMXRA8 and CHIKV virus-like particles (VLPs). The authors used SPR to measure the binding between CHIKV E and hMXRA8, showing an affinity of $6.02 \pm 0.39\,\mu M$. In order to identify the key amino acids of interface, the authors performed a series of mutations at different sites on hMXRA8: Q63, D64, R69, Y92, E96, R98, E227, and R228. SPR results revealed that single mutations at Q63, D64, E96, and the double mutation at E227/R228 slightly reduced the binding of hMXRA8 to CHIKV E. A single mutation at Y92 partially decreased the affinity of the binding. However, single mutations at R69 and R98 completely abolished the binding. These findings suggest that amino acids at these different sites contribute to the interaction to varying degrees, and these results are fully consistent with the corresponding crystal structure data. Notably, they also discovered using SPR that the absence of an internal disulfide bond in hMXRA8 (C144A/C279A or C144S/C279S) could also disrupt its binding to CHIKV E, indicating that this disulfide bond is crucial for the stability of hMXRA8. Based on the critical data from SPR, the interaction mechanism between the envelope protein E of Chikungunya virus and its receptor MXRA8 have been perfectly verified [50], providing new targets for the development of antiviral drugs and the design of novel vaccines.

6.2.2 SPR Application in Ternary PPI Systems

Ternary PPI systems refer to specific binding events involving three molecules. These molecules are typically proteins but can also include other small molecules or biomacromolecules. Ternary PPIs play a crucial regulatory role within cells, implicated in complex biological processes and pathways. In the study of ternary PPIs, SPR is a common and powerful technique used for various applications, such as antibody epitope mapping, investigation of PPI modulators, and the development of targeted protein degraders.

6.2.2.1 SPR-Based Epitope Competition Assays

In epitope competition experiments, three common detection methods are typically employed: the sandwich method, the premixing method, and the tandem method, as illustrated in Figure 6.3. In the sandwich method, the first antibody is immobilized on the sensor chip surface, followed by the sequential introduction of the antigen and the second antibody. If the second antibody can bind to the antigen, it indicates that the two antibodies have different epitopes; otherwise, they share similar epitopes. The premixing method involves premixing an excess of the second antibody with the antigen before introducing them together to the sensor surface where the first antibody is immobilized. This method is suitable for detecting multivalent antigens and can validate the results of the sandwich method. The tandem method involves immobilizing the antigen on the sensor surface first, then continuously introducing the two antibodies. By observing the binding of the second antibody, it can be determined whether the two antibodies compete for the same epitope.

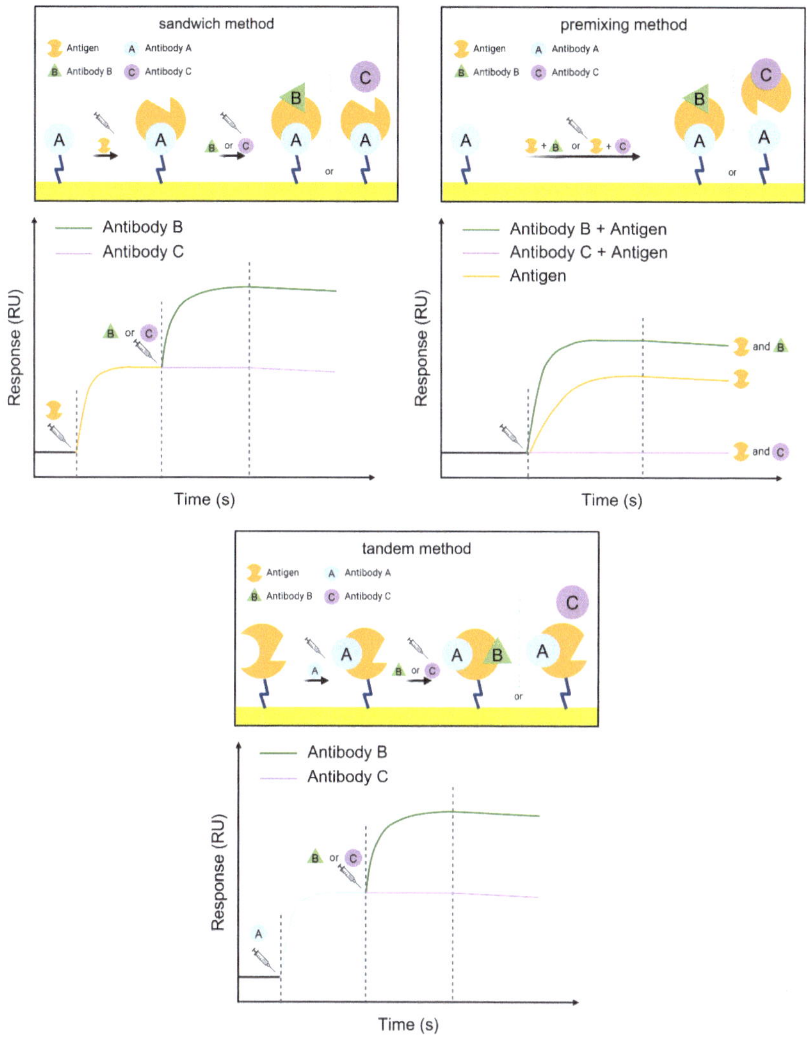

Figure 6.3 The typical sensorgrams for SPR-based binning assays: the sandwich method, the premixing method, and the tandem method. Source: Yuanyuan Xie and Jianrong Xu. In the sandwich method (left panel), the first antibody (Antibody A) is immobilized on the sensor chip surface, followed by sequentially introducing the antigen and the second antibody (Antibody B or Antibody C). If the second antibody can continue to bind to the antigen, it indicates that these two antibodies have different epitopes, e.g. Antibody A and Antibody B; otherwise, they could share similar epitopes, e.g. Antibody A and Antibody C. The premixing method (middle panel) involves premixing the second antibody (Antibody B or Antibody C) with the antigen before introducing them to the sensor surface where the first antibody (Antibody A) is immobilized. If the antigen can continue to bind to the immobilized antibody, it indicates that these two antibodies have different epitopes, e.g. Antibody A and Antibody B; otherwise, they could share similar epitopes, e.g. Antibody A and Antibody C. The tandem method (right panel) involves immobilizing the antigen on the sensor surface first, then tandemly introducing two different antibodies. If the second antibody can bind to the antigen on the chip, it indicates that these two antibodies have different epitopes, e.g. Antibody A and Antibody B; otherwise, they could share similar epitopes, e.g. Antibody A and Antibody C.

To mitigate viral escape, Zhang et al. developed a cocktail of neutralizing antibodies (Nabs) targeting epitopes located on different domains of the spike protein. To gain a better understanding of the synergy observed during the neutralization of SARS-CoV-2 by a cocktail of Nabs, researchers performed competitive SPR assays. The results of the assays were expected to reveal whether the Nabs recognize the same or different patches of the epitopes. As expected, the binding of the N-terminal domain-specific FC05 does not affect the attachment of any of the three RBD-specific Nabs to the SARS-CoV-2 spike trimer, explaining the cooperativity between the antibodies of the cocktail as they bind simultaneously to distinct domains. Conversely, the three RBD-targeting Nabs competed with each other for binding to the SARS-CoV-2 S trimer, which may imply that these RBD-targeting antibodies recognize similar epitopes or their epitopes overlap partially [51]. SPR results demonstrated that the synergy of cocktail antibodies is due to their simultaneous binding to different domains. This suggests that the enhanced efficacy observed in antibody combinations may arise from their ability to engage distinct regions on the viral spike protein, thereby potentiating neutralization and potentially blocking multiple pathways of viral entry.

6.2.2.2 SPR-Based Drug Discovery of PPI Modulators

Inhibition and stabilization of PPIs are two different strategies, resulting in two types of PPI modulator, inhibitors and stabilizers, both of which play important roles by altering the binding affinity, stability, or conformation of the target proteins [52]. One protein is typically immobilized on chip surface, followed by passing a mixture of the modulator and a known protein over the chip to observe response change. By comparing the SPR response sensorgrams of PPIs before and after the addition of the modulator, an increase in response after adding the modulator may indicate a stabilizer; a decrease may suggest an inhibitor, as illustrated in Figure 6.4.

PPI Inhibitors Galectin-3 (Gal-3) is associated with pancreatic ductal adenocarcinoma (PDAC) and can serve as a potential therapeutic target. A polysaccharide, RN1, extracted from *Panax notoginseng* flowers by L Zhang and colleagues can bind to both Gal-3 and the epidermal growth factor receptor (EGFR). As a novel Gal-3 inhibitor, RN1 can be considered a potential candidate for the treatment of human PDAC. Researchers first confirmed the binding of Gal-3 to EGFR through SPR experiments, with an affinity of 7.8 μM. Moreover, when different concentrations of RN1 were mixed with 50 μg/ml of Gal-3, it was found that the response value was inversely related to the concentration of RN1, indicating that RN1 impeded the binding of Gal-3 to EGFR. To further verify this competitive relationship, the interaction between RN1 and EGFR was detected by SPR assay, showing an affinity of 4.8 μM, corroborating the negative correlation between the response value and RN1 concentration observed in the competition experiment, which was due to RN1 competitively displacing Gal-3 originally bound to EGFR. Subsequent cell experiment results also aligned with the SPR experiment findings [53].

Tang and colleagues explored the mechanism of heparin inhibiting LPS cytoplasmic transport at the cellular and molecular levels, discovering that heparin (Hep)

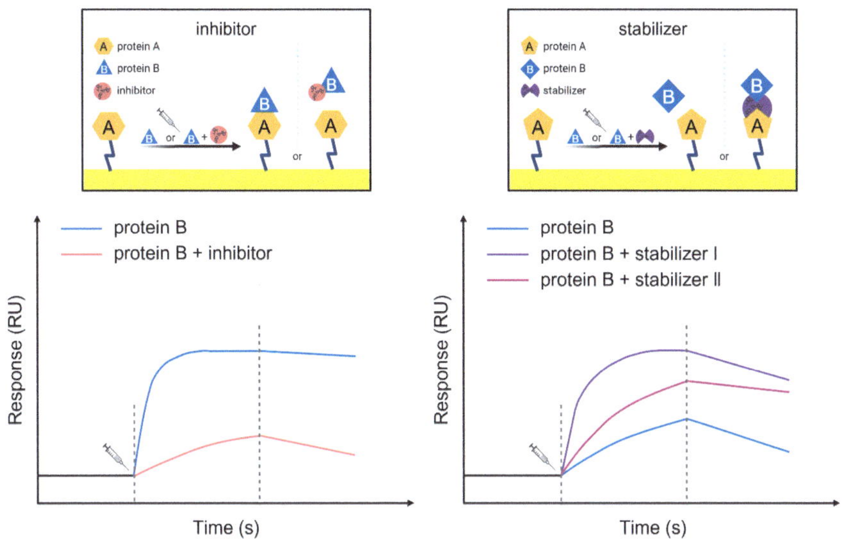

Figure 6.4 The typical SPR sensorgrams in PPI inhibitor and PPI stabilizer characterization. Source: Yuanyuan Xie and Jianrong Xu. Typically, one protein is immobilized on the chip surface as a ligand, followed by passing a mixture of the modulator and the bait protein over the chip surface to observe changes in response. By comparing the SPR response of PPIs in the absence or presence of the modulator, a decrease in response indicates that the modulator is an inhibitor (left); an increase suggests that the modulator is a stabilizer (right). Type I stabilizer increased the binding rate constant, while type II stabilizer reduced the dissociation rate constant of the PPI.

and non-anticoagulant heparin (NAH) primarily block LPS cytoplasmic transport through two targets: heparanase and high-mobility group protein-1 (HMGB1). Utilizing the SPR assay, the team verified that heparin directly inhibits the binding of HMGB1 to LPS. First, they confirmed that both heparin and NAH exhibit binding for HMGB1, with affinity constants of 5.59×10^{-6} M and 2.27×10^{-4} M, respectively. Subsequently, the SPR assay was further employed to verify that both heparin and NAH could inhibit the binding of HMGB1 to LPS. Thus, heparin suppresses HMGB1-mediated cytosolic delivery of LPS, at least in part, through direct inhibition of the HMGB1-LPS binding [54].

PPI Stabilizers Dysregulation of RAS–MAPK (RAS–RAF–MEK–ERK) pathway signaling is one of the most frequently occurring events in tumor development, presenting a promising and challenging clinical target in precision oncology. Ryan et al. developed NST-628, a potent pan-RAF–MEK molecular glue that prevents the phosphorylation and activation of MEK by RAF, overcoming the limitations of traditional RAS–MAPK inhibitors and leading to deep durable inhibition of the pathway. To elucidate the molecular glue mechanism, SPR–based ternary complex assays were performed. The binding affinities between MEK1–BRAF and MEK1–CRAF were determined to be 250 and 310 nM, respectively. These affinities are consistent with the previously reported values from SPR assays. The addition of NST-628 increased the affinity between MEK1 and BRAF from 250 to 50 nM and increased the affinity between MEK1 and CRAF from 309 to 39 nM. The increase in affinity for both systems was driven solely by a reduction in kd, indicating that the presence of NST-628 slows the dissociation of MEK1 from RAFs [55].

6.2.2.3 SPR Applications in Targeted Protein Degradation

Targeted protein degradation (TPD) is an emerging drug development strategy that utilizes two of the cell's most important degradation pathways, the proteasome and lysosome, to break down proteins associated with disease, thereby achieving a therapeutic effect. The advantage of targeted protein degradation technology lies in its ability to overcome the limitations of traditional drug development, such as difficult-to-target protein surfaces and PPIs. In addition, since the protein is completely degraded, this method has the potential to achieve long-lasting effects. Some early studies have already demonstrated its potential in treating cancer, neurodegenerative diseases, and other areas [56, 57]. Currently, proteolysis-targeting chimera (PROTAC) and molecular glue are two very promising types of protein degraders [58]. In the TPD research, the formation of highly stable ternary complexes is crucial for the mechanism of action of PROTAC/molecular glue. SPR can provide precise quantitative data, including affinity and kinetics, and calculate the cooperativity factor α and $t_{1/2}$ values of the complexes. This allows for predicting the rate and activity of PROTAC/molecular glue-mediated target degradation, offering key data for the development of PROTAC/molecular glue drugs.

PROTAC A PROTAC molecule consists of three components: a portion that binds to the target protein, a portion that recruits an E3-ubiquitin ligase, and a linker that connects these two parts. When a PROTAC molecule binds to both the target protein and the E3-ubiquitin ligase, it induces the ubiquitination of the target protein, marking it for degradation. The ubiquitinated target protein is recognized by the proteasome and then degraded [59].

Roy and colleagues have developed an SPR-based assay to quantify the stability of PROTAC-induced ternary complexes. This involves measuring the kinetics of their formation and dissociation *in vitro* using purified proteins. Authors benchmark their assay using four PROTACs that target the bromodomains (BDs) of bromodomain and extraterminal domain proteins Brd2, Brd3, and Brd4 to the Von Hippel–Lindau (VHL) E3 ligase. The results reveal significant differences in ternary complex off-rates for different PROTACs, which exhibit either positive or negative cooperativity for ternary complex formation relative to binary binding. These findings establish a novel assay to measure the kinetics of PROTAC ternary complexes and elucidate the important kinetic parameters that drive effective target degradation [60].

In an effort to design efficient PROTACs, Wurz and colleagues sought to describe the relationship between the initial step (ternary complex formation) and final step (target degradation) of the PROTAC-mediated degradation pathway. In this study, the authors employed SPR to quantify the formation of ternary complexes and assessed the cellular efficacy and degradation kinetics for two series of PROTACs. They characterized the PROTAC-mediated degradation process by utilizing VHL recruiting PROTACs targeting two distinct proteins, SMARCA2 and BRD4 [61].

To expand the roster of E3 ligases that can be utilized for targeted protein degradation, Hickey and colleagues describe the discovery and biochemical characterization of small-molecule ligands targeting the E3 ligase KLHDC2. Researchers utilized computer-aided drug design to synthesize analogue small molecules. Using the SPR assay, with KLHDC2 (E3 ligase) as the ligand, the binding activity and key binding sites of the analogue molecule KDRLKZ were detected. The researchers synthesized a series of KLHDC2–BRD4 PROTAC molecules, with BRD4 as the ligand and KLHDC2–BRD4 PROTAC, either premixed with KLHDC2 or not, as the analyte for the binary SPR assay and ternary SPR assay. Experimental data revealed that K2-B4-3 exhibited strong ternary complex formation effects, and the efficiency of formation was dependent on the recruitment of KLHDC2. To better understand the mechanism of the PROTAC molecule-induced proximity effect, the researchers explored the length and composition of the linker, with experimental designs identical to the binary SPR assays and ternary SPR assays. The results indicated that the formation of the ternary complex also depended on the length and composition of the linker: a longer, structured linker was crucial [62].

Molecular Glue Degraders Molecular glue degraders (including intrinsic degraders) induce or stabilize PPIs between a ubiquitin ligase and a target protein (substrate), leading to protein ubiquitination and subsequent proteasome-based degradation. Such compounds generally recruit target proteins for degradation by enhancing pre-existing PPIs or inducing the formation of new PPI complexes with the substrate receptors of E3-ubiquitin ligases [63].

Research has shown that the degradation of IKZF2 in regulatory T (Treg) cells and other T cells enhances the antitumor immune response, making IKZF2 a target for cancer immunotherapy [64]. Bonazzi et al. modified the IKZF1 degrader pomalidomide to develop the specific IKZF2 degrader NVP-DKY709, which demonstrated good CRBN binding capability and selective IKZF2 degradation ability. To understand this interaction, the authors synthesized different zinc finger structures of the transcription factor separately and used SPR to measure the binary or ternary interactions between proteins, discovering that ZF2-3 is the minimal binding component for IKZF2 to interact with CRBN in the presence of DKY709. The authors also used SPR to detect the interaction between IKZF2 and CRBN in the presence of DKY709, finding a significant enhancement in ZF2 binding compared to the control group without DKY709, which strongly supports the synergistic behavior of the complex [65].

6.3 Advantages and Limitations of SPR Application for PPIs

The SPR assay offers a robust tool for the study of PPIs, allowing real-time analysis of binding affinities and kinetics without the need for labeling. This method is particularly useful in identifying PPIs that are critical for understanding biological functions and potential disease treatments. SPR can detect interactions ranging from weak to strong, which makes it versatile for various molecular interactions, including protein–protein, protein–DNA, and receptor–drug. Its ability to monitor association and dissociation rates provides insights into the mechanisms and stability of these interactions. Furthermore, SPR can evaluate the efficacy of competitive inhibitors by calculating IC_{50} values. In qualitative analyses, SPR confirms protein interactions and aids in identifying key residues or domains affecting these interactions. Overall, SPR is widely regarded as a gold standard for molecular interaction studies and has broad applications in life sciences, drug discovery, and medical diagnostics.

The SPR assay requires high-purity samples and stable experimental conditions. High-purity samples are essential to minimize background noise and interference from nonspecific binding, while the experimental conditions, such as temperature and pH must be precisely controlled to ensure the accuracy of results. The high cost of SPR instruments and their complex operational procedures, which require professional technicians for operation and data interpretation, limit their accessibility and application in laboratories with limited resources. Although SPR can provide kinetic and affinity information on PPIs, it does not offer three-dimensional structural information about complexes. Researchers seeking detailed binding sites and interaction mechanisms need to rely on other techniques, such as X-ray crystallography and nuclear magnetic resonance, for more in-depth studies.

6.4 Future Directions

The future development of the SPR assay will focus on improving the speed of data acquisition and processing. By optimizing hardware design, enhancing automation, and developing more efficient data analysis algorithms, SPR assay can achieve faster data collection and analysis, enabling high-throughput screening. With the advancement of artificial intelligence and machine learning, their application in processing and interpreting SPR data can significantly enhance the accuracy and efficiency of data analysis and lower the barrier to adoption. These advanced technologies can help researchers extract more valuable insights from complex SPR data, such as automatically identifying and removing noise signals and more accurately fitting kinetic models. The application scope of the SPR assay for PPIs is expected to expand further. In clinical diagnostics, based on specific PPIs, SPR can be used to detect biomarkers, enabling early disease diagnosis and real-time monitoring. In drug development, SPR can accelerate the process of drug screening and optimization by monitoring the binding of drugs to targets in real time, quickly assessing the efficacy and safety. The integration of the SPR assay with other scientific disciplines will be an important trend in the future. For instance, combining with nanotechnology can enhance detection sensitivity and expand functionalities. This interdisciplinary innovation will bring new possibilities and broader prospects for the application of the SPR assay.

References

1 Durham, J., Zhang, J., Humphreys, I.R. et al. (2023). Recent advances in predicting and modeling protein-protein interactions. *Trends in Biochemical Sciences* 48 (6): 527–538.

2 Cheng, F., Zhao, J., Wang, Y. et al. (2021). Comprehensive characterization of protein-protein interactions perturbed by disease mutations. *Nature Genetics* 53 (3): 342–353.

3 Kim, M., Park, J., Bouhaddou, M. et al. (2021). A protein interaction landscape of breast cancer. *Science* 374(6563):eabf3066.

4 Thompson, T.B., Chaggar, P., Kuhl, E. et al. (2020). Protein-protein interactions in neurodegenerative diseases: a conspiracy theory. *PLoS Computational Biology* 16 (10): e1008267.

5 Ghosh, S.K. and Pal, T. (2007). Interparticle coupling effect on the surface plasmon resonance of gold nanoparticles: from theory to applications. *Chemical Reviews* 107 (11): 4797–4862.

6 Lee, T.H., Hirst, D.J., Kulkarni, K. et al. (2018). Exploring molecular-biomembrane interactions with surface plasmon resonance and dual polarization interferometry technology: expanding the spotlight onto biomembrane structure. *Chemical Reviews* 118 (11): 5392–5487.

7 Berggard, T., Linse, S., and James, P. (2007). Methods for the detection and analysis of protein-protein interactions. *Proteomics* 7 (16): 2833–2842.

8 Park, S.H., Ko, W., Lee, H.S., and Shin, I. (2019). Analysis of protein-protein interaction in a single live cell by using a FRET system based on genetic code expansion technology. *Journal of the American Chemical Society* 141 (10): 4273–4281.

9 Cui, Y., Zhang, X., Yu, M. et al. (2019). Techniques for detecting protein-protein interactions in living cells: principles, limitations, and recent progress. *Science China. Life Sciences* 62 (5): 619–632.

10 Douzi, B. (2017). Protein-protein interactions: surface plasmon resonance. *Methods in Molecular Biology* 1615: 257–275.

11 Berggard, T., Linse, S., and James, P. (2007). Methods for the detection and analysis of protein-protein interactions. *Proteomics* 7 (16): 2833–2842.

12 Ohlson, S., Strandh, M., and Nilshans, H. (1997). Detection and characterization of weak affinity antibody antigen recognition with biomolecular interaction analysis. *Journal of Molecular Recognition* 10 (3): 135–138.

13 Tao, L., Yang, G., Sun, T. et al. (2023). Capsaicin receptor TRPV1 maintains quiescence of hepatic stellate cells in the liver via recruitment of SARM1. *Journal of Hepatology* 78 (4): 805–819.

14 Nguyen, H.H., Park, J., Kang, S., and Kim, M. (2015). Surface plasmon resonance: a versatile technique for biosensor applications. *Sensors (Basel)* 15 (5): 10481–10510.

15 Wojtowicz, W.M., Vielmetter, J., Fernandes, R.A. et al. (2020). A human IgSF cell-surface interactome reveals a complex network of protein-protein interactions. *Cell* 182 (4): 1027–43 e17.

16 Verschueren, E., Husain, B., Yuen, K. et al. (2020). The immunoglobulin superfamily receptome defines cancer-relevant networks associated with clinical outcome. *Cell* 182 (2): 329–44 e19.

17 Hu, Y., Hu, Q., Li, Y. et al. (2023). gammadelta T cells: origin and fate, subsets, diseases and immunotherapy. *Signal Transduction and Targeted Therapy* 8 (1): 434.

18 Mensurado, S., Blanco-Dominguez, R., and Silva-Santos, B. (2023). The emerging roles of gammadelta T cells in cancer immunotherapy. *Nature Reviews. Clinical Oncology* 20 (3): 178–191.

19 Silva-Santos, B., Mensurado, S., and Coffelt, S.B. (2019). gammadelta T cells: pleiotropic immune effectors with therapeutic potential in cancer. *Nature Reviews. Cancer* 19 (7): 392–404.

20 Willcox, C.R., Vantourout, P., Salim, M. et al. (2019). Butyrophilin-like 3 directly binds a human $V\gamma4^{+}$ T cell receptor using a modality distinct from clonally-restricted antigen. *Immunity* 51 (5): 813–25 e4.

21 Tanaka, Y., Morita, C.T., Tanaka, Y. et al. (1995). Natural and synthetic non-peptide antigens recognized by human gamma delta T cells. *Nature* 375 (6527): 155–158.

22 Constant, P., Davodeau, F., Peyrat, M.A. et al. (1994). Stimulation of human gamma delta T cells by nonpeptidic mycobacterial ligands. *Science* 264 (5156): 267–270.

23 Rigau, M., Ostrouska, S., Fulford, T.S. et al. (2020). Butyrophilin 2A1 is essential for phosphoantigen reactivity by gammadelta T cells. *Science* 367 (6478): eaay5516.

24 Tao, L., Ma, W., Wu, L. et al. (2019). Glial cell line-derived neurotrophic factor (GDNF) mediates hepatic stellate cell activation via ALK5/Smad signalling. *Gut* 68 (12): 2214–2227.

25 Shang, J., Ye, G., Shi, K. et al. (2020). Structural basis of receptor recognition by SARS-CoV-2. *Nature* 581 (7807): 221–224.

26 Toelzer, C., Gupta, K., Yadav, S.K.N. et al. (2020). Free fatty acid binding pocket in the locked structure of SARS-CoV-2 spike protein. *Science* 370 (6517): 725–730.

27 Han, P., Su, C., Zhang, Y. et al. (2021). Molecular insights into receptor binding of recent emerging SARS-CoV-2 variants. *Nature Communications* 12 (1): 6103.

28 Gong, X., Qian, H., Zhou, X. et al. (2016). Structural insights into the Niemann-Pick C1 (NPC1)-mediated cholesterol transfer and Ebola infection. *Cell* 165 (6): 1467–1478.

29 Restelli, C., Ruella, M., Paruzzo, L. et al. (2024). Recent advances in immune-based therapies for acute myeloid leukemia. *Blood Cancer Discovery* 5 (4): 234–248.

30 Rubahamya, B., Dong, S., and Thurber, G.M. (2024). Clinical translation of antibody drug conjugate dosing in solid tumors from preclinical mouse data. *Science Advances* 10 (22): eadk1894.

31 Milazzo, F.M., Vesci, L., Anastasi, A.M. et al. (2019). ErbB2 targeted epigenetic modulation: anti-tumor efficacy of the ADC trastuzumab-HDACi ST8176AA1. *Frontiers in Oncology* 9: 1534.

32 Chen, L., Wang, L., Shion, H. et al. (2016). In-depth structural characterization of Kadcyla(R) (ado-trastuzumab emtansine) and its biosimilar candidate. *MAbs* 8 (7): 1210–1223.

33 Healey, G.D., Frostell, A., Fagge, T. et al. (2019). A RAGE-targeted antibody-drug conjugate: surface plasmon resonance as a platform for accelerating effective ADC design and development. *Antibodies (Basel)* 8 (1): 7.

34 Soderberg, L., Johannesson, M., Nygren, P. et al. (2023). Lecanemab, aducanumab, and gantenerumab - binding profiles to different forms of amyloid-beta might explain efficacy and side effects in clinical trials for Alzheimer's disease. *Neurotherapeutics* 20 (1): 195–206.

35 Dai, L., Zheng, T., Xu, K. et al. (2020). A universal design of betacoronavirus vaccines against COVID-19, MERS, and SARS. *Cell* 182 (3): 722–33 e11.

36 Shi, R., Shan, C., Duan, X. et al. (2020). A human neutralizing antibody targets the receptor-binding site of SARS-CoV-2. *Nature* 584 (7819): 120–124.

37 Wrapp, D., De Vlieger, D., Corbett, K.S. et al. (2020). Structural basis for potent neutralization of betacoronaviruses by single-domain camelid antibodies. *Cell* 181 (5): 1004–15 e15.

38 Xie, M., Liu, Y.U., Zhao, S. et al. (2022). TREM2 interacts with TDP-43 and mediates microglial neuroprotection against TDP-43-related neurodegeneration. *Nature Neuroscience* 25 (1): 26–38.

39 Ghosh, D.K. and Ranjan, A. (2022). HYPK coordinates degradation of polyneddylated proteins by autophagy. *Autophagy* 18 (8): 1763–1784.

40 Honma, M., Ikebuchi, Y., and Suzuki, H. (2021). RANKL as a key figure in bridging between the bone and immune system: Its physiological functions and potential as a pharmacological target. *Pharmacology & Therapeutics* 218: 107682.

41 Lacey, D.L., Boyle, W.J., Simonet, W.S. et al. (2012). Bench to bedside: elucidation of the OPG-RANK-RANKL pathway and the development of denosumab. *Nature Reviews. Drug Discovery* 11 (5): 401–419.

42 Wada, T., Nakashima, T., Hiroshi, N., and Penninger, J.M. (2006). RANKL-RANK signaling in osteoclastogenesis and bone disease. *Trends in Molecular Medicine* 12 (1): 17–25.

43 Luo, J., Yang, Z., Ma, Y. et al. (2016). LGR4 is a receptor for RANKL and negatively regulates osteoclast differentiation and bone resorption. *Nature Medicine* 22 (5): 539–546.

44 Lowe, C., Yoneda, T., Boyce, B.F. et al. (1993). Osteopetrosis in Src-deficient mice is due to an autonomous defect of osteoclasts. *Proceedings of the National Academy of Sciences of the United States of America* 90 (10): 4485–4489.

45 Horne, W.C., Sanjay, A., Bruzzaniti, A., and Baron, R. (2005). The role(s) of Src kinase and Cbl proteins in the regulation of osteoclast differentiation and function. *Immunological Reviews* 208: 106–125.

46 Li, Z., Yang, X., Fu, R. et al. (2024). Kisspeptin-10 binding to Gpr54 in osteoclasts prevents bone loss by activating Dusp18-mediated dephosphorylation of Src. *Nature Communications* 15 (1): 1300.

47 Del Toro, D., Carrasquero-Ordaz, M.A., Chu, A. et al. (2020). Structural basis of Teneurin-Latrophilin interaction in repulsive guidance of migrating neurons. *Cell* 180 (2): 323–39 e19.

48 Wang, Q., Zhang, Y., Wu, L. et al. (2020). Structural and functional basis of SARS-CoV-2 entry by using human ACE2. *Cell* 181 (4): 894–904 e9.

49 Yurkovetskiy, L., Wang, X., Pascal, K.E. et al. (2020). Structural and functional analysis of the D614G SARS-CoV-2 spike protein variant. *Cell* 183 (3): 739–51 e8.

50 Song, H., Zhao, Z., Chai, Y. et al. (2019). Molecular basis of arthritogenic alphavirus receptor MXRA8 binding to Chikungunya virus envelope protein. *Cell* 177 (7): 1714–24 e12.

51 Zhang, L., Cao, L., Gao, X.S. et al. (2021). A proof of concept for neutralizing antibody-guided vaccine design against SARS-CoV-2. *National Science Review* 8 (8): nwab053.

52 Wu, D., Li, Y., Zheng, L. et al. (2023). Small molecules targeting protein-protein interactions for cancer therapy. *Acta Pharmaceutica Sinica B* 13 (10): 4060–4088.

53 Zhang, L., Wang, P., Qin, Y. et al. (2017). RN1, a novel galectin-3 inhibitor, inhibits pancreatic cancer cell growth in vitro and in vivo via blocking galectin-3 associated signaling pathways. *Oncogene* 36 (9): 1297–1308.

54 Tang, Y., Wang, X., Li, Z. et al. (2021). Heparin prevents caspase-11-dependent septic lethality independent of anticoagulant properties. *Immunity* 54 (3): 454–67 e6.

55 Ryan, M.B., Quade, B., Schenk, N. et al. (2024). The pan-RAF-MEK nondegrading molecular glue NST-628 is a potent and brain penetrant inhibitor of the RAS-MAPK pathway with activity across diverse RAS- and RAF-driven cancers. *Cancer Discovery* 14 (7): 1190–1205.

56 Dale, B., Cheng, M., Park, K.S. et al. (2021). Advancing targeted protein degradation for cancer therapy. *Nature Reviews. Cancer* 21 (10): 638–654.

57 Li, X. and Song, Y. (2020). Proteolysis-targeting chimera (PROTAC) for targeted protein degradation and cancer therapy. *Journal of Hematology & Oncology* 13 (1): 50.

58 Schapira, M., Calabrese, M.F., Bullock, A.N., and Crews, C.M. (2019). Targeted protein degradation: expanding the toolbox. *Nature Reviews. Drug Discovery* 18 (12): 949–963.

59 Zhao, L., Zhao, J., Zhong, K. et al. (2022). Targeted protein degradation: mechanisms, strategies and application. *Signal Transduction and Targeted Therapy* 7 (1): 113.

60 Roy, M.J., Winkler, S., Hughes, S.J. et al. (2019). SPR-measured dissociation kinetics of PROTAC ternary complexes influence target degradation rate. *ACS Chemical Biology* 14 (3): 361–368.

61 Wurz, R.P., Rui, H., Dellamaggiore, K. et al. (2023). Affinity and cooperativity modulate ternary complex formation to drive targeted protein degradation. *Nature Communications* 14 (1): 4177.

62 Hickey, C.M., Digianantonio, K.M., Zimmermann, K. et al. (2024). Co-opting the E3 ligase KLHDC2 for targeted protein degradation by small molecules. *Nature Structural & Molecular Biology* 31 (2): 311–322.

63 Maniaci, C. and Ciulli, A. (2019). Bifunctional chemical probes inducing protein-protein interactions. *Current Opinion in Chemical Biology* 52: 145–156.

64 Mognol, G.P., Spreafico, R., Wong, V. et al. (2017). Exhaustion-associated regulatory regions in CD8(+) tumor-infiltrating T cells. *Proceedings of the National Academy of Sciences of the United States of America* 114 (13): E2776–E2785.

65 Bonazzi, S., d'Hennezel, E., Beckwith, R.E.J. et al. (2023). Discovery and characterization of a selective IKZF2 glue degrader for cancer immunotherapy. *Cell Chemical Biology* 30 (3): 235–47 e12.

7

Computational Methods for Protein–Protein Interactions

Hao Li, Yurui Li, and Sheng-You Huang

Huazhong University of Science and Technology, School of Physics, 1037 Luoyu Rd, Wuhan 430074, China

7.1 Introduction

Proteins are among the most important biological macromolecules in living organisms. As principal mediators of vital biological processes, they play central roles in almost all life processes, including gene replication, transcription, translation, and regulation [1–4], signal transduction [5, 6], immune response [7], enzyme catalysis [8], and other functions. Proteins rarely function in isolation; instead, they typically exert their biological effects through interactions with other molecules such as proteins, nucleic acids, carbohydrates, and small molecules [9, 10]. Among these interactions, PPIs account for the largest proportion and have received the most extensive attention and research [11, 12]. Dysregulation of PPIs is implicated in numerous diseases. For example, in the brains of Alzheimer's disease patients, abnormal PPIs between β-amyloid protein (Aβ) molecules lead to aggregate formation, causing neuronal damage [13]. A major feature of Parkinson's disease is the aggregation of α-synuclein in neurons [14]. Abnormal interactions between α-synuclein molecules cause them to form toxic aggregates that damage neuronal function. p53 is a tumor suppressor protein that inhibits tumor formation by promoting cell cycle arrest and apoptosis. However, the MDM2 protein inhibits the function of p53 by interacting with p53. In certain cancers, overexpression of MDM2 leads to p53 dysfunction, thereby promoting tumor development. Designing drugs to disrupt the abnormal interaction between MDM2 and p53 has become an important strategy in cancer therapy [15, 16]. Therefore, studying PPIs plays an important role in our in-depth understanding of life activities, clarifying the mechanism behind them and intervening and treating related diseases.

The study of PPIs is so important that it has always been a focal point in biology and related fields. Researchers have analyzed and studied the principles of PPIs from both experimental and theoretical calculation aspects [11]. The interaction between proteins depends on the mutual recognition of the protein surface. Seeing the PPI clearly at the structural level will help us understand its molecular mechanism so that we can design drugs to regulate this interaction in a targeted

Targeting Protein–Protein Interactions for Drug Discovery, First Edition. Edited by Jian Zhang.

manner. Experimental biologists can resolve the structure of protein–protein complexes through nuclear magnetic resonance (NMR), X-ray crystallography, small-angle X-ray scattering (SAXS), cryo-electron microscopy (Cryo-EM), and other means. SAXS can only obtain low-resolution protein structures, while X-ray crystallography and NMR can offer high-resolution structures at the atomic level. However, protein–protein complexes are often difficult to crystallize and have large molecular weights, making it challenging to obtain their structures using X-ray crystallography and NMR. With the breakthrough progress of Cryo-EM technology, Cryo-EM has gradually become the main tool for researchers to resolve the structure of large protein complexes. Despite these advancements, experimental methods for resolving protein structures still face some limitations. First, they require high costs to maintain expensive experimental equipment and for sample preparation. Second, experimental methods typically take a long time to prepare samples, conduct experiments, and process data, often taking weeks or even months to yield results. Furthermore, experimental methods cannot resolve a large number of protein complexes in a high-throughput manner.

Due to the limitations of experimental methods in resolving protein–protein complex structures, the number of protein–protein structures in the Protein Data Bank (PDB) database remains very limited. However, with the continuous growth of experimental data and advancements in computational fields, computational methods are increasingly valued for studying PPIs.

Before the widespread application of deep learning, protein–protein docking was the mainstream method for predicting protein–protein complex structures. Protein–protein docking algorithms can be divided into template-based docking and ab initio docking methods. Template-based docking algorithms can generate high-quality complex structures if reliable templates can be found in the PDB database. However, due to the limited number of complexes in the PDB database, many PPIs cannot find complex templates in the PDB database, necessitating the use of ab initio docking methods. Ab initio docking treats the larger protein as the receptor and the smaller protein as the ligand, searches for possible binding modes, and uses scoring functions to rank the binding modes. With the breakthrough developments in deep learning technology and its application in biology, deep learning-based computational predictions of PPIs have become a hot research area. Below, we first introduce protein–protein docking algorithms and end-to-end deep learning structure prediction algorithms used for predicting protein–protein complex structures, as shown in Figure 7.1. Then, we discuss the Critical Assessment of PRedicted Interactions (CAPRI) experiment, which has significantly advanced the development of PPI prediction methods. Finally, we discuss the current challenges and future directions in the development of PPI prediction.

7.2 Protein–Protein Docking

Since Janin and Wodak pioneered the use of the first automated algorithm to predict the three-dimensional structure of the interaction between bovine pancreatic trypsin and its natural inhibitor [17], the field of protein–protein docking has entered

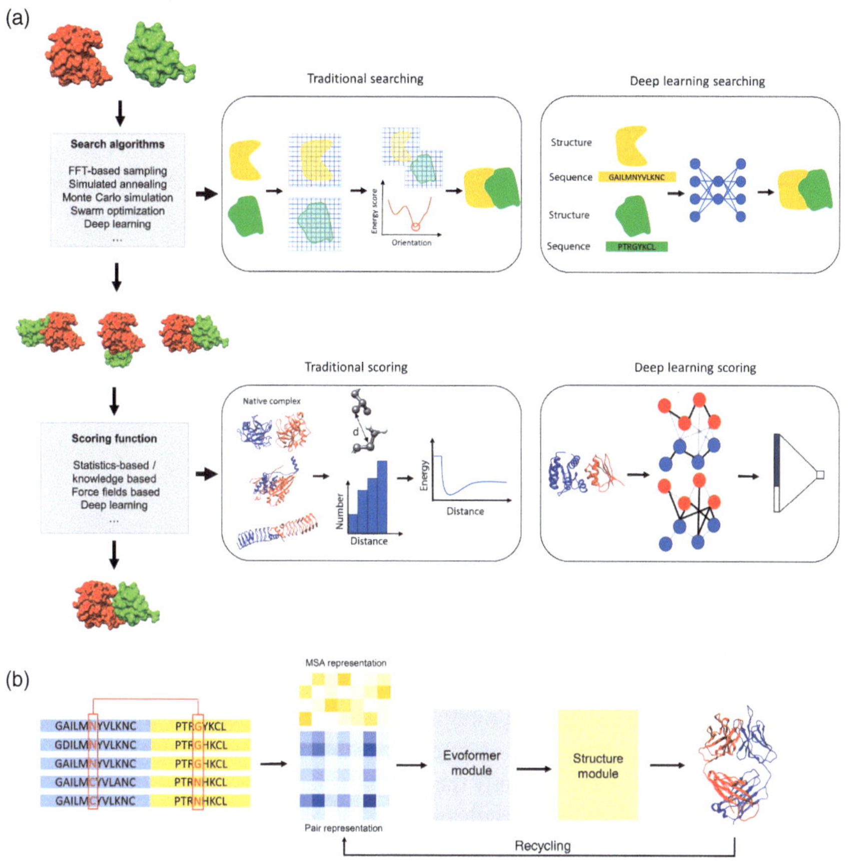

Figure 7.1 Computational methods for protein–protein complex structure prediction. (a) Protein–protein docking algorithms include two steps: sampling and scoring. Both sampling algorithms and scoring functions include traditional and deep learning ones. (b) End-to-end structure prediction algorithms represented by AlphaFold input the MSA of the sequence, then extract and enhance features through Evoformer module, and then generate the structure through structure module.

a period of rapid development. Over the past four decades, numerous docking algorithms have emerged, employing a variety of strategies to predict the structures of protein complexes. Protein–protein docking algorithms have become an important tool for analyzing PPIs and understanding their interaction mechanisms at the structural level.

Protein–protein docking involves predicting the structure of a complex (the bound state) starting from the individual structures of its components (the unbound state). Initially, the monomer structures used for protein–protein docking could only be obtained from experimentally resolved protein structures. Later, as more and more protein structures were solved experimentally, these resolved structures could be used as templates for homology modeling to obtain the monomer structures for docking. Currently, with the tremendous success of end-to-end structure prediction algorithms based on deep learning, exemplified by AlphaFold2 [18], predicting protein monomer structures is no longer a significant challenge. Additionally, the DeepMind team has released the predicted structures of the entire proteome using AlphaFold [19], making these structures easily accessible and downloadable for users.

Given the monomer structures of two proteins, protein–protein docking first explores the potential binding modes between the two proteins through a search algorithm and then evaluates these potential binding modes using a scoring function. The most promising complex conformations are selected as the final prediction results. Therefore, traditional protein–protein docking is divided into two steps: sampling and scoring. This docking approach is also known as free docking or ab initio docking and is the most longstanding and widely applied type of docking algorithm [20, 21]. Sampling and scoring can be performed simultaneously or independently at different stages [22].

7.2.1 Sampling

Sampling includes the search for the relative orientations between two proteins and the search for conformations of the two proteins themselves. During the binding process, proteins often undergo conformational changes, and searching within the protein's conformational space allows these changes to be taken into account. However, the conformational space of proteins is vast, making exhaustive sampling impractical. Even coarse consideration of protein conformational changes can significantly increases computational costs. Consequently, many docking algorithms treat proteins as rigid bodies, considering only the relative orientations between them. Since many proteins exhibit relatively small conformational changes during binding, this approach can still yield good prediction results. Depending on whether deep learning techniques are employed during sampling, search algorithms can be classified into traditional search algorithms and deep learning-based search algorithms.

7.2.1.1 Traditional Search Algorithms

Traditional search algorithms have been developed over a long period and remain the most widely used. Currently, mainstream protein–protein docking algorithms predominantly utilize traditional search algorithms. Common traditional search algorithms include those based on local shape feature matching, FFT, and heuristic search algorithms.

In local shape feature matching algorithms, the molecular shape representation of the protein is first derived, and for example, the Connolly surface is obtained by rolling the probe water molecule around the protein surface. The derived shape surface is then divided into different regions (patches). The algorithm calculates the shape complementarity between different patches of the two proteins to find the two regions with the best shape match, thereby identifying the binding mode with good local shape complementarity between the two proteins. Due to the focus on local shape matching, the resulting binding modes often have many atomic clashes, necessitating subsequent filtering to remove those binding modes with excessive atomic clashes. In addition, this type of method tends to perform more intensive searches in areas with good local shape matches, resulting in high redundancy in the binding modes, which requires clustering to remove redundant binding modes. Typical examples of local shape feature matching algorithms include DOCK [23], PatchDock [24], LZerD [25, 26], 3D-Garden [27], and GAPDOCK [28].

The docking algorithm based on Fast Fourier Transform (FFT) is the most widely used type of docking algorithm. Katchalski-Katzir et al. pioneered the application of FFT to protein–protein binding mode searches in 1992, significantly improving the computational efficiency by using three-dimensional FFT to accelerate the sampling of the three degrees of freedom in the translation space [29]. This type of method first discretizes the two input proteins into three-dimensional grids and assigns values to the grids based on the occupancy of protein atoms. The ligand (smaller protein) grids are then translated relative to the receptor (larger protein) grids, achieving sampling of the translational space. The calculation of the overlap score between receptor and ligand grids can be expedited by computing the three-dimensional Fourier transform, thus accelerating the translational sampling. Due to the high efficiency of FFT-based sampling, many teams have developed their own docking algorithms based on this approach. For instance, ZDOCK2.1 introduced the Pairwise Shape Complementarity (PSC) score by considering the influence of neighboring atoms on surface grids. In this way, the score of the surface grid points with more surrounding atoms is higher, which reflects the information of the protein surface curvature and better describes the shape of the protein surface [30]. Yan Y et al. proposed the Long-range Shape Complementarity (LSC) score, which considers multiple layers of neighboring atoms for surface grids and assigns values using a Gaussian function, enhancing the accuracy of geometric matching [31]. In addition to geometric complementarity, other biophysical factors, such as electrostatic interactions, atomic desolvation effects, hydrophobic interactions, atomic pair potentials, binding site information, and other factors have been integrated into FFT-based docking algorithms, resulting in various FFT-based docking algorithms, such as HDOCK [32, 33], ZDOCK [34–36], DOT [37], GRAMM [38], MOLFIT [39], FTDOCK [40], PIPER [41], ASPDOCK [42], SDOCK [43], CoDOCKPP [44], and others.

Unlike the local shape feature matching algorithms, which require the protein to be represented as a shape, and the FFT-based algorithms, which require the protein to be represented as a grid, the heuristic search algorithms do not require special representations of proteins. Heuristic search algorithms typically use the all-atom models or coarse-grained models of proteins, allowing better consideration of internal conformational changes. Due to computational limitations, heuristic search algorithms cannot explore the entire conformational space and have a degree of randomness in their searches. However, because they use atomic representations, it is easier to incorporate experimental or computational site information or distance information to assist in complex structure prediction. Methods in this category include HADDOCK [45, 46] using simulated annealing, RosettaDock [47] employing Monte Carlo simulation for conformational searches, ATTRACT [48] using quasi-Newton optimization algorithms, SwarmDock [49] and JabberDock [50] utilizing particle swarm optimization algorithms, and LightDock [51, 52] employing glowworm swarm optimization algorithms.

7.2.1.2 Deep Learning-Based Search Algorithms

In addition to traditional search algorithms, new search algorithms that leverage deep learning to generate binding modes have also emerged in recent years. Depending on whether they search for conformational changes of the constituent proteins during binding or not, deep learning search algorithms can be divided into rigid-body search and flexible search. Rigid-body search treats proteins as rigid bodies and only focuses on exploring their rotation and translation relative to each other. In contrast, flexible search, in addition to searching for the relative position between two proteins, also considers changes in the backbone and side-chain dihedral angles of the proteins during the binding process.

In the past few years, several new rigid-body search algorithms based on deep learning have been proposed. Ganea et al. developed EQUIDOCK, which utilizes a pairwise-independent SE(3)-equivariant graph matching network to predict the relative rotational and translational transformations between two proteins [53]. Although the complex prediction results of EQUIDOCK are not as accurate as HDOCK based on the FFT search, it offers a remarkable computational speed, taking just an average of five seconds to predict a complex. DIFFDOCK-PP proposed by Ketata et al. is the first to apply the diffusion generative model to rigid-body protein–protein docking [54]. With comparable computational efficiency, DIFFDOCK-PP demonstrates significantly improved complex structure prediction performance compared to EQUIDOCK. xTrimoDock, developed by the BioMap team, employs the Evoformer architecture inspired by AlphaFold2 to predict the distance map [55]. It then derives the relative rotational and translational transformations between proteins by minimizing the difference between the predicted distance map and the distance map corresponding to the sampling structure. Given that the distance map prediction network is specially trained on antibody–antigen complexes, xTrimoDock has achieved better results than AlphaFold-Multimer [56] in antibody–antigen docking.

Beyond rigid-body search approaches, researchers have also investigated how to incorporate protein flexibility into the sampling process. One such method, DockGPT, developed by McPartlon and Xu, utilizes a generative protein transformer specifically designed for flexible, site-specific protein–protein docking [57]. DockGPT processes the unbound structures and sequences of the proteins involved, optionally including binding site details. The neural network directly produces the complex structure, effectively sidestepping the computational challenges associated with flexible sampling. Similarly, GeoDock, introduced by Chu et al., feeds the unbound structures and sequence features embedded by ESM-2 into a streamlined version of the AlphaFold2 framework to generate the complex structures [58]. Both DockGPT and GeoDock rely on the sequences and structures of the proteins to directly predict the complex structure, akin to AlphaFold-Multimer, but without requiring multiple sequence alignments (MSAs). However, Chu et al. noted that both GeoDock and DockGPT have relatively low success rates in blind unbound docking, highlighting the ongoing difficulties that deep learning-based docking algorithms face in accurately modeling backbone conformational changes. While deep learning approaches that directly generate complex structures can bypass the intensive orientational and conformational searches, they tend to reduce the diversity of sampling. To address this, various techniques have been implemented to increase sampling diversity in such end-to-end methods. For instance, Wallner improved the diversity of structures generated by AlphaFold-Multimer by employing different versions, increasing the number of recycles, and activating dropout during inference [59].

7.2.2 Scoring

Once a large number of candidate complex structures have been generated by traditional or deep learning sampling methods, it is very important to develop an efficient scoring function to identify the most accurate structures. Ideally, we want the scores assigned by the scoring function to be highly correlated with the quality of the models, enabling the identification of the correct binding mode from a large number of incorrect ones. However, despite continuous improvements, current scoring functions are far from perfect. Similar to sampling methods, scoring functions for PPIs can be categorized into traditional scoring functions and deep learning scoring functions.

7.2.2.1 Traditional Scoring Function

Traditional scoring functions commonly used for PPIs can be divided into two main types. The first type of method is to obtain the total score of PPIs by weighted summing various physical or empirical scoring terms, such as shape complementarity, buried surface area, van der Waals potentials, and electrostatic potentials [43, 60–64]. According to the theory of the lock-and-key model, protein–protein recognition depends on the shape complementarity of the interacting proteins. Experimentally resolved PPI interfaces often exhibit significant shape complementarity, making it crucial for evaluating the interface of protein complexes. Therefore, most scoring functions include a shape complementarity term [34]. The buried surface area can reflect the binding energy between two proteins to a certain extent. Generally, the binding energy will increase with the increase in the buried surface area between the two proteins. In addition, the buried surface area also indicates the hydrophobic effect, which plays a very important role in the protein–protein binding process, as hydrophobic regions often correspond to interaction sites. Many scoring functions use buried surface area to calculate the solvation free energy in the protein–protein binding energy [65]. Van der Waals and electrostatic interactions play a crucial role in PPI and recognition. Most force fields provide formulas and corresponding parameters for these two interactions. Researchers can use these theoretical formulas and parameters to quantitatively calculate the van der Waals interaction and electrostatic interaction between two proteins. Van der Waals interactions are typically calculated using the Lennard–Jones potential of the 6–12 form, while the electrostatic interactions are often calculated using the Coulomb potential of point charges. The weights for the combined terms are usually optimized to enhance the recognition success rate of the correct binding mode from a training set containing many candidate poses.

The second type of traditional scoring functions is statistics-based or knowledge-based scoring functions [66–69], which learn the knowledge of PPIs based on the experimental structures of protein complexes that have been solved in the PDB database. This interaction knowledge is typically presented in the form of distance-dependent potentials or contact potentials for atom–atom or residue–residue pairs, which are derived from statistical occurrences observed at the PPI interface in the experimental structure database. According to the inverse Boltzmann relation, more frequently observed distances or contacts represent lower corresponding energies. Knowledge-based statistical potentials are usually derived by dividing the actual observed distribution of a specific residue–residue or atom–atom pair in a non-redundant protein complex structure database by their expected distribution in the reference state. The differences among various statistical potentials lie in the categorization of atom or residue pairs, the selection of the reference state, and the dataset used for deriving the actual distribution. In addition to being divided into atom pair potential and residue pair potential according to the degree of coarse-graining, statistical potential can also be divided into contact potential and distance-dependent potential. Taking atom pair potential as an example, if the contact potential is considered, when the actual distance between two atoms located on two proteins is within a given cutoff distance (generally 6 Å), the two atoms are considered to be in contact with each other and their atomic contact potential is calculated [35]. The total contact potential is obtained by summing the contact potentials of all contacting atom pairs. The distance-dependent potential is more refined than the contact potential. It divides the distances of atomic pairs within a specific distance cutoff into several bins, each with a corresponding potential. The total energy is obtained by summing the potentials of all atom pairs at different distances within the cutoff [66].

However, the two methods mentioned above are not mutually exclusive. Some scoring functions combine these approaches to form a composite scoring function [70–72]. For example, some scoring functions integrate statistical potentials with shape complementarity and electrostatic terms to leverage the strengths of both approaches [35].

7.2.2.2 Deep Learning-Based Scoring Function

In recent years, as deep learning has been well applied in many research fields, various scoring functions based on deep learning have also been proposed. Convolutional neural network (CNN)-based methods include DOVE [73], TRScore [74], and DeepRank [75]. DOVE is the pioneering scoring algorithm that utilizes CNN to evaluate docking models. TRScore leveraged a 3D RepVGG framework and demonstrated superior prediction performance compared to DOVE. DeepRank maps both atomic and residue-level features onto 3D grids to train the network for ranking docking models. However, a drawback of these methods is that they are not rotation invariant, making them sensitive to the orientation of the input structure. To address this limitation, DLA-Ranker [76] considers the local environment of each interface residue by using a local coordinate frame defined by each residue. This local frame definition grants DLA-Ranker rotational invariance to the input structures. Another drawback of these CNN-based methods is that discretizing the structures onto 3D grids will cause a loss of information.

Considering the limitations of CNN-based methods, one viable solution is to use graphs to represent complex structures. Graph neural networks (GNNs) have been increasingly used to evaluate protein docking models. GNN-DOVE [77], the GNN adaptation of DOVE, is the first algorithm to apply GNNs for ranking protein docking models. Thanks to the integration of GNNs, GNN-DOVE significantly outperforms its CNN-based counterpart, DOVE. Bonvin and colleagues also updated their DeepRank to DeepRank-GNN [78], which exhibits notable improvements in terms of classification of correct and incorrect docking models and computational efficiency compared to DeepRank. In the latest version of DeepRank-GNN-esm [79], the Bonvin team employs ESM-2 embedding to replace PSSM features, reducing the computational demands of DeepRank-GNN to generate PSSM. Kim et al. proposed G-RANK [80], which uses a geometric vector perceptron-graph neural network (GVP-GNN) to score protein–protein docking models. When assessed on the CAPRI score set, G-RANK achieved better performance than DeepRank-GNN and GNN-DOVE. DProQA [81] introduced by Cheng's team uses a novel gated neighborhood-modulating graph transformer to evaluate the quality of predicted protein complexes. DProQA ranked 3rd in the CASP15 model quality assessment experiment. Shuvo et al. have developed PIQLE [82], including evolutionary information and multimeric interaction geometries into a multi-head graph attention network for scoring protein complex interfaces. When evaluated on the docking benchmark set of Dockground version 1, PIQLE significantly outperforms DProQA [82].

While the aforementioned methods perform well on specific test sets, a substantial portion of them have not been evaluated in the blind test of CASP15. Therefore, their performance in real-world applications remains uncertain. Furthermore, many of these methods are trained to identify acceptable quality structures generated by traditional docking algorithms, potentially limiting their effectiveness in identifying high-quality models produced by deep learning approaches, such as AlphaFold-Multimer. Remarkably, in the CASP15 experiment, the Venclovas team used AlphaFold with different parameters and versions to sample and incorporated their developed VoroIF-jury to rank the models sampled by AlphaFold [83]. Thanks to VoroIF-jury's excellent scoring capability, they achieved results that are significantly superior to the standard AlphaFold-Multimer pipeline, and they also ranked first in the estimation of model accuracy (EMA) for multimeric structures in CASP15. VoroIF-jury is a multi-model method that groups and ranks models selected by different scoring methods based on the interface Contact Area Difference (CAD) score. A key component of Voro-jury is VoroIF-GNN [84], which employs an attention-based graph neural network to evaluate protein–protein interfaces.

7.2.3 Template-based Docking

Template-based docking involves two main steps: searching for reliable complex templates and constructing complex structures based on these templates. The first step entails identifying complex structures in the PDB database that are similar to both input proteins. Although the search primarily targets structurally similar templates, two proteins with high sequence similarity are generally similar in structure. In addition, in many cases, only the protein sequence is input instead of the structure. Therefore, sequence similarity is used instead of structure similarity, leading to a broader focus on sequence similarity. In order to find templates with high similarity to the input protein sequence, we often use sequence alignment algorithms, such as BLAST [85], FASTA [86], PSI-BLAST [87], HHblits [88], etc. Among them, HHblits uses the alignment between the hidden Markov models of sequences to identify homologous sequences, which can better identify remote homologs. Alternatively, structural alignment tools such as Foldseek [89] and US-align [90] can also be used to directly search for templates with similar structures. Once suitable templates are identified, the next step is to build a complex structure based on the template structure. Generally, the two component proteins are built separately and then superimposed on the complex template.

Template-based docking algorithms effectively utilize existing PPI data from the PDB structure database, circumventing challenges associated with protein flexibility encountered in free docking methods. However, due to the limited availability of resolved protein complex structures in the PDB, many PPIs lack reliable complex structure templates. Consequently, some research groups propose hybrid strategies combining both free docking and template-based docking approaches, exemplified by algorithms like HDOCK [32, 33], GalaxyHomomer [91], InterEvDock [92, 93], ClusPro [94], HADDOCK [45], and ZING [95]. These algorithms vary in their utilization of complex template information. HADDOCK and ClusPro extract residue distance constraints from complex templates to guide sampling in free docking, and filter resultant poses based on these constraints. GalaxyHomomer and HDOCK directly construct complex structure models from identified templates, followed by dissociating and re-docking the constituent proteins, thereby merging homology modeling and free docking results. ZING adopts a similar approach by integrating predictions from the template-based algorithm SPRING and free docking algorithm ZDOCK, employing a regression model to rank and score the combined results. In contrast, InterEvDock solely performs free docking using monomer structures derived from complex templates, excluding template-based modeling outputs from its results.

7.3 End-to-end Structure Prediction

With the continuous application and evolution of deep learning technology in protein structure prediction, especially in achieving unprecedented accuracy in end-to-end predictions of monomeric protein structures, the field of end-to-end protein complex structure prediction has also experienced rapid development. In the CASP14 competition, the performance of the end-to-end deep learning-based protein structure prediction algorithm AlphaFold2 reached an unprecedented level, with its predictions significantly surpassing those of other participants in terms of structural accuracy. In July 2021, the DeepMind team released the inference code and models for AlphaFold2 [18], further promoting the popularization and application of protein structure prediction technology. Considering the excellent performance of AlphaFold2 in protein monomer structure prediction, researchers began to explore its application in predicting complex structures. Since AlphaFold2 only supports the input of a single protein sequence, some researchers attempted to connect the amino acid sequences of different protein chains using poly-glycine linkers [96, 97], thus concatenating multiple chains into one sequence that can be input into AlphaFold2. Others modified the code of AlphaFold2 to add relative position gaps between concatenated chains, enabling AlphaFold2 to directly predict protein complex structures [98]. Additionally, AF2Complex stacked the MSAs of different chains in a diagonal form, filling the non-diagonal regions with gaps to create an MSA for the entire complex, which was then directly input into AlphaFold2 to achieve high-precision protein complex structure prediction [99]. Apart from modifying the input MSA, Ghani U et al. used ClusPro to separate and re-dock the complexes predicted by AlphaFold2 with poly-glycine linkers, and then replaced the default templates searched by AlphaFold2 with the docked structures, thus enhancing the performance of protein complex structure prediction based on AlphaFold2 [100].

Inspired by some work on protein complex structure prediction based on AlphaFold2, the DeepMind team launched AlphaFold-Multimer, an end-to-end structure prediction algorithm specifically retrained for protein complexes [56]. AlphaFold-Multimer is based on AlphaFold2, with adaptive modifications to the network framework and training process to cater to the characteristics of protein complexes. Although AlphaFold-Multimer has further improved the performance of protein complex structure prediction, it still faces several challenges that need to be addressed. Like AlphaFold2, AlphaFold-Multimer requires the input of MSAs for the complex, so the first challenge is how to make the input MSA of the complex more useful. In order to enable the network to more effectively extract evolutionary information between different chains, we need to pair the sequences in the MSAs of different chains. Since each protein chain has a large number of homologous sequences, effectively pairing the homologous sequences of different chains has been a long-standing challenge. In the past, some researchers attempted to use genome and species information for pairing [101, 102]. For instance, using genomic distance-based pairing strategies under the assumption that interacting genes are often located close to each other on the chromosome into operons, thereby inferring the pairing relationship through genomic distance. Others tried to pair interacting sequences within the same species based on phylogenetic trees [103], grouping sequences according to the species' phylogenetic tree and sorting them by sequence similarity for pairing. Although phylogenetic tree-based methods can enhance the extraction of co-evolutionary signals between proteins, sequence similarity-based pairing strategies are still not very reliable.

To improve the accuracy and effectiveness of MSA pairing, researchers began to explore pairing methods based on the protein language model (PLM). PLM is trained on large-scale protein sequence data and can capture deeper sequence relationships and evolutionary information. This method can more effectively identify homology relationships and co-evolutionary signals between different chains by processing the sequences of each chain and using the features generated by PLM for pairing. This PLM-based pairing method has shown great potential in predicting protein complex structures. For example, ESMPair uses attention scores generated by ESM-MSA-1b for pairing and combines it with AlphaFold-Multimer for structure prediction, significantly improving the accuracy of protein complex structure predictions [104]. However, some proteins lack homologous sequences, referred to as orphan proteins. The absence of homologous sequences makes it difficult to construct effective MSAs, resulting in poor quality of both monomeric structures predicted by AlphaFold2 and complex structures containing orphan proteins predicted by AlphaFold-Multimer. To overcome this challenge, researchers have introduced single-sequence PLMs, which do not rely on MSAs and can directly predict the structure of individual proteins [105–107]. In addition, by adding a linker between two proteins, researchers successfully extended protein monomer structure prediction models to the modeling of protein complexes [108, 109]. This approach not only fills the gap in predicting orphan proteins but also provides an effective solution for other proteins for which homologous sequences are difficult to find.

For interaction between two proteins, AlphaFold-Multimer's predictions are relatively accurate. However, as the number of interacting proteins increases, the accuracy of AlphaFold-Multimer in predicting the structure of the large protein complex they form will gradually decrease because the complexity of the interaction patterns increases exponentially, gradually exceeding the capacity of AlphaFold-Multimer to capture the interaction information. To address this challenge, Bryant P et al. proposed a Monte Carlo tree search-based method, MolPC, which assembles the entire large protein complex by progressively adding chains from predicted subcomponents [110]. Additionally, Shor et al. proposed a method, CombFold, which predicts large protein complexes through a combinatorial and hierarchical assembly algorithm. This algorithm uses pairwise interaction data between proteins and employs a hierarchical method to effectively combine multiple interacting subunits, generating high-quality complex prediction models [111].

Recently, the DeepMind team further developed AlphaFold2 into a new algorithm for predicting biomolecular interactions, named AlphaFold3 [112]. This algorithm expands the prediction scope from PPIs to include interactions among proteins, nucleic acids, small molecules, and ions. AlphaFold3 has made significant advancements in predicting PPIs by incorporating a diffusion model and an improved architecture, enhancing both accuracy and applicability. Unlike the attention mechanism used in AlphaFold2, AlphaFold3 emplys a diffusion model. This model predicts the native structure of proteins by progressively denoising from noisy data, allowing the model to handle complex interactions with greater flexibility and precision. The new architecture includes a Pairformer module, which focuses on pairwise representations and simplifies the processing of MSAs. This reduction in computational complexity increases prediction efficiency. Moreover, AlphaFold3 employs a cross-distillation method, integrating structures predicted by AlphaFold-Multimer for training, which enables better modeling of disordered regions. AlphaFold3 excels in predicting antigen–antibody complex structures, significantly outperforming AlphaFold-Multimer v2.3. However, to obtain high-quality antigen–antibody prediction models, AlphaFold3 needs to generate numerous predictions and rank them based on the protein–protein interface-predicted Template Modeling (ipTM) score. Currently, the DeepMind team provides a web server version of AlphaFold3, allowing users to access and use the algorithm online, with a submission limit of 20 prediction tasks per day.

7.4 CAPRI Experiments

Generally speaking, the reliability of protein complex structures predicted by various computational methods is often questioned by users. Protein complex structure prediction algorithms are trained and validated using a variety of datasets during their development and evolution. Most of the protein complex structure prediction algorithms that have been developed will evaluate their performance on common test sets. However, the data in these common test sets might have already been used to train these algorithms, and this evaluation lacks a unified process and standard, resulting in insufficient credibility. In order to objectively evaluate the existing protein structure prediction algorithms, the Critical Assessment of PRediction of Interactions (CAPRI) was launched by EMBL-EBI in 2001. This community-wide blind experiment aims to objectively assess the existing structure prediction algorithms to advance the development of protein structure prediction algorithms [113].

The CAPRI competition was designed after the Critical Assessment of protein Structure Prediction (CASP) competition [114]. Both are community-wide protein structure prediction competitions. The difference lies in their focus: CASP predicts the three-dimensional structure of single-chain proteins from amino acid sequences, while CAPRI predicts the three-dimensional structure of protein complexes from multi-chain amino acid sequences or unbound structures. The CAPRI challenge includes two tracks: prediction and scoring, and the two tracks are further divided into human groups and web server groups.

The first event is the prediction contest, in which all participants predict the structure of the protein complex based on the protein sequence and related supplementary information provided by the organizers, and submit a maximum of 100 prediction models within a set timeframe. Initially, the top 10 submitted models were used for prediction quality evaluation, but recent CAPRI rounds only evaluate the top 5 models. Both human and server predictors use the same information, but server predictors have less time for predictions. After the prediction track is over, the CAPRI organizers provide all submitted models to scoring track participants. The scorers use their own scoring functions and customized screening strategies to score and rank these models, and select 10 models for quality evaluation. The CAPRI challenge is a blind competition, ensuring that participants do not know the correct complex structures during the prediction and scoring stages. The blind setup ensures the reliability of the evaluation results and has earned CAPRI widespread recognition.

With advances in experimental techniques for resolving protein complex structures and computational methods for predicting protein monomer structures, protein complex structure prediction has entered a new stage. To integrate the existing protein monomer structure prediction and protein–protein docking algorithms and combine the expertise and advantages of CASP and CAPRI participants, the first CASP-CAPRI joint experiment was held in the summer of 2014. Since then, it has been held every two years, and the 6th CASP-CAPRI joint competition was held in the summer of 2024. Below we introduce the results of recent CASP-CAPRI competitions.

7.4.1 CASP13-CAPRI

The third joint CASP-CAPRI experiment was held in 2018, corresponding to the CASP13 competition and the 46th round of the CAPRI competition. A total of 20 targets were released in this round, including 14 homo-oligomers and 6 hetero-complexes [115]. About 30 groups participated in this CAPRI competition, including 9 web server predictors. Additionally, 18 teams of scorers participated in the scoring track. The top predictor in this round was the Venclovas team. They correctly predicted 13 of 20 targets in the top 5 predictions, achieving a success rate of 65%. Venclovas et al. obtained high-quality models for 6 targets, medium-quality models for 6 targets, and acceptable quality models for 1 target. The top server predictor, HDOCK, correctly predicted 10 of 20 targets. Among them, 5 targets contain high-quality models and the other 5 targets contain medium-quality models. The top scoring group, Fernandez-Recio team, generated correct models for 12 targets in the top 5 predictions, 1 less than the first place in the prediction track. They submitted high-quality models for 5 targets, medium-quality models for 6 targets, and acceptable quality models for the remaining 1 target.

7.4.2 CASP14-CAPRI

The fourth CASP-CAPRI experiment corresponded to the CASP14 competition and the 50th round of CAPRI competition. This round released 12 targets, including 6 dimers (5 homodimers and 1 heterodimer), 3 trimers (2 homotrimers and 1 heterotrimer), and 3 higher-order oligomers (2 homo-oligomers and 1 hetero-oligomer) [116]. A total of 25 groups competed in the prediction track, including 8 automated servers. Twenty scorers competed in the scoring track. Compared with the previous CASP-CAPRI experiments, the performance of this experiment has been significantly improved. The best-performing predictors achieved a success rate of 70–75%, but few models reached high accuracy standards. The scoring participants performed better than the predictors, with more groups correctly predicting 70–80% of the targets. The top predictors, Seok team, correctly predicted 9 of 12 targets in the top 5 predictions, achieving a success rate of 75%. They generated medium-quality models for 4 targets, and acceptable quality models for the remaining 5 targets. When considering the top 1 prediction, they also successfully generated correct models for 8 targets, with a success rate of 66.7%. However, only 2 of the 8 correct models met the medium quality standard. The Zou team, which ranked first in the scoring group, submitted correct models for 10 targets when considering the top 5 predictions, with a success rate of 83.3%.

7.4.3 CASP15-CAPRI

The fifth joint CASP-CAPRI challenge corresponded to the CASP15 competition and the 54th round of CAPRI. The number of targets released in this round far exceeds that of the previous rounds. A total of 37 targets were released in this round, including 14 homodimers, 3 homotrimers, 13 heterodimers, and 7 higher-order complexes [117]. This round had a higher proportion of hetero-complexes than in previous rounds and included 3 antibody–antigen complexes and 5 nanobody–antigen complexes, making the predictions more challenging than in previous rounds. Thanks to the release of AlphaFold and AlphaFold-Multimer, generating accurate protein monomers and complex structures has become easier. Researchers could modify their structure prediction algorithms based on AlphaFold and AlphaFold-Multimer or adjust the parameters of AlphaFold-Multimer to customize the generation of more candidate protein complex structures. Consequently, the number of participants in this round far exceeded that of the previous rounds. About 70 CASP and CAPRI predictor groups participated in this challenge, including over 20 automated server predictors. Additionally, 15 scoring groups participated in the scoring track. The quality of the models produced in this round is significantly better than the previous one, with high-quality models submitted for about 40% of the targets, compared to only 8% in the previous round. This improvement was largely due to the extensive use of AlphaFold and AlphaFold-Multimer. The predictors customized the input MSA and used dropout during inference and other methods to enhance the sampling of AlphaFold-Multimer, achieving performance that was better than the default version of AlphaFold-Multimer.

The CAPRI experiment has promoted the development of PPI prediction methods. CAPRI provides a platform for evaluating various PPI prediction methods through blind testing, which allows researchers to compare the accuracy and performance of different computational methods and identify the most effective techniques and continuously improve their tools. In addition, the CAPRI competition encourages algorithm developers to continuously improve and optimize their structure prediction tools. The results of the competition provide researchers with feedback, pointing out the strengths and weaknesses of their methods, prompting them to make improvements in future research. The CAPRI competition also promotes collaboration and communication among the global research community. Participants can share their insights and technologies to promote progress in the entire field. Finally, accurate PPI predictions are crucial to understand biological processes and disease mechanisms. Progress in the CAPRI competition will help improve the accuracy of biomedical research, thereby promoting drug development and disease treatment research.

7.5 Challenges and Future Directions

Although great progress has been made in predicting protein–protein complex structures, there are still many challenges. The mainstream end-to-end method predicts PPI interfaces based on co-evolutionary signals, so it is very important to effectively pair the MSAs of different protein chains. Despite some pairing strategies having been developed, there are still limitations in dealing with some specific cases. For example, some proteins lack sufficient homologous sequences, making it difficult to construct high-quality paired MSAs. In addition, a large portion of PPIs are weak and transient [118], with very weak interchain co-evolutionary signals, making it challenging to extract interchain co-evolutionary information from their MSAs. Antigen–antibody interactions are a typical example of interactions lacking co-evolutionary signals, and the success rate of AlphaFold-Multimer in predicting antigen–antibody complex structures is also low. Although AlphaFold3 has further improved the accuracy of predicting antigen–antibody complex structures, it still lags the performance of general protein–protein complex predictions. One potential solution to this problem is to train large-scale models on extensive sequence and structure data, enabling them to directly learn the mapping relationship from single sequences to PPI interfaces.

Intrinsically disordered regions (IDRs), which constitute a significant proportion of the proteome, also participate in numerous PPIs [119]. Unlike common folded proteins, IDRs are characterized by high flexibility and dynamicity, posing significant challenges for predicting their associated interactions. IDRs do not form stable three-dimensional structures under physiological conditions but exist in multiple conformations. This dynamic nature makes it difficult for general structure prediction methods to accurately capture their conformational states. Furthermore, IDRs can adopt various conformations by binding to different molecular partners, further increasing the complexity of structure prediction. Predicting a single static structure cannot adequately reflect the true behavior of IDRs. Moreover, most structure prediction algorithms rely on conserved evolutionary signals to predict interaction interfaces, but due to their high variability, IDRs usually lack these conserved signals. This limitation renders algorithms based on co-evolutionary information less effective in predicting interactions involving IDRs. Additionally, due to the high dynamicity of IDRs, obtaining high-resolution structural data using traditional experimental methods (such as X-ray crystallography and cryo-EM) is extremely challenging, resulting in a lack of IDR-related structures in the PDB database. The scarcity of experimental structures limits the data available for training and validating prediction models, thereby restricting model accuracy and generalizability. To address the insufficient experimental structural information of IDRs, combining physics-based molecular dynamics simulations with deep learning methods offers a promising approach. Molecular dynamics simulation can generate more IDR-related data for training and verifying deep learning models, thereby improving the generalization ability of deep learning models [120].

In the future, protein–protein complex structure prediction will need to integrate multiple data sources, including sequence data, experimental data, and literature data. Combining these different types of data can improve the accuracy and reliability of predictions. For example, combining low-resolution experimental data with high-resolution structure prediction data can enable more precise modeling of protein complex structures. Additionally, with advances in computational technology, developing more powerful computational models will become feasible. These models should not only predict static interaction interfaces but also capture protein flexibility and conformational changes. Integrating traditional physicochemical methods with deep learning, leveraging the efficiency of deep learning and the accuracy of physicochemical methods, is another research direction for developing new prediction models.

Acknowledgments

This work was supported by the National Natural Science Foundation of China (Grant Nos. 62072199 and 32161133002) and the startup grant of Huazhong University of Science and Technology.

Author Contributions

S.H. conceived and supervised the project. H.L. and Y.L. wrote the draft. S.H. and H.L. edited and finalized the manuscript.

References

1 Okazaki, R., Okazaki, T., Sakabe, K. et al. (1968). Mechanism of DNA chain growth. I. Possible discontinuity and unusual secondary structure of newly synthesized chains. *National Academy of Sciences of the United States of America* 59 (2): 598–605.

2 Luger, K., Mäder, A.W., Richmond, R.K. et al. (1997). Crystal structure of the nucleosome core particle at 2.8 Å resolution. *Nature* 389 (6648): 251–260.

3 Schimmel, P., Giegé, R., Moras, D., and Yokoyama, S. (1993). An operational RNA code for amino acids and possible relationship to genetic code. *National Academy of Sciences of the United States of America* 90 (19): 8763–8768.

4 Orphanides, G., Lagrange, T., and Reinberg, D. (1996). The general transcription factors of RNA polymerase II. *Genes & Development* 10 (21): 2657–2683.

5 Groves, J.T. and Kuriyan, J. (2010). Molecular mechanisms in signal transduction at the membrane. *Nature Structural & Molecular Biology* 17 (6): 659–665.

6 Sousa, J.S., Calisto, F., Langer, J.D. et al. (2018). Structural basis for energy transduction by respiratory alternative complex III. *Nature Communications* 9 (1): 1728.

7 Sompayrac, L.M. (2022). *How the Immune System Works*. Wiley.

8 Litman, Z.C., Wang, Y., Zhao, H., and Hartwig, J.F. (2018). Cooperative asymmetric reactions combining photocatalysis and enzymatic catalysis. *Nature* 560 (7718): 355–359.

9 Bader, G.D., Betel, D., and Hogue, C.W. (2003). BIND: the biomolecular interaction network database. *Nucleic Acids Research* 31 (1): 248–250.

10 Yamada, T. and Bork, P. (2009). Evolution of biomolecular networks—lessons from metabolic and protein interactions. *Nature Reviews. Molecular Cell Biology* 10 (11): 791–803.

11 Rao, V.S., Srinivas, K., Sujini, G.N., and Kumar, G.S. (2014). Protein-protein interaction detection: methods and analysis. *International Journal of Proteomics* 2014 (1): 147648.

12 Jones, S. and Thornton, J.M. (1996). Principles of protein-protein interactions. *National Academy of Sciences of the United States of America* 93 (1): 13–20.

13 Sisodia, S.S. and Price, D.L. (1995). Role of the β-amyloid protein in Alzheimer's disease. *The FASEB Journal.* 9 (5): 366–370.

14 Lücking, C.B. and Brice, A. (2000). Alpha-synuclein and Parkinson's disease. *Cellular and Molecular Life Sciences: CMLS* 57: 1894–1908.

15 Moll, U.M. and Petrenko, O. (2003). The MDM2-p53 interaction. *Molecular Cancer Research* 1 (14): 1001–1008.

16 Chène, P. (2003). Inhibiting the p53–MDM2 interaction: an important target for cancer therapy. *Nature Reviews Cancer* 3 (2): 102–109.

17 Wodak, S.J. and Janin, J. (1978). Computer analysis of protein-protein interaction. *Journal of Molecular Biology* 124 (2): 323–342.

18 Jumper, J., Evans, R., Pritzel, A. et al. (2021). Highly accurate protein structure prediction with AlphaFold. *Nature* 596 (7873): 583–589.

19 Varadi, M., Anyango, S., Deshpande, M. et al. (2022). AlphaFold protein structure database: massively expanding the structural coverage of protein-sequence space with high-accuracy models. *Nucleic Acids Research* 50 (D1): D439–D444.

20 Huang, S.Y. (2014). Search strategies and evaluation in protein–protein docking: principles, advances and challenges. *Drug Discovery Today* 19 (8): 1081–1096.

21 Halperin, I., Ma, B., Wolfson, H., and Nussinov, R. (2002). Principles of docking: an overview of search algorithms and a guide to scoring functions. *Proteins: Structure, Function, and Bioinformatics* 47 (4): 409–443.

22 Vajda, S., Hall, D.R., and Kozakov, D. (2013). Sampling and scoring: a marriage made in heaven. *Proteins: Structure, Function, and Bioinformatics* 81 (11): 1874–1884.

23 Kuntz, I.D., Blaney, J.M., Oatley, S.J. et al. (1982). A geometric approach to macromolecule-ligand interactions. *Journal of Molecular Biology* 161 (2): 269–288.

24 Duhovny, D., Nussinov, R., Wolfson, H.J. (2002). Efficient unbound docking of rigid molecules. *Proceedings of the 2nd Workshop on Algorithms in Bioinformatics (WABI) Rome*, Italy, Lecture Notes in Computer Science (LNCS) 2452, Springer Verlag, Berlin Heidelberg, 185–200.

25 Venkatraman, V., Yang, Y.D., Sael, L., and Kihara, D. (2009). Protein-protein docking using region-based 3D Zernike descriptors. *BMC Bioinformatics* 10: 1–21.

26 Esquivel-Rodríguez, J., Yang, Y.D., and Kihara, D. (2012). Multi-LZerD: multiple protein docking for asymmetric complexes. *Proteins: Structure, Function, and Bioinformatics* 80 (7): 1818–1833.

27 Lesk, V.I. and Sternberg, M.J. (2008). 3D-garden: a system for modelling protein–protein complexes based on conformational refinement of ensembles generated with the marching cubes algorithm. *Bioinformatics* 24 (9): 1137–1144.

28 Gardiner, E.J., Willett, P., and Artymiuk, P.J. (2001). Protein docking using a genetic algorithm. *Proteins: Structure, Function, and Bioinformatics* 44 (1): 44–56.

29 Katchalski-Katzir, E., Shariv, I., Eisenstein, M. et al. (1992). Molecular surface recognition: determination of geometric fit between proteins and their ligands by correlation techniques. *National Academy of Sciences of the United States of America* 89 (6): 2195–2199.

30 Chen, R. and Weng, Z. (2003). A novel shape complementarity scoring function for protein-protein docking. *Proteins: Structure, Function, and Bioinformatics* 51 (3): 397–408.

31 Yan, Y. and Huang, S.Y. (2019). Pushing the accuracy limit of shape complementarity for protein-protein docking. *BMC Bioinformatics* 20: 1–10.

32 Yan, Y., Zhang, D., Zhou, P. et al. (2017). HDOCK: a web server for protein–protein and protein–DNA/RNA docking based on a hybrid strategy. *Nucleic Acids Research* 45 (W1): W365–W373.

33 Yan, Y., Tao, H., He, J., and Huang, S.Y. (2020). The HDOCK server for integrated protein–protein docking. *Nature Protocols* 15 (5): 1829–1852.

34 Chen, R., Li, L., and Weng, Z. (2003). ZDOCK: an initial-stage protein-docking algorithm. *Proteins: Structure, Function, and Bioinformatics* 52 (1): 80–87.

35 Mintseris, J., Pierce, B., Wiehe, K. et al. (2007). Integrating statistical pair potentials into protein complex prediction. *Proteins: Structure, Function, and Bioinformatics* 69 (3): 511–520.

36 Pierce, B.G., Hourai, Y., and Weng, Z. (2011). Accelerating protein docking in ZDOCK using an advanced 3D convolution library. *PLoS One* 6 (9): e24657.

37 Roberts, V.A., Thompson, E.E., Pique, M.E. et al. (2013). DOT2: macromolecular docking with improved biophysical models. *Journal of Computational Chemistry* 34 (20): 1743–1758.

38 Vakser, I.A. (1997). Evaluation of GRAMM low-resolution docking methodology on the hemagglutinin-antibody complex. *Proteins: Structure, Function, and Bioinformatics* 29 (S1): 226–230.

39 Berchanski, A., Shapira, B., and Eisenstein, M. (2004). Hydrophobic complementarity in protein–protein docking. *Proteins: Structure, Function, and Bioinformatics* 56 (1): 130–142.

40 Gabb, H.A., Jackson, R.M., and Sternberg, M.J. (1997). Modelling protein docking using shape complementarity, electrostatics and biochemical information. *Journal of Molecular Biology* 272 (1): 106–120.

41 Kozakov, D., Brenke, R., Comeau, S.R., and Vajda, S. (2006). PIPER: an FFT-based protein docking program with pairwise potentials. *Proteins: Structure, Function, and Bioinformatics* 65 (2): 392–406.

42 Li, L., Guo, D., Huang, Y. et al. (2011). ASPDock: protein-protein docking algorithm using atomic solvation parameters model. *BMC Bioinformatics* 12: 1–9.

43 Zhang, C. and Lai, L. (2011). SDOCK: a global protein-protein docking program using stepwise force-field potentials. *Journal of Computational Chemistry* 32 (12): 2598–2612.

44 Kong, R., Wang, F., Zhang, J. et al. (2019). CoDockPP: a multistage approach for global and site-specific protein–protein docking. *Journal of Chemical Information and Modeling* 59 (8): 3556–3564.

45 De Vries, S.J., Van Dijk, M., and Bonvin, A.M. (2010). The HADDOCK web server for data-driven biomolecular docking. *Nature Protocols* 5 (5): 883–897.

46 Dominguez, C., Boelens, R., and Bonvin, A.M. (2003). HADDOCK: a protein–protein docking approach based on biochemical or biophysical information. *Journal of the American Chemical Society* 125 (7): 1731–1737.

47 Lyskov, S. and Gray, J.J. (2008). The RosettaDock server for local protein–protein docking. *Nucleic Acids Research* 36 (suppl_2): W233-8.

48 de Vries, S.J., Schindler, C.E., de Beauchêne, I.C., and Zacharias, M. (2015). A web interface for easy flexible protein-protein docking with attract. *Biophysical Journal* 108 (3): 462–465.

49 Torchala, M., Moal, I.H., Chaleil, R.A. et al. (2013). SwarmDock: a server for flexible protein–protein docking. *Bioinformatics* 29 (6): 807–809.

50 Rudden, L.S. and Degiacomi, M.T. (2019). Protein docking using a single representation for protein surface, electrostatics, and local dynamics. *Journal of Chemical Theory and Computation* 15 (9): 5135–5143.

51 Jiménez-García, B., Roel-Touris, J., Romero-Durana, M. et al. (2018). LightDock: a new multi-scale approach to protein–protein docking. *Bioinformatics* 34 (1): 49–55.

52 Jiménez-García, B., Roel-Touris, J., and Barradas-Bautista, D. (2023). The LightDock server: artificial intelligence-powered modeling of macromolecular interactions. *Nucleic Acids Research* 51 (W1): W298–W304.

53 Ganea, O.E., Huang, X., Bunne, C. et al. (2021). Independent se (3)-equivariant models for end-to-end rigid protein docking. arXiv preprint arXiv:2111.07786.

54 Ketata, M.A., Laue, C., Mammadov, R. et al. (2023). Diffdock-pp: rigid protein-protein docking with diffusion models. arXiv preprint arXiv:2304.03889.

55 Luo, Y., Li, S., Sun, Y. et al. (2023). xTrimoDock: rigid protein docking via cross-modal representation learning and spectral algorithm. *bioRxiv* 6: 2023-02.

56 Evans, R., O'Neill, M., Pritzel, A. et al. (2021). Protein complex prediction with AlphaFold-multimer. *bioRxiv* 4: 2021-10.

57 McPartlon, M. and Xu, J. (2023). Deep learning for flexible and site-specific protein docking and design. *bioRxiv* 2: 2023-04.

58 Chu, L.S., Ruffolo, J.A., Harmalkar, A., and Gray, J.J. (2024). Flexible protein–protein docking with a multitrack iterative transformer. *Protein Science* 33 (2): e4862.

59 Wallner, B. (2023). AFsample: improving multimer prediction with AlphaFold using massive sampling. *Bioinformatics* 39 (9): btad573.

60 Feng, T., Chen, F., Kang, Y. et al. (2017). HawkRank: a new scoring function for protein–protein docking based on weighted energy terms. *Journal of Cheminformatics* 9: 1–5.

61 Kynast, P., Derreumaux, P., and Strodel, B. (2016). Evaluation of the coarse-grained OPEP force field for protein-protein docking. *BMC Biophysics* 9: 1–7.

62 Gray, J.J., Moughon, S., Wang, C. et al. (2003). Protein–protein docking with simultaneous optimization of rigid-body displacement and side-chain conformations. *Journal of Molecular Biology* 331 (1): 281–299.

63 Cheng, T.M., Blundell, T.L., and Fernandez-Recio, J. (2007). pyDock: electrostatics and desolvation for effective scoring of rigid-body protein–protein docking. *Proteins: Structure, Function, and Bioinformatics* 68 (2): 503–515.

64 Gong, X., Wang, P., Yang, F. et al. (2010). Protein–protein docking with binding site patch prediction and network-based terms enhanced combinatorial scoring. *Proteins: Structure, Function, and Bioinformatics* 78 (15): 3150–3155.

65 Fernandez-Recio, J., Totrov, M., and Abagyan, R. (2004). Identification of protein–protein interaction sites from docking energy landscapes. *Journal of Molecular Biology* 335 (3): 843–865.

66 Huang, S.Y. and Zou, X. (2008). An iterative knowledge-based scoring function for protein–protein recognition. *Proteins: Structure, Function, and Bioinformatics* 72 (2): 557–579.

67 Andreani, J., Faure, G., and Guerois, R. (2013). InterEvScore: a novel coarse-grained interface scoring function using a multi-body statistical potential coupled to evolution. *Bioinformatics* 29 (14): 1742–1749.

68 Kowalsman, N. and Eisenstein, M. (2009). Combining interface core and whole interface descriptors in postscan processing of protein-protein docking models. *Proteins: Structure, Function, and Bioinformatics* 77 (2).

69 Lu, H., Lu, L., and Skolnick, J. (2003). Development of unified statistical potentials describing protein-protein interactions. *Biophysical Journal* 84 (3): 1895–1901.

70 Vreven, T., Hwang, H., and Weng, Z. (2011). Integrating atom-based and residue-based scoring functions for protein–protein docking. *Protein Science* 20 (9): 1576–1586.

71 Li, H., Yan, Y., Zhao, X., and Huang, S.Y. (2022). Inclusion of desolvation energy into protein–protein docking through atomic contact potentials. *Journal of Chemical Information and Modeling* 62 (3): 740–750.

72 Andrusier, N., Nussinov, R., and Wolfson, H.J. (2007). FireDock: fast interaction refinement in molecular docking. *Proteins: Structure, Function, and Bioinformatics* 69 (1): 139–159.

73 Wang, X., Terashi, G., Christoffer, C.W. et al. (2020). Protein docking model evaluation by 3D deep convolutional neural networks. *Bioinformatics* 36 (7): 2113–2118.

74 Guo, L., He, J., Lin, P. et al. (2022). TRScore: a 3D RepVGG-based scoring method for ranking protein docking models. *Bioinformatics* 38 (9): 2444–2451.

75 Renaud, N., Geng, C., Georgievska, S. et al. (2021). DeepRank: a deep learning framework for data mining 3D protein-protein interfaces. *Nature Communications* 12 (1): 7068.

76 Mohseni Behbahani, Y., Crouzet, S., Laine, E., and Carbone, A. (2022). Deep local analysis evaluates protein docking conformations with locally oriented cubes. *Bioinformatics* 38 (19): 4505–4512.

77 Wang, X., Flannery, S.T., and Kihara, D. (2021). Protein docking model evaluation by graph neural networks. *Frontiers in Molecular Biosciences* 25 (8): 647915.

78 Réau, M., Renaud, N., Xue, L.C., and Bonvin, A.M. (2023). DeepRank-GNN: a graph neural network framework to learn patterns in protein–protein interfaces. *Bioinformatics* 39 (1): btac759.

79 Xu, X. and Bonvin, A.M. (2024). DeepRank-GNN-esm: a graph neural network for scoring protein–protein models using protein language model. *Bioinformatics Advances* 4 (1): vbad191.

80 Kim, H.Y., Kim, S., Park, W.Y., and Kim, D. (2023). G-RANK: an equivariant graph neural network for the scoring of protein–protein docking models. *Bioinformatics Advances* 3 (1): vbad011.

81 Chen, X., Morehead, A., Liu, J., and Cheng, J. (2023). A gated graph transformer for protein complex structure quality assessment and its performance in CASP15. *Bioinformatics* 39 (Supplement_1): i308-17.

82 Shuvo, M.H., Karim, M., Roche, R., and Bhattacharya, D. (2023). PIQLE: protein–protein interface quality estimation by deep graph learning of multimeric interaction geometries. *Bioinformatics Advances* 3 (1): vbad070.

83 Olechnovič, K., Valančauskas, L., Dapkūnas, J., and Venclovas, Č. (2023). Prediction of protein assemblies by structure sampling followed by interface-focused scoring. *Proteins: Structure, Function, and Bioinformatics* 91 (12): 1724–1733.

84 Olechnovič, K. and Venclovas, Č. (2023). VoroIF-GNN: Voronoi tessellation-derived protein–protein interface assessment using a graph neural network. *Proteins: Structure, Function, and Bioinformatics* 91 (12): 1879–1888.

85 Camacho, C., Coulouris, G., Avagyan, V. et al. (2009). BLAST+: architecture and applications. *BMC Bioinformatics* 10: 1–9.

86 Pearson, W.R. and Lipman, D.J. (1988). Improved tools for biological sequence comparison. *National Academy of Sciences of the United States of America* 85 (8): 2444–2448.

87 Altschul, S.F., Madden, T.L., Schäffer, A.A. et al. (1997). Gapped BLAST and PSI-BLAST: a new generation of protein database search programs. *Nucleic Acids Research* 25 (17): 3389–3402.

88 Remmert, M., Biegert, A., Hauser, A., and Söding, J. (2012). HHblits: lightning-fast iterative protein sequence searching by HMM-HMM alignment. *Nature Methods* 9 (2): 173–175.

89 van Kempen, M., Kim, S.S., Tumescheit, C. et al. (2022). Foldseek: fast and accurate protein structure search. *bioRxiv* 9: 2022-02.

90 Zhang, C., Shine, M., Pyle, A.M., and Zhang, Y. (2022). US-align: universal structure alignments of proteins, nucleic acids, and macromolecular complexes. *Nature Methods* 19 (9): 1109–1115.

91 Baek, M., Park, T., Heo, L. et al. (2017). GalaxyHomomer: a web server for protein homo-oligomer structure prediction from a monomer sequence or structure. *Nucleic Acids Research* 45 (W1): W320–W324.

92 Quignot, C., Rey, J., Yu, J. et al. (2018). InterEvDock2: an expanded server for protein docking using evolutionary and biological information from homology models and multimeric inputs. *Nucleic Acids Research* 46 (W1): W408–W416.

93 Quignot, C., Postic, G., Bret, H. et al. (2021). InterEvDock3: a combined template-based and free docking server with increased performance through explicit modeling of complex homologs and integration of covariation-based contact maps. *Nucleic Acids Research* 49 (W1): W277–W284.

94 Kozakov, D., Hall, D.R., Xia, B. et al. (2017). The ClusPro web server for protein–protein docking. *Nature Protocols* 12 (2): 255–278.

95 Vangaveti, S., Vreven, T., Zhang, Y., and Weng, Z. (2020). Integrating ab initio and template-based algorithms for protein–protein complex structure prediction. *Bioinformatics* 36 (3): 751–757.

96 Ko, J. and Lee, J. (2021). Can AlphaFold2 predict protein-peptide complex structures accurately? *bioRxiv* 27: 2021-07.

97 Tsaban, T., Varga, J.K., Avraham, O. et al. (2022). Harnessing protein folding neural networks for peptide–protein docking. *Nature Communications* 13 (1): 176.

98 Bryant, P., Pozzati, G., and Elofsson, A. (2022). Improved prediction of protein-protein interactions using AlphaFold2. *Nature Communications* 13 (1): 1265.

99 Gao, M., Nakajima An, D., Parks, J.M., and Skolnick, J. (2022). AF2Complex predicts direct physical interactions in multimeric proteins with deep learning. *Nature Communications* 13 (1): 1744.

100 Ghani, U., Desta, I., Jindal, A. et al. (2021). Improved docking of protein models by a combination of AlphaFold2 and cluspro. *bioRxiv* 7: 2021-09.

101 Hopf, T.A., Schärfe, C.P., Rodrigues, J.P. et al. (2014). Sequence co-evolution gives 3D contacts and structures of protein complexes. *eLife* 25 (3): e03430.

102 Ovchinnikov, S., Kamisetty, H., and Baker, D. (2014). Robust and accurate prediction of residue–residue interactions across protein interfaces using evolutionary information. *eLife* 1 (3): e02030.

103 Zeng, H., Wang, S., Zhou, T. et al. (2018). ComplexContact: a web server for inter-protein contact prediction using deep learning. *Nucleic Acids Research* 46 (W1): W432–W437.

104 Chen, B., Xie, Z., Qiu, J. et al. (2023). Improved the heterodimer protein complex prediction with protein language models. *Briefings in Bioinformatics* 24 (4): bbad221.

105 Chowdhury, R., Bouatta, N., Biswas, S. et al. (2022). Single-sequence protein structure prediction using a language model and deep learning. *Nature Biotechnology* 40 (11): 1617–1623.

106 Wu, R., Ding, F., Wang, R. et al. (2022). High-resolution de novo structure prediction from primary sequence. *bioRxiv* 22: 2022-07.

107 Jing, X., Wu, F., Luo, X., and Xu, J. (2024). Single-sequence protein structure prediction by integrating protein language models. *National Academy of Sciences of the United States of America* 121 (13): e2308788121.

108 Wang, W., Peng, Z., and Yang, J. (2022). Single-sequence protein structure prediction using supervised transformer protein language models. *Nature Computational Science* 2 (12): 804–814.

109 Lin, Z., Akin, H., Rao, R. et al. (2023). Evolutionary-scale prediction of atomic-level protein structure with a language model. *Science* 379 (6637): 1123–1130.

110 Bryant, P., Pozzati, G., Zhu, W. et al. (2022). Predicting the structure of large protein complexes using AlphaFold and Monte Carlo tree search. *Nature Communications* 13 (1): 6028.

111 Shor, B. and Schneidman-Duhovny, D. (2024). CombFold: predicting structures of large protein assemblies using a combinatorial assembly algorithm and AlphaFold2. *Nature Methods* 21 (3): 477–487.

112 Abramson, J., Adler, J., Dunger, J. et al. (2024). Accurate structure prediction of biomolecular interactions with AlphaFold 3. *Nature* 8: 1–3.

113 Vajda, S., Vakser, I.A., Sternberg, M.J., and Janin, J. (2002, 2002). Modeling of protein interactions in genomes. *Proteins* 47 (4): 444–446.

114 Moult, J., Pedersen, J.T., Judson, R., and Fidelis, K. (1995). A large-scale experiment to assess protein structure prediction methods. *Proteins: Structure, Function, and Bioinformatics* 23 (3): ii–v.

115 Lensink, M.F., Brysbaert, G., Nadzirin, N. et al. (2019). Blind prediction of homo-and hetero-protein complexes: the CASP13-CAPRI experiment. *Proteins: Structure, Function, and Bioinformatics* 87 (12): 1200–1221.

116 Lensink, M.F., Brysbaert, G., Mauri, T. et al. (2021). Prediction of protein assemblies, the next frontier: the CASP14-CAPRI experiment. *Proteins: Structure, Function, and Bioinformatics* 89 (12): 1800–1823.

117 Lensink, M.F., Brysbaert, G., Raouraoua, N. et al. (2023). Impact of AlphaFold on structure prediction of protein complexes: the CASP15-CAPRI experiment. *Proteins: Structure, Function, and Bioinformatics* 91 (12): 1658–1683.

118 Chichili, V.P., Kumar, V., and Sivaraman, J. (2013). A method to trap transient and weak interacting protein complexes for structural studies. *Intrinsically Disordered Proteins* 1 (1): e25464.

119 Teilum, K., Olsen, J.G., and Kragelund, B.B. (2021). On the specificity of protein–protein interactions in the context of disorder. *The Biochemical Journal* 478 (11): 2035–2050.

120 Lotthammer, J.M., Ginell, G.M., Griffith, D. et al. (2024). Direct prediction of intrinsically disordered protein conformational properties from sequence. *Biophysical Journal* 123 (3): 43a.

8

Foldamers as Inhibitors of Aberrant Protein–Protein Interactions

Nicholas H. Stillman, Ryan A. Dohoney, Charles Z. Baysah, and Sunil Kumar

University of Denver, Knoebel Institute of Healthy Aging, Department of Chemistry and Biochemistry, 2155 E Wesley Ave., Denver, CO, 80210, USA

8.1 Introduction

Protein–protein interactions (PPIs) are phenomena involving contact between two or more protein units. These interactions are primarily driven by the electrostatic and hydrophobic nature of the interfaces, which can extend over an area between 1500 and 3000 Å^2. These surfaces typically contain charged and hydrophobic residues that facilitate their interactions [1, 2]. PPIs are critical elements for numerous molecular mechanisms, including the formation of protein complexes and networks. PPIs play a pivotal role in regulating key physiological processes such as signal transduction, cell cycle regulation, metabolism, and immune response [3–6]. In signal transduction, PPIs are crucial for transmitting signals from cell surface receptors to intracellular targets, thereby eliciting cellular responses [7]. These interactions facilitate the formation of signaling complexes, ensuring the specificity and regulation of pathways such as the mitogen-activated protein kinase (MAPK) and phosphatidylinositol 3-kinase/protein kinase B (PI3K/Akt) pathways [8–10]. In cell cycle regulation, PPIs govern the activation and inhibition of cyclins and cyclin-dependent kinases (CDKs), thereby ensuring appropriate cell division and growth [11–14]. During metabolism, PPIs facilitate the organization of enzyme complexes that regulate metabolic pathways, thus optimizing substrate channeling and pathway efficiency [15, 16]. In the immune response process, PPIs are implicated in antigen recognition and the formation of immune complexes, thus orchestrating the activation and coordination of immune cells to mount defense against pathogens [17–19].

Despite their critical roles in fundamental biological processes, aberrant protein–protein interactions (aPPIs) have been identified as key factors in the onset of several diseases, including cancer, diabetes, and neurodegenerative diseases such

Targeting Protein–Protein Interactions for Drug Discovery, First Edition. Edited by Jian Zhang.

as Alzheimer's disease (AD), Parkinson's disease (PD), and Amyotrophic Lateral Sclerosis (ALS) [2, 3, 20–27]. The vital role of transcription factor p53 as a tumor suppressor, among its many other roles in cellular processes, is impeded by mutations in at least 50% of human cancers [28–30]. The more promiscuous DNA binding and non-physiological PPIs of mutant p53 are correlated with an increase in cancer aggressiveness and progression [29]. The aggregation of islet amyloid polypeptide (IAPP), which regulates and maintains glucose homeostasis, significantly contributes to β-cell dysfunction and death, a key feature of type II diabetes [31]. In other amyloid diseases such as AD, PD, and ALS, the respective aggregation-prone proteins, amyloid beta 42 ($A\beta_{42}$), α-synuclein (αS), and TDP-43 accumulate as amyloid plaques, which are the key pathological hallmarks of neurodegeneration [2, 20–27, 32–34]. In light of this, the modulation of aPPIs has been identified as a strategic approach to disease mitigation, as highlighted by the extensive research in this field in recent years. Small-molecule drug discovery for the modulation of aPPIs has seen immense progress over the years, with several candidates reaching clinical trials [3, 35–39]. However, despite the headway, several challenges from design and synthesis to efficient targeting still exist [5, 36]. Here, we highlight recent advancements made in the field of aPPI modulation using a class of protein-mimetics called foldamers.

8.2 The Evolution of Hamilton's Oligopyridylamides

aPPIs are valuable targets for therapeutics as they are the root of an array of pathological diseases, including amyloid diseases, viral infection, neurodegeneration, and cancer [3, 20, 22–25, 27, 32, 36, 40]. Disrupting these interactions is challenging due to the large surface area and unique topographical distribution of hydrophobic and hydrophilic residues at the interface between the interacting proteins [22, 41]. Sixty percent of PPIs in the protein databank involve an α-helix, and furthermore, 60% of these bind one face of a helical peptide [3]. In 2001, Hamilton et al. first introduced a class of molecules known as α-helical mimetics with the 3,2′,2″-trisubstituted terphenyl scaffold [42]. These molecules are designed so their projected functional groups mimic and interact with the i, $i+3/4$, and $i+7$ amino acid residues (Figure 8.1) on one face of an α-helical segment of the target protein [41–43]. The design of these molecules was confirmed through X-ray crystallography experiments, which showed the spacing of the projected functional groups to be between 5.1 and 8.83 Å, which reasonably coincides with the i, $i+3/4$, and $i+7$ β carbons of an α-helix [42]. To demonstrate the ability of the terphenyl protein mimetics to interrupt PPIs, Hamilton's group synthesized a mimic of the known calmodulin binding face of smMLCK [41, 42]. Through competition experiments, they showed the terphenyl mimetic could displace C20W, a protein known to bind in the same region as smMLCK [42]. In another example, Ernst et al. developed another terphenyl protein mimetic capable of impeding the gp41 fusion-active conformation consisting of six helical proteins, consequently preventing cellular entry of the HIV-1 virus [41, 43].

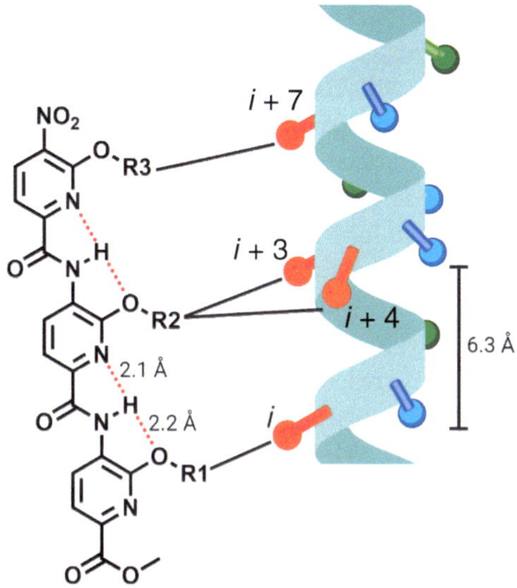

Figure 8.1 Schematic of an oligopyridylamide interacting with the *i*, *i* + 3/4, and *i* + 7 amino acid residues on one face of an α-helical peptide. Red dashed lines indicate intramolecular bifurcated hydrogen bonding. Created in BioRender. Kumar, S. (2024) BioRender.com/f43d880. Source: Nicholas H. Stillman.

More recently, Hamilton's group proposed a new scaffold of protein mimetics known as the oligopyridylamide (OP) foldamer [44]. Foldamers are defined as molecules whose conformations are constrained by internal, noncovalent interactions including hydrogen bonding, hydrophobic interactions, and π–π stacking [45–47]. This novel scaffold maintains the key spacing of the projected functional groups (Figure 8.1) with the added advantage of structural rigidity due to inherent bifurcated intramolecular hydrogen bonding within the backbone [44]. X-ray crystallography of a tri-substituted isopropyl OP confirmed that the molecule was planar with all three isopropyl groups on one side (Figure 8.1), and the distances from the amide proton to the pyridyl nitrogen and the oxygen atom of the sidechains were 2.1 and 2.2 Å, respectively [44]. To verify the potential therapeutic abilities of the OP scaffold, Ernst et al. synthesized a small library of OPs to inhibit Bak BH3/Bcl-xL complex formation [44]. This protein complex formation prevents apoptotic pathways in the cell and imbalances are linked to cancer progression and metastasis [44, 48, 49]. By implementing a fluorescence polarization (FP) assay, they found an OP capable of inhibiting the interaction between Bak BH3 and Bcl-xL with a K_i value of 2.3 μM [44].

8.3 Limitations of a Tedious Synthetic Route

Until recently, the libraries of OPs synthesized have been relatively small primarily due to the tedious synthetic pathways, numerous chromatography steps required, and low total % yield for one tripyridyl [44, 47, 50–58]. First, 2,6-dichloro-3-nitropyridine is reacted with a primary alcohol and sodium hydride to substitute the chlorine in the second position on the ring [51, 58]. The remaining chlorine is then substituted with a terminal alkene in a reaction involving tri-*n*-butyl(vinyl)tin and Tetra(triphenylphosphine)palladium [51, 58]. The resulting alkene is then oxidized to a carboxylic acid with potassium permanganate [51, 58]. Some amount of each functionalized picolinic acid is then converted to a methyl ester with methyl iodide and potassium carbonate in order to be in the first (bottom) position on the final tripyridyl [51, 58]. The coupling reaction to form the dipyridyl and tripyridyl generally involves a reduction with palladium on activated carbon and hydrogen gas, followed by the amide coupling with 2-chloro-1-methylpyridinium iodide and triethylamine [51, 58]. Altogether, one tripyridyl required 14 synthesis steps and 11 chromatography purification steps [51, 58]. These reactions are repeated to generate a library of monopyridyls, which are then repeatedly reduced and coupled together to generate a small tripyridyl library [50, 51, 58, 59]. Libraries of OPs up to this point were typically limited to 10 different functional groups consisting of a positive charge, a negative charge, a couple of aromatic hydrocarbons, and several branched and unbranched aliphatic chains [50, 51, 53, 56]. The total % yield of this outdated synthesis ranged from 0.78% to 3.61% [58].

8.4 OPs as Antagonists of Neurodegeneration

Since the debut of the OP scaffold protein mimetic, derivatives of the molecule have been shown to disrupt aberrant protein–protein and nucleic acid–protein interactions related to several diseases, including AD, HIV, diabetes, and cancer [51, 52, 54, 60]. AD is the most common form of dementia and the sixth leading cause of death in the United States and is estimated to affect 15 million Americans by the year 2060 [33, 61–63]. The neuronal dysfunction and cognitive decline associated with AD have been linked to the accumulation of oligomers and β-rich proteinaceous fibrils of $A\beta_{42}$, the transmembrane cleavage product of amyloid precursor protein [51, 64]. There is currently no cure for AD, and the most effective modulators of $A\beta_{42}$ aggregation lack cellular stability and exhibit a reduced ability to cross the blood–brain barrier [65]. In 2017, Kumar and Hamilton showed through thioflavin T (ThT) kinetics experiments (Figure 8.2a–c) that a tripyridyl OP (ADH-41) with two sequential amines and a benzyl functional group could selectively inhibit the self-assembly of $A\beta_{42}$ *in vitro* [51]. This same molecule was proven to be relatively ineffective toward the aggregation of IAPP, a process associated with type II diabetes, which contains two sequentially similar segments to that of $A\beta_{42}$ (Figure 8.2c) [51]. The inhibition of $A\beta_{42}$ aggregation by this novel OP protein mimetic was confirmed through dynamic light scattering, atomic force microscopy, transmission electron microscopy, and dot blot experiments with A11 and 6E10 antibodies (Figure 8.2d–j) [51]. Furthermore, circular dichroism (CD) experiments revealed that ADH-41 induced and stabilized an α-helical secondary structure in $A\beta_{42}$ up to 24 hours under aggregation conditions, compared to the protein, which exhibited a β-sheet structure within 12 hours (Figure 8.2k,l) [51]. Analysis of ^{1}H-^{15}N HSQC NMR data indicated that ADH-41 induced an α-helical

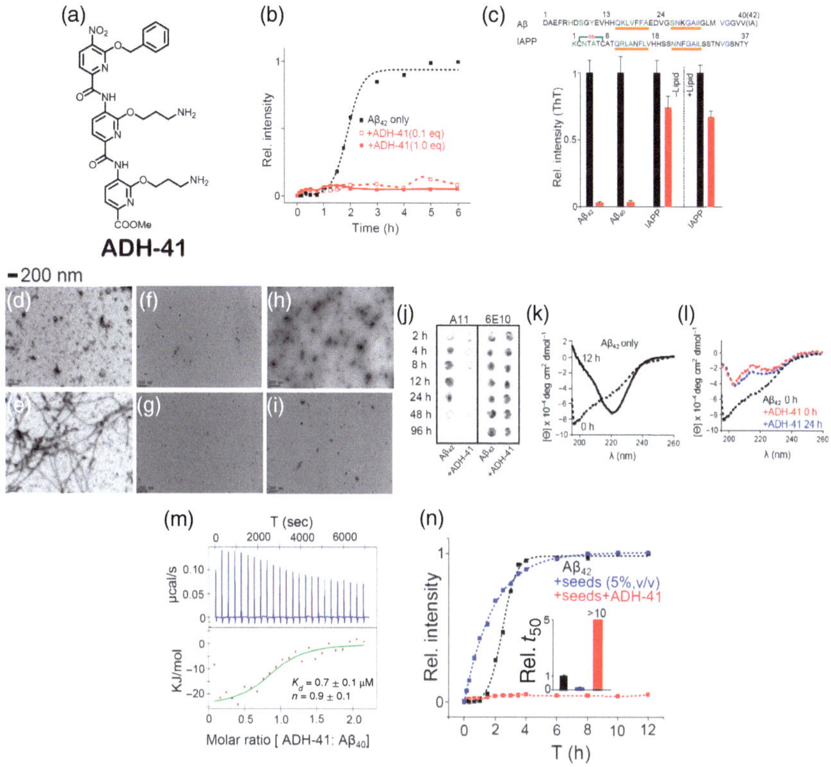

Figure 8.2 (a) Chemical structure of ADH-41; (b) Representative ThT kinetic traces of 20 μM Aβ42 aggregation in the absence and presence of ADH-41 at the indicated equivalents; (c) Sequence comparison of Aβ40, Aβ42, and IAPP. Identical and chemically similar residues are shown in blue and green, respectively. Sequences of greatest similarity are underlined in orange. (Below) Relative change in the fluorescence intensity of the aggregation of Aβ40, Aβ42, and IAPP (with and without lipids) in the absence (black) and presence (red) of ADH-41; TEM images of 5 μM Aβ42 at 4 hours (d) and 24 hours (e) of aggregation in the absence of ADH-41; TEM images of Aβ42 in the presence of ADH-41 at 4 hours (f) and 24 hours (g); TEM images of Aβ42 after 24 hours in the presence of ADH-41 at a sub-stoichiometric ratio of (h) 1 : 0.1 and (i) 1 : 0.5 (Aβ42:ADH-41); (j) Kinetics of 5 μM Aβ42 aggregation probed using oligomer-specific A11 and sequence-specific 6E10 antibodies in the absence and presence of ADH-41 at an equimolar ratio over 96 hours; (k) Overlayed CD spectra of 15 μM Aβ42 at 0 and 12 hours; (l) Overlayed CD spectra of 15 μM Aβ42 in the absence (black) and presence of ADH-41 at an equimolar ratio at 0 (red) and 24 hours (blue); (m) ITC thermogram for the titration of a solution of Aβ40 with ADH-41. The upper panel represents the heat burst curves and the lower panel is the corrected injection heat associated with each injection plotted as a function of the molar ratio; (n) ThT curve of the seed-catalyzed aggregation of 20 μM Aβ42 in the absence and presence of ADH-41 at an equimolar ratio. Source: [51]/with permission of American Chemical Society.

conformation in 2 segments spanning from residues 15-24 and 28-35 [51]. The authors concluded that the benzyl functional group likely interacts with the Phe residue at positions 19 and/or 20, potentially disrupting the π–π interactions known to lead to Aβ_{42} fibril formation [51]. Isothermal titration calorimetry (ITC) further determined that ADH-41 binds to Aβ_{40} at a 1 : 1 stoichiometry, with a K_d of $0.7 \pm 0.1\,\mu M$ affinity (Figure 8.2m) [51]. Lastly, since it is known that Aβ_{42} aggregation spreads in a prion-like manner, ThT and TEM experiments showed that ADH-41 was able to prevent the seed-catalyzed aggregation of Aβ_{42} (Figure 8.2n) [51]. Altogether, the data presented indicates the extraordinary potential of OP ADH-41 as a potential therapeutic for AD.

8.5 OPs Inhibit HIV Infection

The OP protein mimetic was also shown to inhibit the interactions between the TAR RNA and TAT protein of the human immunodeficiency virus (HIV) [60]. The complex formation of these two biomolecules significantly accelerates the transcription of the HIV virus through the recruitment of the host cell's transcription proteins [60, 66–70]. In this work, Maity et al. conducted fluorescent intercalator displacement (FID) assays with a library of OPs (Figure 8.3a,b), where the largest decrease in fluorescence corresponds to the highest affinity [60]. ADH-19, ADH-41, and ADH-40 decreased the fluorescence of the reporter dye, thiazole orange (TO), from 100% to 7%, 10%, and 14%, respectively (Figure 8.3b) [60]. ADH-19, with primary amines for all three functional groups, had the highest affinity for the TAR RNA, whereas ADH-41 and ADH-40 have benzyl and butyl groups in the third position, respectively, and retain functionality (Figure 8.3a) [60]. Interestingly, when the same functional groups as ADH-41 were presented in a different arrangement (benzyl group in the second position, ADH-37), the binding affinity was lost (Figure 8.3a) [60]. Spermidine, a molecule containing three primary amines, showed no effect on fluorescence intensity in the same experiment, indicating the activity is due to the spatial arrangement of the functional groups on the OP scaffold and not the presence of three positive charges (Figure 8.3a) [60]. The authors also administered competitive FRET experiments using a 16 amino acid peptide derived from TAT, with fluorescein and TAMRA fluorescent tags on its N- and C-termini, respectively [60]. When unbound, the peptide folds so that the fluorophores are in close proximity, quenching the fluorescence [60]. When the peptide was bound to the TAR RNA, the peptide was unfolded, and a fluorescence increase was observed [60]. Using a variety of doses of the OPs in the FRET experiments, an IC_{50} of $3.9 \pm 1.0\,\mu M$ was calculated for ADH-19 (Figure 8.3b) [60]. IC_{50} values for ADH-41 and ADH-40 followed the same trend as the previous FID experiments (Figure 8.3b) [60]. The authors then used ^1H-^1H 2D TOCSY NMR to characterize the binding of TAR RNA and ADH-19 (Figure 8.3c) [60]. Changes in the chemical shifts and peak volumes of the H5/H6 pyrimidine protons of the RNA indicated ADH-19 was binding to the lower bulge region of the TAR RNA hairpin [60]. To determine the inhibitory effects of ADH-19

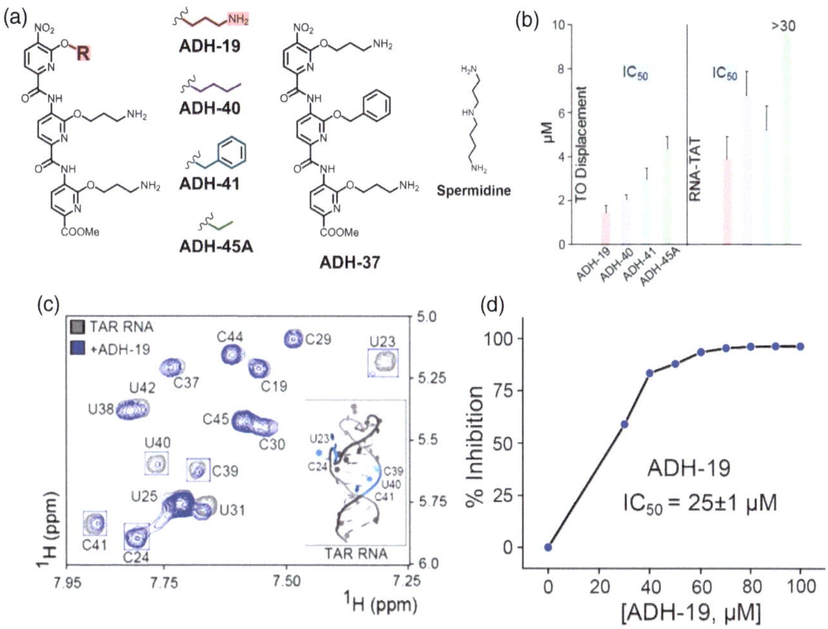

Figure 8.3 (a) Chemical structures of ADH-19, ADH-40, ADH-41, ADH-45A, ADH-37, and Spermidine; (b) A comparison of the IC50 values of the oligopyridylamides against TAR RNA using thiazole orange (TO) displacement assay and the RNA-TAT FRET experiments; (c) Overlayed 2D TOCSY NMR spectra of 500 μM TAR RNA in the absence (black) and presence (blue) of 0.6 eq of ADH-19; (d) A plot of the % inhibition of the TZM-bl cells toward HIV-1HXB-2 pseudovirus infection at various doses of ADH-19. Source: Adapted from Maity et al. [60]. All rights reserved, including rights for text and data mining and training of artificial intelligence technologies or similar technologies.

on HIV infection, the HIV-1 pseudovirus HIV-1HXB-2 was pretreated with ADH-19 before treating TZM-bl cells in a single-cell infection assay (Figure 8.3d) [60]. The IC_{50} was determined to be 25 μM in the dose-dependent study (Figure 8.3d) [60]. Furthermore, ADH-19 was not inherently toxic to the TMB-bl under exact conditions, with a determined cytotoxic concentration of 172 ± 10 μM [60]. Overall, the authors have clearly shown that OPs are also capable of disrupting nucleic acid–protein interactions that could prevent HIV-1 infection.

8.6 OPs Targeting Type II Diabetes

IAPP is a small, unstructured hormonal peptide that is co-secreted with insulin by the β-cells of the pancreas and plays a role in gastric emptying, glucose homeostasis, adiposity control, and other metabolic functions [54, 71, 72]. IAPP takes on an α-helical conformation upon binding to the cell membrane with the positive charges presented on one face of the helix [54, 72]. There is evidence to support

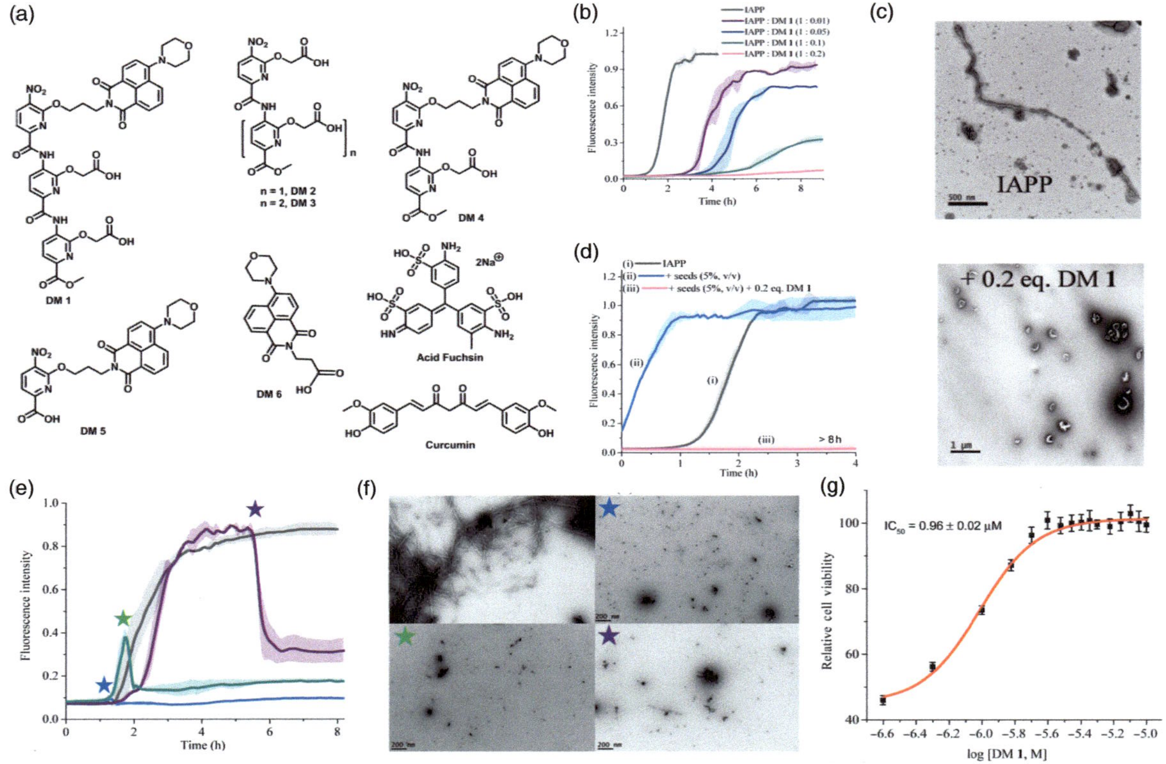

Figure 8.4 (a) Chemical structures of DM 1 and its derivatives; (b) ThT Kinetic traces of lipid-catalyzed IAPP aggregation in the absence and presence of DM 1 at the indicated sub-stoichiometric ratios; (c) TEM images of 15 μM IAPP after 2 hours in the absence of DM 1 (top) and after 8 hours in the presence of 3 μM DM 1 (bottom) in the presence of lipids; (d) ThT kinetic profile of (i) 25 μM IAPP, (ii) in the presence of 5% v/v aggregated IAPP seeds, and (iii) in the presence of 5% v/v IAPP seeds and 0.2 M equivalents of DM 1; (e) ThT kinetic profile of 25 μM IAPP in the absence (grey) and presence of 0.2 equivalents of DM 1 added at the time points indicated by stars; (f) TEM images of the solutions from (e) after 8 hours; (g) Dose-dependent effect of DM 1 on 10 μM IAPP-induced toxicity in RIN-m cells. Source: [54]/with permission of The Royal Society of Chemistry.

that membrane binding induces the formation of oligomeric species that cause pancreatic β-cell loss and result in type II diabetes [54, 71, 73]. Based on the positively charged face of the membrane-bound protein and evidence that aromatic dyes inhibit IAPP aggregation, in 2020 Maity et al. synthesized an OP (DM 1) with carboxylate and naphthalimide-based hydrophobic dye functionalities (Figure 8.4a) [54]. DM 1 and its derivatives were screened in a ThT kinetic assay of 15 μM IAPP in 3 : 7 DOPG:DOPC (750 μM, $d = 100$ nm) as a plasma membrane model [54]. DM 1 increased the t_{50} from 1.8 to 3.8 hours at a 0.1 M ratio and prevented IAPP fibril formation completely over 8 hours when present at 0.2 equivalents (Figure 8.4a,b) [54]. Derivatives of DM 1 which contained only one or two of its functionalities did not affect the aggregation of IAPP, except DM 4, containing the naphthalimide dye and one carboxylate, which showed a moderate effect on IAPP aggregation (Figure 8.4a) [54]. Additionally, DM 1 was shown to inhibit IAPP amyloidogenesis in 1 : 1 DOPG:DOPC containing 30% cholesterol as a model for mammalian plasma membranes [54]. TEM imaging showed lipid membrane-bound mature IAPP fibrils after two hours of incubation in the absence of DM 1 and no mature fibrils in the presence of DM 1 (Figure 8.4c) [54]. In agreeance with the ThT and TEM results, circular dichroism revealed that IAPP adopts a β-sheet-rich structure after just 30 minutes of incubation in the absence of DM 1; however, the protein remains α-helical after eight hours in the presence of DM 1 [54]. ^{1}H-^{15}N HSQC NMR analysis of ^{15}N-IAPP with equimolar DM 1 indicated binding at the N-terminal residues of the protein, the region of the protein that has been suggested to take on a helical conformation upon membrane binding [54]. Furthermore, the authors demonstrated that DM 1 was capable of inhibiting the aggregation of IAPP, even when catalyzed by the presence of pre-formed fibrils (pFFs) (Figure 8.4d) [54]. Even more impressive, DM 1 was able to disaggregate mature fibrils of IAPP as the ThT fluorescence decreased significantly upon the addition of DM 1 at different time points in the aggregation of the protein (Figure 8.4e) [54]. TEM images of these solutions revealed a complete lack of mature fibrils (Figure 8.4e,f) [54]. Finally, cytotoxicity experiments in rat insulinoma cells (RINm) revealed that DM 1 completely reversed the cytotoxicity associated with IAPP pFFs [54]. Cell viability was reduced to approximately 44% after incubation with 10 μM IAPP; however, adding increasing concentrations of DM 1 rescued the toxicity at sub-stoichiometric levels, and the IC_{50} was determined to be 0.96 ± 0.02 mM (Figure 8.4g) [54]. In this work, the authors designed an OP that not only inhibits the aggregation of IAPP at sub-stoichiometric concentrations but further demonstrates its ability to disaggregate mature pFFs; this is a huge milestone in this field!

8.7 OPs Targeting and Reactivating Mutant Protein in Cancer

In another example of the exceptional properties of OP protein mimetics, Palanikumar et al. discovered a tripyridylamide, which inhibits the aggregation of and restores the function of mutant p53, a protein that when mutated forms inactive protein aggregates in more than half of all human cancers [52, 74–77]. Under normal conditions, p53 exists at a low concentration in the cell as a result of targeted protein degradation by MDM2, an E3 ubiquitin ligase [52, 76, 78–82]. Upon cellular stress including DNA damage, oxidative stress, etc., p53 is accumulated and activated to initiate DNA repair, apoptosis, and autophagy, among several other pathways to prevent cancer cell proliferation [52, 76, 78, 81]. A majority of the cancer-associated mutations in p53 occur within the DNA-binding domain (DBD), which spans from residues 102-292 [52, 76, 78, 81]. In this work, the authors focused on one of the most common p53 mutations, R248W (pR248W) [52, 83, 84]. By performing a ThT kinetics assay using a small library of 10 OPs, the tripyridylamide presenting two primary amines and a butyl group, ADH-6, completely inhibited the aggregation of the R248W mutated segment of the p53 DBD compared to the inactive OP, ADH-1 (Figure 8.5a,b) [52]. When added at the midpoint of the ThT aggregation assay, ADH-6 significantly reduced the ThT fluorescence and disaggregated the fibrils of the mutant peptide, as supported by the TEM images (Figure 8.5c) [52]. MIA PaCa-2 cells expressing pR248W were co-stained with thioflavin S (specific to intracellular aggregates) and PAb 240 anti-p53 antibody (specific to unfolded p53) and treated with 10 μM ADH-6. Approximately 24% of cells were positive for p53 aggregates compared to the control [52]. Using a FITC-labeled analog of ADH-6 (ADH-6_{FITC}) and confocal fluorescence microscopy, researchers were able to see a strong co-localization of FITC and PAb 240 that dissipated as the PAb 240 signal diminished with the disaggregation of the protein (Figure 8.5d,e) [52]. Furthermore, cellular thermal shift assays revealed an increase in the melting point of the MIA PaCa-2 cells, indicating a stabilization of the mutant p53 by ADH-6 [52]. The viability of MCF-7 (expressing WT p53) and p53 null Saos-2 human bone cancer cell lines was not affected by ADH-6 treatment; however, after transfection with mutant p53, ADH-6 induced significant toxicity in the Saos-2 cells [52]. Transcriptome analysis (Figure 8.5f) revealed that the observed cellular toxicity was due to the upregulation of p53 target genes essential to cell cycle arrest and apoptosis, e.g. *CDKN1A*, *FOS*, *EGR1*, etc. [52]. Finally, ADH-6 was administered through intraperitoneal injection of mice bearing MIA PaCa-2 (mutant R248W p53) xenografts on one flank and MCF-7 (WT p53) xenografts on the other flank [52]. Treatment with ADH-6 had no significant effect on the MCF-7 xenografts; however, ADH-6 treatment reduced the tumor mass of the MIA PaCa-2 xenografts to ~3% of that of the controls (Figure 8.5g–i) [52]. ADH-6 was found to reduce the volume of the MIA PaCa-2 tumors to a greater extent than ReACp53, a peptide recently found to reactivate mutant p53 in castration-resistant prostate cancer (Figure 8.5g–i) [52, 85]. Altogether, these researchers have demonstrated excellent diversity in the application of OPs as therapeutics. In the collection of work presented here, the protein mimicry of OPs has been used to abrogate the aggregation of a neuronal protein, prevent HIV-1 viral infection, prevent and reverse aggregation of a metabolic peptide, and reactivate a mutant protein in more than half of human cancers [51, 52, 54, 60].

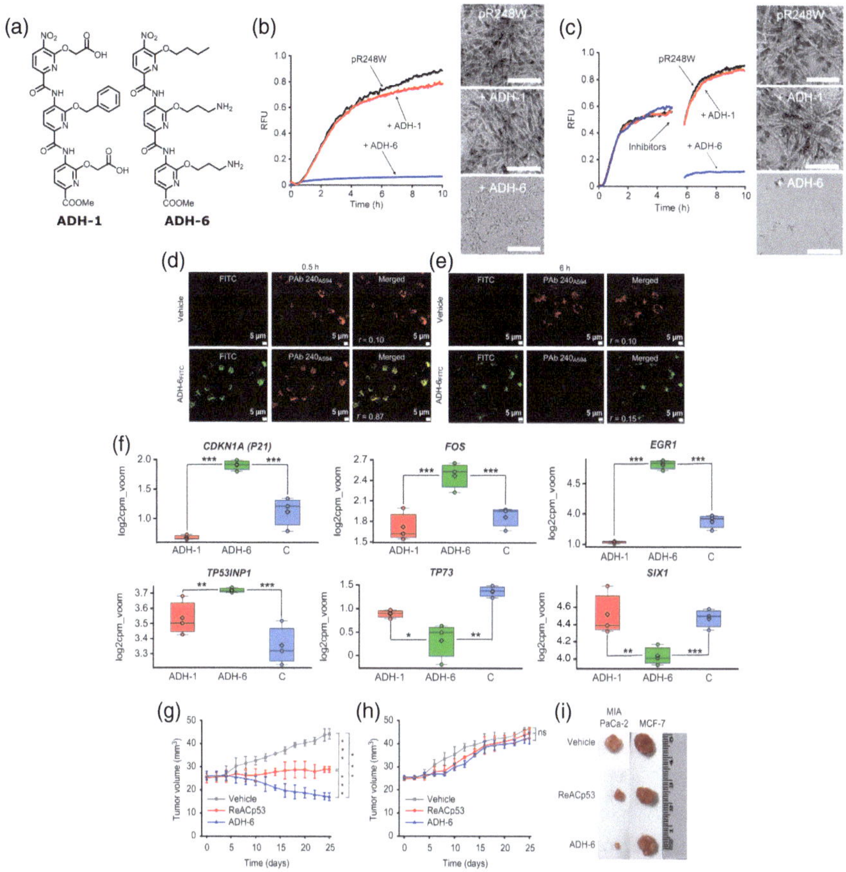

Figure 8.5 (a) Chemical structures of ADH-1 and ADH-6; ThT Kinetic profiles and representative TEM images for the aggregation of 25 μM pR248W in the absence or presence of equimolar ADH-1 or ADH-6 added at the start of the experiment (b) or added during the growth phase after 5 hours (c); Representative confocal images of FITC-labeled ADH-6 with PAb 240-stained pR248W aggregates at 0.5 hours (d) and 6 hours (e) following incubation with the ADH-6FITC (5 μM); (f) Differences in the expression of gene products of p53 in control cells versus those treated with ADH-1 and ADH-6; Tumor volume growth curves for the MIA PaCa-2 (g) and MCF-7 (h) xenografts in the different treatment groups over 25 days (mean ± SD; n = 8); (i) Representative images of the MIA PaCa-2 (g) and MCF-7 (h) tumor tissues from mice after 25 days of treatment. Source: [52]/Springer Nature/CC BY 4.0.

8.8 Novel Synthesis of OPs and Alzheimer's Disease

Up to this point, libraries of OPs were relatively small due to the tedious synthetic methodologies and numerous chromatography steps required to produce them. Recently, Dohoney et al. developed a novel synthetic route to synthesize OPs from a 'common precursor' that eliminates the need for chromatography purification (Figure 8.6) [58]. First, the functionalized monopyridyl is synthesized by reacting 2-chloro-6-methyl-3-nitropyridine (monopyridyl common precursor) with a primary alcohol and sodium hydride in anhydrous toluene [58]. The nitro group of the monopyridyl is then reduced in ethyl acetate with palladium on carbon and hydrogen gas and further coupled to 6-chloro-5-nitro-picolinic acid in a one-pot acid chloride coupling using thionyl chloride and triethylamine in dry dichloromethane [58]. The aromatic chlorine of the dipyridyl common precursor is then substituted with a primary alcohol using sodium metal [58]. By repeating the reduction, coupling, and alcohol substitution reactions, the final tripyridyl OPs are made without a single purification step and a total % yield ranging from 53.3% to 79.8% [58]. A final deprotection step is required to generate the active functional groups on Ops with primary amine(s) or carboxylic acid(s).

We then used this new synthetic route to create an OP with the same functional groups as ADH-41 (RD242), previously shown to inhibit Aβ_{42} aggregation (Figures 8.2b and 8.7a) [51, 58]. The only difference between ADH-41 and RD242 is the C-terminal methyl ester and methyl groups of the respective molecules (Figure 8.7a) [58]. ThT kinetic assays showed comparable inhibition of Aβ_{42} aggregation as indicated by low ThT fluorescence in the presence of the ligands (Figure 8.7b) [58]. The ThT results were corroborated by the complete lack of Aβ_{42} fibrils observed in the TEM images (Figure 8.7c,d) [58]. ^{1}H-^{15}N HSQC NMR indicated that RD242 interacts with Aβ_{42} residues between E11 and F20, similar to the binding site of ADH-41 [51, 58]. Using ITC, a K_d of $0.84 \pm 0.11\,\mu$M was determined for the binding RD242 and Aβ_{42}. This is similar to that of ADH-41 and Aβ_{42} ($K_d = 0.7 \pm 0.1\,\mu$M) [51, 58]. To test RD242 in a living system, we employed a well-established AD model *C. elegans* strain, GMC101, which expresses human Aβ_{42} protein in their body wall muscle cells [58]. *Caenorhabditis elegans* are the ideal model for studying neurodegenerative diseases due to their relatively short lifespan, genetic similarity to humans, and their ability to imitate neuropathological phenotypes [86]. Fluorescence-based experiments demonstrated that RD242 reduced Aβ_{42} aggregation, as well as reduced the formation of reactive oxygen species associated with AD progression in the worms when treated at the early L4 stage (Figure 8.7e–i) [58]. Additionally, RD242 rescued the paralysis associated with Aβ_{42} aggregation in the disease worm model (Figure 8.7j,k) at 10 and 30 μM concentrations [58]. Since treatment for neurodegenerative diseases often begins post-disease onset, we tested the effects of RD242 on late-stage GMC101 *C. elegans* (Figure 8.7l) [58]. Late-stage treatment of RD242 inhibited the fibers-catalyzed progression of Aβ_{42} aggregation in the worms (Figure 8.7m–o) [58]. Clearly, in this work, we have simplified the synthesis of OPs and demonstrated their ability to attenuate AD phenotypes in a *C. elegans* model of the disease.

8.9 2D-FAST

Since the antagonist activity of these molecules increases with additional sidechains, the OP protein mimetic scaffold is perfect for a fragment-based approach.

(a)

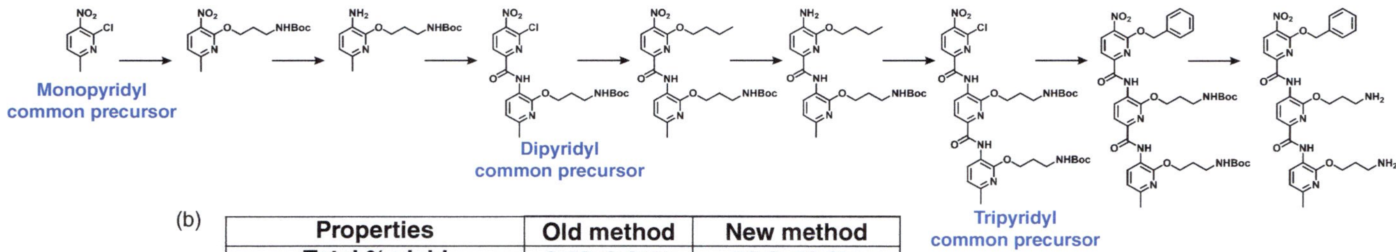

(b)

Properties	Old method	New method
Total % yield	0.78–3.61%	53.3–79.8%
Chromatography steps	11	0
Synthetic steps	14	8
Total side chains	10	20
Total tripyridyls	1 tripyridyl	Tripyridyl library

Figure 8.6 (a) Synthetic scheme for the production of RD242 using the common precursor method; (b) Table comparing the properties of the old synthetic method and the new common precursor method. Source: Nicholas H. Stillman.

Stillman et al. developed a 2-dimensional fragment-assisted structure-based technique (2D-FAST, Figure 8.8) which combines fragment and structure-based techniques to systematically optimize the antagonist activity toward the aggregation of α-synuclein (αS), a process known to impair the dopaminergic neurons and lead to Parkinson's disease (PD) [59, 87–91]. The primary physiological function

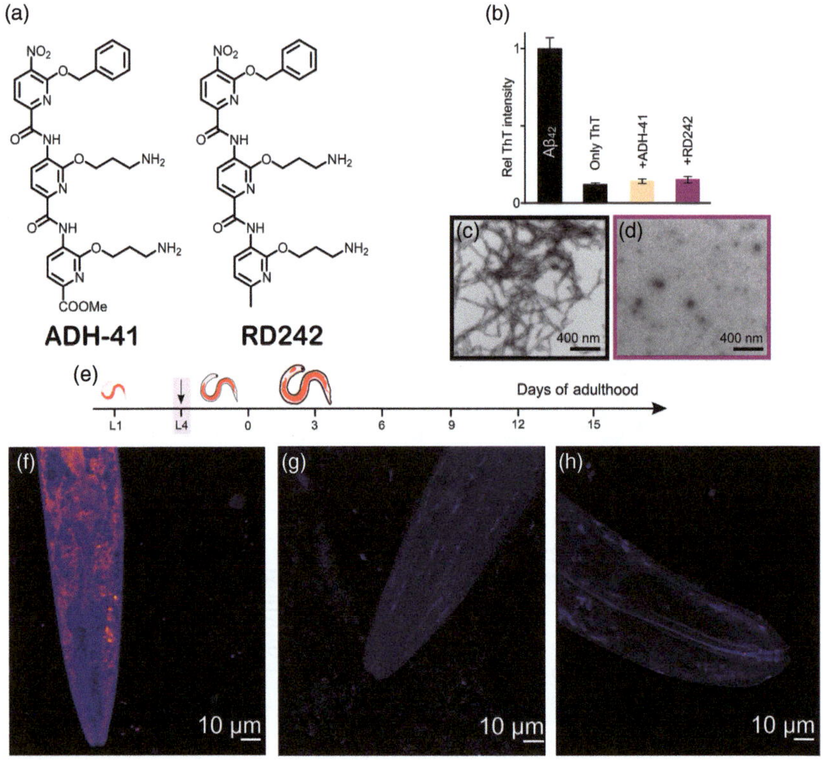

Figure 8.7 (a) Chemical structures of ADH-41 and RD242; (b) Relative ThT fluorescence intensity of 10 μM Aβ42 after 24 hours under aggregating conditions in the absence and presence of the indicated OPs at an equivalent concentration; Representative TEM images of 10 μM Aβ42 in the absence (c) and presence (d) of one equivalent RD242; (e) Schematic of the aging process of GMC101 worms with the arrow indicating the stage of treatment with ligands; Representative confocal images of GMC101 worms in the absence (f) and presence of RD242 (g) and bexarotene (h); (i) Comparison of the ROS levels in GMC101 worms in the absence and presence of the indicated ligands over 2 hours; (j) Comparison of the motility rate of GMC101 and N2 worms in the absence and presence of the indicated ligands; (k) The relative motility rate of GMC101 and N2 worms on day nine in the absence and presence of 30 μM (left) and 10 μM (right) ligands; (l) Schematic of the aging process of GMC101 worms with the arrow indicating the late stage (day five of adulthood) treatment with ligands; Representative confocal images of GMC101 worms on day five (m) and day 10 (n) of adulthood; (o) Representative confocal image of GMC101 worms on day 10 after late-stage treatment (day five) with 30 μM RD242. Source: [58]/with permission of American Chemical Society.

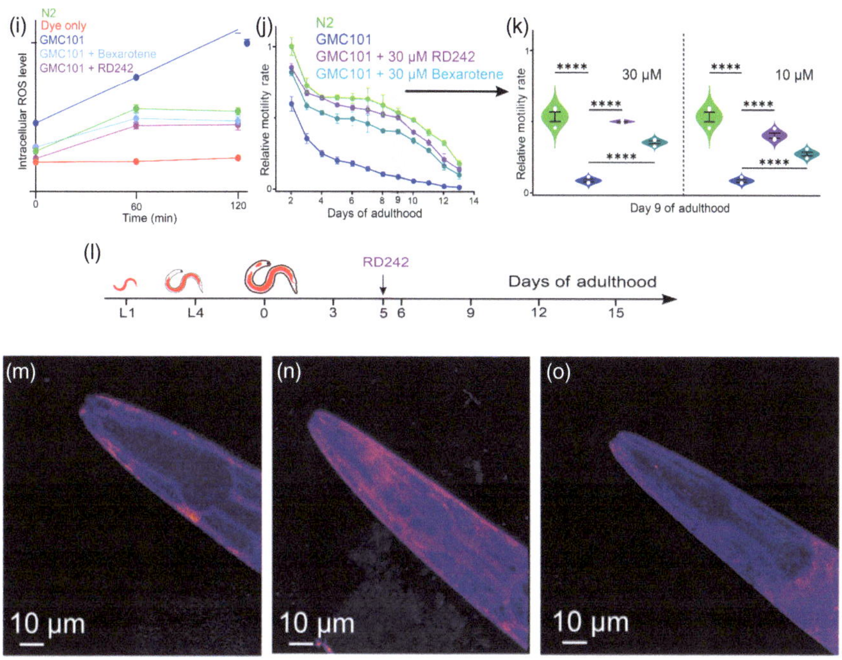

Figure 8.7 (*Continued*)

of αS is synaptic vesicle pooling and neurotransmitter release by facilitating the assembly of the SNARE complex [92, 93]. Aggregation of this protein, likely in the form of smaller oligomers, causes neuron death through fragmentation of the mitochondria, excess mitophagy, and the formation of reactive oxygen species [94, 95]. The 2D-FAST begins with the synthesis and subsequent screening of a library of monopyridyls (Figures 8.8 and 8.9a). Monopyridyls were synthesized by reacting 2-chloro-6-methyl-3-nitropyridine with a variety of primary alcohols to generate a library of molecules with polar, nonpolar, positive, and negatively charged sidechains (Figure 8.9a) [59]. The library of monopyridyls is screened in a ThT assay, and the most active monopyridyl becomes the precursor for the library of dipyridyls (Figure 8.9a–d) [59]. A one-pot coupling reaction generates the chloro-dipyridyl, which in turn is then reacted with a variety of primary amines and thiols to create a diverse library of functionalized dipyridyls (Figure 8.9c) [59]. After screening the library of dipyridyls, the most active dipyridyl is expanded to a library of tripyridyls in the same way (Figure 8.9c–e) [59]. Screening the library of tripyridyls led to the identification of NS132 as the most potent antagonist of αS aggregation (Figure 8.9e,f) [59].

Removing the indole and protecting the carboxylic acid functional group of NS132 diminished the activity of the molecule, further proving that the side chains are crucial for the molecule's activity [59]. ^{1}H-^{15}N HSQC NMR and ITC indicated two binding sites spanning residues 10-20 and 36-43 on the N-terminus of the protein with a K_d of 1.81 ± 0.33 μM and 1 : 1 binding stoichiometry (αS:NS132) [59].

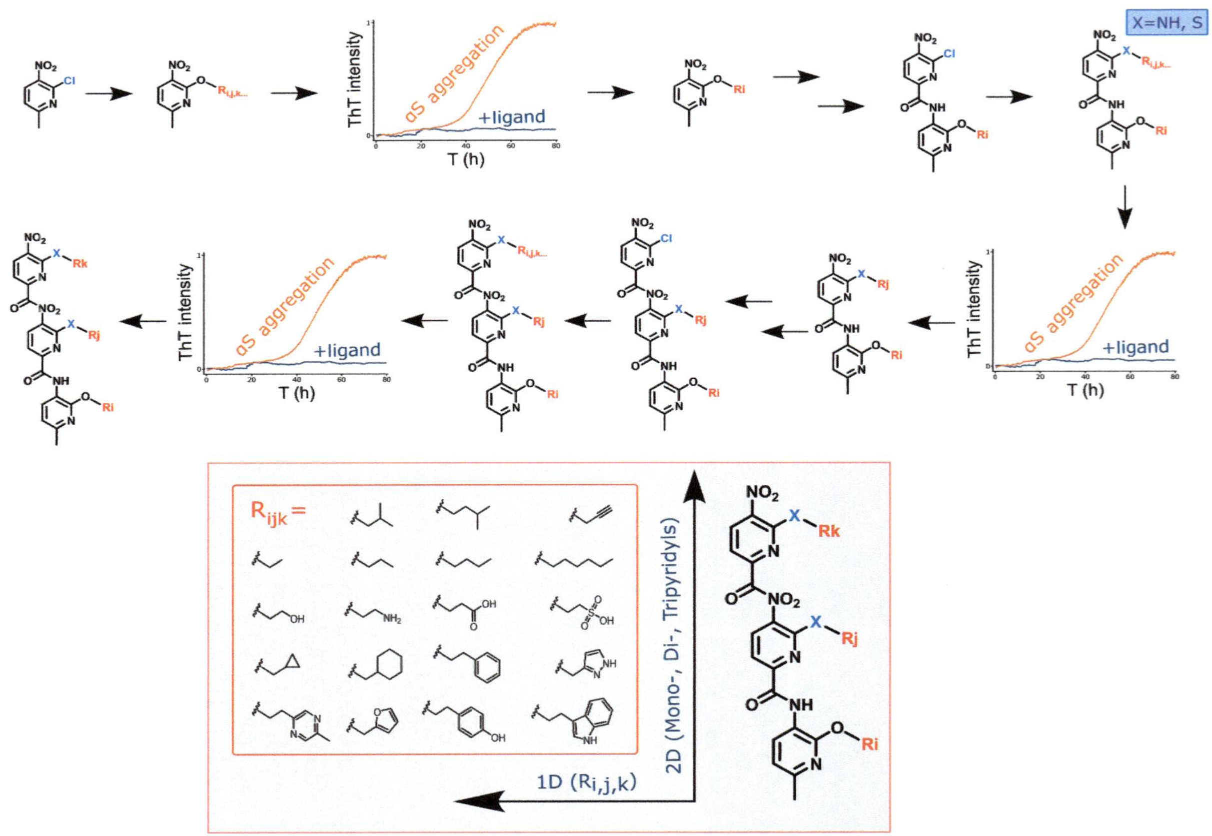

Figure 8.8 The 2-dimensional Fragment-Assisted Structure-based Technique (2D-FAST) for the synthesis of OPs to target αS aggregation. Source: Nicholas H. Stillman.

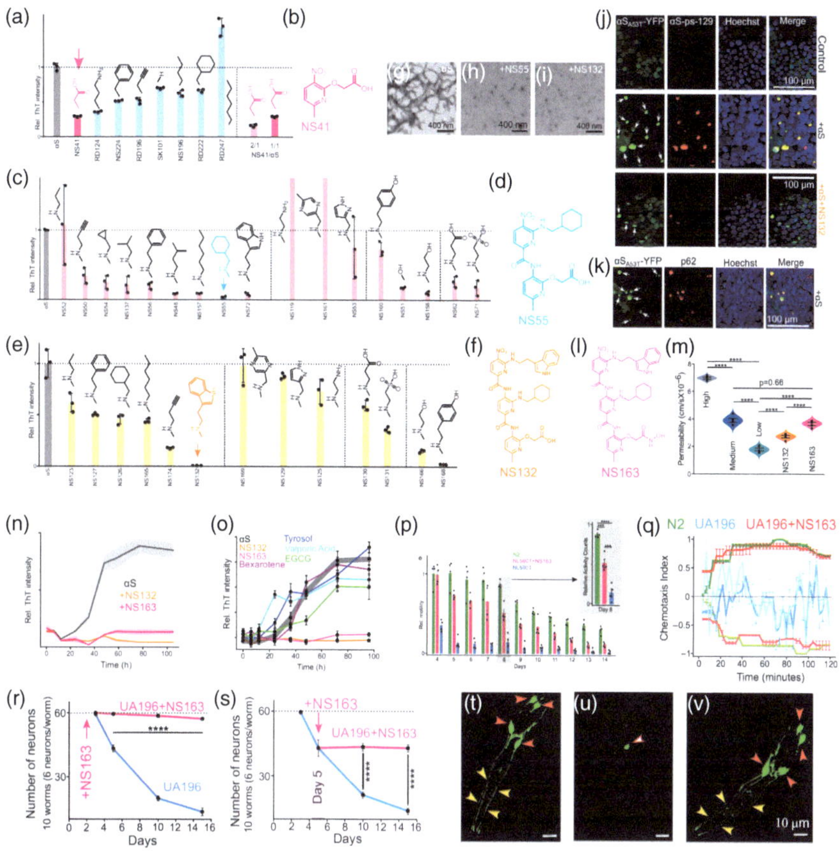

Figure 8.9 The relative ThT intensity of 100 μM αS after aggregation for 96 hours in the absence and presence of monopyridyls (a), dipyridyls (c), and tripyridyls (e) at an equimolar ratio; The chemical structures of the most active mono- (NS41, b), di- (NS55, d), and tripyridyls (NS132, f); Representative TEM images of 100 μM αS in the absence (g) and presence of NS55 (h) and NS132 (i) at one equivalent; (j) Representative confocal images of HEK cells treated with 5 μM αS aggregated in the absence and presence of equimolar NS132. αSA53T-YFP = green, αS-pS-129 = red, Hoechst = blue, merge = αSA53T-YFP, αS-pS-129, and Hoechst; (k) Representative confocal images of HEK cells treated with 5 μM aggregated αS. αSA53T-YFP = green, p62 = red, Hoechst = blue, merge = αSA53T-YFP, p62, Hoechst; The chemical structure of NS163 (l) and a comparison of the cell permeability of NS132 and NS163 with the high, medium, and low permeability standards (m); (n) Overlayed ThT kinetic profiles of 100 μM αS aggregation in the absence and presence of NS132 and NS163; (o) Comparison of the ThT kinetic traces of 100 μM αS aggregation in the absence and presence of commercially available antagonists; (p) The motility of N2 (green bar) and NL5901 *C. elegans* in the absence (blue bar) and presence (pink bar) of 50 μM NS163. The inset provides statistics for day eight; (q) The chemotaxis index graph of N2 and UA196 worms (with and without treatment of 50 μM NS163) on day 10 of adulthood; The number of intact neurons in 10 UA196 worms following treatment of 50 μM NS163 at the early stage (days two and four) (r) and late-stage (day five) (s) of the disease progression; Representative confocal images of untreated UA196 worms on day three (t) and day 15 (u), and in the presence of 50 μM NS163 (v) on day 15. Where relevant, statistics were determined by one-way ANOVA with Tukey's multiple comparisons test (***p* < 0.01, ****p* < 0.001, *****p* < 0.0001). Source: [59]/Springer Nature/CC BY 4.0.

Transfection of HEK293T cells, which stably express YFP-tagged A53T mutant αS (αS_{A53T}-YFP), with pFFs of wildtype αS led to a drastic increase in puncta within the cells as the fibrils template and catalyze the aggregation of the endogenously expressed protein (Figure 8.9j) [59]. The observed puncta were co-localized with p62 and αS-pS-129 (αS phosphorylated at Serine residue 129), indicating the formation of Lewy-body-like structures, the hallmark of PD (Figure 8.9j,k) [59]. Contrarily, far fewer puncta were observed when the cells were transfected with αS aggregated in the presence of NS132, which suggests that NS132 does not generate fiber-competent cytotoxic structures (Figure 8.9j) [59]. It was anticipated that the cell permeability of NS132 would be less than ideal given the carboxylic acid functional group; therefore we synthetically modified the carboxylic acid to make the hydroxamic acid analog, NS163 (Figure 8.9f,l) [59]. Using a parallel artificial membrane permeability assay (PAMPA), we demonstrated that NS163 did in fact have a higher cell permeability than NS132 (Figure 8.9m) [59]. ThT assays, TEM imaging, ^{1}H-^{15}N HSQC NMR, and ITC experiments all indicated that NS163 maintained its antagonist activity toward αS aggregation, binding to residues 9-21 and 36-43 with a K_d of 1.81 ± 0.33 μM and 1 : 1 binding stoichiometry (Figure 8.9n) [59]. Both NS132 and NS163 outperformed the commercially available antagonists of αS aggregation (including Bexarotine, EGCG, Tyrosol, and Valporic acid) in a ThT assay (Figure 8.9o) [59]. Furthermore, pretreating the HEK293T cells with OPs NS163 and NS132, followed by transfection with αS pFFs, reduced the puncta formation by 80% and 57%, respectively [59]. We ascertain that the increased activity of NS163 in the cells is due to the increased cell permeability of the molecule. Next, we tested the rescuing effects of NS163 in the NL5901 strain PD model *C. elegans*, which expresses human αS with a yellow fluorescent protein tag in their body wall muscle cells [59]. The motility of the worms becomes significantly hindered as the protein aggregates during the aging process (Figure 8.9p, blue bars) [59]. Fewer inclusions and significantly improved motility were observed in the diseased strain of worms after treatment with NS163, compared to the untreated (Figure 8.9p, pink bars) [59]. A second strain of *C. elegans*, UA196, was used to measure the neuroprotective effects of NS163 [59]. This strain of *C. elegans* expresses both human αS and green fluorescent protein in each of their six dopaminergic neurons [59]. As the disease progresses, there is a gradual loss of the dopaminergic neurons, which drives the observed behavioral deficits in the worms. Using the WMicroTracker ARENA plate reader, we showed that UA196 *C. elegans* treated with NS163 maintain a preference for their food over ethanol, whereas the untreated diseased worms appear to have no preference (Figure 8.9q) [59]. The gradual loss of the neurons in the UA196 worms was observed through confocal microscopy (Figure 8.9r,t,u) [59]. There was no significant decline or change in the morphology of the dopaminergic neurons up to day 15 in the diseased worms treated with NS163 (Figure 8.9r,v) [59]. Since treatment for neurodegenerative diseases often begins after the onset of symptoms, it is important to test the neuroprotective effects of therapeutics post-disease onset. Treatment of UA196 worms on day four of adulthood, when ~28% of the dopaminergic neurons had degraded, prevented further loss of the remaining neurons compared to the untreated worms (Figure 8.9s)

[59]. Altogether, we have developed and applied a novel fragment-based method to systematically optimize the antagonist activity of OPs and demonstrated its application toward rescuing PD phenotypes in an early and post-disease onset disease model.

8.10 OQ Foldamers – Structure and Discovery

The design and synthesis of foldamers targeting disease-related aPPIs often focuses specifically on the mimicry of helical domains, as they are key structural motifs present in biopolymers such as proteins and DNA [22, 41–43, 96, 97]. α-Helical domains are the most prevalent secondary structure observed in proteins, and greater than 60% of protein–protein complexes in the PDB exhibit a helical interface, making them attractive targets for the development of therapeutics toward aPPI-related diseases [98, 99]. This interface, however, remains a challenging therapeutic target due to the complex chemical topology and surface area present at the interface of PPIs [22, 41, 97, 100, 101]. Numerous helical foldamer mimetics comprised of unnatural amino acids have been designed and synthesized to target these interactions, such as stapled peptides, β-peptides, peptoids, and sulfono-γ-AApeptides, yet stabilization of their helical conformation, tunability, and stability in the cellular milieu remain impending challenges for these foldamers in biologically relevant applications [97, 102–105].

Meta-substituted pyridyl and benzyl oligoamide foldamers with a bond angle of around 120° (Figure 8.10a) between amide coupling functionalities were found to form helical complexes varying between 4.5 and 8 units per turn, notably much larger than peptide helices with an average of 3.6 units per turn [100, 106, 107]. In 2002, the Ivan Huc group introduced a new class of quinoline-derived synthetic foldamers called OQs (previously δ-peptides) that, unlike other planar oligoamide scaffolds, exhibited the ability to fold into stable helical conformations [108]. In designing the OQ scaffold, the authors correctly presumed that the synthesis of aromatic oligoamides from functionalized monomeric quinoline units, bearing nitro and methyl ester substituents (reduced or saponified for amide coupling, respectively) oriented at 60° rather than 120°, (Figure 8.10b) would yield a more compact helix with ~2.5 units per turn [108]. Each monomer also contains a side chain functionality in the fourth position of the quinoline, allowing for the introduction of chemical diversity on the surface of the helix. The folding of the OQ helix is driven by hydrogen bonding interactions between the amide proton and neighboring endocyclic nitrogen atoms (Figure 8.10c), and its stability is further increased through π–π interactions within the helix [109]. Crystallographic and two-dimensional NMR studies demonstrate that side chain functionalities in the folded conformation orient residue $i+5$ above i with a rise of ~7.6 Å, closely mimicking the structure and surface topology of biomolecules or peptide α-helices (Figure 8.10c) [108–110]. This pivotal design afforded a dynamic and synthetically tunable scaffold capable of participating in the complex chemical interactions seen

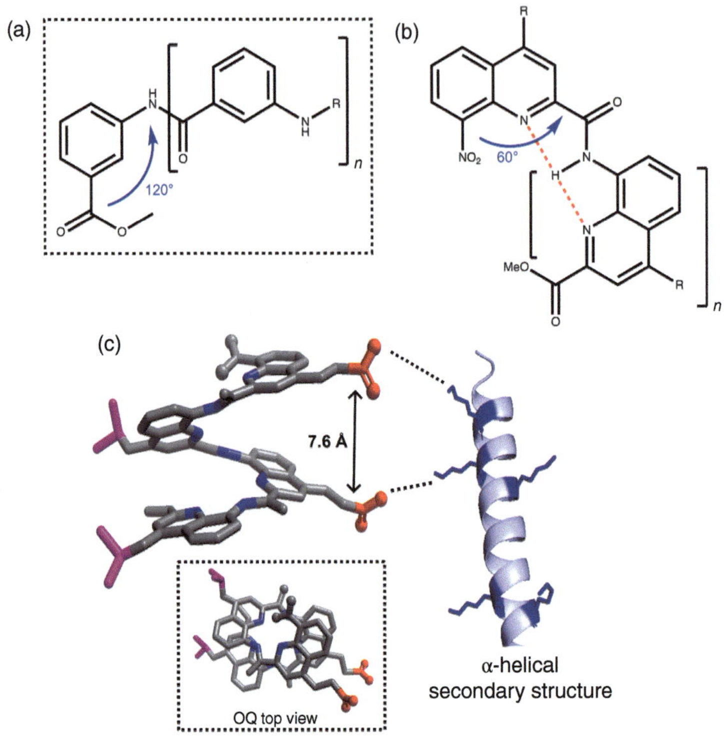

Figure 8.10 (a) Representative chemical structure of helical oligobenzamides. (b) Representative chemical structure of oligoquinolines. Dotted lines indicate hydrogen bonding that drives helix folding and stability (c) Crystal structure of a tetrameric OQ representing the chemical interactions with protein α-helices. Source: Ryan A. Dohoney.

at the interface of PPIs and opened the door to ongoing studies on the applicability of OQs as potential therapeutics.

8.11 Synthesis of OQ Foldamers

OQ foldamers demonstrate numerous advantageous properties that make them strong candidates for the targeting of aPPIs [47, 111–115]. Their rather facile synthesis allows for the production of an array of chemically diverse ligands as potential therapeutic agents. The synthesis begins with the addition of diisopropyl azodicarboxylate to 2-nitroaniline, followed by thermal ring closure of the pyridine ring, yielding the precursor methyl 4-hydroxy-8-nitroquinoline-2-carboxylate (**Q1**, Figure 8.11a) [108, 116]. Under Mitsunobu conditions [109, 115] or Williamson alkylation [113], the alkyl-aryl ether of **Q1** can be formed with an array of side chains that mimic those of naturally occurring amino acids. (Figure 8.11a) These monomeric quinoline building blocks can then undergo either reduction of the nitro or saponification of the methyl ester to be coupled in an iterative sequence-specific manner. The coupling of the monomeric units is facilitated by the activation of the carboxylic acid using thionyl chloride ($SOCl_2$), or a one-pot coupling procedure with a variety of activation agents employed. While standard peptide coupling reagents (HBTU/HOBT, HCTU, PyBOP, PyBrOP, etc.) tend to be more inefficient in the amide coupling, there have been other reagents successfully employed in a one-pot reaction without the need for pre-activation of the carboxylic monomers [109, 114, 115]. It is also important to note that the pre-activation and formation of acid chlorides using $SOCl_2$ is not compatible with many BOC-protected functionalities, and thus a major improvement to these synthetic methods was the use of 1-chloro-*N,N,2*-propenylamine [115, 117], an activating agent compatible with the *t*-butyl ester and carbamate protecting groups of most α-amino acid side chains. With this same synthetic methodology, it is also possible to produce OQs of significantly greater chain lengths in fewer steps by the coupling of longer segments (di, tri, tetrameric, etc.) to one another, i.e. the coupling of two dimers with alternating hydrophobic and hydrophilic residues yields a tetramer with the same alternating sequence (Figure 8.11b) [109]. This is especially useful in the production of amphipathic OQs and can significantly expand the chemical diversity and activity toward aPPIs of the folded structures. Further, the coupling of monomeric quinolines has successfully been adapted and performed on solid-phase resins, allowing for a synthetic route that requires limited chromatography [118].

8.12 OQs as Modulators of Type II Diabetes-Related aPPIs

Following the development of the OQ foldamer, most studies surrounding the scaffold focused primarily on its structural attributes as well as material and photochemical properties, yet biochemical applications relating to the prevention of aPPIs remained sparse. In 2013, Kumar et al. published one of the first investigations on the use of OQs to disrupt biologically relevant aPPIs [119]. As previously described, the peptide hormone IAPP co-secreted with insulin in pancreatic β-cells is implicated in type I and type II diabetes pathologies through the formation of toxic oligomeric complexes [54, 71, 72]. The α-helical membrane interacting segment of IAPP is shown to be amphipathic and largely cationic on one face, facilitating the formation of oligomers and leading to disruption of the cell membrane and dysregulation of mitochondrial function [54, 72, 120, 121]. Previously, the Hamilton lab had reported on a pentacarboxylate OP mimetic (Figure 8.12a) capable of modulating the lipid-catalyzed IAPP oligomerization and fibrilization [121, 122]. In the 2013 report on OQ foldamers, a pentacarboxylate OQ (Figure 8.12a) was synthesized, and the activity was compared to the previously described oligopyridyl, as well as an unstructured pentacarboxylate peptoid mimetic (Figure 8.12a) for its ability to abrogate the toxic oligomerization of IAPP. To achieve this, the authors carried out a lipid-catalyzed fibril formation assay in which 10 μM IAPP was aggerated in a buffer solution containing 20 μM ThT and 100 μM of each pentacarboxylate ligand with 630 μM unilamellar liposomes (1 : 1 DOPG:DOPC) [119]. The kinetics of the fibril

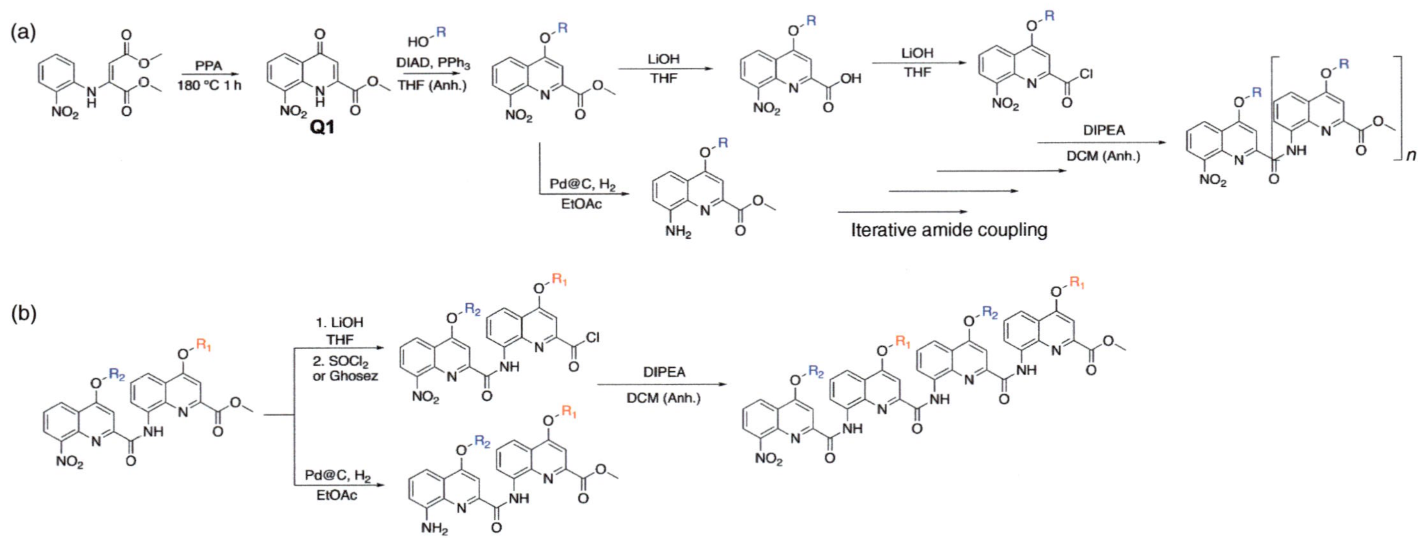

Figure 8.11 (a) Synthetic scheme of OQs bearing side chain functionalities varying in number of repeat quinoline units; (b) Representative synthetic scheme of amphipathic quinolines from the coupling of two dimeric OQ sub-units baring alternating hydrophobic and hydrophilic R-groups. Source: Ryan A. Dohoney.

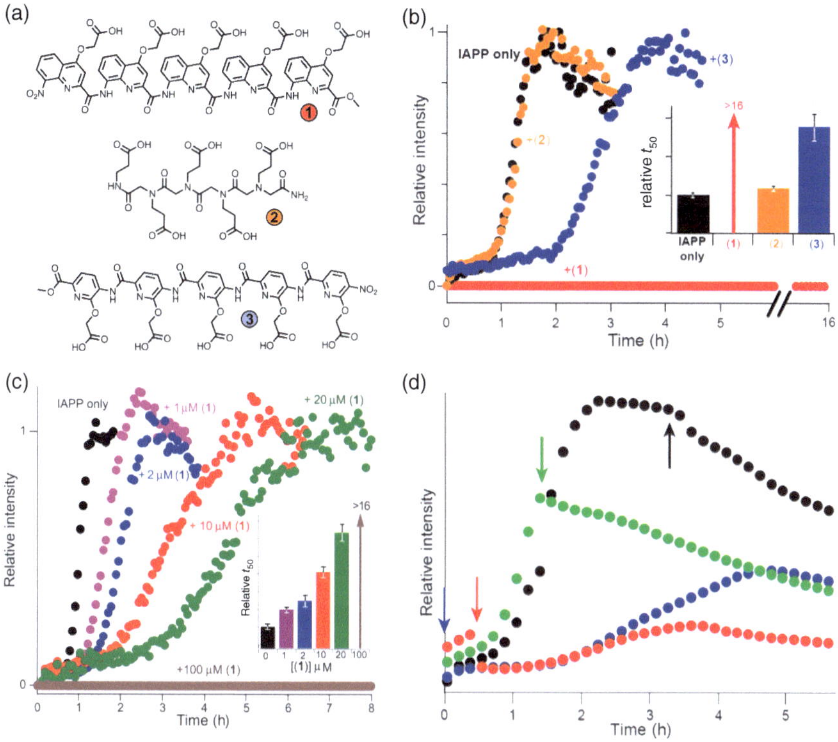

Figure 8.12 (a) Chemical structures of OQ, peptoid, and OP pentacarboxylate ligands (1-3); (b) Lipid-catalyzed fibril formation assay in the absence and presence of indicated ligands (1-3), Inset depicts the quantification of t_{50} extracted from the fibril formation assay for each ligand; (c) Dose-dependent effect of OQ (1) on the lipid-catalyzed fibril formation of IAPP, Inset represents quantification of the t_{50} of (1) at varying molar concentrations; (d) Timed-dependent addition of (1) to the lipid-catalyzed fiber formation assay at an equimolar ratio (IAPP/Ligand), arrows indicate time of addition of (1) in each experiment. Source: Adapted from Kumar and Miranker [119].

formation (or conversion of IAPP to a β-sheet-rich species) are determined by monitoring the changes in fluorescence intensity of the ThT dye over 16 hours, where an increase in fluorescence intensity is indicative of fibril or amyloid formation. The kinetic profile of liposome-catalyzed IAPP aggregation is nucleation dependent, and therefore the fluorescence data can be fit to extract the time required for 50% IAPP precursor conversion (t_{50}) (Figure 8.12b) [119]. In the presence of a 10 : 1 M ratio of the unstructured pentacarboxylate peptoid (**2**), there was little to no effect on the aggregation kinetics of IAPP relative to the control ($t_{50} = 1.1 \pm 0.1$ h). In marked contrast, the structured pyridyl scaffold (**3**) increased the t_{50} to 2.9 ± 0.3 h correlating to an increase in amyloid formation time by a factor of 2.7. Remarkably, in the presence of the OQ ligand (**1**) at the same molar ratio, IAPP aggregation was wholly inhibited yielding a $t_{50} > 16$ h [119]. This demonstrated the first success of OQ foldamers as potential therapeutic ligands toward aPPIs.

The authors further tested the inhibition of IAPP aggregation by (**1**) in a dose-dependent manner. A lipid-catalyzed fiber formation assay was carried out with the addition of various concentrations of (**1**) ranging from 0.1 to 10 M equivalents (Ligand : IAPP), and the relative t_{50} was quantified for each experiment (Figure 8.12c). Notably, this experiment demonstrated that at a molar equivalent of 1 : 1 (Ligand:IAPP), the OQ ligand (**1**) achieved the same level of inhibition as the OP ligand (**3**) at a molar ratio of 10 : 1 suggesting the potency of the OQ ligand toward IAPP aggregation is far greater than the other mimetic ligands [119]. To further investigate the mechanism by which the OQ inhibits IAPP amyloid formation, under matched conditions to the previously described assay (10 μM IAPP, 630 μM liposomes), an equimolar aliquot of (**1**) was added at various time points and was found to wholly inhibit further aggregation following its addition (Figure 8.12d, arrows). This is in marked contrast to the OP ligand (**3**) that reached a saturation point even at very high (~40 : 1) molar equivalents where the t_{50} and aggregation were no longer reduced further [119, 123]. Altogether, it is evident that though the pentacarboxylate ligands share similar chemical functionalities, the OQ ligand (**1**) demonstrates a far greater effect and higher potency than the OP (**3**), and their mechanism of inhibition may differ significantly. The time-dependent study also suggests that (**1**) is capable of effectively modulating both the nucleation of protein into new fibrils, as well as the elongation of existing fibrils, two of the main pathogenic events in amyloid formation [119]. Together, this study is one of the first reports on the successful modulation of disease-related aPPIs with the OQ foldamer scaffold.

8.13 Mechanistic Insights into OQ Manipulation of aPPIs

A follow-up report from Kumar et al. further explored the mechanism by which OQs influence the IAPP aggregation pathway [114]. Previously, protein mimetic ligands have been shown to bind to stabilized membrane-bound IAPP structures and inhibit further aPPIs [22, 119, 121]. However, the dynamic structure and chemical diversity of OQs allow for different types of interactions with the protein, in which the OQ causes a stabilizing conformational change to a target protein structure rather than direct interference with aggregation-prone surfaces [114, 119]. Hence, OQs modulate IAPP aggregation through a distinct mechanism when compared to the OP ligands targeting IAPP. In the absence of foldamers, circular dichroism (CD) spectroscopy demonstrates that under lipid-catalyzed conditions, IAPP undergoes a rapid transition from a random coil conformation to predominantly α-helical in nature. Following this event, IAPP converts to a β-sheet-rich species within one hour of the interaction, indicative of fibrillization [114, 119, 121]. However, in the presence of equimolar OQ_5, the formation of fibrils is significantly delayed while retaining the helicity of IAPP intermediates (Figures 8.12c and 8.13a,b) [114, 119].

The authors further reveal the differences in the OP versus OQ inhibition mechanism through the solution phase and seed-catalyzed conditions. It is shown that IAPP can undergo self-assembly in the absence of liposomes at higher concentrations (40 μM). In buffer-only conditions, the t_{50} for fibrillization of IAPP at 40 μM is

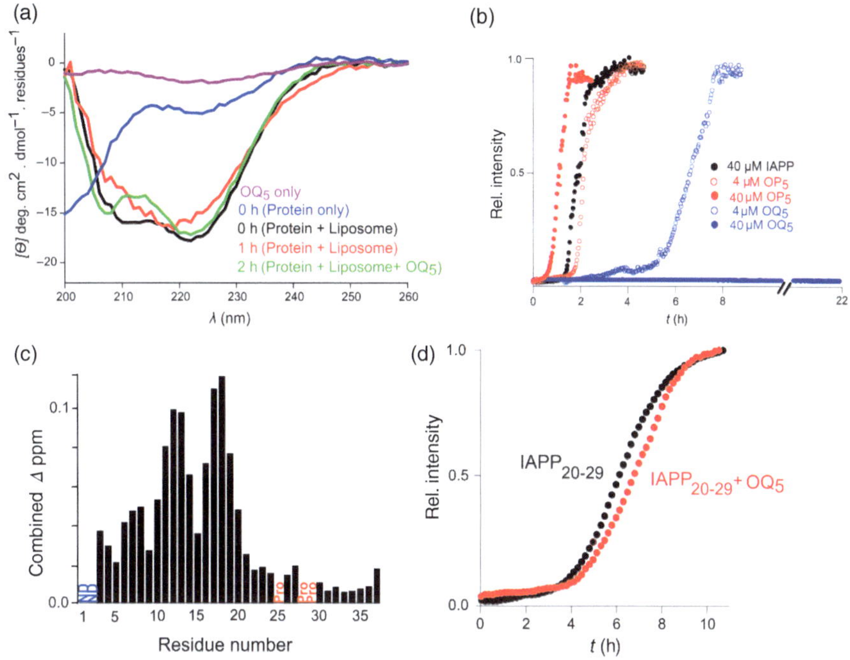

Figure 8.13 (a) Circular dichroism spectra of IAPP at various time points in the absence and presence of liposomes and/or OQ5; (b) Liposome-catalyzed fibril formation assay of IAPP in the absence and presence of 0.1 or 1 M equivalents of OP5 or OQ5 ligands; (c) Graphical representation of the calculated chemical shift perturbation obtained by 1H-15 N HSQC for individual residues on the IAPP peptide when titrated with OQ5; (d) ThT Kinetic aggregation profile of the IAPP20-29 subregion in the absence and presence of OQ5. Source: Kumar et al. [114]/with permission of Elsevier.

1.9 ± 0.1 h. In the presence of equimolar (40 μM) OP_5 in buffer only, the t_{50} is found to be 1.1 ± 0.1 h, suggesting no effect or a slightly agonistic effect on IAPP fibrillization (Figure 8.13b) [114]. In marked contrast, the presence of 0.1 M equivalents of OQ_5 under the same conditions yielded a threefold to fourfold increase in the inhibitory effect (Figure 8.13b), with no detectable conversion at an equimolar ratio [114]. Under seed-catalyzed conditions, (1% mol/mol) preformed IAPP fibrils were added to the liposome-catalyzed and uncatalyzed IAPP reactions, where a fivefold to sixfold increase in fibrillization was observed in untreated conditions. Little to no effect was observed on the rate of fibrillization when treated with OP_5; however, in the presence of equimolar OQ_5, the increase in seed-catalyzed fibrillization is negated, and at 10 M equivalents, fibril formation is wholly inhibited [114].

^{1}H-^{15}N HSQC NMR (chemical shift perturbation) studies revealed that OQ_5 interacts between residues 3-20 (Figure 8.13c), which is the region of IAPP shown to adopt α-helical structure upon membrane binding [124]. The subregion (residues 20-29) of IAPP implicated in amyloid formation showed little change in the NMR studies and was further tested in a kinetic aggregation assay (Figure 8.13d). The IAPP fragment consisting of residues 20-29 was aggregated at

200 μM and demonstrated a t_{50} of 1.9 ± 0.1 h. In the presence of equimolar OQ_5, no change was observed in the kinetic profile of aggregation, further suggesting the OQ does not interact directly with the aggregation-prone surface [114].

Taken together, the data suggest that OQ foldamers exhibit a unique mechanism for the inhibition of aPPIs as compared to other small molecules or foldamer-based ligands. Unlike other scaffolds, the dynamic OQ scaffold is capable of targeting intermediate species and inducing a stabilizing conformational change during amyloid formation to successfully modulate protein aggregation.

8.14 Chemical Diversity and Structure Modulate Efficacy of OQs

Having demonstrated the remarkable effects of OQs as potential therapeutics in biophysical IAPP studies, the focus is then shifted to optimization of the chemical diversity of OQs to develop ligands with improved efficacy and drug-like properties [46, 47, 114, 119]. In 2016, follow-up studies on IAPP by Kumar et al. explored the functionalities of OQ ligands to further identify potent modulators of pre-amyloid IAPP toxicity [47, 113]. The authors synthesized a library of more than 40 OQ compounds consisting of monomers, dimers, trimers, tetramers, and pentamers to be screened against IAPP aggregation. These ligands, unlike the previously identified pentacarboxylate (OQ_5), introduced a variety of OQs with alternating sequences bearing alkyl and carboxylate functionalities (Figure 8.14a). Similar to previous experiments, the ligands were screened against IAPP in a lipid-catalyzed fibril formation assay (Figure 8.14b) to determine the t_{50}. In this work, the authors identified a new OQ tetramer (Figure 8.14a, ADM-116) bearing two alternating alkyl and carboxylate functionalities that was a highly potent antagonist of IAPP aggregation [47, 113].

Compared to the previously reported OQ_5 (referred to as ADM-118 in this study), ADM-116 completely suppressed liposome-catalyzed fibrillization of IAPP over 45 hours (Figure 8.14b, t_{50} > 40 hours) [47]. This translates to a ligand that is 10-fold

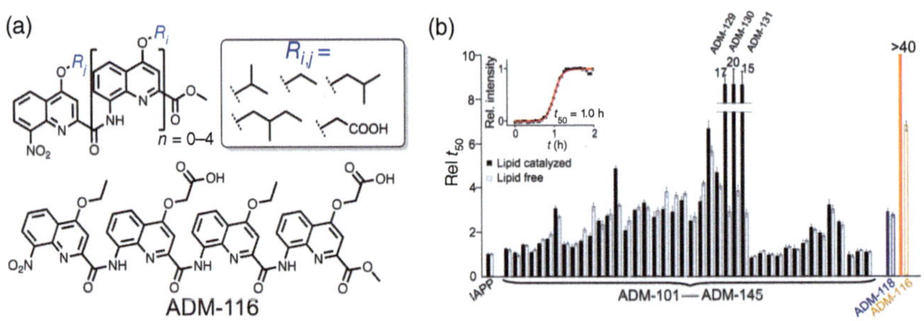

Figure 8.14 (a) Representative chemical structures of OQ analogs ranging from monomeric to pentameric including side chain functionalities in the library of ligands and the chemical structure of ADM-116; (b) Quantification of the t_{50} of IAPP aggregation in the presence of a library of OQ analogs extracted from lipid-catalyzed fiber formation assays. Source: Adapted from Kumar et al. [47].

more effective than the previously identified potent inhibitor. The authors also analyzed structural changes to ADM-116 in which the ethyl functionalities were replaced with isopropyl, isobutyl, and isopentyl alkyl groups (ADM-129, ADM-130, and ADM-131, respectively). It is shown that while the analogs maintained activity toward IAPP, the efficacy of each was significantly reduced in comparison to ADM-118 (Figure 8.14b, t_{50} around 15–20 hours). Similarly, when the number of quinoline units is increased or decreased, there is also an observed loss of activity toward IAPP for OQ analogs. This supports that the inhibition by OQ foldamers relies heavily on structure and sequence specific interactions with target intermediate protein conformations [47].

The effect of ADM-116 was also studied for the first time in a pancreatic cell line, INS-1, for its ability to rescue IAPP-mediated cytotoxicity [113]. Upon incubation of INS-1 cells with 13 μM IAPP for 48 hours, viability was reduced by 78%. In contrast, when treated with IAPP and ADM-116 at an equimolar ratio, cell viability of INS-1 was wholly restored [113]. The OQ ligand was also demonstrated to restore viability to 48% when the cells were treated 24 hours after the initial addition of IAPP. When cells were treated with ADM-116 alone, no cytotoxicity was observed. Confocal microscopy of a fluorescently-labeled ADM-116 analog further demonstrates that the OQ foldamer was capable of penetrating the cell membrane at 200 nM concentration and internalizing in INS-1 cells, suggesting internalization further contributes to the OQs mechanism of rescue [113]. Though this passive membrane permeation is surprising as ADM-116 is di-anionic in nature, it suggests that the dynamic nature and reversibility of folding may play an important role in its permeability. These remarkable findings support that OQ foldamers exhibit numerous advantageous properties in biological systems toward the development of therapeutic ligands and operate through sequence and structure specific targeting of aPPIs.

8.15 Modulation of Alzheimer's Disease-Related Aβ

Alzheimer's disease is the most common neurodegenerative disease, and its pathology is related to the self-assembly and fibrilization of the Aβ peptide into β-sheet-rich neuronal plaques, eventually leading to neuronal death [33, 51, 61–65]. To date, there are no known cures or therapeutics to slow the progression of AD [65]. Since AD is a progressive disease involving aPPIs, ligands that can modulate the aggregation pathway of Aβ may be considered as viable therapeutic interventions [33, 34, 112, 125]. As OQs have been previously shown to modulate the self-assembly of IAPP, they also demonstrate the potential to rescue other neurodegenerative disease phenotypes such as AD. In 2017, Kumar et al. reported on the use of OQ foldamers to attenuate $A\beta_{42}/A\beta_{40}$ aggregation *in vitro* and explore mechanistic insights into the interaction of OQs with Aβ [112]. Similar to the reports on IAPP aggregation [47, 113, 114, 119], a library of OQ foldamers was screened against $A\beta_{42}$ aggregation using a ThT kinetic aggregation assay. The authors identified a different OQ ligand (Figure 8.15a, **5**) from the library that was capable of wholly inhibiting $A\beta_{42}$ aggregation at an equimolar ratio ($A\beta_{42}$:**5**). At sub-stoichiometric

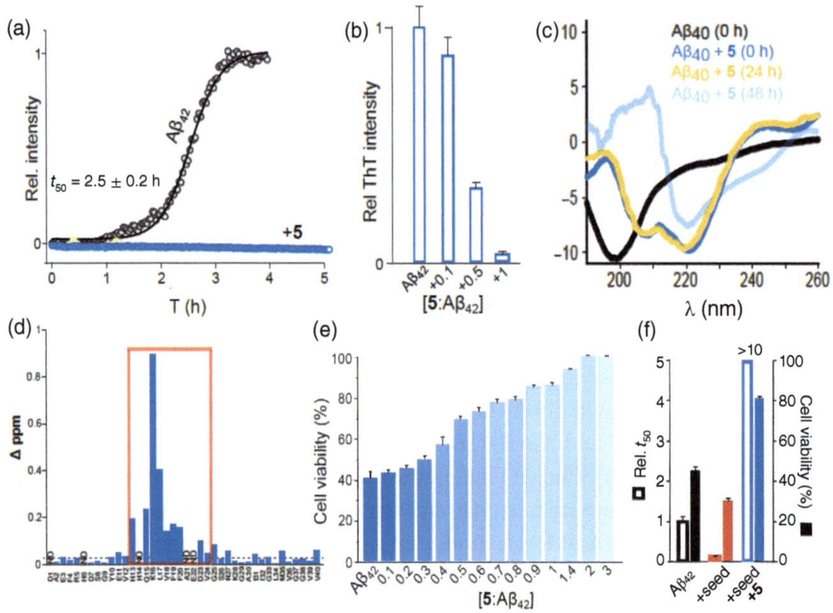

Figure 8.15 (a) ThT kinetic aggregation assay in the absence and presence of (**5**); (b) Relative change in ThT fluorescence intensity of Aβ42 samples after 6 hours in the absence and presence of (**5**) at 0.1, 0.5, and 1 equivalents (5:Aβ42); (c) Circular dichroism spectra of Aβ40 in the absence and presence of (**5**) over 48 hours; (d) Quantification by HSQC of chemical shift perturbations of Aβ40 when titrated with (**5**); (e) Dose-dependent viability of mouse neuroblastoma (N2a) cells when treated with (**5**) at various molar equivalents; (f) Quantification of seed-catalyzed (10% v/v Aβ42 seeds) aggregation and cytotoxicity of 5 μM Aβ42 in the absence and presence of (**5**). Source: Adapted from Kumar et al. [112].

ratios, the ligand was able to maintain activity and attenuate the ThT intensity by ~70% (Figure 8.15b, 0.5 eq) and confirmed by a lack of fibrillization in TEM imaging [112]. Fluorescence titrations between (**5**) and fluorescein-labeled $A\beta_{40}$ allowed for the determination of a K_d of 2.2 ± 0.3 μM supporting that (**5**) binds to $A\beta_{40}$ with high affinity [112]. Immunoassays and TEM imaging demonstrated that (**5**) is capable of preventing the formation of oligomeric $A\beta_{42}$ species, which are thought to be key neurotoxic species [126], as well as inhibiting their fibrillization.

The structural attenuation of $A\beta_{40}$ aggregation by (**5**) was investigated using CD and ^{1}H-^{15}N HSQC NMR spectroscopy. It is shown that over a 24-hour period, 20 μM $A\beta_{40}$ transitions from a random coil conformation to one that is β-sheet rich. However, in the presence of (**5**), $A\beta_{40}$ adopts an α-helical conformation that remains for up to 24 hours [112]. Interestingly, (**5**) demonstrates no CD bands in the absence of $A\beta_{40}$, but upon combining $A\beta_{40}$ with (**5**) in solution at an equimolar ratio, a strong positive CD band is observed at ~390 nm (Figure 8.15c) [112]. This suggests that (**5**) folds into a right-handed helix upon interacting with $A\beta_{40}$, and therefore the conformational changes observed for both peptide and ligand are driven by their interaction with one another. HSQC NMR studies provide insights into the potential binding site of the ligand on $A\beta_{40}$. Chemical shift perturbation was quantified by ^{1}H-^{15}N HSQC NMR for the interaction of (**5**) with $A\beta_{40}$ and revealed that significant changes in the peptide structure occurred between residues 13-24 (Figure 8.15d), with the largest changes occurring between His^{13}-Lys^{15} and Leu^{17}-Phe^{20} [112]. It is hypothesized that the carboxylates on (**5**) form salt bridges with the positively charged His^{13} and Lys^{15}, while the aliphatic groups disrupt hydrophobic interactions within Leu^{17}-Phe^{20} [112]. This is supported by previous reports that aggregation is promoted by interactions involving π–π and electrostatic interactions between Lys^{16} and Phe^{19}/Phe^{20} [127, 128]. Overall, the OQ foldamer (**5**) is capable of inducing and stabilizing an α-helical conformation of $A\beta_{40}$ to potently disrupt aggregation-prone interactions.

Mouse neuroblastoma (N2a) cellular assays were also used to elucidate the toxic effects of Aβ aggregation. When N2a cells were treated with 5 μM $A\beta_{42}$, their viability was reduced to 45% of that of the control cells. When incubated with $A\beta_{42}$ in the presence of (**5**) at various molar ratios, a dose-dependent response is observed with nearly complete rescue of cell viability at two molar equivalents of ligand (Figure 8.15e) [112]. Seed-catalyzed $A\beta_{42}$ experiments also exhibited a rapid increase in aggregation kinetics and cytotoxicity when aggregation occurred in the presence of 10% (v/v) preformed $A\beta_{42}$ fibrils. In the presence of (**5**) under seed-catalyzed conditions, aggregation was almost wholly inhibited and a significant rescue of cell viability was observed [112]. This study demonstrates the remarkable ability of OQ foldamers to target neurodegenerative-related aPPIs and serves to provide a deeper mechanistic insights into the further development of OQ therapeutics targeting neurodegeneration (Figure 8.15f).

8.16 OQs for the Modulation of Synucleinopathies

As previously described, the aggregation and accumulation of αS is one of the main pathogenic hallmarks in the progression of Parkinson's disease, leading to the impairment and death of dopaminergic neurons [59, 87–91]. An extensive report by Ahmed et al. utilized the OQ foldamer scaffold to identify a potent OQ capable of modulating aggregation of αS, while also investigating the underlying mechanistic factors involved in disease pathology [111]. This report is one of the first that successfully demonstrated the efficacy of OQ ligands in both *in vitro* and *in vivo* models targeting PD phenotypes, while revealing pertinent pathogenic associations between PD and αS aggregation. Using a ThT kinetic assay, the authors identified a potent tetrameric inhibitor of αS aggregation (SK-129) from a previously established library of OQs (Kumar Aβ) bearing alternating carboxylate and isopropyl modalities (Figure 8.16a). FP assays confirmed a fluorescent SK-129 analog is a high-affinity ligand for αS with a $K_d = 0.81 \pm 0.09$ μM [111]. In cellular models, HEK cell lines stably expressing fluorescently labeled WT αS (αS-YFP) and familial mutant αS (αS-$_{A53T}$-YFP) exhibit fluorescent puncta formation as aggregation progresses following transfection with preformed αS fibrils [129, 130]. In the presence of SK-129, a significant reduction in puncta formation was observed (Figure 8.16b), and cell viability was improved from 68% to 94% relative to control cells as measured by an MTT assay (Figure 8.16c) [111].

To elucidate underlying mechanisms involved in the aggregation of αS and the prevention thereof by SK-129, ^{1}H-^{15}N HSQC experiments were carried out using ^{15}N-labeled αS. The relative percent change in backbone amide peak intensity was then calculated over a series of SK-129 titrations (Figure 8.16c). Residues 6-12, 15-23, 36-45, and 48-53 demonstrated the greatest changes in NMR peak intensity upon the addition of SK-129 and are therefore hypothesized to be potential binding sites or regions influenced by a conformational change upon binding of the ligand (Figure 8.16d) [111]. To further probe the involvement of these residues in the aggregation of αS, the authors expressed four αS mutants (Figure 8.16e, αS1-4) with deletions of each span of residues individually. Each mutant was aggregated in the absence and presence of SK-129, and it is shown by ThT, TEM, and CD experiments that mutants with deletions of residues 6-12 or 15-23 (αS1 and αS2, respectively) aggregate under seed-catalyzed and uncatalyzed conditions (Figure 8.1f); however, a deletion of the remaining segments in mutants αS3 and αS4 displayed no propensity to aggregate. In all mutants, treatment with SK-129 inhibited aggregation [111]. This demonstrates that SK-129 may be binding to several regions of αS and/or inducing conformational change in the N-terminus to prevent protein aggregation, while also identifying that residues 36-46 and 48-53 are necessary for protein aggregation to occur (Figure 8.1g) [111]. Together, these studies support that the OQ scaffold not only exhibits favorable therapeutic potential but also provides mechanistic insights into targeting disease-related aPPIs.

Seed-catalyzed aggregation (prion-like spread) of αS also plays a significant role in the progression of PD phenotypes [131–133]. To investigate the effect of seed-catalyzed αS aggregation, *in vitro* and ex vivo experiments were carried out with αS seeds extracted from brain samples of postmortem PD patients. Healthy (control) brain samples demonstrated no significant increase in αS aggregates or toxicity in HEK cell models and primary rat hippocampal neuron models; however, when incubated with samples extracted from PD brains, a significant increase in aggregate formation (inclusions) and neurotoxicity in hippocampal neurons was observed (approximately eightfold greater than the control) [111]. In contrast, when incubated with PD brain extracts and SK-129 treatment, no significant inclusions or neurotoxicity was reported in HEK cells or hippocampal neurons [111]. This supports that SK-129 is capable of modulating the disease-relevant seed-catalyzed aggregation of αS.

To evaluate the rescuing effect of SK-129 *in vivo*, a well-established *C. elegans* model (NL5901) was used, in which the worms express WT αS-YFP in their body wall muscles [111]. NL5901 worms exhibit PD phenotypes upon aging, such as the aggregation of αS in the body wall muscle cells and a decline in motility rate in comparison to non-diseased strains (N2) [59]. The diseased *C. elegans* demonstrated a drastic decline in motility (Figure 8.16h,i) and significant formation of aggregates (Figure 8.16j) over 10 days. When treated with SK-129, the number of aggregates was drastically reduced, and a motility rate similar to the control was restored, suggesting

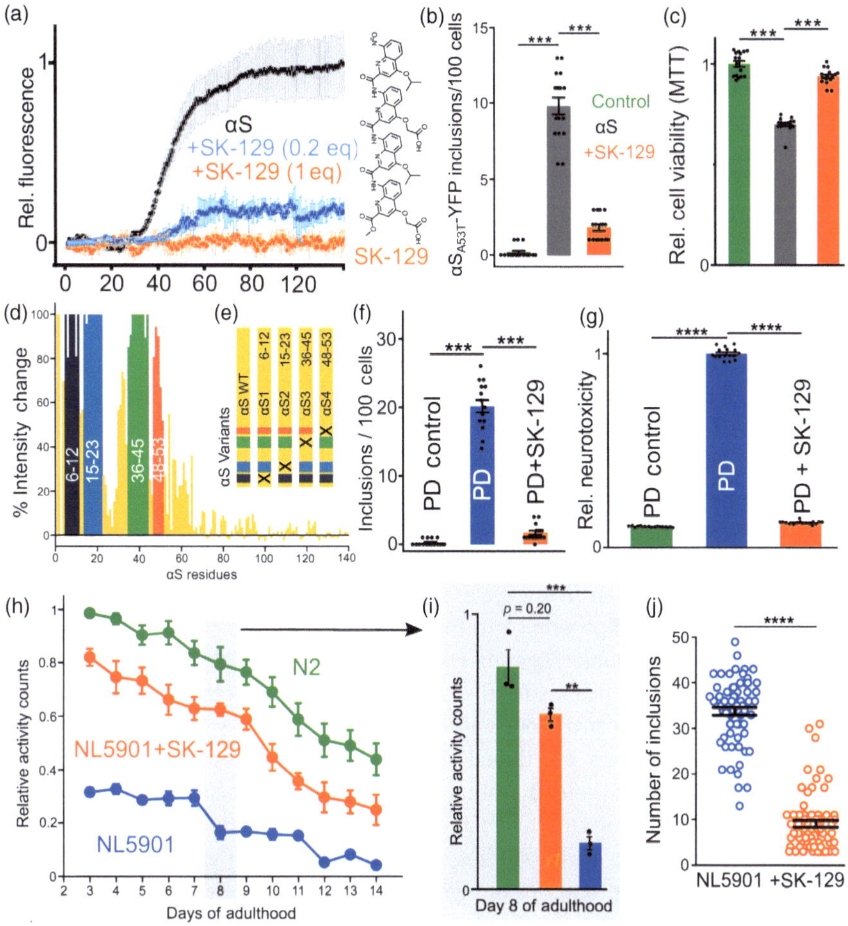

Figure 8.16 (a) ThT aggregation assay of αS in the absence and presence of SK-129 at various molar equivalents and the chemical structure of SK-129; (b) Quantification of inclusions/100 cells in HEK cells expressing αS-A53T-YFP transfected with αS preformed fibrils in the absence and presence of equimolar SK-129; (c) Relative cell viability as quantified by MTT of HEK cells transfected with αS preformed fibrils in the absence and presence of SK-129; (d) Percent intensity change of amide backbone HSQC resonances of 15 N-labeled αS in the presence of SK-129; (e) Representation of αS1-4 mutations denoting sequence residues removed in each mutant; (f) Quantification of inclusions/100 cells in HEK cells treated with healthy (control) and PD brain extracts in the absence and presence of SK-129 at an equimolar ratio; (g) The relative neurotoxicity of primary neurons incubated with healthy and PD brain extracts in the absence and presence of equimolar SK-129; The relative activity counts for control (N2), diseased (NL5901) and treated (NL5901+SK-129) *C. elegans* from day three to day 14 of adulthood (h) and statistical analysis of activity counts on day eight (i); (j) Quantification and statistical analysis of inclusions quantified by confocal microscopy of NL5901 worms in the absence and presence of SK-129. The statistical analysis was performed using one-way analysis of variance (ANOVA) with Tukey's multiple comparison test. *$p < 0.05$, **$p < 0.01$, ***$p < 0.001$. Source: Ahmed et al. [111]/Springer Nature/CC BY 4.0.

SK-129 effectively rescues PD phenotypes in an *in vivo* model [111]. This comprehensive study is the first report of an OQ foldamer with the capability of potently modulating disease-related aPPIs in several PD model systems.

8.17 Epilogue

Unlike enzymes and other traditional small-molecule targets with small, well-defined pockets (300–1000 $Å^2$) as binding sites, the PPI interfaces are significantly larger, (1500–3000 $Å^2$) predominantly hydrophobic flat surfaces with few pockets, rendering them a challenging target for small molecule design [96, 134–136]. Numerous peptide-based PPI inhibitors with significant inhibitory potency have been reported in recent years [137]. However, the persistent issues of conformational stability, bioavailability, and susceptibility to proteolytic degradation remain key hindrances to this approach despite efforts employing techniques such as stapling to mitigate these drawbacks [58, 59, 138, 139]. The rigidity from the intramolecular hydrogen bonds of OPs allows them to mimic the secondary structure of target proteins and modulate disease-related aPPIs, while resisting enzymatic degradation in living systems. The dynamic nature of the OQ scaffold allows for a sampling of protein conformations to stabilize.

Several of these foldamer ligands (RD242, SK-129, and NS132 and its analog NS163) have exhibited promising therapeutic efficacy, as demonstrated in assays conducted in a lower *in vivo* model, specifically *C. elegans* [58, 59, 111]. Advancing development necessitates preclinical studies aimed at assessing properties that will reflect measures of their suitability for the clinical trial stage. Considering the demonstrated efficacy of OP and OQ foldamer-based ligands toward modulation of aPPIs in *in vitro* and *in vivo* studies, a logical progression would be to evaluate their pharmacokinetic properties in higher level mouse models and ex vivo human studies. A crucial prerequisite for developing treatments for neurological conditions is the capacity of the candidate molecules to cross the blood–brain barrier (BBB). Numerous potential therapeutic interventions for neurological disorders have not progressed to clinical trials due to their inability to traverse the BBB [140–142], hence the necessity to further explore these capabilities of OPs and OQs. As demonstrated with NS163, foldamers are synthetically tunable to increase their therapeutic properties without sacrificing aPPI modulation. The dynamic nature of the OQ scaffold allows for induced conformational changes in its target proteins to combat aberrant interactions while providing mechanistic insights into amyloid formation [47, 111–114, 119]. The marked success of both OP and OQ foldamer scaffolds in the abrogation of aPPIs is a promising step toward the development of potent therapeutics for diseases once thought to be "undruggable."

Acknowledgement

The authors would like to thank the Department of Chemistry and Biochemistry, The Knoebel Institute for Healthy Aging, and the University of Denver for the startup funds. The authors would like to thank all the former and current member of Sunil Kumar's lab who contributed to some of work cited in this article. We would also like to thank the American Parkinson Disease Association Research Grant R79, The Stanley Fahn Junior Faculty Award from Parkinson's Foundation PF-SF-JFA-1046268, and the NIH grant 1R61NS137824-01 (to S.K.) for the support of the research conducted in this manuscript. The authors would also like to thank Dr. Anza Darehshouri at CU Anschutz for TEM assistance.

References

1 Conte, L.L., Chothia, C., and Janin, J. (1999). The atomic structure of protein-protein recognition sites. *Journal of Molecular Biology* 285: 2177–2198.

2 Jones, S. and Thornton, J.M. (1996). Principles of protein-protein interactions. *Proceedings of the National Academy of Sciences* 93: 13–20.

3 Arkin, M.R., Tang, Y., and Wells, J.A. (2014). Small-molecule inhibitors of protein-protein interactions: progressing toward the reality. *Chemistry & Biology* 21: 1102–1114.

4 Rual, J.-F., Venkatesan, K., Hao, T. et al. (2005). Towards a proteome-scale map of the human protein–protein interaction network. *Nature* 437: 1173–1178.

5 Sheng, C., Dong, G., Miao, Z. et al. (2015). State-of-the-art strategies for targeting protein–protein interactions by small-molecule inhibitors. *Chemical Society Reviews* 44: 8238–8259.

6 Stelzl, U., Worm, U., Lalowski, M. et al. (2005). A human protein-protein interaction network: a resource for annotating the proteome. *Cell* 122: 957–968.

7 Lee, M.J. and Yaffe, M.B. (2016). Protein regulation in signal transduction. *Cold Spring Harbor Perspectives in Biology* 8: a005918.

8 Cargnello, M. and Roux, P.P. (2011). Activation and function of the MAPKs and their substrates, the MAPK-activated protein kinases. *Microbiology and Molecular Biology Reviews: MMBR* 75: 50–83.

9 Zhang, W. and Liu, H.T. (2002). MAPK signal pathways in the regulation of cell proliferation in mammalian cells. *Cell Research* 12: 9–18.

10 Lee, E.-R., Kim, J.-Y., Kang, Y.-J. et al. (2006). Interplay between PI3K/Akt and MAPK signaling pathways in DNA-damaging drug-induced apoptosis. *Biochimica et Biophysica Acta (BBA)-Molecular Cell Research* 1763: 958–968.

11 Malumbres, M. (2014). Cyclin-dependent kinases. *Genome Biology* 15: 122.

12 Łukasik, P., Baranowska-Bosiacka, I., Kulczycka, K., and Gutowska, I. (2021). Inhibitors of cyclin-dependent kinases: types and their mechanism of action. *International Journal of Molecular Sciences* 22: 2806.

13 Lucas, M., Morris, A., Townsend-Teague, A. et al. (2023). Inferring cell cycle phases from a partially temporal network of protein interactions. *Cell Reports Methods* 3: 100397.

14 Stanyon, C.A., Liu, G., Mangiola, B.A. et al. (2004). A drosophila protein-interaction map centered on cell-cycle regulators. *Genome Biology* 5: R96.

15 Welch, G.R. (1978). On the role of organized multienzyme systems in cellular metabolism: a general synthesis. *Progress in Biophysics and Molecular Biology* 32: 103–191.

16 Schmitt, D.L. and An, S. (2017). Spatial organization of metabolic enzyme complexes in cells. *Biochemistry* 56: 3184–3196.

17 Rebsamen, M., Kandasamy, R.K., and Superti-Furga, G. (2013). Protein interaction networks in innate immunity. *Trends in Immunology* 34: 610–619.

18 Fukuyama, H., Verdier, Y., Guan, Y. et al. (2013). Landscape of protein–protein interactions in drosophila immune deficiency signaling during bacterial challenge. *Proceedings of the National Academy of Sciences* 110: 10717–10722.

19 Peng, R. and Deng, M. (2023). Mapping the protein–protein interactome in the tumor immune microenvironment. *Antibody Therapeutics* 6: 311–321.

20 Biza, K.V., Nastou, K.C., Tsiolaki, P.L. et al. (2017). The amyloid interactome: exploring protein aggregation. *PLoS One* 12: e0173163.

21 Cheng, F., Zhao, J., Wang, Y. et al. (2021). Comprehensive characterization of protein–protein interactions perturbed by disease mutations. *Nature Genetics* 53: 342–353.

22 Cummings, C.G. and Hamilton, A.D. (2010). Disrupting protein–protein interactions with non-peptidic, small molecule α-helix mimetics. *Current Opinion in Chemical Biology* 14: 341–346.

23 Fry, D.C. (2015). Targeting protein-protein interactions for drug discovery. In: *Protein-Protein Interact* (ed. C.L. Meyerkord and H. Fu), 93–106. New York, NY: Springer.

24 Gonzalez, M.W. and Kann, M.G. (2012). Chapter 4: protein interactions and disease. *PLoS Computational Biology* 8: e1002819.

25 Lage, K. (2014). Protein–protein interactions and genetic diseases: the interactome. *Biochimica et Biophysica Acta (BBA) - Molecular Basis of Disease* 1842: 1971–1980.

26 Mintseris, J. and Weng, Z. (2005). Structure, function, and evolution of transient and obligate protein–protein interactions. *Proceedings of the National Academy of Sciences* 102: 10930–10935.

27 Modell, A.E., Blosser, S.L., and Arora, P.S. (2016). Systematic targeting of protein–protein interactions. *Trends in Pharmacological Sciences* 37: 702–713.

28 Joerger, A.C. and Fersht, A.R. (2008). Structural biology of the tumor suppressor p53. *Annual Review of Biochemistry* 77: 557–582.

29 Silva, J.L., Gallo, C.V.D.M., Costa, D.C.F., and Rangel, L.P. (2014). Prion-like aggregation of mutant p53 in cancer. *Trends in Biochemical Sciences* 39: 260–267.

30 Pitolli, C., Wang, Y., Mancini, M. et al. (2019). Do mutations turn p53 into an oncogene? *International Journal of Molecular Sciences* 20: 6241.

31 Abedini, A. and Schmidt, A.M. (2013). Mechanisms of islet amyloidosis toxicity in type 2 diabetes. *FEBS Letters* 587: 1119–1127.

32 Brown, J. and Horrocks, M.H. (2020). A sticky situation: aberrant protein–protein interactions in Parkinson's disease. *Seminars in Cell & Developmental Biology* 99: 65–77.

33 Sadigh-Eteghad, S., Sabermarouf, B., Majdi, A. et al. (2015). Amyloid-beta: a crucial factor in Alzheimer's disease. *Medical Principles and Practice* 24: 1–10.

34 Sehar, U., Rawat, P., Reddy, A.P. et al. (2022). Amyloid beta in aging and Alzheimer's disease. *International Journal of Molecular Sciences* 23: 12924.

35 Jennings, D., Huntwork-Rodriguez, S., Henry, A.G. et al. (2022). Preclinical and clinical evaluation of the LRRK2 inhibitor DNL201 for Parkinson's disease. *Science Translational Medicine* 14: eabj2658.

36 Lu, H., Zhou, Q., He, J. et al. (2020). Recent advances in the development of protein–protein interactions modulators: mechanisms and clinical trials. *Signal Transduction and Targeted Therapy* 5: 213.

37 Pérez-Arancibia, R., Cisternas-Olmedo, M., Sepúlveda, D. et al. (2023). Small molecules to perform big roles: the search for Parkinson's and Huntington's disease therapeutics. *Frontiers in Neuroscience* 16: 1084493.

38 Wagner, J., Ryazanov, S., Leonov, A. et al. (2013). Anle138b: a novel oligomer modulator for disease-modifying therapy of neurodegenerative diseases such as prion and Parkinson's disease. *Acta Neuropathol (Berl)* 125: 795–813.

39 Wang, S., Zhao, Y., Aguilar, A. et al. (2017). Targeting the MDM2–p53 protein–protein interaction for new cancer therapy: progress and challenges. *Cold Spring Harbor Perspectives in Medicine* 7: a026245.

40 Vabulas, R.M. and Hartl, F.U. (2011). Aberrant protein interactions in amyloid disease. *Cell Cycle* 10: 1512–1513.

41 Fletcher, S. and Hamilton, A.D. (2005). Protein surface recognition and proteomimetics: mimics of protein surface structure and function. *Current Opinion in Chemical Biology* 9: 632–638.

42 Orner, B.P., Ernst, J.T., and Hamilton, A.D. (2001). Toward proteomimetics: terphenyl derivatives as structural and functional mimics of extended regions of an α-helix. *Journal of the American Chemical Society* 123: 5382–5383.

43 Ernst, J.T., Kutzki, O., Debnath, A.K. et al. (2002). Design of a protein surface antagonist based on α-helix mimicry: inhibition of gp41 assembly and viral fusion. *Angewandte Chemie, International Edition* 41: 278.

44 Ernst, J.T., Becerril, J., Park, H.S. et al. (2003). Design and application of an α-helix-mimetic scaffold based on an oligoamide-foldamer strategy: antagonism of the Bak BH3/Bcl-xL complex. *Angewandte Chemie, International Edition* 42: 535–539.

45 Gellman, S.H. (1998). Foldamers: a manifesto. *Accounts of Chemical Research* 31: 173–180.

46 Goodman, C.M., Choi, S., Shandler, S., and DeGrado, W.F. (2007). Foldamers as versatile frameworks for the design and evolution of function. *Nature Chemical Biology* 3: 252–262.

47 Kumar, S., Birol, M., and Miranker, A.D. (2016). Foldamer scaffolds suggest distinct structures are associated with alternative gains-of-function in a preamyloid toxin. *Chemical Communications* 52: 6391–6394.

48 Trisciuoglio, D., Tupone, M.G., Desideri, M. et al. (2017). BCL-XL overexpression promotes tumor progression-associated properties. *Cell Death & Disease* 8: 3216.

49 Choi, S., Chen, Z., Tang, L.H. et al. (2016). Bcl-xL promotes metastasis independent of its anti-apoptotic activity. *Nature Communications* 7: 10384.

50 Kumar, S., Schlamadinger, D.E., Brown, M.A. et al. (2015). Islet amyloid-induced cell death and bilayer integrity loss share a molecular origin targetable with oligopyridylamide-based α-helical mimetics. *Chemistry & Biology* 22: 369–378.

51 Kumar, S. and Hamilton, A.D. (2017). α-helix mimetics as modulators of Aβ self-assembly. *Journal of the American Chemical Society* 139: 5744–5755.

52 Palanikumar, L., Karpauskaite, L., Al-Sayegh, M. et al. (2021). Protein mimetic amyloid inhibitor potently abrogates cancer-associated mutant p53 aggregation and restores tumor suppressor function. *Nature Communications* 12: 3962.

53 Kumar, S., Henning-Knechtel, A., Magzoub, M., and Hamilton, A.D. (2018). Peptidomimetic-based multidomain targeting offers critical evaluation of Aβ structure and toxic function. *Journal of the American Chemical Society* 140: 6562–6574.

54 Maity, D., Kumar, S., AlHussein, R. et al. (2020). Sub-stoichiometric inhibition of IAPP aggregation: a peptidomimetic approach to anti-amyloid agents. *RSC Chemical Biology* 1: 225–232.

55 Kulikov, O.V., Kumar, S., Magzoub, M. et al. (2015). Amphiphilic oligoamide α-helix peptidomimetics inhibit islet amyloid polypeptide aggregation. *Tetrahedron Letters* 56: 3670–3673.

56 Bavinton, C.E., Sternke-Hoffmann, R., Yamashita, T. et al. (2022). Rationally designed helical peptidomimetics disrupt α-synuclein fibrillation. *Chemical Communications* 58: 5132–5135.

57 Hebda, J.A., Saraogi, I., Magzoub, M. et al. (2009). A peptidomimetic approach to targeting pre-amyloidogenic states in type II diabetes. *Chemistry & Biology* 16: 943–950.

58 Dohoney, R.A., Joseph, J.A., Baysah, C. et al. (2023). "Common-precursor" protein mimetic approach to rescue Aβ aggregation-mediated Alzheimer's phenotypes. *ACS Chemical Biology* 18: 1510–1522.

59 Stillman, N.H., Joseph, J.A., Ahmed, J. et al. (2024). Protein mimetic 2D FAST rescues alpha synuclein aggregation mediated early and post disease Parkinson's phenotypes. *Nature Communications* 15: 3658.

60 Maity, D., Kumar, S., Curreli, F. et al. (2019). α-helix-mimetic foldamers for targeting HIV-1 TAR RNA. *Chemistry - A European Journal* 25: 7265–7269.

61 Jakob-Roetne, R. and Jacobsen, H. (2009). Alzheimer's disease: from pathology to therapeutic approaches. *Angewandte Chemie, International Edition* 48: 3030–3059.

62 Mao, Y., Fisher, D.W., Yang, S. et al. (2020). Protein-protein interactions underlying the behavioral and psychological symptoms of dementia (BPSD) and Alzheimer's disease. *PLoS One* 15: e0226021.

63 Matthews, K.A., Xu, W., Gaglioti, A.H. et al. (2019). Racial and ethnic estimates of Alzheimer's disease and related dementias in the United States (2015–2060) in adults aged ≥65 years. *Alzheimers Dement* 15: 17–24.

64 Trejo-Lopez, J.A., Yachnis, A.T., and Prokop, S. (2022). Neuropathology of Alzheimer's disease. *Neurotherapeutics* 19: 173–185.

65 Pagano, K., Tomaselli, S., Molinari, H., and Ragona, L. (2020). Natural compounds as inhibitors of Aβ peptide aggregation: chemical requirements and molecular mechanisms. *Frontiers in Neuroscience* 14: 619667.

66 Berkhout, B., Silverman, R.H., and Jeang, K.-T. (1989). Tat trans-activates the human immunodeficiency virus through a nascent RNA target. *Cell* 59: 273–282.

67 Charnay, N., Ivanyi-Nagy, R., Soto-Rifo, R. et al. (2009). Mechanism of HIV-1 tat RNA translation and its activation by the tat protein. *Retrovirology* 6: 74.

68 Dingwall, C., Ernberg, I., Gait, M.J. et al. (1990). HIV-1 tat protein stimulates transcription by binding to a U-rich bulge in the stem of the TAR RNA structure. *The EMBO Journal* 9: 4145–4153.

69 Dingwall, C., Ernberg, I., Gait, M.J. et al. (1989). Human immunodeficiency virus 1 tat protein binds trans-activation-responsive region (TAR) RNA in vitro. *Proceedings of the National Academy of Sciences* 86: 6925–6929.

70 Hamy, F., Felder, E.R., Heizmann, G. et al. (1997). An inhibitor of the tat/TAR RNA interaction that effectively suppresses HIV-1 replication. *Proceedings of the National Academy of Sciences* 94: 3548–3553.

71 Westermark, P., Andersson, A., and Westermark, G.T. (2011). Islet amyloid polypeptide, islet amyloid, and diabetes mellitus. *Physiological Reviews* 91: 795–826.

72 Akter, R., Cao, P., Noor, H. et al. (2016). Islet amyloid polypeptide: structure, function, and pathophysiology. *Journal Diabetes Research* 2016: 1–18.

73 Wirth, F., Heitz, F.D., Seeger, C. et al. (2023). A human antibody against pathologic IAPP aggregates protects beta cells in type 2 diabetes models. *Nature Communications* 14: 6294.

74 Bieging, K.T., Mello, S.S., and Attardi, L.D. (2014). Unravelling mechanisms of p53-mediated tumour suppression. *Nature Reviews. Cancer* 14: 359–370.

75 Khoo, K.H., Verma, C.S., and Lane, D.P. (2014). Drugging the p53 pathway: understanding the route to clinical efficacy. *Nature Reviews. Drug Discovery* 13: 217–236.

76 Hu, J., Cao, J., Topatana, W. et al. (2021). Targeting mutant p53 for cancer therapy: direct and indirect strategies. *Journal of Hematology & Oncology* 14: 157.

77 Kim, M.P. and Lozano, G. (2018). Mutant p53 partners in crime. *Cell Death and Differentiation* 25: 161–168.

78 Hafner, A., Bulyk, M.L., Jambhekar, A., and Lahav, G. (2019). The multiple mechanisms that regulate p53 activity and cell fate. *Nature Reviews. Molecular Cell Biology* 20: 199–210.

79 Kubbutat, M.H.G., Jones, S.N., and Vousden, K.H. (1997). Regulation of p53 stability by mdm2. *Nature* 387: 299–303.

80 Koo, N., Sharma, A.K., and Narayan, S. (2022). Therapeutics targeting p53-MDM2 interaction to induce cancer cell death. *International Journal of Molecular Sciences* 23: 5005.

81 Wang, H., Guo, M., Wei, H., and Chen, Y. (2023). Targeting p53 pathways: mechanisms, structures and advances in therapy. *Signal Transduction and Targeted Therapy* 8: 92.

82 Haupt, Y., Maya, R., Kazaz, A., and Oren, M. (1997). Mdm2 promotes the rapid degradation of p53. *Nature* 387: 296–299.

83 Klemke, L., Fehlau, C.F., Winkler, N. et al. (2021). The gain-of-function p53 R248W mutant promotes migration by STAT3 deregulation in human pancreatic cancer cells. *Frontiers in Oncology* 11: 642603.

84 Chasov, V., Mirgayazova, R., Zmievskaya, E. et al. (2020). Key players in the mutant p53 team: small molecules, gene editing, immunotherapy. *Frontiers in Oncology* 10: 1460.

85 Zhang, Y., Xu, L., Chang, Y. et al. (2020). Therapeutic potential of ReACp53 targeting mutant p53 protein in CRPC. *Prostate Cancer and Prostatic Diseases* 23: 160–171.

86 Teschendorf, D. and Link, C.D. (2009). What have worm models told us about the mechanisms of neuronal dysfunction in human neurodegenerative diseases? *Molecular Neurodegeneration* 4: 38.

87 Dawson, T.M. and Dawson, V.L. (2003). Molecular pathways of neurodegeneration in Parkinson's disease. *Science* 302: 819–822.

88 Dettmer, U., Selkoe, D., and Bartels, T. (2016). New insights into cellular α-synuclein homeostasis in health and disease. *Current Opinion in Neurobiology* 36: 15–22.

89 Goedert, M. (2001). Alpha-synuclein and neurodegenerative diseases. *Nature Reviews. Neuroscience* 2: 492–501.

90 Kingwell, K. (2017). Zeroing in on neurodegenerative α-synuclein. *Nature Reviews. Drug Discovery* 16: 371–373.

91 Spillantini, M.G., Schmidt, M.L., Lee, V.M.-Y. et al. (1997). α-synuclein in Lewy bodies. *Nature* 388: 839–840.

92 Gao, V., Briano, J.A., Komer, L.E., and Burré, J. (2023). Functional and pathological effects of α-synuclein on synaptic SNARE complexes. *Journal of Molecular Biology* 435: 167714.

93 Yoo, G., Shin, Y.-K., and Lee, N.K. (2023). The role of α-synuclein in SNARE-mediated synaptic vesicle fusion. *Journal of Molecular Biology* 435: 167775.

94 Stefanis, L. (2012). α-synuclein in Parkinson's disease. *Cold Spring Harbor Perspectives in Medicine* 2: a009399–a009399.

95 Lurette, O., Martín-Jiménez, R., Khan, M. et al. (2023). Aggregation of alpha-synuclein disrupts mitochondrial metabolism and induce mitophagy via cardiolipin externalization. *Cell Death & Disease* 14: 729.

96 Fry, D. (2008). Drug-like inhibitors of protein-protein interactions: a structural examination of effective protein mimicry. *Current Protein & Peptide Science* 9: 240–247.

97 Sang, P. and Cai, J. Unnatural helical peptidic foldamers as protein segment mimics. *Chemical Society Reviews* 52: 4843–4877.

98 Mahon, A.B. and Arora, P.S. (2012). End-capped α-helices as modulators of protein function. *Drug Discovery Today: Technologies* 9: e57–e62.

99 Pace, C.N. and Scholtz, J.M. (1998). A helix propensity scale based on experimental studies of peptides and proteins. *Biophysical Journal* 75: 422–427.

100 Keskin, O., Gursoy, A., Ma, B., and Nussinov, R. (2008). Principles of protein–protein interactions: what are the preferred ways for proteins to interact? *Chemical Reviews* 108: 1225–1244.

101 Scott, D.E., Bayly, A.R., Abell, C., and Skidmore, J. (2016). Small molecules, big targets: drug discovery faces the protein–protein interaction challenge. *Nature Reviews. Drug Discovery* 15: 533–550.

102 Kaffy, J., Berardet, C., Mathieu, L. et al. (2020). Helical γ-peptide foldamers as dual inhibitors of amyloid-β peptide and islet amyloid polypeptide oligomerization and fibrillization. *Chemistry - A European Journal* 26: 14612–14622.

103 Liu, H., Cui, Y., Zhao, X. et al. (2024). Helical sulfonyl-γ-AApeptides modulating Aβ oligomerization and cytotoxicity by recognizing Aβ helix. *Proceedings of the National Academy of Sciences* 121: e2311733121.

104 Cheng, R.P., Gellman, S.H., and DeGrado, W.F. (2001). β-peptides: from structure to function. *Chemical Reviews* 101: 3219–3232.

105 Simon, R.J., Kania, R.S., Zuckermann, R.N. et al. (1992). Peptoids: a modular approach to drug discovery. *Proceedings of the National Academy of Sciences* 89: 9367–9371.

106 Zhu, J., Parra, R.D., Zeng, H. et al. (2000). A new class of folding oligomers: crescent Oligoamides. *Journal of the American Chemical Society* 122: 4219–4220.

107 Nguyen, K.T., Le Clair, S.V., Ye, S., and Chen, Z. (2009). Orientation determination of protein helical secondary structures using linear and nonlinear vibrational spectroscopy. *The Journal of Physical Chemistry B* 113: 12169–12180.

108 Jiang, H., Léger, J.-M., and Huc, I. (2003). Aromatic δ-peptides. *Journal of the American Chemical Society* 125: 3448–3449.

109 Gillies, E.R., Dolain, C., Léger, J.-M., and Huc, I. (2006). Amphipathic helices from aromatic amino acid oligomers. *The Journal of Organic Chemistry* 71: 7931–7939.

110 Dolain, C., Grélard, A., Laguerre, M. et al. (2005). Solution structure of quinoline- and pyridine-derived oligoamide foldamers. *Chemistry - A European Journal* 11: 6135–6144.

111 Ahmed, J., Fitch, T.C., Donnelly, C.M. et al. (2022). Foldamers reveal and validate therapeutic targets associated with toxic α-synuclein self-assembly. *Nature Communications* 13: 2273.

112 Kumar, S., Henning-Knechtel, A., Chehade, I. et al. (2017). Foldamer-mediated structural rearrangement attenuates Aβ oligomerization and cytotoxicity. *Journal of the American Chemical Society* 139: 17098–17108.

113 Kumar, S., Birol, M., Schlamadinger, D.E. et al. (2016). Foldamer-mediated manipulation of a pre-amyloid toxin. *Nature Communications* 7: 11412.

114 Kumar, S., Brown, M.A., Nath, A., and Miranker, A.D. (2014). Folded small molecule manipulation of islet amyloid polypeptide. *Chemistry & Biology* 21: 775–781.

115 Gillies, E.R., Deiss, F., Staedel, C. et al. (2007). Development and biological assessment of fully water-soluble helical aromatic amide foldamers. *Angewandte Chemie, International Edition* 46: 4081–4084.

116 Qi, T., Deschrijver, T., and Huc, I. (2013). Large-scale and chromatography-free synthesis of an octameric quinoline-based aromatic amide helical foldamer. *Nature Protocols* 8: 693–708.

117 Ghosez, L., Haveaux, B., and Viehe, H.G. (1969). Alkyl- und Aryl-α-chlorenamine. *Angewandte Chemie* 81: 468–469.

118 Baptiste, B., Douat-Casassus, C., Laxmi-Reddy, K. et al. (2010). Solid phase synthesis of aromatic oligoamides: application to helical water-soluble foldamers. *The Journal of Organic Chemistry* 75: 7175–7185.

119 Kumar, S. and Miranker, A.D. (2013). A foldamer approach to targeting membrane bound helical states of islet amyloid polypeptide. *Chemical Communications* 49: 4749–4751.

120 Magzoub, M. and Miranker, A.D. (2012). Concentration-dependent transitions govern the subcellular localization of islet amyloid polypeptide. *The FASEB Journal* 26: 1228–1238.

121 Hebda, J.A. and Miranker, A.D. (2009). The interplay of catalysis and toxicity by amyloid intermediates on lipid bilayers: insights from type II diabetes. *Annual Review of Biophysics* 38: 125–152.

122 Saraogi, I., Hebda, J.A., Becerril, J. et al. (2010). Synthetic α-helix mimetics as agonists and antagonists of islet amyloid polypeptide aggregation. *Angewandte Chemie, International Edition* 49: 736–739.

123 Wolfe, L.S., Calabrese, M.F., Nath, A. et al. (2010). Protein-induced photophysical changes to the amyloid indicator dye thioflavin T. *Proceedings of the National Academy of Sciences* 107: 16863–16868.

124 Ruschak, A.M. and Miranker, A.D. (2007). Fiber-dependent amyloid formation as catalysis of an existing reaction pathway. *Proceedings of the National Academy of Sciences* 104: 12341–12346.

125 Chen, G., Xu, T., Yan, Y. et al. (2017). Amyloid beta: structure, biology and structure-based therapeutic development. *Acta Pharmacologica Sinica* 38: 1205–1235.

126 Tolar, M., Hey, J., Power, A., and Abushakra, S. (2021). Neurotoxic soluble amyloid oligomers drive Alzheimer's pathogenesis and represent a clinically validated target for slowing disease progression. *International Journal of Molecular Sciences* 22: 6355.

127 Sinha, S., Lopes, D.H.J., Du, Z. et al. (2011). Lysine-specific molecular tweezers are broad-spectrum inhibitors of assembly and toxicity of amyloid proteins. *Journal of the American Chemical Society* 133: 16958–16969.

128 Gazit, E. (2002). A possible role for π-stacking in the self-assembly of amyloid fibrils. *The FASEB Journal* 16: 77–83.

129 Sangwan, S., Sahay, S., Murray, K.A. et al. (2020). Inhibition of synucleinopathic seeding by rationally designed inhibitors. *eLife* 9: e46775.

130 Sanders, D.W., Kaufman, S.K., DeVos, S.L. et al. (2014). Distinct tau prion strains propagate in cells and mice and define different Tauopathies. *Neuron* 82: 1271–1288.

131 Guerrero-Ferreira, R., Taylor, N.M., Arteni, A.-A. et al. (2019). Two new polymorphic structures of human full-length alpha-synuclein fibrils solved by cryo-electron microscopy. *eLife* 8: e48907.

132 Strohäker, T., Jung, B.C., Liou, S.-H. et al. (2019). Structural heterogeneity of α-synuclein fibrils amplified from patient brain extracts. *Nature Communications* 10: 5535.

133 Bousset, L., Pieri, L., Ruiz-Arlandis, G. et al. (2013). Structural and functional characterization of two alpha-synuclein strains. *Nature Communications* 4: 2575.

134 Mullard, A. (2012). Protein–protein interaction inhibitors get into the groove. *Nature Reviews. Drug Discovery* 11: 173–175.

135 Whitty, A. and Kumaravel, G. (2006). Between a rock and a hard place? *Nature Chemical Biology* 2: 112–118.

136 Chène, P. (2006). Drugs targeting protein–protein interactions. *ChemMedChem* 1: 400–411.

137 Wójcik, P. and Berlicki, Ł. (2016). Peptide-based inhibitors of protein–protein interactions. *Bioorganic & Medicinal Chemistry Letters* 26: 707–713.

138 Pereira, A.J., de Campos, L.J., Xing, H., and Conda-Sheridan, M. (2024). Peptide-based therapeutics: challenges and solutions. *Medicinal Chemistry Research* 33: 1275–1280.

139 Philippe, G., Huang, Y., Cheneval, O. et al. (2016). Development of cell-penetrating peptide-based drug leads to inhibit MDMX:p53 and MDM2:p53 interactions. *Peptide Science* 106: 853–863.

140 Pardridge, W.M. (2009). Alzheimer's disease drug development and the problem of the blood-brain barrier. *Alzheimers Dement* 5: 427–432.

141 Pardridge, W.M. (2020). Treatment of Alzheimer's disease and blood–brain barrier drug delivery. *Pharmaceuticals* 13: 394.

142 Palmer, A.M. (2011). The role of the blood brain barrier in neurodegenerative disorders and their treatment. *Journal of Alzheimer's Disease* 24: 643–656.

9

Application of Sulfonyl-γ-AApeptides for PPI Drug Discovery

Jarais Fontaine and Jianfeng Cai

University of South Florida, Department of Chemistry, Fl, Tampa, 33620, USA

9.1 Introduction

In accordance with their merit, peptidomimetics with a folding propensity, so-called foldamers, have distinguished themselves as potential biologically active entities thanks to their promising application in terms of molecular recognition [1]. Over the last 26 years after Gellman first introduced β-peptides as foldamers, researchers have developed a plethora of diverse foldamers such as α/β-peptides, γ-peptides, peptoids, oligourea, etc. [2] [3]. In 2011, our group reported a new class of foldameric peptidomimetics known as γ-AApeptides. Such peptides are comprised of γ-substituted-*N*-acylated-*N*-aminoethyl amino acids that were derived from the γ-chiral peptide nucleic acids. Even in their infancy, γ-AApeptides illustrated excellent proteolytic stability and potential in biomedical applications [4]. As of 2020, our lab successfully obtained crystal structures of a series of sulfonyl-γ-AApeptides that illustrated how our unique scaffold folded into well-defined helices stabilized by their own set of parameters. Based on the crystal structure, we determined that the sulfonyl-γ-AApeptides possess an intramolecular 14-hydrogen-bonding pattern, a radius of 2.8 Å, and a helical pitch of 5.1 Å, as well as exactly four side chains per turn [5] (Figure 9.1a-h). Compared to canonical α-helices, sulfonyl-γ-AApeptides demonstrated higher helical stability, presumably owing to both intramolecular bonding and constrained tertiary-sulfonamide bonds leading to greater helical stability [6].

As this chapter aims to illustrate the rationale of the design and application of sulfonyl-γ-AApeptides, the detailed synthesis of the building blocks and peptide mimics is not included but can be accessed from our previous reports [7]. Therefore, throughout the course of this chapter, we intend to highlight the biomedical applications and rational design of sulfonyl-γ-AApeptides targeting a variety of different protein–protein interactions.

Targeting Protein–Protein Interactions for Drug Discovery, First Edition. Edited by Jian Zhang.

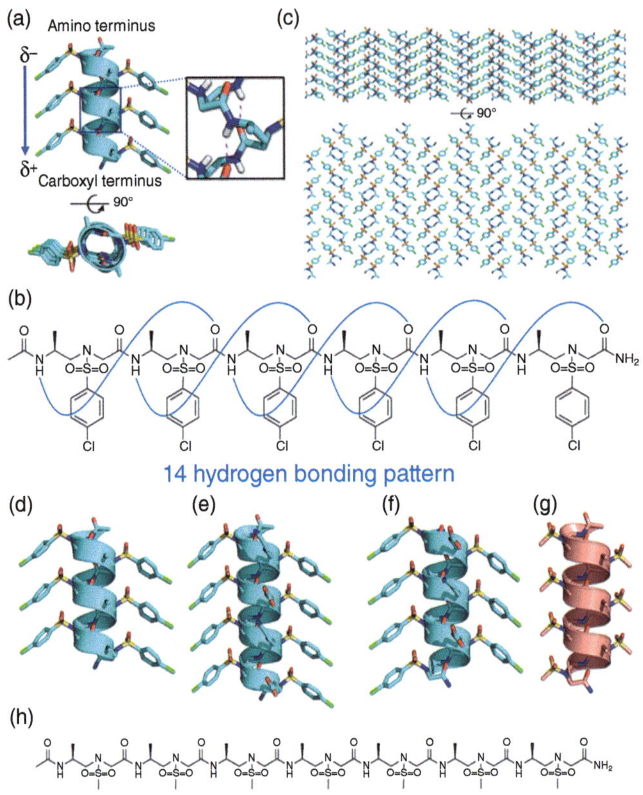

Figure 9.1 (a) Side and top views of the crystal structure of **1a**. Hydrogen bonding is shown in red. (b) The intramolecular 14-hydrogen-bonding pattern of **1a** detected in the crystal structure. (c) Crystal packing of **1a** viewed perpendicular to and down the helix axis, (d–g) comparison of crystal structures of **1a**, **3a**, **4b**, and **6a**. (h) Sequence structure of oligomer **6a**. Source: [5]/with permission of John Wiley & Sons.

9.2 Application of Sulfonyl-γ-AApeptides

9.2.1 Modulation of PPIs Involved in Cancer

9.2.1.1 Inhibition of β-catenin/B-cell lymphoma 9 PPIs

In 2019, our lab reported the first series of helical sulfonyl-γ-AApeptides that disrupted protein–protein interactions. Previous studies illustrate that B-cell lymphoma 9 (BCL9) interacts with β-catenin within an α-helix in its HD2 domain. The formation of this complex is the crucial factor leading to the Wnt/β-catenin signaling pathway, thus leading to cell growth and proliferation [8]. Therefore, we believed that by disrupting the protein–protein interactions (PPIs) between BCL9 and β-catenin, we could inhibit the Wnt/β-catenin signaling pathway, thus resulting in a unique anticancer agent. As previously mentioned, sulfonyl-γ-AApeptides exhibit excellent folding propensity by adopting a helical secondary structure, as well as having enhanced proteolytic degradation stability. Henceforth, we utilized

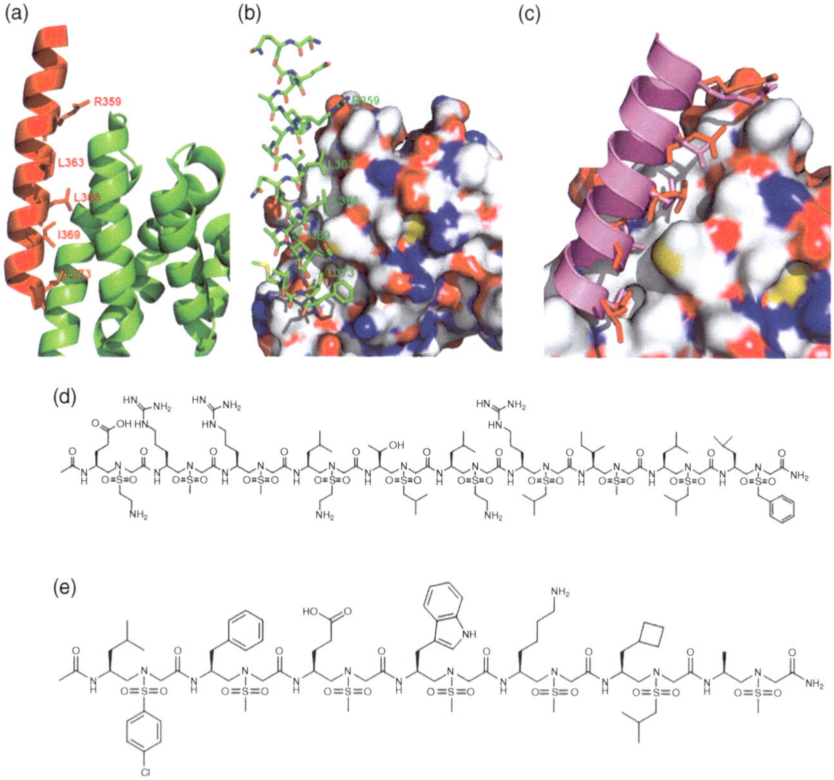

Figure 9.2 (A+B) The alpha-helical domain of HD2 domain of BCL9. (C) Overlay of 6 with critical residues of BCL9 on the binding surface of β-catenin (PDB ID code 2GL7) using PyMOL software. (D) Structure of Peptide 2. (E) Structure of lead compound PS10. A-C were [9]/with permission of National Academy of Sciences.

this unique scaffold to mimic the critical side chains given by the already solved crystal structure of the BLC9/β-catenin/TCF-4 complex (Figure 9.2a and b). It is to our delight that the initial design led to the peptidomimetic 2 with a binding affinity as low as 0.43 μM. An overlay of the sulfonyl-γ-AApeptides with the BCL-9 peptide suggests that their side chains involved in the interaction with β-catenin largely overlapped with each other (Figure 9.2c and d).

The structure-activity relationship (SAR) study led to a few sequences with excellent binding activities such as 2, 3, 4, 6 [9]. One obstacle that many peptidomimetics must overcome is cell permeability [10]. After examining the cellular reuptake of our FITC-labeled peptides using SW480 cells by confocal fluorescence microscopy, it was observed that our peptides were cell-permeable at concentrations as low as 1 μM, whereas the native helical peptide 1 showed no cell permeability. We next examined our peptide's ability to inhibit the proliferation of cancer cells. The cell visible assay illustrated that in hyperactive Wnt/β-catenin signaling cells, our peptidomimetics possessed excellent inhibitory activity and good selectivity over the cells bearing normal Wnt/β-catenin signaling. Thus, we further investigated their selectivity through the TOPFlash/FOPLash luciferase reporter assay. The results indicated that our sulfonyl-γ-AApeptides were able to selectively inhibit the Wnt/β-catenin signaling pathway without inhibiting other transcriptional pathways. When compared to the native BCL9 peptide, our sulfonyl-γ-AApeptides possess superior binding affinity, cell permeability, selective inhibition, as well as proteolytic resistance. The discovery of these unnatural helical mimetics as potential anti-cancer agents despite the previously known challenges of peptidomimetics illustrates the potential of synthetic foldamers as modulators of protein-protein interactions, and to date, we continue to work toward obtaining a co-crystal structure of our peptidomimetics with β-catenin to confirm the binding mechanism of our sulfonyl-γ-AApeptides to inhibit PPIs.

9.2.1.2 p53-MDM2/MDMX PPIs Inhibitor

The PPIs between protein 53 (p53) and human double minute 2 (MDM2), as well as MDMX, have been extensively researched due to the crucial role they play in cancer development and progression. MDM2 is an E3 ubiquitin ligase that is abundant in cancer cells and specifically binds to tumor suppressant p53 inducing proteasomal degradation. MDMX also binds to p53 in a similar fashion but differs in the mechanism of regulation. MDMX has not been shown to promote the degradation of p53 but rather inhibits DNA-binding and transcriptional activity. Due to these inherent differences in action, the necessity for finding selective inhibitors for p53/MDM2, p53/MDMX, or both remains medicinally relevant [11].

The crystal structure of the p53/MDM2 complex has revealed that there are three crucial hydrophobic side chains that align to make deep contact with the MDM2 cleft (Phe19, Trp23, and Leu26) [12]. Therefore, we set out to use sulfonyl-γ-AApeptide where we based our design on our previously reported γ-AApeptide. After careful examination of the position of the critical side chains, as well as some optimization of non-key residues, we synthesized lead compound PS10 (Figure 9.2e). The overlay of PS10 with the p53 peptide suggests that PS10 could mimic the p53 helix very well (Figure 9.2i and j). When compared to the pure p53 compound, our sulfonyl-γ-AApeptides exhibited a binding affinity and IC_{50} five times more potent (26 nM and 0.891 μM, respectively). In order to evaluate PS10 potential as an anti-cancer therapeutic, $^{15}N/^{1}H$ HSQC was performed, illustrating the slow binding exchange between MDM2 and our lead compound PS10. Next, we looked to see if PS10 was able to bind and inhibit MDMX. Our sulfonyl-γ-AApeptides that bind to MDM2 also bind to MDMX with good activity to which PS10 performed best with K_d and IC_{50} values of 221 nM and 3.6 μM, respectively [11]. Enzymatic degradation poses as another obstacle within peptidomimetics [13]. In accordance with previous studies of our sulfonyl-γ-AA peptides, lead compound PS10 showed no enzymatic degradation in the presence of protease, while native p53 peptide was completely degraded which once again demonstrated the enhanced proteolytic stability of sulfonyl-γ-AA peptides. As a potential alternative strategy toward targeting MDM2/MDMX-p53 protein–protein interactions, our unique scaffold was able to bind to MDM2 and MDMX with good activity with present-day studies geared toward obtaining the crystal structure to confirm our initial findings and predicted binding modes.

9.2.1.3 HIF-1α PPI's Inhibitor

One of the main characteristics of solid tumors is their low-oxygen environment or hypoxia. As a result of this hypoxic state, tumor cells adapt leading to a glycolytic metabolism, promotion of proliferation, apoptosis resistance, angiogenesis, and migration to less hypoxic areas [14]. Hypoxia-inducible factor 1 (HIF-1) is a transcription factor responsible for the regulation of over 100 genes that ultimately promotes rapid tumor progression and mediates the adaptive response brought by hypoxic environments. HIF-1 is a heterodimer consisting of two subunits (α and β) in which the former is responsible for the oxygen-sensing and transcriptional effects of HIF-1. Under homeostatic conditions, HIF-1α is readily degraded; however, under hypoxic conditions, it is stable and able to form a complex with HIF-1β. This complex then interacts with p300/calmodulin-binding peptide (CBP) resulting in the transcription of hypoxia-induced response genes that are crucial for tumor growth. Due to previous reports of elevated levels of HIF-1α in an abundance of different cancer cells, PPIs inhibitors toward HIF-1α have been of continued interest. We decided to use sulfonyl-γ-AA peptide scaffold to potentially inhibit the HIF-1α signaling pathway. In lieu of the solved crystal structure of the p300/CBP complex, many insights regarding the critical interactions have been solved due to solution nuclear magnetic resonance (NMR) studies. The NMR studies indicate that the binding interactions are confined to the C-terminal activation domain of HIF-1α (776–826) and the transcriptional adapter zinc-binding domain 1 (TAZ1) of p300/CBP (345–439). Further studies indicate that the C-terminal of HIF-1α is comprised of three helices (A, B, and C). Of the three helices, it was shown through mutagenesis studies that B and C are critical for the binding of HIF-1α to p300/CPB. By examining the solution NMR studies of helix C (815–824), it was determined that Leu-818, Leu-822, and Val-825 penetrate deep into the hydrophobic cleft. Therefore, we rationally designed a set of left-handed sulfonyl-γ-AA peptides with critical residues on the same face resulting in lead compound **HC13** (Figure 9.3a) that has a strong binding affinity of 97 nM as well as the IC_{50} value of 0.83 μM. In western blot analysis and pull-down studies, **HC13** possessed the best activity for selectively inhibiting the transcriptional activity of HIF-1α. Previous studies on sulfonyl-γ-AA peptides as previously stated within this chapter illustrated excellent cell permeability and enhanced proteolytic stability which is typically one of the limitations of peptides. Using FITC-labeled peptidomimetics, we were able to visualize **HC13** excellent cell permeability which is consistent with our prior findings, suggesting that sulfonyl-γ-AApeptide scaffold possesses enhanced cell permeability. To assess **HC13's** ability to regulate the HIF-1α-mediated signaling pathway, we performed a gene expression profile study. Within this study, we determined that many hypoxia-related transcripts were significantly downregulated by our lead compound at least twofold when compared to the normoxia control group. Specifically, our findings suggest that sulfonyl-γ-AApeptides are able to suppress the signaling pathway, namely, by downregulating eight specific genes that are typically upregulated under hypoxic conditions (LDHB, LDHA, PFKL, MKNK2, VHL, ALDOA, GAPDH, and PKFM). We further analyzed the permeability potential of our compounds through parallel artificial membrane permeability assay-blood–brain barrier (PAMPA-BBB), and, to our delight, our scaffold exhibited superior permeability, with lead compound HC13 exhibiting a permeability coefficient of 145×10^{-6} cm/s. Thus, it suggested that our peptidomimetics could cross the blood–brain barrier effectively. Lead compound HC13 in the presence of pronase over the course of 24 hours showed no degradation, whereas the canonical peptide hybridization chain reaction (HCR) was completely degraded within the allotted time. Once again, the immense potential the sulfonyl-γ-AApeptide backbone possesses in terms of overcoming boundaries that regular amino acid sequences have is illustrated [15].

9.2.2 Anti-Viral

9.2.2.1 HIV Fusion Inhibitor

To combat HIV, one of the most dangerous diseases, scientists have attempted to develop highly active antiretroviral therapeutics. Despite its major prevalence and threat, deaths resulting from AIDS have significantly decreased because of the development of a variety of different therapeutics [16]. The challenge remains; however, to eradicate the disease, researchers have to continue to look for novel antiviral agents. HIV type 1 (HIV-1) Env spike protein is a membrane protein that resides on the cell membrane that induces infection via the fusion of the viral and healthy cell membranes. Through a careful examination of the mode of action of cell fusion, the viral particle and cell membrane are brought into proximity for fusion after a series of interactions that form a hair-pin-like six-helix structure within the gp41 structure. The helical bundle contains a deep hydrophobic pocket that is penetrated by three residues (Trp628, Trp631, and Ile635) from the protein-binding domain of the C-terminal heptad repeat (CHR) helix in gp41. Upon binding, the six-helix core is stabilized, which provides researchers with an ample target for PPI's inhibitors. Thus, our lab set out to utilize the advantages provided by sulfonyl-γ-AApeptides.

We initially based our design on a previously reported amino acid sequence (MTSC22EK). It was anticipated that by introducing six chiral hydrophobic residues along the same face of the helix, we could bind deeply into the hydrophobic pocket. Based on this rationale, we synthesized a series of left-handed sulfonyl-γ-AApeptides to investigate the importance of the critical hydrophobic side chain positions. It has also been shown that by the conjugation of a lipid, it is possible to increase the activity of potential HIV-1 fusion inhibitors by anchoring the lipopeptide to the cell membrane, raising the local concentration of the inhibitor. This ultimately resulted in the synthesis of lead compounds **2′** and **12′** (Figure 9.3b and c) with IC_{50} and selectivity index values of 0.98 and 0.18 μM and > 40 and 108, respectively. To understand the depth of **2′** and **12′** potential as HIV-1 fusion inhibitors, we tested their activity against five pseudo-types and two authentic HIV-1 clinical isolates. Our results illustrate that both sequences possessed potent inhibitory activity, with IC_{50} values ranging from 46.7 to 154.0 nM, which is comparable to the current HIV fusion inhibitor T20. We then tested our peptides' inhibitory activity against two authentic clinical isolates, where they performed with similar efficiency to the other isolate types with

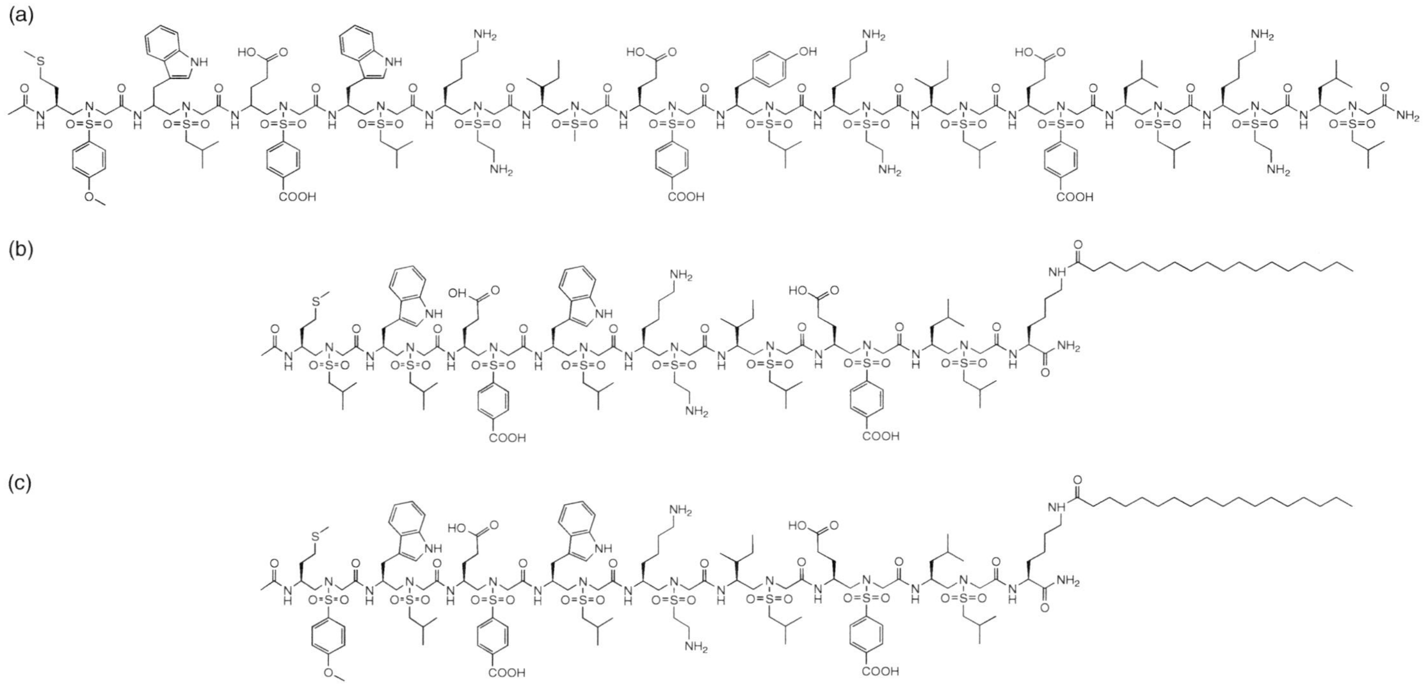

Figure 9.3 (a) Structure of lead compound HC13. Structures of **2′** (b) and **12′** (c). Source: Jarais Fontaine and Jianfeng Cai.

IC_{50} values ranging from 81.3 to 940.2 nM. The consistent potency of our unnatural sulfonyl-γ-AA foldamer mimics could potentially be potent anti-HIV-1 agents and overcome the barriers brought about by drug-resistant mutations of HIV-1.

To increase our comprehension of the mechanism of how **2′** and **12′** behave as anti-viral agents, we examined the circular dichroism (CD) spectra of the complexes formed. The six-helix bundle is formed when the N-terminal heptad repeat (NHR) region of gp41 interacts with the CHR regions. Therefore, we functionalized the NHR-peptide mimic N36 and the CHR-peptide mimic C34 and illustrated that on their own they do not form a helical secondary structure; however, upon mixing, we observed the pronounced signatures at 208 and 222 nm, thus indicating a helical secondary structure. Upon mixing N36 with **12′**, we observed an even more pronounced conformational change in its secondary structure, suggesting that our peptide scaffold strongly interacts with N36 by significantly enhancing its helical stability. Our findings were then confirmed by fluorescence polarization assay, demonstrating **12′** and N36 to have an excellent binding affinity ($K_d = 0.67\ \mu M$). Once C34 was added as a binding competitor, we observed a significant decrease in binding affinity suggesting that our compound competes with C34, confirming that our scaffold can mimic the binding mode of C34.

Another challenge that previous anti-HIV-1 struggles to overcome is the ability to pass through the blood–brain barrier. Once again, we tested our compounds using PAMPA to assess our compounds' potential as drug candidates. To assess oral administration, we functionalized the gastrointestinal variant of PAMPA (PAMPA-GIT) followed by PAMPA-BBB to determine our lead compound's ability to pass blood–brain barrier. To our delight, we were happy to report that in both instances, **2′** and **12′** possessed favorable oral bioavailability independently of pH and could easily penetrate the blood–brain barrier. The last obstacle hindering peptide mimics is their susceptibility to enzymatic degradation. Upon subjecting **2′** and **12′** to pronase, we observed no degradation, which is consistent with previous reports that the sulfonyl-γ-AApeptides possess enhanced resistance to enzymatic degradation [17]. Taken together, our sulfonyl-γ-AA peptides show great promise as potential candidates for anti-HIV-1 therapeutics.

9.2.2.2 Pan-Coronavirus Fusion inhibitor

As a result of the COVID-19 virus, the overwhelming need for anti-viral therapeutics has brought out the development of new anti-viral drugs. As of 2023, there have been several small molecule anti-viral drugs and 11 monoclonal antibodies marketed for the treatment against COVID-19 [18]. As the virus mutates it only becomes more imperative that advances are made in medicinal chemistry to combat the worldwide pandemic. The severe acute respiratory syndrome coronavirus 2 (SARS-CoV-2) virus is comprised of 4 structural components in which the spike (S) protein facilitates the viral entry into targeted cells by recruiting cellular serine protease TMPRSS2 as well as angiotensin-converting enzyme 2 (ACE-2). Following the binding of S1 to ACE-2, there is a conformational change in S2 permitting the fusion peptide domain to insert into the target cell. Finally, the heptad-repeating domain 1 (HR1), which resides in S2, assembles into a homotrimeric structure, exposing three hydrophobic pockets that HR2 interacts with, forming a six-helical bundle that brings the cell membranes into proximity to begin the fusion process [19]. As the HR1 domain is conserved throughout the different CoVs, it has become a promising target for pan-coronavirus fusion inhibitors [20]. Previous efforts in peptidomimetics have proven to be highly effective; however, due to their canonical amino acid backbone structure, they are found lacking in terms of bioavailability and biostability. Therefore, our lab decided that it may be possible to use sulfonyl-γ-AApeptides to mimic the critical residues within HR2 to interrupt the interactions between HR1 and HR2. To design our helical mimetics, we looked at the crystal structure of the six-helical bundle, which illustrates that there are six hydrophobic residues along the same face of the HR2 domain that tightly bind to a deep hydrophobic pocket within HR1. Contrary to previously discussed peptide mimics throughout the course of this chapter, we designed a series of right-handed sulfonyl-γ-AApeptides by utilizing D-amino acid as the chiral side chain rather than L-amino acids [21]. After careful modifications to the side chain position as well as the lipidation of high-performing peptides, the lead compound XY4-C7 (Figure 9.4a) demonstrated excellent potency and selectivity for SARS-CoV-2 virus exhibiting an IC_{50} value of 0.24 μM with an SI index of 62.04. As **XY4-7C** was our lead compound, we tested its binding affinity postlipidation to ensure binding toward HR1 using fluorescence polarization assay and isothermal titration calorimetry, which showed the addition of cholesterol did not alter the binding affinity (K_d = 0.1 and 0.8 μM). In order to assess the mechanism behind inhibitory activity, we utilized circular dichroism. On their lonesome, both HR1 and HR2 possessed characteristics of the helical secondary structure, which only became more pronounced upon mixing. In the presence of **XY4-7C**, the definite CD signature of HR1 decreased dramatically, indicating a substantial change in conformation due to the interactions between **XY4-7C**. Another promising feature of **XY4-7C** is that it showed effective inhibition in the SARS-CoV-2 delta variant, three pseudotype HCoVs and one pseudotype bat SARSr-CoV. Congruent with previous findings regarding the advantages of sulfonyl-γ-AApeptides, **XY4-7C** was highly stable showing no degradation in the presence of pronase and extremely favorable permeability into the blood–brain barrier and gastrointestinal tract suggesting excellent bioavailability in vivo. To confirm **XY4-7C**'s ability as a potential orally administrated drug, we performed proteinase K (PK) studies with both intraperitoneal and oral routes of administration. Our studies find that effective blood exposure is as high as 350-fold that of the IC_{50} value of **XY4-7C** with a half-life as long as 28.9 hours. As a result, **XY4-7C** once again proves its oral bioavailability capabilities indicating its potential as an orally administered drug in the future with the suggested binding mode to HR1.

9.2.3 Aβ-Oligomerization Modulation

Amyloidosis is a complex process characterized by the accumulation, and deposition of amyloid fibers can lead to pathological conditions such as Alzheimer's and Parkinson's disease, type II diabetes, and dialysis-related amyloidosis [21]. One of the hallmarks in the pathogenic pathway of Alzheimer's disease (AD) is the

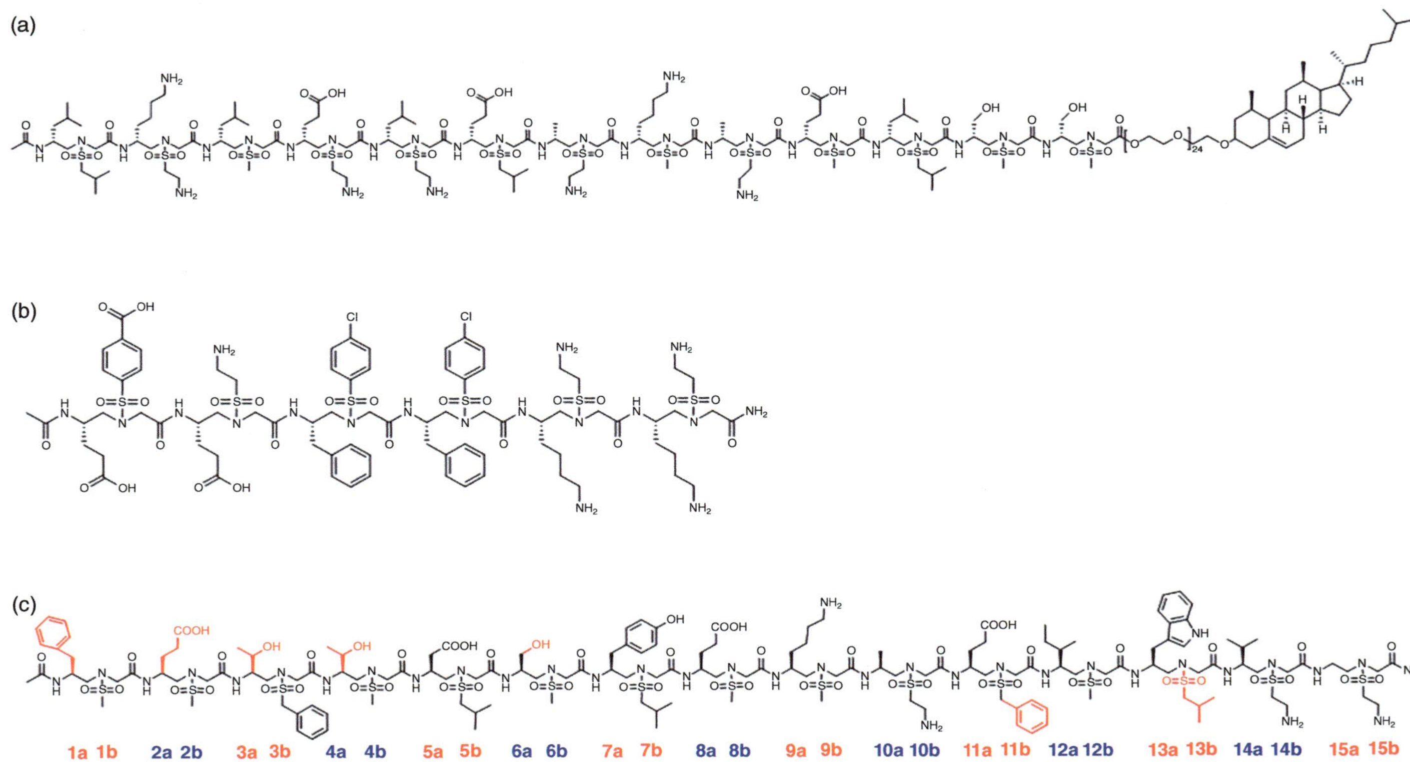

Figure 9.4 (a) Structure of lead compound XY4-7C. (b) Structure of lead compound Ab-6. (c) Chemical structure of the foldamer 3. Source: Jarais Fontaine and Jianfeng Cai.

aggregation and fibrillar formation of β-amyloid peptide Aβ-42. Due to the prevalent nature of AD affecting roughly 50 million people as of 2023 and rising projections, therapeutics targeting AD/AD pathogenesis are becoming more readily needed. Current efforts are focused on preventing the β-sheet growth of Aβ aggregates or stabilizing Aβ in its α-helical secondary structure through protein antibodies, β-sheet mimics, N-methylated and N-amino peptides, cyclic α-peptides, and helical foldamers [22]. We therefore theorized that sulfonyl-γ-AA peptides could potentially recognize the face of the Aβ to stabilize its helical formation. Even though both Aβ40 and Aβ42 are intrinsically unstructured within aqueous media, they could adopt a secondary structure upon external environmental stimulation. From our careful analysis of the α-helix of Aβ40, we selected three potentially crucial domains for the design of ligands that may recognize and stabilize the Aβ40 helix. We speculated that by targeting all three domains simultaneously, we would best be able to recognize the Aβ surface through hydrophobic and electrostatic interactions. To this end, we designed a series of sulfonyl γ-AA peptides that targeted the cationic domain 1, the hydrophobic domain 2, and the anionic domain 3 (Figure 9.4b).

The initial set of foldamer mimics performed well by inhibiting aggregation by 60% when compared to control with the ThT-based assay. We then altered our approach by extending the length of our peptide by one to have further interactions with the $A\beta_{40}$ surface, resulting in lead compound **Ab-6**, which inhibited aggregation up to 80%. The binding affinity of **Ab-6** was assessed through titrating N^{α}-fluorescein-labeled $A\beta_{40}$ and was determined to be 0.32 μM, which corresponds to our finding in ThT-based assays. Given the knowledge that preformed fibers can catalyze the formation of neurotoxic oligomers, we then tested **Ab-6** inhibitory activity against $A\beta_{42}$ in the presence of preformed fibers. It was observed that the fibrillation process was almost completely inhibited up to 24 hours. We next employed CD spectroscopy to assess whether **Ab-6** could induce a helical secondary structure in $A\beta_{42}$. On its own, $A\beta_{42}$ aggregates into amyloids characterized by a single minimum at 214 nm indicative of β-sheets. The introduction of **Ab-6**, in a seemingly immediate fashion, was able to show the characteristic double minimum at 208 and 222 nm, indicating a pure α-helical secondary structure. It provided the insight that our sulfonyl-γ-AApeptides could potentially serve as a therapeutic in the prevention of β-sheet oligomerization and fibrillation.

One of the key characteristics of the oligomerization process in $A\beta_{42}$ is a dramatic increase in cell toxicity. Due to **Ab-6**'s ability to inhibit oligomerization, we investigated the cytotoxicity profile of $A\beta_{42}$ in occupancy with **Ab-6**. Using N2a cells, the control group showed a cell viability of 78.6% after 24 hours. When added in a dose-dependent manner of up to the equimolar ratio, we observed that the cell viability increases to 99.6% with an EC_{50} value of 0.46 μM. Based on this result, we functionalized confocal microscopy, which revealed that both $A\beta_{42}$ and **Ab-6** could enter the cell and localize within the mitochondria, leading us to believe that **Ab-6** is capable of rescuing cell viability through the protection of mitochondrial function that typically is impaired by $A\beta_{42}$ aggregation.

Overall, our sulfonyl-γ-AA peptides foldamers could induce and stabilize an α-helix secondary structure with Aβ. Our studies demonstrated new insights within the recognition of potential Aβ antagonists to which sulfonyl-γ-AApeptide can serve as a template for the exploration of PPI inhibitors of other amyloid proteins [22].

9.2.4 Diabetes Therapeutics

One of the leading methods of combating type II diabetes and obesity is through the usage of helical peptide ligands glucagon-like peptide-1 (GLP-1) analogs. One of the main limitations of GLP-1 is its susceptibility to rapid degradation from proteases, resulting in the increased need to stabilize GLP-1 to regulate diabetes as it is projected that 7079 individuals in 100,000 will be diagnosed with type II diabetes by 2030. There have been several efforts to use foldamers; however, as of 2020, there had yet to be a synthetic foldamer mimic that functionalized an entire unnatural backbone. It was therefore reasonable to believe that we could mimic the long helical nature of GLP-1 using sulfonyl-γ-AApeptides (Figure 9.4c) and overcome the limitation of a natural peptide backbone that was previously reported. Examination of the interactions between GLP-1 and the GLP-1 receptor suggested that both N- and C-terminus are engaged resulting in strong binding interactions. Within the N-terminus, polar residues H7, E9, T11, T13, and S17 are most crucial, and within the C-terminus, hydrophobic residues F28 and L32 are imperative to the interaction of GLP-1 and GLP-1R extracellular domain. We designed a series of sulfonyl-γ-AApeptides extending the entire length of GLP-1 peptides 3–16, with peptides 1 and 2 being the canonical GLP-1 peptides. Upon examination of our peptide's agonist activity, we found that our best-performing peptidomimetic 3 is only two orders of magnitude lower than the canonical GLP-1 peptide 1 at 39.79 nM. We then observed the CD spectra of lead peptide 3 to see if it could mimic the helical secondary structure of GLP-1, to which we observed a strong cotton effect at around 208 nm, indicating a well-defined left-handed helical structure consistent with our previously reported homogenous sulfonyl-γ-AA peptides. Also, in conjunction with the previously mentioned peptide mimics within this chapter, peptide 3 exhibited no enzymatic degradation in the presence of pronase once again, illustrating that the sulfonyl-γ-AApeptides enhanced stability against enzymatic degradation when compared to canonical α-amino acid mimics. As a potential therapeutic, we tested peptide 3's blood glucose-lowering effect. At both given doses (4 and 40 mg/kg), there was a dramatic decrease in blood glucose levels with the results indicating good pharmacodynamics in vivo in a dose-dependent manner (Figure 9.4) [23].

9.3 Future Directions/Conclusion

Throughout the course of this chapter, sulfonyl-γ-AApeptides have been employed to modulate PPIs of multiple different protein targets. Despite the promise our peptides show in inhibiting key interactions, we lack co-crystal structures illustrating that our peptides are behaving in the manner we theorize. Today's efforts are designated toward such a purpose as well as continuing to pursue the full length of the potential of sulfonyl-γ-AA peptides. It is our hope that our unique unnatural amino acid backbone will serve as a template for others to explore alternative avenues of modulating PPIs.

Acknowledgments

This work was supported by NIH 2R01AG056569-06 and NIH 1R01GM150196.

References

1 Guichard, G. and Huc, I. (2011). Synthetic foldamers. *Chemical Communications* 47 (21): 5933. https://doi.org/10.1039/c1cc11137j.
2 Gellman, S.H. (1998). Foldamers: a manifesto. *Accounts of Chemical Research* 31 (4): 173–180. https://doi.org/10.1021/ar960298r.
3 Sang, P. and Cai, J. (2023). Unnatural helical peptidic foldamers as protein segment mimics. *Chemical Society Reviews* 52 (15): 4843–4877. https://doi.org/10.1039/d2cs00395c.
4 Niu, Y., Hu, Y., Li, X. et al. (2011). γ-AApeptides: design, synthesis and evaluation. *New Journal of Chemistry* 35 (3): 542. https://doi.org/10.1039/c0nj00943a.
5 She, F., Teng, P., Peguero-Tejada, A. et al. (2018). De novo left-handed synthetic peptidomimetic foldamers. *Angewandte Chemie (International Ed.)* 57 (31): 9916–9920. https://doi.org/10.1002/anie.201805184.
6 Sang, P., Shi, Y., Huang, B. et al. (2020). Sulfono-γ-AApeptides as helical mimetics: crystal structures and applications. *Accounts of Chemical Research* 53 (10): 2425–2442. https://doi.org/10.1021/acs.accounts.0c00482.
7 Cai, J.F.J. (2023). Sulfonyl-γ-AApeptide tools for modulating biology. *Methods in Enzymology* 698: 243–262.
8 van Tienen, L.M., Mieszczanek, J., Fiedler, M. et al. (2017). Constitutive scaffolding of multiple Wnt enhanceosome components by Legless/BCL9. *eLife* 6. https://doi.org/10.7554/eLife.20882.
9 Sang, P., Zhang, M., Shi, Y. et al. (2019). Inhibition of beta-catenin/B cell lymphoma 9 protein-protein interaction using alpha-helix-mimicking sulfono-gamma-AApeptide inhibitors. *Proceedings of the National Academy of Sciences of the United States of America* 116 (22): 10757–10762. https://doi.org/10.1073/pnas.1819663116 From NLM Medline.
10 Nero, T.L., Morton, C.J., Holien, J.K. et al. (2014). Oncogenic protein interfaces: small molecules, big challenges. *Nature Reviews Cancer* 14 (4): 248–262. https://doi.org/10.1038/nrc3690.
11 Sang, P., Shi, Y., Lu, J. et al. (2020). α-Helix-mimicking sulfono-γ-AApeptide inhibitors for p53–MDM2/MDMX protein–protein interactions. *Journal of Medicinal Chemistry* 63 (3): 975–986. https://doi.org/10.1021/acs.jmedchem.9b00993.

12 Murray, J.K. and Gellman, S.H. (2007). Targeting protein–protein interactions: lessons from p53/MDM2. *Peptide Science* 88 (5): 657–686. https://doi.org/10.1002/bip.20741.

13 Kremsmayr, T., Aljnabi, A., Blanco-Canosa, J.B. et al. (2022). On the utility of chemical strategies to improve peptide gut stability. *Journal of Medicinal Chemistry* 65 (8): 6191–6206. https://doi.org/10.1021/acs.jmedchem.2c00094.

14 Ruan, K., Song, G., and Ouyang, G. (2009). Role of hypoxia in the hallmarks of human cancer. *Journal of Cellular Biochemistry* 107 (6): 1053–1062. https://doi.org/10.1002/jcb.22214.

15 Jiang, W., Abdulkadir, S., Zhao, X. et al. (2023). Inhibition of hypoxia-inducible transcription factor (HIF-1α) signaling with sulfonyl-γ-AApeptide helices. *Journal of the American Chemical Society* 145 (36): 20009–20020. https://doi.org/10.1021/jacs.3c06694.

16 Arts, E.J. and Hazuda, D.J. (2012). HIV-1 antiretroviral drug therapy. *Cold Spring Harbor Perspectives in Medicine* 2 (4): a007161–a007161. https://doi.org/10.1101/cshperspect.a007161.

17 Xue, S., Xu, W., Wang, L. et al. (2023). Rational design of sulfonyl-γ-AApeptides as highly potent HIV-1 fusion inhibitors with broad-spectrum activity. *Journal of Medicinal Chemistry* 66 (18): 13319–13331. https://doi.org/10.1021/acs.jmedchem.3c01412.

18 Li, G., Hilgenfeld, R., Whitley, R., and De Clercq, E. (2023). Therapeutic strategies for COVID-19: progress and lessons learned. *Nature Reviews Drug Discovery* 22 (6): 449–475. https://doi.org/10.1038/s41573-023-00672-y.

19 Xue, S., Xu, W., Wang, L. et al. (2023). An HR2-mimicking Sulfonyl-γ-AApeptide is a potent pan-coronavirus fusion inhibitor with strong blood–brain barrier permeability, long half-life, and promising oral bioavailability. *ACS Central Science* 9 (5): 1046–1058. https://doi.org/10.1021/acscentsci.3c00313.

20 Xia, S., Yan, L., Xu, W. et al. (2019). A pan-coronavirus fusion inhibitor targeting the HR1 domain of human coronavirus spike. *ScienceAdvances* 5 (4): eaav4580. https://doi.org/10.1126/sciadv.aav4580.

21 Teng, P., Ma, N., Cerrato, D.C. et al. (2017). Right-handed helical foldamers consisting of de Novo d-AApeptides. *Journal of the American Chemical Society* 139 (21): 7363–7369. https://doi.org/10.1021/jacs.7b03007.

22 Liu, H., Cui, Y., Zhao, X. et al. (2024). Helical sulfonyl-γ-AApeptides modulating Aβ oligomerization and cytotoxicity by recognizing Aβ helix. *Proceedings of the National Academy of Sciences* 121 (6): https://doi.org/10.1073/pnas.2311733121.

23 Sang, P., Zhou, Z., Shi, Y. et al. (2020). The activity of sulfono-γ-AApeptide helical foldamers that mimic GLP-1. *ScienceAdvances* 6 (20): eaaz4988. https://doi.org/10.1126/sciadv.aaz4988.

10

Introduction of the Application of Stapled Peptides in Protein–Protein Interactions Drug Discovery and Their Successful Examples

Maxwell J. Austin and Danny Hung-Chieh Chou

Department of Pediatrics-Endocrinology, Stanford University, 1701 Page Mill Rd, Palo Alto, CA, USA

10.1 Introduction

Protein–protein interactions (PPIs) are essential and ubiquitous for nearly every biological process, including signal transduction, cell cycle regulation, immune responses, and metabolic control [1]. These interactions form the molecular basis of many cellular functions, where proteins bind to one another in a specific manner to regulate biological pathways. Given their central role in biology, aberrant PPIs are implicated in numerous diseases, including cancer, neurodegenerative disorders, and infectious diseases. As such, modulating these interactions represents a promising strategy in drug discovery.

However, targeting PPIs for therapeutic intervention presents significant challenges [2]. Unlike enzyme active sites or receptor binding pockets, the interaction surfaces in PPIs are often large and flat, making it difficult to design small molecules that can effectively disrupt these interactions. Traditional drug discovery approaches, including small molecule inhibitors and biologics, often fall short in addressing these challenges. Small molecules are typically too small to cover the extensive binding surfaces involved in PPIs, while biologics, such as monoclonal antibodies, can be too large to penetrate cells and access intracellular targets.

10.1.1 Stapled Peptides as a Solution to PPI Challenges

In response to these challenges, stapled peptides have emerged as a promising class of molecules for modulating PPIs [3]. Peptides, being derived from natural proteins, have the potential to mimic the structure of protein interaction surfaces and therefore offer a more suitable scaffold for targeting PPIs. However, native peptides suffer from several limitations, including poor stability, rapid proteolytic degradation, and low cell permeability, which significantly hinder their therapeutic potential.

Stapled peptides address many of these limitations [4]. By chemically constraining the peptide into an α-helical structure through the introduction of "staples," researchers can enhance the peptide's stability, increase its binding affinity to target proteins, and improve its cell permeability. These modifications make stapled

Targeting Protein–Protein Interactions for Drug Discovery, First Edition. Edited by Jian Zhang.

peptides an attractive solution for targeting intracellular PPIs that were previously deemed "undruggable." The development of stapled peptides represents a major innovation in the field of peptide therapeutics, aiming to bridge the gap between small molecules and biologics.

10.1.2 Growing Importance of PPIs in Drug Discovery

As our understanding of cellular processes has expanded, so too has the appreciation for the importance of PPIs in disease. Traditionally, drug discovery efforts have focused on enzymes, receptors, and other proteins with well-defined active sites that can be targeted by small molecules. However, it has become increasingly clear that many diseases are driven by aberrant PPIs, which play a key role in the progression of cancer, neurodegenerative diseases, viral infections, and more [5]. For instance, in cancer, PPIs involved in cell cycle regulation, apoptosis, and growth signaling pathways are often dysregulated, leading to uncontrolled cell growth. Targeting these PPIs could provide new avenues for therapeutic intervention. In neurodegenerative diseases, such as Alzheimer's and Parkinson's, protein misfolding and aggregation, driven by abnormal PPIs, are key pathological features. Modulating these interactions could help prevent or slow the progression of these diseases. Similarly, in infectious diseases, host–pathogen interactions are often mediated by PPIs, making them attractive targets for antiviral and antibacterial therapies.

Despite the therapeutic potential of targeting PPIs, the large and often flat surfaces involved in these interactions have made them difficult to drug. Traditional approaches, including small molecule inhibitors and biologics, have struggled to achieve success in this area. This challenge has spurred the development of new approaches, such as stapled peptides, which are designed to better mimic the structure and function of natural protein interaction surfaces.

10.1.3 Early Development and Success of Stapled Peptides

The concept of stapled peptides was pioneered by Gregory L. Verdine and his colleagues at Harvard University in the early 2000s [6]. Their goal was to develop a new class of therapeutics that could overcome the limitations of traditional peptides while maintaining their ability to target PPIs. By incorporating a hydrocarbon "staple" into the peptide sequence, they were able to stabilize the α-helical structure of the peptide, enhancing its ability to bind to target proteins and resist degradation.

One of the first successful application of stapled peptides is in targeting the interaction between MDM2 and p53, a key PPI involved in the regulation of apoptosis [7]. The tumor suppressor protein p53 is often inactivated in cancer due to its interaction with the oncoprotein MDM2, which leads to its degradation. Stapled peptides designed to disrupt this interaction were shown to restore p53 function, leading to the induction of apoptosis in cancer cells. This breakthrough demonstrated the potential of stapled peptides as a new class of therapeutics for targeting PPIs.

Since then, stapled peptides have been applied to a wide range of PPIs, including those involved in cancer, viral infections, and autoimmune diseases. Their ability to modulate intracellular interactions, combined with their enhanced stability and permeability, has made them a promising platform for drug discovery.

10.1.4 Overview of the Chapter

In the following sections, we will explore the design and structural features of stapled peptides, their advantages over traditional molecules, and key examples of their successful application in drug discovery. We will also discuss the challenges and limitations associated with stapled peptides and highlight potential future directions for their development.

10.2 Stapled Peptides: Structure Features and Benefits

Stapled peptides represent a class of chemically modified peptides designed to stabilize the α-helical structure, a critical secondary structure in proteins involved in PPIs. By "stapling" peptides through covalent bonds, these molecules maintain structural rigidity and show enhanced therapeutic properties, such as increased stability, binding affinity, and cell permeability. This section explores structural design, stapling techniques, and molecular mechanisms that determine the effectiveness of stapled peptides in modulating PPIs.

10.2.1 Importance of α-Helical Structures in PPIs

α-helices play a critical role in the structure and function of proteins. Many PPIs are mediated by α-helical regions that bind to complementary surfaces on their target proteins [8]. In particular, the helix serves as an interaction motif, where the outward-facing residues engage in specific interactions with the binding partner. This makes α-helical peptides appealing scaffolds for mimicking the structure of protein interaction domains.

However, synthetic peptides based on natural sequences often fail to maintain a stable helical structure in solution or the cellular environment. Native peptides are inherently flexible, rapidly lose their secondary structure, and are susceptible to proteolytic degradation. This structural instability limits their therapeutic potential because it reduces binding affinity to the target protein and diminishes the overall effectiveness of the peptide. Stapling peptides address this issue by locking the peptide into a pre-organized α-helical conformation, enhancing its structural integrity and improving its biological activity [7].

10.2.2 Designing Stapled Peptides: Mechanism of Stapling

The core innovation in stapled peptides is the introduction of covalent chemical bonds (staples) to stabilize the α-helix. These staples are typically hydrocarbon bridges that span between two amino acid side chains within the peptide sequence. In the seminal examples, this process involves the use of non-natural amino acids containing olefinic (carbon-carbon double bond) side chains, which are strategically incorporated into the peptide at specific positions along the helical face.

The most common approach involves linking these side chains via a chemical reaction known as ring-closing metathesis (RCM) [9]. This reaction forms a covalent hydrocarbon bond that "staples" the peptide into its α-helical conformation [4].

- Stapling chemistry: The design of stapled peptides often involves the substitution of two residues on the same face of the α-helix with α,α-disubstituted non-natural amino acids, which possess the reactive olefinic side chains. After incorporating these amino acids into the peptide sequence, RCM forms the staple, resulting in a structurally constrained helix.
- Staple positioning: The placement of the staple is crucial for achieving optimal stabilization. A helical turn has 3.6 residues on average. Therefore, the staple can be introduced at different positions within the peptide sequence, typically spanning i, i + 4, i, i + 7, or i, i + 11 residues. The choice of position depends on the desired degree of helical stabilization and the specific requirements of the PPI target.
  - i, i + 4 stapling: This is the most commonly used strategy, which covalently links two amino acids that are four residues apart in the peptide sequence. It reinforces the natural tendency of the peptide to adopt an α-helical structure by stabilizing one turn of the helix.
  - i, i + 7 and i, i + 11 stapling: These approaches stabilize two or three helical turns, respectively. These longer staples may be used when a more extended or rigid structure is required.

10.2.3 Additional Stapling Strategies

While the traditional hydrocarbon staple has proven effective, other stapling methods have also been developed to further expand the toolkit of stapled peptides (Figure 10.1) [10]. These include:

10.2.3.1 Lactamization Between Lysine and Glutamate/Aspartate

One of the earliest peptide stapling methods involves creating a lactam bridge between Lys and Asp or Glu residues [11]. Numerous studies focused on lactamization have examined the formation of amide bonds with various side chain lengths and positions to enhance helix stability in different peptide systems [10–12]. While both Asp–Lys and Lys–Asp bridges improve helical stability, Lys–Asp lactamization is preferred due to its superior helix-inducing effect. The optimal configuration for helix constraint in stapled peptides, using lactam bridges between Lys and Asp/Glu residues, is the i, i + 4 spacing, which cannot be achieved through single-component stapling reactions at the i, i + 7 position due to the length. Additionally, two Lys residues can be used for peptide stapling via a two-component lactamization process. In this approach, NHS ester-containing cross-linkers are attached to the amino groups of Lys–Lys residues, resulting in bi-lactam stapled peptides [13]. The

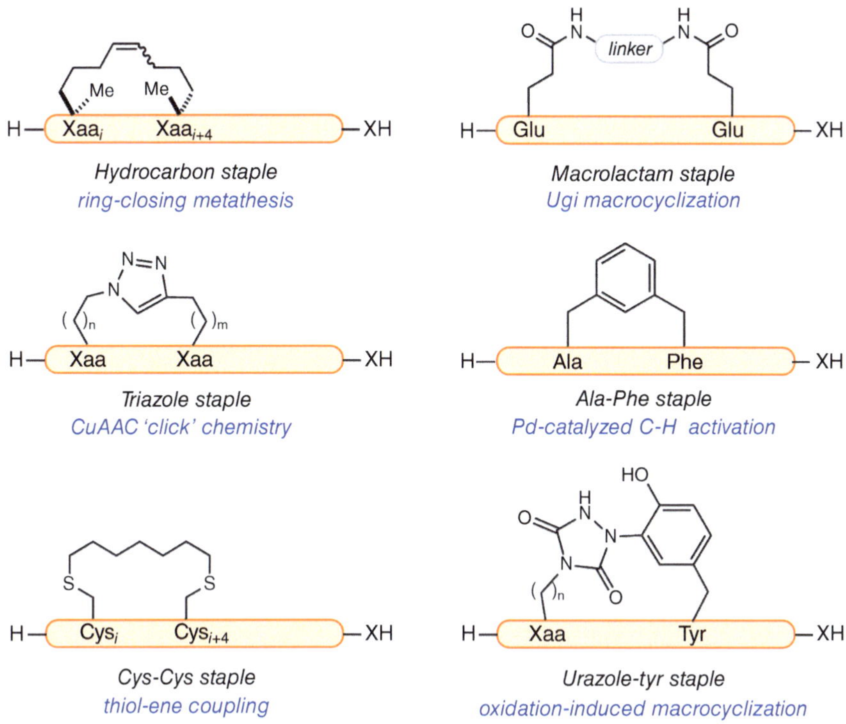

Figure 10.1 Development of stapling strategies using various chemical methodologies. Source: MAXWELL J. AUSTIN; DANNY HUNG-CHIEH CHOU.

helix stability achieved through this Lys–Lys stapling method correlates with the length and rigidity of the cross-linkers. However, the efficiency of the reaction and minimizing byproduct formation are key challenges in this approach. Similar to Lys–Lys, Glu–Glu residues can also generate a bi-lactam staple through bis-amide [14]. Additionally, Ugi four-component reactions have been used to synthesize Asp-Asp or Glu–Glu stapled peptides, with Glu–Glu peptides showing greater helical stability and binding activity compared to Asp–Asp stapled peptides [14].

10.2.3.2 Azide–Alkyne Cycloaddition

Cu-assisted azide–alkyne cycloaddition (CuAAC), also known as Huisgen cycloaddition or click chemistry, is a key reaction in biochemistry. It requires unnatural stapling amino acids, such as ω-azido- and ω-yl-α-amino acids, to form a triazole staple [15]. Scrima et al. were the first to synthesize the i–to–i + 4 side chain to side chain cyclization, with variations in the size of the disubstituted triazolyl-containing bridge [16]. Structural studies showed that the triazolyl flanked by a total of five or six methylene groups appropriately accommodated α-helical structures and reproduced the helical structure stabilized by a lactam bridge. CuAAC has the advantage of being compatible with a wide range of functional groups, enabling unprotected peptides to serve as linear precursors for triazole stapling.

10.2.3.3 C—H Activation

C—H activation involves forming a C(sp3)—C(sp2) bond between the aryl group of iodine-containing Phe and the β-carbon of Ala, using a phthaloyl-protected amino acid [17–19]. It is noteworthy that this reaction can be efficiently performed in the solid phase, yielding staples with 6- or 7-atom lengths at the i, i + 3 or i, i + 4 positions, even when Ala–Phe residues are not located at the N- or C-terminal ends. This strategy lays the groundwork for novel peptide stapling via C—H activation.

10.2.3.4 Cys–Cys Initiated Stapling Strategy

The Cys–Cys stapling approach primarily involves Cys-alkylation and Cys-arylation, which is achieved by cross-linking two thiols using bifunctional agents [18, 19]. The synthesis typically begins with conventional solid phase assembly and peptide cleavage, followed by two thiol cyclizations in solution with bifunctional linkers. Reducing agents like TCEP, GSH, and DTT are often co-incubated with the linear peptide precursor containing Cys residues to prevent oxidation before stapling. Although this method may require an additional synthesis and purification step, it eliminates the need for costly non-natural amino acids or orthogonally protected amino acids. Another key advantage of Cys–Cys stapling is the ability to easily modify and functionalize stapled peptides by using customizable cross-linkers.

Using this strategy, our lab developed a thiol-ene reaction method for peptide stapling [20]. Thiol-ene coupling occurs under ultraviolet light and requires a radical initiator. The initiator generates a thiyl radical from the sulfhydryl group, which then selectively adds to an alkene, forming a stable thioether bond. We expanded the thiol-ene conjugation method to cyclize peptides containing two cysteine residues positioned at (i, i + 4), (i, i + 7), or (i, i + 8) locations and demonstrated that unprotected peptides could be efficiently cross-linked using diene linkers of varying lengths, rigidity, and heteroatom composition, with the radical initiator 2,2-dimethoxy-2-phenylacetophenone (DMPA) in *N*-methyl-2-pyrrolidone (NMP). To investigate whether thiol-ene stapling could be applied to biologically relevant peptides, we synthesized and studied a thioether-stapled version of a previously reported p53 hydrocarbon-stapled peptide. In this case, two cysteines at the (i, i + 7) positions in the 16-residue peptide were cross-linked with 1,9-nonadiene, yielding a stapled peptide with similar enhanced helical structure, p53/mdm2 inhibitory activity, and cancer cell toxicity as the linear version. In a subsequent paper, we further expanded the application scope into aqueous conditions and thiol-yne reactions, where we can introduce two sets of thiol-ene coupling [21]. These reactions provide useful tools for biological applications.

10.2.3.5 Tyrosine Stapling

Peptide stapling strategies based on tyrosine crosslinks may be able to leverage native tyrosine functionality for structure stabilization and may also take advantage of potential bioactivities of biaryl-like macrocyclic peptide components. Biaryl crosslinks are well-characterized structural motifs in cyclic peptide natural products and are known to impart structural stability and bioactivity. For instance, the arylomycin family of antimicrobial peptides are hydroxyphenylglycine (Hpg)-tyrosine linked macrocycles that, with synthetic optimization, exhibit broad-spectrum antimicrobial activity [22]. Total synthesis efforts to construct the arylomycin core have led to the development of biaryl cross-linking techniques via C—H functionalization and oxidative copper-catalyzed crosslinking reactions, for example [23, 24].

Albericio, Lavilla, and coworkers developed a Pd-catalyzed C—H activation method to staple between tryptophan and tyrosine residues and applied this method to stabilize bioactive RGD epitope-containing peptides [25].

In the context of larger peptides, recent developments in tyrosine stapling include a few recent developments. Zhang et al. reported a method using tethered 1,3,5-triazines for peptide stapling between two tyrosine residues [26]. This strategy may allow for easily varying the length of the staple for further optimization of stapled peptides. Li et al. reported a lysine–tyrosine stapling method using aqueous formaldehyde, which after imine formation with lysine undergoes a mannich-type reaction with a neighboring tyrosine to afford a lysine–tyrosine linkage [27].

Click-like chemistries for selective tyrosine modifications have been extended for selective cyclization at tyrosine. Originally reported by Barbas in 2010, 1,2,4-triazoline-3,5-dione (TAD) reagents have been developed and used to selectively label tyrosine residues in peptides and proteins [28]. Inspired by this chemoselective method for modifying tyrosine residues and the desire to access biaryl scaffold-mimics, Roberts and coworkers developed a TAD-based peptide cyclization protocol in which an *N4*-substituted urazole is installed on an N-terminal or side-chain amine during solid-phase peptide synthesis (SPPS) and subsequently oxidized to the reactive TAD species upon which spontaneous cyclization at tyrosine occurs [29]. The afforded urazole–tyrosine moiety may be useful for biaryl natural product scaffold mimicry or stabilization of bioactive peptides as a facile and less hydrophobic stapling method.

10.2.4 Advantages of Stapled Peptides in Drug Discovery

Stapled peptides are increasingly recognized for their potential to address critical challenges in drug discovery, particularly in targeting PPIs, a class of biological targets that has traditionally been considered difficult to drug. Their unique properties make them a promising therapeutic platform, bridging the gap between small molecule inhibitors and biologics. In this section, we explore the key advantages of stapled peptides that make them highly attractive candidates in drug discovery efforts.

10.2.4.1 Enhanced Proteolytic Stability

One of the major hurdles in developing peptide-based therapeutics is their susceptibility to proteolytic degradation. Native peptides are prone to rapid breakdown by proteases in the blood and within cellular environments, significantly reducing their biological half-life. This limits their therapeutic utility, as high doses and frequent administration may be required to achieve and maintain therapeutic concentrations.

Stapled peptides, however, demonstrate remarkable resistance to proteolysis due to the structural rigidity imparted by the staple [30]. By constraining the peptide into a stable α-helical conformation, the backbone becomes less flexible and less accessible to proteases, which often require an unstructured or dynamic substrate to cleave. This proteolytic resistance translates into an extended biological half-life, allowing stapled peptides to remain active in the bloodstream or within cells for longer periods.

In drug discovery, enhanced stability offers several key benefits, including reduced dosage and frequency of administration. With longer half-lives, stapled peptides can maintain therapeutic concentrations over extended periods, allowing for lower doses to be used. Additionally, increased stability ensures that the peptide remains intact and functional long enough to exert its therapeutic effects, thereby improving efficacy. Furthermore, enhanced proteolytic stability might enable a wider range of delivery routes, such as oral administration, which is typically less feasible for traditional peptides.

10.2.4.2 Improved Cell Permeability

Another critical limitation of traditional peptide therapeutics is their poor ability to cross cellular membranes. Many peptide-based drugs are large, hydrophilic molecules that cannot efficiently penetrate the lipid bilayer of cells. This limits their application, particularly when targeting intracellular PPIs, which account for a significant proportion of therapeutic targets in oncology, immunology, and infectious diseases.

Stapled peptides, in contrast, demonstrate significantly enhanced cell permeability. The hydrocarbon staple not only stabilizes the peptide's secondary structure but also increases its overall hydrophobicity, allowing the peptide to more readily diffuse through the lipid-rich membranes of cells. This property is crucial for drug discovery, particularly when targeting intracellular protein interactions that are inaccessible to large biologics, such as antibodies [31].

Studies have shown that stapled peptides can efficiently enter cells without the need for delivery vectors or membrane-disrupting agents [4–7]. This opens up new opportunities for targeting intracellular PPIs that were previously considered undruggable. By enhancing cell permeability, stapled peptides can potentially treat diseases with intracellular targets, such as cancer, neurodegenerative diseases, and viral infections.

10.2.4.3 Ability to Target Previously "Undruggable" PPIs

PPIs are involved in many essential cellular processes, and disrupting these interactions has the potential to treat a wide variety of diseases. However, PPIs are notoriously difficult to target using conventional small molecules or biologics, largely due to their extensive and often featureless binding surfaces [31]. Traditional drug discovery efforts have focused on proteins with well-defined active sites, such as enzymes or receptors, but many critical PPIs lack such obvious druggable pockets.

Stapled peptides have emerged as a solution to this challenge. Unlike small molecules, stapled peptides are large enough to cover the extensive interaction surfaces involved in PPIs. Their structured, helical conformation mimics the natural interaction domains of proteins, enabling them to engage with and modulate

PPIs effectively. This unique ability to mimic protein structures and disrupt large surface area interactions allows stapled peptides to target PPIs that were previously considered undruggable.

10.2.4.4 Specificity and Affinity Considerations

One of the most significant advantages of stapled peptides in drug discovery is their high specificity and binding affinity for target proteins. Unlike small molecules, which often rely on nonspecific interactions and may lead off-target effects, stapled peptides are designed to mimic specific regions of protein interaction surfaces, allowing them to engage with their targets in a highly selective manner. The stabilization of the α-helical conformation in stapled peptides leads to improved molecular recognition, as key binding residues are preorganized in an optimal orientation. This pre-organized structure minimizes the entropic cost of binding, resulting in higher affinity interactions with the target protein.

Specificity and affinity are critical factors in drug discovery, as they influence both the efficacy and safety of the therapeutic agent. High-affinity binding ensures that the stapled peptide can effectively compete with natural ligands for the target protein, while specificity reduces the risk of off-target interactions and associated side effects. By combining both of these attributes, stapled peptides offer a robust platform for the development of highly selective therapeutics with fewer unwanted interactions.

10.3 Successful Applications of Stapled Peptides in Drug Discovery

Stapled peptides have emerged as an innovative therapeutic tool, especially for modulating PPIs, which are often considered difficult targets for small molecules or biologics. Their ability to stabilize specific peptide structures has broadened the scope of drug discovery, providing new avenues for treating a range of diseases. From oncology to infectious and autoimmune diseases, stapled peptides have demonstrated a unique versatility in addressing complex biological challenges [32]. We discuss various successful applications of stapled peptides, highlighting their potential in drug discovery and development.

10.3.1 Targeting the MDM2–p53 Interaction

p53 is a crucial tumor-suppressor gene, and its deletion or mutation is a significant factor in tumor development [33, 34]. MDM2 and MDMX act as negative regulators of p53, suppressing its activity and contributing to tumorigenesis. The peptide sequence $p53_{14-29}$, located at the N-terminus of p53, interacts with the hydrophobic cleft of MDM2 [35]. Verdine's team designed 10 stapled peptides based on this sequence using an all-hydrocarbon stapling technique, which enhanced α-helicity [36]. The lead peptide, SAH-p53-8, was able to penetrate cell membranes and induce apoptosis in cancer cells. Further studies revealed that SAH-p53-8 also effectively inhibited MDMX, making it a promising dual-target inhibitor for cancer therapy [37]. Building on this, our team modified SAH-p53-8 by incorporating cysteine residues and using thiol-ene stapling, which led to the creation of stapled peptides SAH-p53-8 and SAH-p53-14 [20]. The latter exhibited a stronger inhibition of p53–MDM2 interactions, outperforming linear peptides by three to four times, illustrating the complementarity of thiol-ene and traditional RCM stapling methods.

Phan et al. used phage display to identify a peptide (named PDI) that was 300 times more potent than $p53_{14-29}$ in disrupting MDM2/MDMX–p53 interactions [38]. Chang et al. further optimized PDI using all-hydrocarbon stapling to create ATSP-7041, which showed superior tumor inhibition in both *in vitro* and *in vivo* studies, along with a sevenfold increase in α-helicity compared to PDI [39]. However, challenges in large-scale production led to the development of ALRN-6924, a modified version of ATSP-7041. ALRN-6924, a clinical candidate, mimics the p53 N-terminal domain and strongly binds MDM2 and MDMX, activating the p53 pathway [40]. Incorporating penta-L-alanyl-D-alaninamide into its C-terminal enhanced its cellular potency threefold. In addition, ALRN-6924 produces a stable metabolite, ALRN-8714, which binds effectively to MDM2/MDMX and has slow dissociation kinetics. ALRN-6924 has dual utility: at high doses, it exhibits anticancer activity, and at low doses, it protects healthy cells during chemotherapy without shielding cancer cells with p53 mutations. This dual function has advanced its clinical development and contributed to the evolution of other stapled peptide therapies [41].

To further improve PDI's properties, Pentelute's group employed the lysine N-arylation stapling method, which led to the creation of SPDI-14, a perfluoroaromatic stapled peptide with enhanced hydrolytic stability and improved binding [42]. This method addresses the chemical stability limitations of cysteine arylation. In another study, Li et al. developed the antagonist Mtide-01, which was improved into the stapled peptide SPMI-2 through all-hydrocarbon stapling [43]. SPMI-2 demonstrated potent and sustained p53 activation and stronger MDMX binding compared to SAH-p53-8. It effectively activated p53 signaling *in vivo*, showing promise in tumor suppression due to the rapid accumulation of MDM2 and MDMX in cancer cells.

10.3.2 Targeting BCL-2 Family Proteins

Intracellular PPIs are critical control points in many signaling pathways, but they are often difficult to target with small molecules due to their extensive, shallow, and hydrophobic interfaces. Stapling techniques, such as hydrocarbon stapling, can improve the properties of native peptides, offering an alternative approach to studying PPIs and influencing biological pathways. The BCL-2 family of proteins play an essential role in regulating apoptosis by balancing pro-apoptotic and anti-apoptotic signals within cells [44]. Dysregulation of this balance is a hallmark of many cancers, where anti-apoptotic proteins like BCL-2, BCL-XL, and MCL-1 are overexpressed, allowing cancer cells to evade programmed cell death. Disrupting these interactions is a promising strategy in cancer therapy.

Bcl-2, an anti-apoptotic protein containing multiple conserved Bcl-2 homology (BH) domains, is highly expressed in certain cancer cells and plays a role in antitumor effects [45]. Research has shown that an amphiphilic α-helical fragment of BH3, a homologous domain of Bcl-2, can bind to the hydrophobic regions of BH1, BH2, and BH3, regulating Bcl-2 and promoting cell apoptosis [46]. However, the instability of this α-helical fragment in its free state makes it vulnerable to proteases and limits its ability to penetrate cells. To address this, Walensky and colleagues used the α-helical peptide BID BH3 from the BID region as a model to synthesize five stapled peptides (SAHBs) using an all-hydrocarbon stapling approach [47]. They identified $SAHB_A$ as a promising candidate, with increased helicity (from 15.7% to 87.5%), improved serum half-life (from 3.1 to 29.4 hours), and stronger binding affinity (K_d reduced from 269 to 38.8 nM) compared to the BID BH3 model peptide.

Mcl-1, another anti-apoptotic protein from the Bcl-2 family, is overexpressed in some cancer cells, protecting them from apoptosis and reducing their sensitivity to treatments [48]. This makes Mcl-1 an attractive target for cancer therapy. Stewart et al. identified that the BH3 peptide sequence of Mcl-1 inhibits the protein [49]. To improve the protease resistance and cellular permeability of this peptide, they employed the hydrocarbon stapling method to create a series of stapled peptides, among which Mcl-1 $SAHB_D$ exhibited the highest inhibitory activity ($K_d = 10 \pm 3$ nM). Mcl-1 $SAHB_D$ neutralizes Mcl-1's interaction with the pro-apoptotic protein BAK, sensitizing cancer cells to apoptosis via the caspase-dependent pathway.

Bcl-xL, which regulates apoptosis by interacting with the BH3 region of pro-apoptotic proteins, is another potential target for cancer treatment [50]. Kneissl et al. designed two stapled peptides, $Bak_{72-87}{}^{i+7}$ and $Bak_{72-87}{}^{i+11}$, based on the BH3 region of the Bcl-xL/Bak complex [51]. They introduced a photoactivatable cross-linker azobenzene at specific cysteine residues. These stabilized stapled peptides exhibited strong binding affinity to Bcl-xL, with K_d values of 42 ± 9 nM and 21 ± 1 nM, respectively, and up to a 20-fold increase in affinity for their dark-adapted states with destabilized helix forms. Additionally, these peptides showed over 200-fold selectivity for Bcl-xL compared to HDM2 binding. This research highlights the potential for using photo-controlled peptide switches to selectively and reversibly modulate biomolecular interactions, providing new tools for investigating and regulating cellular processes.

10.3.3 Targeting the β-Catenin/TCF Interaction

The Wnt/β-catenin signaling pathway is crucial in regulating cellular processes like proliferation and differentiation [52, 53]. β-Catenin is a key player in the Wnt signaling pathway, which is crucial for tumor invasion, metastasis, proliferation, and apoptosis. The interaction between T-cell factor (TCF) and β-catenin is a critical downstream point in the Wnt pathway, and regulating this interaction can inhibit tumor growth [54]. Dysregulation of this pathway, often due to mutations in key regulatory proteins such as adenomatous polyposis coli (APC), leads to the accumulation of β-catenin, which subsequently interacts with TCF to drive the expression of genes involved in tumorigenesis, particularly in colorectal cancer. Verdine's group initially explored the β-catenin binding domain (CBD) in TCF4 as a template peptide but found that its extended linear structure made it vulnerable to proteolysis [55]. They then identified that the binding domain of Axin, another component in the Wnt pathway, overlaps with TCF4's binding site on β-catenin. Since Axin's CBD consists of a more stable α-helix, Verdine's group used this sequence to create three stapled peptides (StAx1-3) by introducing two unnatural amino acids at specific sites to improve structural stability and protein binding. Among them, StAx-3 displayed the highest α-helicity and the strongest affinity for β-catenin. Through further optimization, they developed StAx-3-35R, a stapled peptide that directly antagonized β-catenin and inhibited Wnt-dependent cancer cell growth. However, despite its effectiveness in inhibiting Wnt signaling at the cellular level, StAx-3-35R faces challenges in efficiently suppressing Wnt-dependent tumor growth *in vivo*, partly due to rapid β-catenin accumulation. To address this limitation, Liao et al. developed a chimeric peptide inhibitor, StAx-3-35R-VHL, by using PROTAC technology [56]. This new inhibitor couples the β-catenin-targeting StAx-3-35R with a peptide ligand that recruits the Von Hippel–Lindau (VHL) ubiquitin ligase, leading to β-catenin degradation. StAx-3-35R-VHL effectively degraded β-catenin within 12 hours and maintained low levels of the protein over time, indicating its potential for sustained β-catenin suppression.

BCL9 interacts with β-catenin through its α-helical homology domain 2 (HD2), promoting tumor cell proliferation, migration, and invasion. Inhibiting this interaction may suppress tumor growth [57]. Takada et al. designed three stapled peptides (SAH-BCL9x) based on BCL9HD2 using an all-hydrocarbon stapling approach [58]. Among them, SAH-BCL9B had the highest α-helical content and showed a fivefold increase in binding affinity and a fourfold increase in anti-proliferative potency compared to the original template. SAH-BCL9B disrupted natural β-catenin/BCL9 complexes, selectively inhibited Wnt transcription, and demonstrated significant anti-proliferative effects against colon cancer cells.

10.3.4 Infectious Diseases

Stapled peptides have shown promise in targeting viral host PPIs, which are critical for the replication and spread of viruses [59]. For instance, in the case of HIV and respiratory syncytial virus (RSV), stapled peptides designed to disrupt viral protein interactions with host factors have been developed. These peptides have been shown to inhibit viral replication, offering a novel approach to antiviral therapy that targets the intracellular interactions essential for viral life cycles.

The capsid domain of the human immunodeficiency virus type 1 (HIV-1) Gag polyprotein is essential for virus assembly, making it a potential target for AIDS treatment [60]. Virus assembly occurs through the controlled polymerization of the Gag polyprotein at the host cell membrane. Zhang et al. developed a cell-penetrating peptide (CPP), NYAD-1, by applying hydrocarbon stapling to a 12-mer α-helical peptide called the capsid assembly inhibitor (CAI), which disrupts capsid particle assembly *in vitro*. NYAD-1 was created by replacing two natural amino acids in the CAI sequence with a non-natural amino acid, enhancing its α-helicity and cell penetration [60]. This CPP effectively inhibited HIV-1 in cell culture at low micromolar concentrations, demonstrating its potential as an antiviral drug for AIDS.

For RSV, the fusion protein (F) is activated into a metastable pre-fusion state before viral entry into the host cell, containing a hydrophobic fusion peptide and two heptad repeats (HR1 and HR2). Following insertion into the target cell membrane, F undergoes a conformational change, forming a six-helix bundle structure that facilitates membrane fusion and viral entry. Gaillard et al. designed double-stapled peptides that disrupted the formation of this post-fusion structure, displaying potent antiviral activity and proteolytic resistance [61]. Intranasal administration of the most effective peptide significantly reduced RSV infection in the respiratory tracts of BALB/c mice.

Coronaviruses also use a six-helix bundle for membrane fusion with host cells, though the N-terminal heptad repeat (NHR) and C-terminal heptad repeat (CHR) sequences vary across viral families. Wang et al. used hydrocarbon stapling to design double-stapled peptides that mimic the CHR helical motifs of the Middle East Respiratory Syndrome (MERS) coronavirus [62]. These peptides inhibited spike (S) protein-mediated fusion, preventing MERS infection. Another example of stapled peptide design involves HIV-1 inhibitors. The 36-mer peptide enfuvirtide blocks HIV-1 cell fusion by disrupting the six-helix bundle formation of the viral fusion apparatus [63]. Despite showing strong efficacy and safety in HIV-1 patients, enfuvirtide has poor oral bioavailability and requires frequent injections due to its low *in vivo* stability, limiting its use to last-resort treatments. Bird et al. improved HIV-1 fusion inhibitors by using all-hydrocarbon cross-linking to create next-generation gp41 HR2 peptides with enhanced protease resistance and antiviral activity compared to enfuvirtide [32]. More recently, Gomara et al. designed stapled peptides based on E1P47, a fragment of the E1 envelope glycoprotein of the human pegivirus GB virus C [64]. These stapled peptides were synthesized using lactamization between Lys and Asp or Glu residues, resulting in greater inhibitory potency against HIV-1 in cellular models and colorectal tissue explants. However, they did not show improved resistance to serum proteases compared to the parent peptide.

10.3.5 Clinical Progress and Future Directions

The success of stapled peptides in preclinical models has generated significant interest in their clinical development, particularly in oncology. Several stapled peptide candidates have entered early-phase clinical trials, showing encouraging results. However, further optimization of these molecules, particularly in terms of delivery, stability, and specificity, will be crucial for their advancement.

10.4 Challenges and Limitations

While stapled peptides hold great promise as a therapeutic modality, there are still significant challenges that need to be addressed in their development and clinical application. As with any emerging technology, stapled peptides face technical hurdles that can limit their broad adoption. We discuss the key challenges and limitations associated with stapled peptides, including issues related to their manufacturing, delivery, off-target effects, and potential drug resistance mechanisms.

10.4.1 Manufacturing and Cost Considerations

Stapled peptides often require specialized synthetic methods due to the incorporation of non-natural amino acids and the stapling process itself, which involves covalent bond formation between side chains. While traditional peptide synthesis techniques, such as SPPS, can be adapted for the creation of stapled peptides, the addition of non-natural amino acids and the need for post-synthesis modifications, such as RCM for hydrocarbon staples, significantly increase the complexity of manufacturing.

Key manufacturing challenges include increased synthesis complexity, where the introduction of non-natural amino acids and subsequent stapling reactions add layers of difficulty compared to the production of linear, unmodified peptides. This heightened complexity can result in lower yields and increased purification challenges. Another major challenge is the cost of raw materials. Non-natural amino acids, particularly those required to introduce orthogonal reacting groups, are often more expensive and less readily available than standard amino acids, thereby driving up production costs, especially for large-scale synthesis. There is also an environmental consideration where SPPS may lead to more waste due to the excess use of catalysts and coupling reagents compared to the case of using recombinant protein expressions.

Moreover, the post-synthesis modifications, such as the stapling process itself, require additional chemical steps that further complicate the manufacturing pipeline and add to the overall cost. Given the current state of peptide synthesis technologies, the cost of manufacturing stapled peptides remains relatively high, which could impact their accessibility and scalability. Therefore, efforts to streamline production methods, enhance the efficiency of stapling reactions, and reduce the cost of raw materials will be crucial to making stapled peptides more feasible for widespread therapeutic use.

10.4.2 Delivery Issues: Overcoming Biological Barriers

Effective delivery of stapled peptides remains a significant challenge, particularly when targeting tissues or organs beyond the bloodstream. Although stapled peptides exhibit enhanced cell permeability compared to traditional peptides, achieving effective delivery to specific tissues, organs, or compartments within the body can still be problematic.

One of the primary delivery challenges is oral bioavailability. Similar to many peptide-based therapeutics, stapled peptides are generally not orally bioavailable due to degradation by proteases in the gastrointestinal tract and poor absorption across the intestinal epithelium. This limitation restricts their administration to injection routes, such as intravenous or subcutaneous injections, which are less convenient for patients. Additionally, tissue targeting poses another challenge. Ensuring that stapled peptides reach their intended target tissue or organ is difficult. For instance, delivering these molecules to solid tumors or crossing the blood–brain barrier in neurological disorders requires overcoming significant biological barriers. Moreover, immune clearance is a concern, as stapled peptides, like other foreign biomolecules, can be recognized by the immune system and rapidly cleared from the bloodstream, potentially reducing their therapeutic efficacy. Overcoming rapid clearance while maintaining a balance between safety and efficacy remains a critical hurdle.

To address these issues, ongoing research is exploring delivery strategies such as nanoparticle-based systems, liposomal encapsulation, and chemical modifications to enhance the distribution and stability of stapled peptides *in vivo*. These approaches aim to improve targeted delivery and minimize the limitations associated with conventional administration routes.

10.4.3 Off-Target Effects and Safety Concerns

While stapled peptides exhibit improved specificity relative to small molecules, the potential for off-target effects still exists. These effects can arise from several factors, including nonspecific binding and immune activation. Nonspecific binding is a concern because, despite their increased specificity for PPI targets, stapled peptides may still interact with unintended proteins or biological surfaces due to their size and hydrophobicity. This unintended binding can lead to off-target pharmacology, potentially causing undesired side effects or toxicities. Another factor is immune activation. As synthetic molecules containing non-natural amino acids, stapled peptides have the potential to be recognized by the immune system as foreign entities, which could trigger immune responses such as hypersensitivity or allergic reactions.

To address these concerns, preclinical safety studies to rule out potential harmful *in vivo* toxicity are essential. Researchers are currently exploring strategies such as modifying peptide sequences to fine-tune specificity, reducing off-target binding, and optimizing dosing regimens to mitigate these risks and improve the safety profile of stapled peptides.

10.4.4 Resistance Mechanisms

One of the concerns with any therapeutic agent is the potential for the development of resistance over time. Although stapled peptides have not yet been widely tested in clinical settings, the possibility of resistance mechanisms emerging remains an important consideration, especially in oncology, where selective pressure on tumor cells can drive resistance to therapies [65]. Potential resistance mechanisms for stapled peptides include mutations of target proteins. Tumor cells, for example, could acquire mutations in the PPI interfaces that prevent stapled peptides from binding effectively. Such mutations would reduce the therapeutic efficacy of these peptides, necessitating the development of next-generation molecules capable of binding to the mutated target proteins (similar to the cases of kinase inhibitor development). Another resistance mechanism involves the overexpression of proteins. Cells might compensate for the disruption of a critical PPI by overexpressing one or both of the interacting proteins, making it more challenging for stapled peptides to achieve complete inhibition. This is analogous to how cancer cells often overexpress anti-apoptotic proteins in response to pro-apoptotic drugs. Additionally, cells could upregulate efflux pumps or other detoxification mechanisms to decrease the intracellular concentration of stapled peptides, leading to reduced efficacy. This mechanism of resistance is already observed with some small molecule inhibitors and chemotherapeutic drugs.

Developing strategies to overcome these resistance mechanisms will be essential to maximizing the long-term therapeutic potential of stapled peptides. Approaches such as combination therapies or designing resistance-resistant peptides will play a critical role in mitigating these challenges and ensuring sustained efficacy.

10.5 Future Directions

The field of stapled peptides is still rapidly evolving, with ongoing research and development aiming to address current limitations and expand the scope of their applications. As stapled peptides continue to advance through preclinical studies and early-stage clinical trials, several key areas of innovation and exploration are likely to shape their future. We highlight the potential future directions for stapled peptides, including their expansion into new therapeutic areas, improvements in stapling techniques, and integration with other therapeutic modalities.

10.5.1 Expansion into New Disease Areas

While much of the current research on stapled peptides has focused on oncology and cancer-related PPIs, there is considerable potential for expanding their application to a broader range of diseases. As our understanding of PPIs in various conditions continues to grow, stapled peptides could be developed to target specific interactions implicated in a wide array of diseases.

One promising area of expansion is in neurodegenerative diseases. Protein misfolding and aggregation are central to disorders such as Alzheimer's, Parkinson's, and Huntington's disease [66]. These pathological processes often involve abnormal PPIs that lead to the formation of toxic protein aggregates. Stapled peptides could be designed to disrupt these interactions, thereby preventing the aggregation of misfolded proteins and offering a new therapeutic approach for neurodegenerative disorders. Furthermore, their potential ability to target intracellular interactions and cross biological barriers, such as the blood–brain barrier, could prove advantageous in this field. Another potential application is in cardiovascular diseases [67]. Stapled peptides could be used to modulate interactions in pathways related to inflammation, apoptosis, or vascular remodeling, providing novel therapeutic interventions for cardiovascular health. Infectious diseases are also a promising target for stapled peptides [59]. While targeting viral host PPIs has already shown promise in the field, there is further potential to address other viral and bacterial infections. For instance, by targeting PPIs critical for the replication or invasion of pathogens, stapled peptides could offer new therapeutic options against diseases such as hepatitis, influenza, and bacterial infections resistant to traditional antibiotics. Additionally, stapled peptides could be applied to metabolic and autoimmune disorders, where PPIs play essential roles in regulating immune responses and metabolic processes. They could be designed to modulate immune cell signaling pathways involved in autoimmune diseases such as rheumatoid arthritis or lupus or to intervene in metabolic disorders like type 2 diabetes by targeting PPIs that regulate insulin sensitivity and glucose homeostasis.

Expanding the application of stapled peptides beyond oncology holds great promise for tackling a diverse range of challenging diseases, offering new therapeutic strategies where conventional approaches may be limited.

10.5.2 Integration with Other Therapeutic Modalities

Stapled peptides hold great potential for integration with other therapeutic platforms, particularly in combination therapies that target multiple disease pathways simultaneously. In the future, stapled peptides may be used synergistically with small molecules, biologics, or gene therapies to enhance their therapeutic efficacy.

One promising avenue is their combination with small molecule inhibitors. In oncology, for example, stapled peptides could be used alongside small molecule inhibitors targeting complementary pathways to overcome drug resistance. For instance, a stapled peptide targeting an oncogenic protein–protein interaction could be combined with a kinase inhibitor that modulates downstream signaling. The multivalent nature of stapled peptides also makes them suitable for targeting multiple PPIs or protein domains within the same pathway, further enhancing their versatility in combination therapies. Another area of potential integration is with antibody or biologic therapies. While biologics such as monoclonal antibodies are highly specific and potent, they are limited by their inability to target intracellular proteins. Stapled peptides, which can penetrate cells and modulate intracellular PPIs, could complement biologic therapies by expanding the range of druggable targets. For instance, in immune checkpoint inhibition, antibodies targeting surface proteins could be paired with stapled peptides to modulate intracellular pathways involved in immune cell activation.

10.5.3 Emerging Delivery Technologies

To overcome some of the current challenges associated with the delivery of stapled peptides, including their limited oral bioavailability and tissue targeting, future research will likely focus on developing novel delivery systems. These systems aim to protect stapled peptides from degradation, enhance their biodistribution, and improve their therapeutic efficacy.

One emerging technology is the use of nanoparticle and liposome carriers. Nanoparticles and liposomes can encapsulate stapled peptides, shielding them from proteolytic degradation and enhancing their delivery to specific tissues or cells. These carriers can be functionalized with targeting ligands to direct stapled peptides to particular tissues, such as tumors or sites of inflammation, thereby improving the precision and effectiveness of the treatment. Another approach is the development of sustained-release formulations. These formulations can gradually release stapled peptides over time, reducing the frequency of administration and maintaining therapeutic concentrations for extended periods. Sustained-release systems may be particularly useful for chronic conditions that require long-term treatment, offering a more consistent therapeutic effect and improved patient compliance. Additionally, peptide-conjugation strategies could enhance the delivery of stapled peptides. Conjugating stapled peptides to other molecules, such as cell-penetrating peptides, can improve their cellular uptake and delivery to specific cells. Furthermore, conjugation with targeting moieties, such as antibodies or ligands, could increase the precision of stapled peptide delivery to specific cell types or tissues, allowing for more targeted therapeutic interventions.

10.5.4 Summary

Stapled peptides represent a groundbreaking advancement in the field of peptide therapeutics and drug discovery, offering a powerful solution to the long-standing challenge of targeting PPIs, which are critical in numerous biological processes and diseases. By stabilizing peptides into α-helical structures, stapling enhances their stability, cell permeability, and binding affinity, making previously "undruggable" PPIs accessible to therapeutic intervention.

The success of stapled peptides in disrupting oncogenic interactions, such as the MDM2–p53 axis, BCL-2 family proteins, and β-Catenin/TCF interactions, has opened up new possibilities in cancer treatment. Furthermore, their potential applications extend well beyond oncology, with growing interest in utilizing stapled peptides for neurodegenerative diseases, infectious diseases, autoimmune disorders, and cardiovascular conditions. As our understanding of PPIs continues to expand, the therapeutic possibilities of stapled peptides will likely broaden even further. While stapled peptides offer many advantages over traditional small molecules and biologics, they also face significant challenges that must be addressed for their widespread clinical application. Issues related to manufacturing complexity, delivery, off-target effects, resistance mechanisms, and regulatory approval must be overcome. Continued advancements in peptide stapling techniques, delivery technologies, and the integration of stapled peptides with other therapeutic modalities will be essential to realizing their full potential.

Looking ahead, the future of stapled peptides is promising. With ongoing research into new stapling chemistries, computational peptide design, personalized medicine approaches, and innovative delivery systems, stapled peptides are poised to become a key component of the next generation of therapeutics. Their ability to target intracellular PPIs and modulate complex signaling pathways offers a unique and versatile platform that bridges the gap between small molecules and biologics, addressing some of the most challenging and critical targets in drug discovery. In conclusion, stapled peptides stand at the frontier of modern drug discovery, representing a transformative step forward in the treatment of diseases driven by PPIs. As researchers continue to refine these molecules and expand their applications, stapled peptides hold the potential to reshape the therapeutic landscape and provide new hope for patients suffering from a wide range of diseases.

Acknowledgments

Maxwell J. Austin thanks NSF Graduate Research Fellowship (DGE 2146755) for funding support. Danny Hung-Chieh Chou thanks National Institute of Health (DK140025) for funding support.

References

1 Fry, D.C. and Vassilev, L.T. (2005). Targeting protein-protein interactions for cancer therapy. *Journal of Molecular Medicine (Berlin, Germany)* 83 (12): 955–963.
2 Shin, W.H., Kumazawa, K., Imai, K. et al. (2020). Current challenges and opportunities in designing protein-protein interaction targeted drugs. *Advances and Applications in Bioinformatics and Chemistry* 13: 11–25.
3 Moiola, M., Memeo, M.G., and Quadrelli, P. (2019). Stapled peptides-a useful improvement for peptide-based drugs. *Molecules* 24 (20): 3654.
4 Walensky, L.D. and Bird, G.H. (2014). Hydrocarbon-stapled peptides: principles, practice, and progress. *Journal of Medicinal Chemistry* 57 (15): 6275–6288.
5 Wang, H., Dawber, R.S., Zhang, P. et al. (2021). Peptide-based inhibitors of protein-protein interactions: biophysical, structural and cellular consequences of introducing a constraint. *Chemical Science* 12 (17): 5977–5993.
6 Schafmeister, C.E., Po, J., and Verdine, G.L. (2000). An all-hydrocarbon cross-linking system for enhancing the helicity and metabolic stability of peptides. *Journal of the American Chemical Society* 122 (24): 5891–5892.
7 Verdine, G.L. and Hilinski, G.J. (2012). Stapled peptides for intracellular drug targets. *Methods in Enzymology* 503: 3–33.
8 Watkins, A.M., Wuo, M.G., and Arora, P.S. (2015). Protein-protein interactions mediated by helical tertiary structure motifs. *Journal of the American Chemical Society* 137 (36): 11622–11630.
9 Blackwell, H.E. and Grubbs, R.H. (1998). Highly efficient synthesis of covalently cross-linked peptide helices by ring-closing metathesis. *Angewandte Chemie (International Ed. in English)* 37 (23): 3281–3284.
10 Lau, Y.H., de Andrade, P., Wu, Y., and Spring, D.R. (2015). Peptide stapling techniques based on different macrocyclisation chemistries. *Chemical Society Reviews* 44 (1): 91–102.

11 Taylor, J.W. (2002). The synthesis and study of side-chain lactam-bridged peptides. *Biopolymers* 66 (1): 49–75.

12 Shepherd, N.E., Hoang, H.N., Abbenante, G., and Fairlie, D.P. (2005). Single turn peptide alpha helices with exceptional stability in water. *Journal of the American Chemical Society* 127 (9): 2974–2983.

13 Fujimoto, K., Oimoto, N., Katsuno, K., and Inouye, M. (2004). Effective stabilisation of alpha-helical structures in short peptides with acetylenic cross-linking agents. *Chemistry Communication (Cambridge, England)* 11: 1280–1281.

14 Ricardo, M.G., Ali, A.M., Plewka, J. et al. (2020). Multicomponent peptide stapling as a diversity-driven tool for the development of inhibitors of protein-protein interactions. *Angewandte Chemie. (International Ed.)* 59 (13): 5235–5241.

15 Li, H., Aneja, R., and Chaiken, I. (2013). Click chemistry in peptide-based drug design. *Molecules* 18 (8): 9797–9817.

16 Scrima, M., Le Chevalier-Isaad, A., Rovero, P. et al. (2010). CuI catalyzed azide-alkyne intramolecular -to-(+4) side-chain-to-side-chain cyclization promotes the formation of helix-like secondary structures. *European Journal of Organic Chemistry* 2010 (3): 446–457.

17 Noisier, A.F.M., García, J., Ionut, I.A., and Albericio, F. (2017). Stapled peptides by late-stage C(sp^3)–H activation. *Angewandte Chemie. (International Ed.)* 56 (1): 314–318.

18 Timmerman, P., Beld, J., Puijk, W.C., and Meloen, R.H. (2005). Rapid and quantitative cyclization of multiple peptide loops onto synthetic scaffolds for structural mimicry of protein surfaces. *Chembiochem* 6 (5): 821–824.

19 Brunel, F.M. and Dawson, P.E. (2005). Synthesis of constrained helical peptides by thioether ligation: application to analogs of gp41. *Chemistry Communication (Cambridge, England)* 20: 2552–2554.

20 Wang, Y. and Chou, D.H. (2015). A thiol-Ene coupling approach to native peptide stapling and macrocyclization. *Angewandte Chemie (International Ed. in English)* 54 (37): 10931–10934.

21 Wang, Y.X., Bruno, B.J., Cornillie, S. et al. (2017). Application of thiol-yne/thiol-ene reactions for peptide and protein macrocyclizations. *Chemistry: A European Journal* 23 (29): 7087–7092.

22 Smith, P.A., Koehler, M.F.T., Girgis, H.S. et al. (2018). Optimized arylomycins are a new class of gram-negative antibiotics. *Nature* 561 (7722): 189–194.

23 Peters, D.S., Romesberg, F.E., and Baran, P.S. (2018). Scalable access to arylomycins via C-H functionalization logic. *Journal of the American Chemical Society* 140 (6): 2072–2075.

24 Ben-Lulu, M., Gaster, E., Libman, A., and Pappo, D. (2020). Synthesis of Biaryl-bridged cyclic peptides via catalytic oxidative cross-coupling reactions. *Angewandte Chemie (International Ed. in English)* 59 (12): 4835–4839.

25 Mendive-Tapia, L., Preciado, S., García, J. et al. (2015). New peptide architectures through C-H activation stapling between tryptophan-phenylalanine/tyrosine residues. *Nature Communications* 6.

26 Zhang, Y., Yin, R.J., Jiang, H. et al. (2023). Peptide stapling through site-directed conjugation of triazine moieties to the tyrosine residues of a peptide. *Organic Letters* 25 (13): 2248–2252.

27 Li, B., Tang, H., Turlik, A. et al. (2021). Cooperative stapling of native peptides at lysine and tyrosine or arginine with formaldehyde. *Angewandte Chemie (International Ed. in English)* 60 (12): 6646–6652.

28 Ban, H., Gavrilyuk, J., and Barbas, C.F. (2010). Tyrosine bioconjugation through aqueous Ene-type reactions: a click-like reaction for tyrosine. *Journal of the American Chemical Society* 132, 5: 1523.

29 Keyes, E.D., Mifflin, M.C., Austin, M.J. et al. (2023). Chemoselective, oxidation-induced macrocyclization of tyrosine-containing peptides. *Journal of the American Chemical Society* 145 (18): 10071–10081.

30 Skowron, K.J., Speltz, T.E., and Moore, T.W. (2019). Recent structural advances in constrained helical peptides. *Medicinal Research Reviews* 39 (2): 749–770.

31 Ali, A.M., Atmaj, J., Van Oosterwijk, N. et al. (2019). Stapled peptides inhibitors: a new window for target drug discovery. *Computational and Structural Biotechnology Journal* 17: 263–281.

32 Bird, G.H., Madani, N., Perry, A.F. et al. (2010). Hydrocarbon double-stapling remedies the proteolytic instability of a lengthy peptide therapeutic. *Proceedings of the National Academy of Sciences of the United States of America* 107 (32): 14093–14098.

33 Kastenhuber, E.R. and Lowe, S.W. (2017). Putting p53 in context. *Cell* 170 (6): 1062–1078.

34 Huang, L., Yan, Z., Liao, X.D. et al. (2011). The p53 inhibitors MDM2/MDMX complex is required for control of p53 activity in vivo. *Proceeding of the National Academy Science of the United States of America* 108 (29): 12001–12006.

35 Vassilev, L.T., Vu, B.T., Graves, B. et al. (2004). In vivo activation of the p53 pathway by small-molecule antagonists of MDM2. *Science* 303 (5659): 844–848.

36 Bernal, F., Tyler, A.F., Korsmeyer, S.J. et al. (2007). Reactivation of the p53 tumor suppressor pathway by a stapled p53 peptide. *Journal of the American Chemical Society* 129 (16): 5298.

37 Bernal, F., Wade, M., Godes, M. et al. (2010). A stapled p53 helix overcomes HDMX-mediated suppression of p53. *Cancer Cell* 18 (5): 411–422.

38 Phan, J., Li, Z.Y., Kasprzak, A. et al. (2010). Structure-based design of high affinity peptides inhibiting the interaction of p53 with MDM2 and MDMX. *The Journal of Biological Chemistry* 285 (3): 2174–2183.

39 Chang, Y.S., Graves, B., Guerlavais, V. et al. (2013). Stapled α-helical peptide drug development: a potent dual inhibitor of MDM2 and MDMX for p53-dependent cancer therapy. *Proceedings of the National Academy of Sciences of the United States of America* 110 (36): E3445–E3454.

40 Guerlavais, V., Sawyer, T.K., Carvajal, L. et al. (2023). Discovery of Sulanemadlin (ALRN-6924), the first cell-permeating, stabilized α-helical peptide in clinical development. *Journal of Medicinal Chemistry* 66 (14): 9401–9417.

41 Saleh, M.N., Patel, M.R., Bauer, T.M. et al. (2021). Phase 1 trial of ALRN-6924, a dual inhibitor of MDMX and MDM2, in patients with solid Tumors and lymphomas bearing wild-type. *Clinical Cancer Research* 27 (19): 5236–5247.

42 Lautrette, G., Touti, F., Lee, H.G. et al. (2016). Nitrogen arylation for macrocyclization of unprotected peptides. *Journal of the American Chemical Society* 138 (27): 8340–8343.

43 Li, C., Pazgier, M., Li, C.Q. et al. (2010). Systematic mutational analysis of peptide inhibition of the p53-MDM2/MDMX interactions. *Journal of Molecular Biology* 398 (2): 200–213.

44 Danial, N.N. (2007). BCL-2 family proteins: critical checkpoints of apoptotic cell death. *Clinical Cancer Research* 13 (24): 7254–7263.

45 Delbridge, A.R.D., Grabow, S., Strasser, A., and Vaux, D.L. (2016). Thirty years of BCL-2: translating cell death discoveries into novel cancer therapies. *Nature Reviews Cancer* 16 (2): 99–109.

46 Muchmore, S.W., Sattler, M., Liang, H. et al. (1996). X-ray and NMR structure of human Bcl-xL, an inhibitor of programmed cell death. *Nature* 381 (6580): 335–341.

47 Walensky, L.D., Kung, A.L., Escher, I. et al. (2004). Activation of apoptosis in vivo by a hydrocarbon-stapled BH3 helix. *Science* 305 (5689): 1466–1470.

48 Germain, M., Nguyen, A.P., Le Grand, J.N. et al. (2011). MCL-1 is a stress sensor that regulates autophagy in a developmentally regulated manner. *The EMBO Journal* 30 (2): 395–407.

49 Stewart, M.L., Fire, E., Keating, A.E., and Walensky, L.D. (2010). The MCL-1 BH3 helix is an exclusive MCL-1 inhibitor and apoptosis sensitizer. *Nature Chemical Biology* 6 (8): 595–601.

50 Sattler, M., Liang, H., Nettesheim, D. et al. (1997). Structure of Bcl-x(L)-Bak peptide complex: recognition between regulators of apoptosis. *Science* 275 (5302): 983–986.

51 Kneissl, S., Loveridge, E.J., Williams, C. et al. (2008). Photocontrollable peptide-based switches target the anti-apoptotic protein Bcl-xL. *Chembiochem* 9 (18): 3046–3054.

52 Hu, L.F., Chen, W., Qian, A.R., and Li, Y.P. (2024). Wnt/β-catenin signaling components and mechanisms in bone formation, homeostasis, and disease. *Bone Research* 12 (1): 39.

53 MacDonald, B.T., Tamai, K., and He, X. (2009). Wnt/β-catenin Signaling: components, mechanisms, and diseases. *Developmental Cell* 17 (1): 9–26.

54 Sampietro, J., Dahlberg, C.L., Cho, U.S. et al. (2006). Crystal structure of a β-catenin/BCL9/Tcf4 complex. *Molecular Cell* 24 (2): 293–300.

55 Grossmann, T.N., Yeh, J.T.H., Bowman, B.R. et al. (2012). Inhibition of oncogenic Wnt signaling through direct targeting of β-catenin. *Proceedings of the National Academy of Sciences of the United States of America* 109 (44): 17942–17947.

56 Liao, H.W., Li, X., Zhao, L.Z. et al. (2020). A PROTAC peptide induces durable β-catenin degradation and suppresses Wnt-dependent intestinal cancer. *Cell Discovery* 6 (1).

57 Deka, J., Wiedemann, N., Anderle, P. et al. (2010). Bcl9/Bcl9l are critical for Wnt-mediated regulation of stem cell traits in colon epithelium and adenocarcinomas. *Cancer Research* 70 (16): 6619–6628.

58 Takada, K., Zhu, D., Bird, G.H. et al. (2012). Targeted disruption of the BCL9/β-catenin complex inhibits oncogenic Wnt Signaling. *Science Translational Medicine* 4 (148): 148a117.

59 Bird, G.H., Patten, J.J., Zavadoski, W. et al. (2024). A stapled lipopeptide platform for preventing and treating highly pathogenic viruses of pandemic potential. *Nature Communications* 15 (1): 274.

60 Zhang, H., Zhao, Q., Bhattacharya, S. et al. (2008). A cell-penetrating helical peptide as a potential HIV-1 inhibitor. *Journal of Molecular Biology* 378 (3): 565–580.

61 Gaillard, V., Galloux, M., Garcin, D. et al. (2017). A short double-stapled peptide inhibits respiratory syncytial virus entry and spreading. *Antimicrob Agents Chemother* 61 (4): e02241-16.

62 Wang, C., Xia, S., Zhang, P.Y. et al. (2018). Discovery of hydrocarbon-stapled short α-helical peptides as promising middle east respiratory syndrome coronavirus (MERS-CoV) fusion inhibitors. *Journal of Medicinal Chemistry* 61 (5): 2018–2026.

63 Kilby, J.M., Hopkins, S., Venetta, T.M. et al. (1998). Potent suppression of HIV-1 replication in humans by T-20, a peptide inhibitor of gp41-mediated virus entry. *Nature Medicine* 4 (11): 1302–1307.

64 Gomara, M.J., Perez, Y., Gomez-Gutierrez, P. et al. (2020). Importance of structure-based studies for the design of a novel HIV-1 inhibitor peptide. *Scientific Reports* 10 (1): 14430.

65 Vasan, N., Baselga, J., and Hyman, D.M. (2019). A view on drug resistance in cancer. *Nature* 575 (7782): 299–309.

66 Soto, C. and Pritzkow, S. (2018). Protein misfolding, aggregation, and conformational strains in neurodegenerative diseases. *Nature Neuroscience* 21 (10): 1332–1340.

67 Kong, P., Cui, Z.Y., Huang, X.F. et al. (2022). Inflammation and atherosclerosis: signaling pathways and therapeutic intervention. *Signal Transduction and Targeted Therapy* 7 (1): 131.

11

Cyclic Peptides for PPI Drug Discovery

Hong-Gang Hu[1] *and Xiang Li*[2]

[1] *Shanghai University, School of Medicine, Shangda Road 99, Shanghai 200444, China*
[2] *Second Military Medical University, School of Pharmacy, Guohe Road 325, Shanghai 200433, China*

11.1 Introduction

The human interactome is estimated to consist of a varying number of distinct protein–protein interactions (PPIs), approximately 1.3×10^5 to 6.5×10^5. However, only about 1% of these PPIs have actually been investigated so far [1]. PPIs are fundamental to the orchestration of various biological processes, including signaling pathways, immune responses, and cellular metabolism [2, 3]. The regulation of these interactions, whether through upregulation or downregulation, has a profound impact on the physiological and pathological state of an organism. It is increasingly clear that aberrations in PPIs are closely associated with the development of various disorders, ranging from cancer to neurodegenerative disorders. Consequently, targeting PPIs has become a crucial approach in drug development, offering the potential to modulate complex disease mechanisms through precise manipulation of these interactions. This approach not only promises novel therapeutic avenues but also provides deeper insights into disease biology, thereby facilitating the design of more effective and selective drugs [4, 5].

Notably, PPIs are characterized as a large interacting surface, which small molecule inhibitors usually fail to target. By contrast, peptides are considered ideal PPI regulators, as they can serve as protein epitope mimetics (PEMs) and bind to target protein by recapitulating their native secondary structures [2, 6, 7]. Peptide drugs have attracted considerable interest over the past few years due to their selectivity, potent efficacy, and comparatively low toxicity relative to traditional small-molecule drugs. A notable example is semaglutide prescribed for type 2 diabetes and obesity, an enduring and effective glucagon-like peptide-1 (GLP-1) receptor agonist. In 2023, semaglutide achieved global sales of approximately $20 billion. Importantly, aided by oral delivering technologies, semaglutide can achieve the oral effectiveness [8, 9].

Nevertheless, peptides exhibit their inherent limitations, including instability against proteolytic degradation and poor membrane permeability, which severely

Targeting Protein–Protein Interactions for Drug Discovery, First Edition. Edited by Jian Zhang.

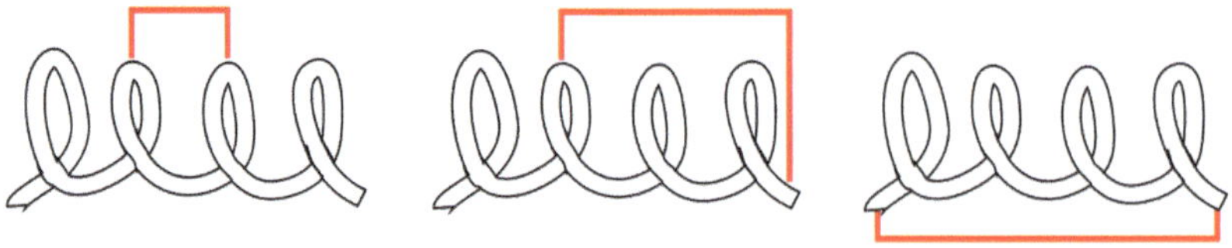

Figure 11.1 Different cyclization modes. Source: Hong-Gang Hu, Xiang Li.

limit the development of peptide drugs. Short linear peptides consisting of natural amino acids without chemical modifications usually adopt the random coils without a stable secondary structure in solution, unless bind to a target protein [10–12]. Cyclic peptides, a kind of conformationally constrained peptides by introducing a rigid chemical linker within the sequence, showcase a great promise to serve as effective peptide-based PPI modulators [13, 14]. It is worth mentioning that out of more than 60 peptides approved by the Food and Drug Administration (FDA) and the European Medicines Agency (EMA), two-thirds were cyclic peptides, indicating the critical role of cyclic peptides in the modern pharmaceutical industry [15]. These introduced macrocycles are proven to maintain the secondary or tertiary structures of peptides by head-to-tail cyclization, side-chain-to-tail/head cyclization, and side-to-side-chain cyclization (Figure 11.1) [16]. The constrained conformations of cyclic peptides include α-helices, β-strands, and loops, turns, or random coils, which depend on the native protein epitope from the vital PPIs.

11.2 α-Helix Cyclic Peptides (Stapled Peptides)

In protein structures, α-helices are the most common secondary structures, characterized by a right-handed coiled structure of amino acid residues. The stability of these helices is maintained through intramolecular hydrogen bonds formed between the carbonyl oxygen of one amino acid and the amide hydrogen of another amino acid located four residues downstream in the sequence. This leads to a repetitive pattern of hydrogen bonds. The α-helix constitutes 40% of the secondary structural patterns found in proteins. Additionally, more than 60% of protein molecules that form complexes have an α-helix at their interface, as recorded in the Protein Data Bank (PDB) database [17].

In this section, α-helical cyclic peptides usually used the side-chain-to-side-chain peptide stapling technologies to construct the stapled helical peptides [11, 12, 18–20]. α-Helices comprise 3.6 residues per turn, with residues positioned on the same helical face at i, $i+4$, $i+7$, and $i+11$, corresponding to one, two, and three turns, respectively. Notably, the side chains at i, $i+4$, and i, $i+7$ positions are the frequent stapling sites for crosslinking single and double helical turns. To date, various peptide stapling technologies have been developed aided by different chemistries, including ring-closing metathesis between olefinic side chains [21], lactamization [22], thiol-ene and thiol-yne [23], thiol/nitrogen-arylation/alkylation [24, 25],

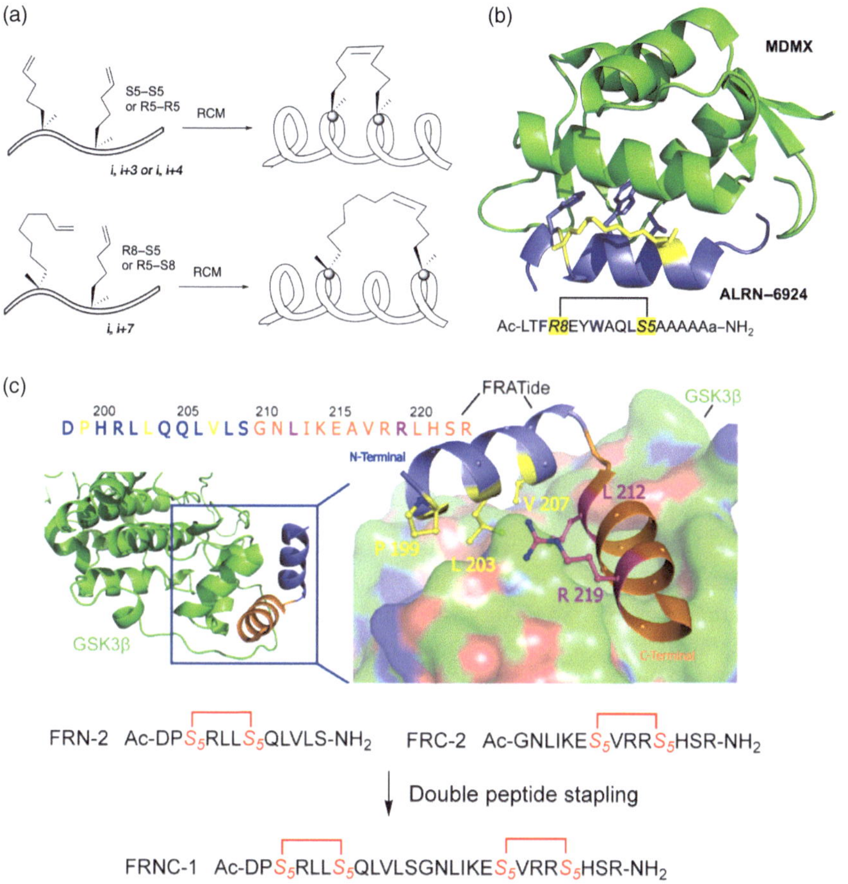

Figure 11.2 Representative peptide all-hydrocarbon stapling technique and its application. (a) S5–S5 or R8–S5 mediated on-resin ring-closing metathesis at i,i + 4 or i,i + 7 position, (b) The structure of clinical stapled peptide drug candidate ALRN-6924 and its complex with MDMX protein, (c) The development of double-stapled peptide targeting phosphorylates GSK-3β. Source: Hong-Gang Hu, Xiang Li.

click-based aide-alkyne cycloaddition [26], C–H activation [27], and so on. Among them, the most representative strategy should be the on-resin ring-closing metathesis mediated all-hydrocarbon peptide stapling, which took advantage of R5–S5 or S5–S5 paired non-natural amino acids at i, $i+3$ or i, $i+4$ position and R8–S5 or R5–S8 residues at i, $i+7$ position to afford one-turn and two-turn helical-stabilizing peptides, respectively [21] (Figure 11.2a).

11.2.1 Antitumor Stapled Peptides

Different PPIs have been proven vital in the development of cancer, with typical examples including p53–MDM2/MDMX interactions, BCL-2 domains, and Wnt/β–catenin interactions. MDM2 and MDMX exert negative regulatory effects on p53, an important tumor suppressor protein. This inhibition ultimately contributes to the development of tumor. Therefore, the p53–MDM2/MDMX interaction has emerged as an important target for the development of various anti-tumor agents. The $p53_{14-29}$ sequence located at the N-terminal was first used as a template, and the stapled peptide termed SAH-p53-8 was then developed using the all-hydrocarbon stapling technique [28] (Table 11.1). Although SAH-p53-8 bound to MDM2 and MDMX with potent affinity, this first-generation stapled peptide failed to kill the cancer cells under normal serum conditions. Subsequently, Chen and colleagues [29] identified the pDI peptide (Table 11.1) utilizing the phage display technique. This peptide demonstrated approximately 300-fold higher affinity for MDM2 and MDMX compared to the $p53_{14-29}$ peptide. Stapled pDI peptide, termed ATSP-7041 (Table 11.1), represents the second-generation stapled peptide targeting p53–MDM2/MDMX interaction [30]. ATSP-7041 exhibited significant tumor growth suppression both *in vitro* and *in vivo*, outperforming SAH-p53-8. Furthermore, the medicinal candidate ALRN-6924 in clinical development was obtained by the introducing a penta-*L*-alanyl-*D*-alaninamide C-terminal moiety and the replacing the unnatural Cba amino acid with the original Leu residue (Table 11.1) in ATSP-7041. ALRN-6294 exhibited the strikingly enhanced cellular potency (EC_{50} = 300 nM) three times more effective than its parent peptide ATSP-7041 (EC_{50} = 900 nM) in the proliferation assay of SJSA1 cells [31]. More importantly, ALRN-6924 exhibited a 75-fold improvement in water solubility compared to ATSP-7041.

As shown in Figure 11.2b, ALRN-6924 is a typical i, $i+7$ two-helix-turn stapled peptide starting from R8 and S5 olefinic amino acids, which exhibits an excellent helical secondary structure. The core binding triad of Phe_{19}, Trp_{23}, and Leu_{26} in the native p53 sequence was retained and the introduced R8–S5 staple could form the additional hydrophobic interaction with the target protein. ALRN-6924 has been evaluated in multiple clinical trials as a potential anticancer treatment. Currently, it is being developed as a chemoprotection drug to decrease or prevent hematological toxicity caused by chemotherapy, like neutropenia, as well as non-hematological toxicities, including alopecia (hair loss).

Of note, using phage display, the Lu group also identified a potent PMI peptide with nanomolar affinities for binding to MDM2 and MDMX [32]. Subsequently, our team reported an ultrahigh-affinity, dual-specificity peptide antagonist of MDM2 and MDMX, obtained through systematic mutational analysis of the PMI peptide and additivity-based molecular design. The yielded PMI-M3 peptide (LTFLEYWAQLMQ) (Table 11.1) demonstrated binding to both MDM2 and MDMX with picomolar K_d values, verified through isothermal titration calorimetry [33]. Unfortunately, both PMI and PMI-M3 peptides exhibit very weak cellular activities probably due to their poor membrane permeability and resistance to proteolysis. Using PMI as the template, the stapled peptide SPMI2 was first developed. SPMI2 activates a distinct p53 response in HCT-116 cells without causing p53-dependent apoptosis [34]. Furthermore, our team developed stapled peptide-based proteolysis-targeting chimeras (SP-PROTACs) that simultaneously exhibited improved cellular uptake and proteolytic stability attributed to the stapled peptides and efficient target protein degradation promoted by the PROTACs. Using this strategy, we designed and synthesized a series of PMI-based SP-PROTACs by introducing the hexapeptide HIF

Table 11.1 The development of stapled p53-derived peptides against p53–MDM2/MDMX interactions.

	N-term	14	15	16	17	18	19	20	21	22	23	24	25	26	27	28	29	30	31	32	33	34		C-term
p53		L	S	Q	E	T	F	S	D	L	W	K	L	L	P	E	N	N	V	L	S	I	W	$-NH_2$
SAH-p53-8	Ac-	Q	S	Q	Q	T	F	R8	N	L	W	R	L	L	S5	Q	N							$-NH_2$
pDI	Ac-				L	T	F	E	H	Y	W	A	Q	L	T	S								$-NH_2$
ATSP-7041	Ac-				L	T	F	R8	E	Y	W	A	Q	Cba	S5	S								$-NH_2$
ALRN-6924	Ac-				L	T	F	R8	E	Y	W	A	Q	L	S5	A	A	A	A	A	a			$-NH_2$
PMI					T	S	F	A	E	Y	W	A	L	L	S	P								
M3					L	T	F	L	E	Y	W	A	Q	L	M	Q								
SPMI2	Ac-				T	S	F	R8	E	Y	W	A	L	L	S5									$-NH_2$
SPMI-HIF2-1	Ac-				T	S	F	R8	E	Y	W	A	L	L	S5	Peg	L	A	Hyp	Y	Hle	P		$-NH_2$
																	VHL binding peptide motif							

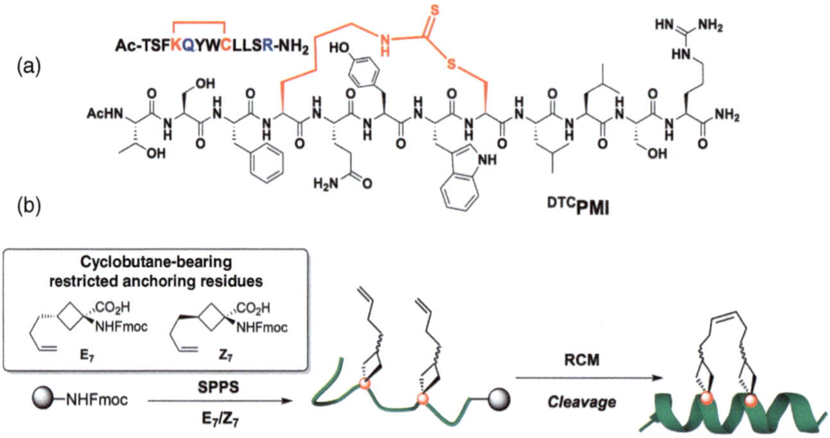

Figure 11.3 Other representative peptide stapling technique and their resultant stapled peptides. (a) The structure of DTCPMI derived from PMI peptide by dithiocarbamate-inspired peptide stapling, (b) Geometry-specific cyclobutane-bearing peptide stapling. Source: Hong-Gang Hu, Xiang Li.

(LA-Hyp-Y-Hle-P) for recruiting Von Hippel-Lindau (VHL) into the PMI peptide sequence, of which SPMI-HIF2-1 exhibited similar binding affinity to MDM2 and MDMX, but obviously higher helical contents, improved proteolytic stability, better cellular permeability, and a better pharmacokinetic profile compared with its linear PMI counterpart. SPMI-HIF2-1 has shown a remarkable ability to cause cancer cells to die and inhibit the formation of tumors in models of colorectal cancer xenografts that are implanted either subcutaneously or orthotopically. This was achieved by simultaneously facilitating the unusual breakdown of both MDM2 and MDMX, resulting in the continuous activation of p53 [35].

In addition to the all-hydrocarbon peptide stapling technique, multiple stapling chemistries have been successfully established in recent years for designing numerous p53–MDM2/MDMX inhibitors [11]. Our team previously reported a new method for stapling peptides using dithiocarbamate chemistry to connect the side chains of Lys(i) as well as Cys($i+4$) residues in unprotected peptides. We then applied this method to a series of peptides obtained from PMIs [36]. The ^{DTC}PMI with ideal characteristics exhibited a high affinity for binding to MDM2 and MDMX that was 50 times better than its linear counterpart (Figure 11.3a). Significantly, the ^{DTC}PMI demonstrated remarkable cell membrane penetration and successfully eliminated HCT116 cancer cells under a laboratory setting through the activation of the suppressor of cancer protein p53. Crystallographic analyses of peptide–MDM2/MDMX complexes confirmed the chemoselectivity for the dithiocarbamate staple, which connects the amino acids Lys and Cys at i, $i+4$ position, as demonstrated by their structural validation [36].

A major concern is the severe side effects of p53-derived stapled peptides. Severe neutropenia (Grade 4) and alopecia associated with ALRN-6924 forced Aileron to terminate the Phase 1b breast cancer trial and halt further development of ALRN-6924. One possible reason is that the introduced staple is extremely hydrophobic, which contrbutes to the cytotoxicity of the developed all-hydrocarbon stapled peptides. Based on the above considerations, more peptide stapling techniques were needed and should be applied in p53-derived peptides.

Other significant targets for combating cancer are the anti-apoptotic BCL-2 family proteins. These proteins play a vital role in enhancing cell survival by capturing the BH3 α-helix of pro-apoptotic BCL-2 proteins, which is responsible for initiating apoptosis [37]. Targeting the BH3-binding pockets of anti-apoptotic proteins remains a crucial objective. To this end, various stabilized α-helix of BCL-2 domains (SAHBs), including BH3 interacting-domain death agonist (BID), Bcl-2 interacting mediator of cell death (BIM), BCL2 associated agonist of cell death (BAD), Phorbol-12-myristate-13-acetate-induced protein 1 (NOXA), and p53 upregulated modulator of apoptosis (PUMA), have been developed as "sensitizers" or "derepressors" of anti-apoptotic BCL-2 proteins. These SAHBs have demonstrated potent anti-tumor activities both *in vitro* and *in vivo* [38, 39]. Notably, SAHBs are all-hydrocarbon stapled peptides. The BH3 α-helix sequences are highly conserved, and their stapled analogs exhibit different functions and specificities. The SAHB derived from the BH3 domain of the BID (Table 11.2) has been shown to adopt a helical conformation, resist proteolytic degradation, and efficiently penetrate cells. These molecules also have a higher affinity for the pockets of multidomain BCL-2 members. Crucially, BID SAHB precisely triggered the process of programmed cell death to eliminate leukemia cells as well as successfully suppressed the leukemia tumors in living organisms [37]. Subsequently, BIM SAHB (Table 11.2) effectively and specifically targeted BCL-2 family members, inhibiting their antiapoptotic interactions and directly activating proapoptotic activity. Consequently, there was a correlation between the dosage and the cell death in hematologic cancer cells, which was specific to the BH3 sequence. The beneficial effects of BIM SAHB have been demonstrated via its capacity to selectively trigger cell death for the abnormal mice lymphoid infiltrates that received BIM-deficient bone marrow, as well as in a xenograft model of human acute myeloid leukemia (AML) [40]. Another BAD SAHB was found to have bifunctional activities in both survival of β cells and production of insulin [41]. Additionally, stapled peptides derived from other BH3-only proteins, such as NOXA and PUMA SAHBs (Table 11.2), have also been shown to induce apoptosis in cancer cells [42, 43].

Multidomain BCL-2 anti-apoptotic proteins also exhibit a self-regulatory mechanism, as observed in the BH3 domain of MCL-1. The MCL-1 BH3 helix-derived MCL-1 SAHB peptide has been found to be a highly effective and specific inhibitor of MCL-1. The MCL-1 SAHB (Table 11.2) selectively inhibits MCL-1, disrupts its interaction with the pro-apoptotic Bcl-2 homologous antagonist killer (BAK), and enhances the susceptibility of cancer cells to caspase-dependent apoptosis [44]. Herein we summarized various SAHBs derived from BCL2 family proteins in Table 11.2, alongside further studies on their specificity and biological functions.

β-Catenin is also a highly effective cellular target for cancer therapy. Aberrant stimulation of the signaling pathway mediated by Wnt/β-catenin is believed to be associated with the initiation and progression of various cancers. Our team previously developed the SP-PROTACs by conjugating the stapled peptide SAHPA1

Table 11.2 The development of SAHBs (stabilized α-helix of BCL-2 domains) modeled after BCL2 family proteins.

SAHB	N-term	1	2	3	4	5	6	7	8	9	10	11	12	13	14	15	16	17	18	19	20	21	22	23	24	25	26	C-term
BID	Ac-			D	I	I	R	N	I	A	R	H	L	A	S5	V	G	D	S5	B	D	R	S	I	R	R		$-NH_2$
BIM	Ac-						I	W	I	A	Q	E	L	R	S5	I	G	D	S5	F	N	A	Y	Y	A	R	R	$-NH_2$
BAD	Ac-	N	L	W	A	A	Q	R	Y	G	R	E	L	R	S5	B	S	D	S5	F	V	D	S	F	K	K		$-NH_2$
NOXA	Ac-				L	E	V	E	S	A	T	Q	L	R	S5	F	G	D	S5	L	N	F	R	Q	K	L		$-NH_2$
PUMA	Ac-			Q	W	A	R	E	I	G	A	Q	L	R	S5	G	A	D	S5	L	N	A	Q	Y				$-NH_2$
BAK	Ac-							Q	V	G	R	Q	L	A	S5	I	G	D	S5	I	N	R	R	Y	D			$-NH_2$
BAX	Ac-			A	S	T	K	K	L	S	E	S	L	K	S5	I	G	D	S5	L	D	S	N					$-NH_2$
BOK	Ac-				R	L	A	E	V	S	A	V	L	L	S5	L	G	D	S5	L	E	B	I	R				$-NH_2$
MCL-1	Ac-							K	A	L	E	T	L	R	S5	V	G	D	S5	V	Q	R	N	H	E	T	A	$F-NH_2$
BCL-2	Ac-							V	V	H	L	T	L	R	S5	A	G	D	S5	F	S	R	R	Y				$-NH_2$
BCL-X_L	Ac-							A	V	K	Q	A	L	R	S5	A	G	D	S5	F	E	L	R	Y				$-NH_2$
BCL-W	Ac-								L	H	Q	A	B	R	S5	A	G	D	S5	F	E	T	R	Y				$-NH_2$
BFL-1-A1	Ac-						K	E	V	E	K	N	L	K	S5	S	L	D	S5	V	N	V	V	S	V			$-NH_2$

or xStAx with the VHL ligand, inspired by the Axin sequence. This design allows for the effective degradation of β-catenin [45]. The xStAx-VHLL facilitated the reduction of β-catenin and demonstrated effective suppression of Wnt signaling across tumor cells as well as $APC^{-/-}$ organoids. Moreover, xStAx-VHLL exhibited the capacity to effectively suppress tumor growth in BALB/C nude mice as well as decrease the size of previous tumors in $APC^{min/+}$ animals. Notably xStAx-VHLL exhibited the potent suppression of the viability of organoids derived from colorectal tumor patients.

The development of stapled peptides against other cancer-related PPIs, including NOTCH ternary complex, ATG5–ATG16L1 complex, LC3/GABARAP, Beclin 1, WASF1, METTL3, and others, was also performed by many researchers [46].

11.2.2 Antiviral Stapled Peptides

During the spike protein-mediated membrane fusion process, the creation of a six-helix bundle (HR1–HR2 complex) has emerged as a potential, conserved candidate for the identification of fusion inhibitors. Accordingly, CoVHR2-0 and EK1 peptides were recently reported to exhibit potent inhibitory activity against coronavirus spike protein-mediated cell–cell fusions [47, 48]. Using the peptide all-hydrocarbon stapling technique, our team developed a series of stapled peptides derived from CoVHR2-0. Two of these peptides, SCH-1-20 and SCH2-1-27, demonstrated enhanced resistance to degradation, increased helical structure, and stronger inhibitory effects against both pseudo- and authentic SARS-CoV-2 in comparison to the linear version. It is worth mentioning that SCH2-1-20 and SCH2-1-27 have been demonstrated to disrupt the process of membrane fusion mediated by the spike protein [49]. For EK1 peptide modification, we synthesized two novel non-natural amino acids: (*E*)-1-amino-3-(but-3-en-1-yl) cyclobutane-1-carboxylic acid (E7) and (*Z*)-1-amino-3-(but-3-en-1-yl)cyclobutane-1-carboxylic acid (Z7). These building blocks enabled ring-closing metathesis (RCM)-mediated peptide stapling to generate geometry-specific stapled peptides (Figure 11.3b). Among the four geometry-specific stapled cyclobutane-bearing peptides, the E7–E7 configuration demonstrated the increased helicity and thus exhibits superior activity compared to canonical hydrocarbon stapled peptides [50].

Recently, Walensky team designed and prepared a stapled lipopeptide library by varying the stapling positions and the length of polyethylene glycol (PEG) linker. The research has shown that the stapling improved the antiviral activity by 1.3–1.5 fold, whereas the combination of an optimal stapling position and the extended PEG linker (PEG8) improved the overall antiviral activity by 24–39 fold. Peptide stapling improves the compound's binding affinity, antiviral potency, stability, and solubility. The leading compound, RQ-01, achieved tissue persistence after intranasal administration to mice and protected hamsters from SARS-CoV-2-induced weight loss and lung damage, whether administered before or after viral inoculation. These results highlighted the potential of RQ-01 as an effective drug to prevent and alleviate symptoms of SARS-CoV-2 infection. The stapled lipopeptide may also be effective against dangerous and potentially deadly viruses such as respiratory syncytial virus (RSV), Ebola, and Nipah viruses [51].

SARS-CoV-2 infection is initiated by the binding of the receptor-binding domain (RBD) in the SARS-CoV-2 spike protein to the host's cellular angiotensin-converting enzyme 2 (hACE2) receptor. Inhibiting the interaction between RBD and hACE2 may block viral entry into human cells, thereby preventing subsequent viral replication [52]. Maas et al. designed three stapled peptides based on the N-terminal α1-helix of hACE2 (referred to as inhibitor 1) to improve the structural stability and binding activity of the corresponding linear peptide [53]. Circular dichroism (CD) analysis revealed that the helicity of stapled peptides increased by 2–5 fold compared to the linear peptide. In addition, binding assays targeting the formation of the RBD–hACE2 complex showed that these stapled peptides exhibited more than a 3-fold improvement in inhibitory activity relative to the linear counterpart. The stapling by forming a lactam bridge at the $i, i+4$ site showed increased cellular uptake, reduced enzymatic degradation, and improved pharmacokinetic properties, resulting in a potent inhibitory effect against the virus.

Human immunodeficiency virus type 1 (HIV-1) capsid proteins play crucial roles in multiple stages of the viral replication cycle and represent promising targets for HIV therapy, as antiviral effects can be achieved by inhibiting capsid assembly. Peptide NYAD-1 can bind to the HIV-1 capsid (CA) to inhibit capsid assembly and exert antiviral activity [54]. To investigate the impact of stapling on bioactivity and conformation of NYAD-1, Pentelute's group modified the binding peptide NYAD-1 of CA [55]. The peptide was stapled using different linkers including carbazide, oxalyl dihydrazide, succinyl dihydrazide, and hexanedihydrazide. The CD spectra showed that the stapled NYAD-1 adopted a typical helical structure. The binding kinetics assays revealed that the K_d value of the staple peptide with succinyl dihydrazide linker was approximately 3-fold higher, compared with that of the linear peptide.

11.2.3 Anti-osteoporosis Stapled Peptides

The Akt/GSK-3β/NFATc1 signaling pathway functions as a negative regulator in osteoclasts, where Akt (protein kinase B) phosphorylates GSK-3β (glycogen synthase kinase-3β), and phosphorylated GSK-3β promotes NFATc1 (nuclear factor of activated T-cells cytoplasmic 1)-mediated osteoclast differentiation. FRATtide, a peptide derived from a GSK-3 binding protein, was initially identified in *Xenopus laevis* embryos as a 26-amino acid peptide (ranging from D198 to R223) segment, which comprised two helical peptides and effectively inhibited osteoclast differentiation by binding to phosphorylated GSK-3β. We established an array of single stapled peptide derivatives by modifying non-interacting residues at i, $i+4$ positions, among which the stapled peptides FRN-2 and FRC-2 exhibited an increased biological activity [56]. Next, we combined the N-terminal FRN-2 and the C-terminal FRC-2 and designed a peptide called FRNC-1 by using a double-stapling technique (Figure 11.2c), which not only exhibited substantial anti-osteoporotic effects both in laboratory experiments and in living organisms but also showed a potent efficacy after oral administration [57].

11.2.4 Anti-inflammation Stapled Peptides

Synaptotagmin-2 is a calcium-sensing protein that forms a specific interface with the Ca^{2+}-binding region of the neuronal soluble *N*-ethylmaleimide-sensitive factor attachment protein receptor (SNARE) complex. This interaction induces Ca^{2+}-triggered membrane fusion and mediates the release of myosin. In chronic airway inflammatory diseases, excessive myosin secretion is an important pathophysiological process. To inhibit this process, Lai's team designed a series of stapled peptides that compete with SNARE for binding to synaptotagmin-2, interfering with the interaction and thereby disrupting Ca^{2+}-triggered membrane fusion. Using the SNAP-25A fragment as a template, a series of monocyclic and bicyclic stapled peptides were synthesized using the all-hydrocarbon peptide stapling strategy [58]. These stapled peptides were more structurally stable and exhibited higher α-helicity than the linear peptide. The study found that the stapled peptide SP9 can effectively inhibit Ca^{2+}-triggered membrane fusion, increasing the inhibitory effect of the template peptide by about 70%. The bicyclic stapled peptide SP9 has the potential to inhibit excessive secretion of myosin, providing a new drug candidate for the treatment of chronic airway inflammatory diseases such as airway inflammation.

11.2.5 Anti-diabetes Stapled Peptides

The PPI between ALMS1 and protein kinase Cα (PKCα) hinders PKCα-mediated glucose uptake in adipocytes. Recently, Schreyer et al. screened α-helical sequences from the structural domain of human PKCα kinase and found that the peptide (ECTMVEKKVLALL) effectively blocked the binding of ALMS1 to PKCα and activated PKCα kinase activity [59]. Subsequently, the linear peptide was stapled and modified by the all-hydrocarbon linkage, and the synthesized stapled peptide was named peptide derived of PKC alpha targeting AlmS (PATAS). The dose required for PKC activation by PATAS was reduced by fivefold, compared with the original sequence. PATAS can target endogenous ALMS1-PKCα PPIs in adipocytes in the absence of insulin to alleviate insulin resistance and its associated comorbidities and effectively reverse type 2 diabetes. It is expected to be a "first-in-class" adipocyte-specific therapeutic agent.

11.3 β-Hairpin Cyclic Peptides

β-Hairpins are vital elements in the protein structure and act as vital binding motifs in several PPIs. β-hairpin is a structural element that describes a reverse turn joining neighboring strands in an antiparallel β-sheet. The concept known as protein epitope mimetics has been instrumental in the design of β-hairpin-based mimics. It began with identifying a significant epitope in molecular identification within the PPI interface. This epitope was added to a hairpin-stabilizing semi-rigid template. The end result was a cyclic β-hairpin structure with a constrained conformation [16, 60]. One of the most known hairpin-stabilizing templates was Pro(D)–Pro(L)

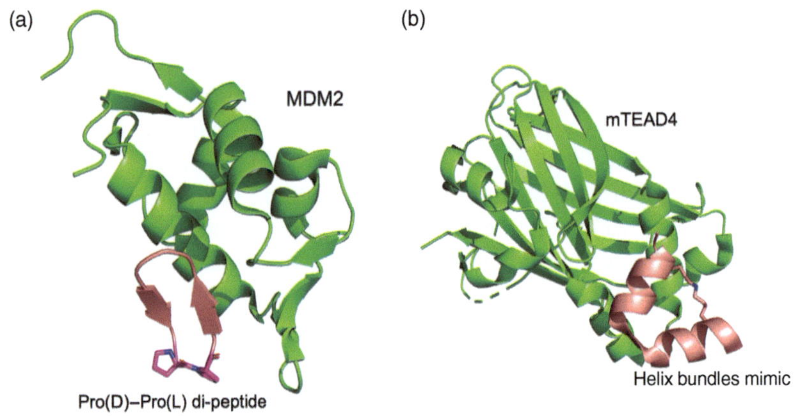

Figure 11.4 Representative head-to-tail cyclic peptides. (a) Change of the p53–MDM2 interaction was accomplished via a stabilized β-hairpin mimic, (b) The macrocyclic peptide mimicking helix bundles targeting mTEAD4. Source: Hong-Gang Hu, Xiang Li.

di-peptide that has been applied in the design of p53–MDM2 interaction inhibitor (Figure 11.4a) [61]. A set of stabilized β-hairpin peptides was synthesized by employing a cyclic antiparallel template as well as a turn inducer Pro(D)–Pro(L), which resulted in the exposure of crucial residues to the peptide surface that fit toward the hydrophobic region of MDM2. Additionally, Trp–Trp crosslinking, backbone disulfide bonds, click cyclization, 4-mercaptoproline, and hydrogen bond surrogates have been developed in the engineering and production of stabilized β-hairpin peptides [62–67].

It is worth noting that although hydrocarbon peptide stapling has proven effective for stabilizing α-helical peptides, its utility is limited when applied to β-sheet peptide structures. Our team recently developed a distinctive double-stapled peptide-based PROTAC targeting the androgen receptor (DSARTC). This peptide effectively stabilized both α-helix and β-sheet structures, as confirmed by molecular dynamics simulations and CD spectroscopy [68]. DSARTC specifically focused on inhibiting the DNA-binding domain of the androgen receptor (AR) as well as MDM2 as an E3 ligase. DSARTC, functioning as a type of SP-PROTAC, demonstrated the capability to break down both the complete AR and AR-V7. DSARTC, a novel peptide therapeutic candidate with a unique structure, demonstrates increased stability, enhanced capacity to enter cells, and efficient suppression of tumor growth in an animal model of enzalutamide-resistant prostate cancer, as compared to its linear version [68]. As synthetic scaffolds that stabilize reverse turns and cross-strand interactions have provided important insights into β-sheet stability and folding, cyclized β-hairpin mimetics capable of blocking PPIs are emerging [69]. Recently, Strømgaard et al. used a 23-amino acid nNOS β-hairpin mimetic as a template peptide and successfully obtained two cyclic nNOS β-hairpin mimetic peptides by native chemical ligation cyclization [70]. Compared with the wild-type (WT) peptide, these cyclic peptides exhibited approximately 100-fold improved binding affinity for PSD-95-PDZ2. Furthermore, these two cyclic peptides can effectively disrupt the nNOS–PSD-95 interaction and serve as the leading compounds for drug development for the treatment of acute ischemic stroke.

11.4 Macrocyclic Peptides

Naturally occurring cyclic peptides possess drug-like characteristics such as metabolic stability and the capacity to pass through cell membranes. They are the tiniest biomolecules that may efficiently identify extensive binding sites on difficult proteins. One of the most pronounced macrocyclic peptides could be cyclosporin A (CSA), an immunosuppressant medication that has been given orally for a period of 40 years [71]. CSA is a cyclic peptide having a non-polar nature and a backbone that is strongly N-methylated, exhibiting moderate membrane permeability by passive diffusion and enhanced proteolytic stability. Non-ribosomal peptide synthetases (NRPSs) are enzymes that commence biosynthesis and add extraordinary chemical diversity to natural cyclic peptides. To accomplish this, they include non-proteinogenic amino acids, performing epimerization, *N*-methylation, and backbone cyclization. Nevertheless, the complex process of creating non-standard cyclic peptides resembling natural products has impeded a methodical investigation of the macrocyclic chemical realm for the purpose of medication development [16, 72].

Native macrocyclic peptides have inspired the development of synthetic macrocyclic peptides as highly effective PPI inhibitors. A typical example is a PRMT5/adaptor protein PPI inhibitor [73]. Based on the mechanism by which PRMT5 interacts with its adaptor protein RioK1, a cyclic peptide library incorporating non-proteinogenic amino acids was constructed, leading to the identification of a potent cyclic peptide that binds PRMT5 with nanomolecular affinity. Another example is a cyclic anti-PD-L1 peptide, which demonstrated up to a 34-fold improvement in PD-1/PD-L1 blockade activity and, importantly, exhibited significant *in vivo* anti-tumor efficacy [74].

Macrocyclic peptides have the advantage of mimicking protein tertiary structures. This is important because mimicking a single secondary structure is often insufficient to disrupt the broad and flat interfaces of PPIs. For example, a stabilized mimic of a protein tertiary structure (helix bundles) was constructed using lactam cyclization, which act as an inhibitor of the interaction between the transcription element mTEAD4 and its corepressor VGL4 (Figure 11.4b) [75]. This macrocyclic peptide was specifically designed to chemically link helix α2 and helix α3 of the VGL4, with the goal of promoting cell growth by modulating the Hippo pathway.

11.5 Summary and Outlook

PPIs as an emerging class of challenging drug targets have attracted an increasing attention both in academia and industries. Characterized as large and flat interacting surfaces, PPIs have always been thought as "undruggable." Due to the rapid development of various techniques, including peptide cyclizing modification and facile PROTAC for protein degradation, several small-molecule and peptide-based drug candidates were rationally designed against different kinds of PPIs. Among them, cyclic peptides showed considerable potential in developing drugs as they offered the advantages of increased proteolytic protection and strengthened membrane permeability over their linear counterparts. More importantly, emerging genetically encoded library methods, including phage, mRNA, as well as DNA display, facilitated fast evaluation and straightforward hit deconvolution of cyclic peptides [76, 77].

Recently, major advances have been made in deep learning models for protein structure prediction and design, exemplified by tools such as Alphafold, Rosetta, and Chroma [78]. By combining artificial intelligence (AI) and computational chemistry, these models are now being translated into practical applications of functional cyclic peptide development for disease treatment and diagnostics. For instance, we previously found AR- and VHL-binding peptides using RFdiffusion and ProteinMPNN. After the completion of the binder sequences, Alphafold2-multimer and ZDOCK were employed for determining the spatial position of the two peptide binders, thus affording a novel peptide PROTAC for AR protein [79]. This proof-of-concept validated the production of peptide-based drug candidates targeting the potential protein by AI technology, and a straightforward workflow for generating the cyclic peptide drug was conducted in our lab.

Although cyclic peptides exhibited significantly enhanced proteolytic stability than their linear counterparts, most of these drug candidates can not be administrated orally. The orally administered cyclic peptides, including CSA and linaclotide, showcased a severely low oral bioavailability. A recently developed tricyclic peptide, which binds to PCSK9 (a crucial controller of plasma low-density lipoprotein [LDL]-cholesterol), has demonstrated a remarkably strong affinity in the near picomolar range. When administered to cynomolgus monkeys along with a permeation enhancer, it obtained an oral availability of 2.9% [80]. Overall, elucidating the design principles and structural features of various cyclization motifs that contribute to proteolytic stability, metabolic resistance, and enhanced cellular permeability remains an urgent need.

References

1 Robertson, N.S. and Spring, D.R. (2018). Using peptidomimetics and constrained peptides as valuable tools for inhibiting protein(-)protein interactions. *Molecules* 23 (4): 959.

2 Rual, J.F., Venkatesan, K., Hao, T. et al. (2005). Towards a proteome-scale map of the human protein-protein interaction network. *Nature* 437 (7062): 1173.

3 Philippe, G.J.B., Craik, D.J., and Henriques, S.T. (2021). Converting peptides into drugs targeting intracellular protein-protein interactions. *Drug Discovery Today* 26 (6): 1521.

4 Bonetta, L. (2010). Interactome under construction. *Nature* 468: 851.

5 Milroy, L.G., Grossmann, T.N., Hennig, S. et al. (2014). Modulators of protein-protein interactions. *Chemical Reviews* 114 (9): 4695.

6 Fosgerau, K. and Hoffmann, T. (2015). Peptide therapeutics: current status and future directions. *Drug Discovery Today* 20 (1): 122.

7 Wang, L., Wang, N., Zhang, W. et al. (2022). Therapeutic peptides: current applications and future directions. *Signal Transduction and Targeted Therapy* 7 (1): 48.
8 Zizzari, A.T., Pliatsika, D., Gall, F.M. et al. (2021). New perspectives in oral peptide delivery. *Drug Discovery Today* 26 (4): 1097.
9 Bucheit, J.D., Pamulapati, L.G., Carter, N. et al. (2020). Oral semaglutide: a review of the first oral glucagon-like peptide 1 receptor agonist. *Diabetes Technology and Therapeutics* 22 (1): 10.
10 Kurrikoff, K., Aphkhazava, D., and Langel, U. (2019). The future of peptides in cancer treatment. *Current Opinion in Pharmacology* 47: 27.
11 Li, X., Chen, S., Zhang, W.-D., and Hu, H.-G. (2020). Stapled helical peptides bearing different anchoring residues. *Chemical Reviews* 120 (18): 10079.
12 Li, X., Zou, Y., and Hu, H.-G. (2018). Different stapling-based peptide drug design: mimicking α-helix as inhibitors of protein–protein interaction. *Chinese Chemical Letters* 29 (7): 1088.
13 Hill, T.A., Shepherd, N.E., Diness, F., and Fairlie, D.P. (2014). Constraining cyclic peptides to mimic protein structure motifs. *Angewandte Chemie (International Ed. in English)* 53 (48): 13020.
14 Pelay-Gimeno, M., Glas, A., Koch, O., and Grossmann, T.N. (2015). Structure-based design of inhibitors of protein-protein interactions: mimicking peptide binding epitopes. *Angewandte Chemie (International Ed. in English)* 54 (31): 8896.
15 Zhang, H. and Chen, S. (2022). Cyclic peptide drugs approved in the last two decades (2001-2021). *RSC Chemical Biology* 3 (1): 18.
16 Cheng, J., Zhou, J., Kong, L. et al. (2023). Stabilized cyclic peptides as modulators of protein-protein interactions: promising strategies and biological evaluation. *RSC Medicinal Chemistry* 14 (12): 2496.
17 Jochim, A.L. and Arora, P.S. (2010). Systematic analysis of helical protein interfaces reveals targets for synthetic inhibitors. *ACS Chemical Biology* 5: 919.
18 Cromm, P.M., Spiegel, J., and Grossmann, T.N. (2015). Hydrocarbon stapled peptides as modulators of biological function. *ACS Chemical Biology* 10 (6): 1362.
19 Reguera, L. and Rivera, D.G. (2019). Multicomponent reaction toolbox for peptide macrocyclization and stapling. *Chemical Reviews* 119: 9836.
20 Lau, Y.H., de Andrade, P., Wu, Y., and Spring, D.R. (2015). Peptide stapling techniques based on different macrocyclisation chemistries. *Chemical Society Reviews* 44 (1): 91.
21 Schafmeister, C.E., Po, J., and Verdine, G.L. (2000). An all-hydrocarbon cross-linking system for enhancing the helicity and metabolic stability of peptides. *Journal of the American Chemical Society* 122: 5891.
22 Vasco, A.V., Brode, M., Mendez, Y. et al. (2020). Synthesis of lactam-bridged and lipidated cyclo-peptides as promising anti-phytopathogenic agents. *Molecules* 25 (4): 811.
23 Wang, Y., Bruno, B.J., Cornillie, S. et al. (2017). Application of thiol-yne/thiol-ene reactions for peptide and protein macrocyclizations. *Chemistry-A European Journal* 23 (29): 7087.

24 Fadzen, C.M., Wolfe, J.M., Cho, C.F. et al. (2017). Perfluoroarene-based peptide macrocycles to enhance penetration across the blood-brain barrier. *Journal of the American Chemical Society* 139 (44): 15628.

25 Spokoyny, A.M., Zou, Y., Ling, J.J. et al. (2013). A perfluoroaryl-cysteine S(N)Ar chemistry approach to unprotected peptide stapling. *Journal of the American Chemical Society* 135 (16): 5946.

26 Lau, Y.H., Wu, Y., Rossmann, M. et al. (2015). Double strain-promoted macrocyclization for the rapid selection of cell-active stapled peptides. *Angewandte Chemie (International Ed. in English)* 54: 15410.

27 Wang, P., Liu, J., Zhu, X. et al. (2023). Modular synthesis of clickable peptides via late-stage maleimidation on C(7)-H tryptophan. *Nature Communications* 14 (1): 3973.

28 Bernal, F.T., Tyler, A.F., Korsmeyer, S.J. et al. (2007). Reactivation of the p53 tumor suppressor pathway by a stapled p53 peptide. *Journal of the American Chemical Society* 129: 2456.

29 Phan, J., Li, Z., Kasprzak, A. et al. (2010). Structure-based design of high affinity peptides inhibiting the interaction of p53 with MDM2 and MDMX. *The Journal of Biological Chemistry* 285 (3): 2174.

30 Chang, Y.S., Graves, B., Guerlavais, V. et al. (2013). Stapled α-helical peptide drug development: a potent dual inhibitor of MDM2 and MDMX for p53-dependent cancer therapy. *Proceedings of the National Academy of Sciences of the United States of America* 110 (36): E3445.

31 Guerlavais, V., Sawyer, T.K., Carvajal, L. et al. (2023). Discovery of sulanemadlin (ALRN-6924), the first cell-permeating, stabilized alpha-helical peptide in clinical development. *Journal of Medicinal Chemistry* 66 (14): 9401.

32 Pazgier, M., Liu, M., Zou, G. et al. (2009). Structural basis for high-affinity peptide inhibition of p53 interactions with MDM2 and MDMX. *Proceedings of the National Academy of Sciences of the United States of America* 106 (12): 4665.

33 Li, X., Gohain, N., Chen, S. et al. (2021). Design of ultrahigh-affinity and dual-specificity peptide antagonists of MDM2 and MDMX for P53 activation and tumor suppression. *Acta Pharmaceutica Sinica B* 11 (9): 2655.

34 Brown, C.J., Quah, S.T., Jong, J. et al. (2013). Stapled peptides with improved potency and specificity that activate p53. *ACS Chemical Biology* 8 (3): 506.

35 Chen, S., Li, X., Li, Y. et al. (2022). Design of stapled peptide-based PROTACs for MDM2/MDMX atypical degradation and tumor suppression. *Theranostics* 12 (15): 6665.

36 Li, X., Tolbert, W.D., Hu, H.G. et al. (2019). Dithiocarbamate-inspired side chain stapling chemistry for peptide drug design. *Chemical Science* 10 (5): 1522.

37 Walensky, L.D., Kung, A.L., Escher, I. et al. (2004). Activation of apoptosis in vivo by a hydrocarbon-stapled BH3 helix. *Science* 305 (5689): 1466.

38 Walensky, L.D., Pitter, K., Morash, J. et al. (2006). A stapled BID BH3 helix directly binds and activates BAX. *Molecular Cell* 24 (2): 199.

39 Bird, G.H., Bernal, F., Pitter, K., and Walensky, L.D. (2008). Synthesis and biophysical characterization of stabilized α-helices of BCL-2 domains. *Methods in Enzymology* 446: 369.

40 LaBelle, J.L., Katz, S.G., Bird, G.H. et al. (2012). A stapled BIM peptide overcomes apoptotic resistance in hematologic cancers. *The Journal of Clinical Investigation* 122 (6): 2018.

41 Danial, N.N., Walensky, L.D., Zhang, C.-Y. et al. (2008). Dual role of proapoptotic BAD in insulin secretion and beta cell survival. *Nature Medicine* 14 (2): 144.

42 Huhn, A.J., Guerra, R.M., Harvey, E.P. et al. (2016). Selective covalent targeting of anti-apoptotic BFL-1 by cysteine-reactive stapled peptide inhibitors. *Cell Chemical Biology* 23 (9): 1123.

43 Edwards, A.L., Gavathiotis, E., LaBelle, J.L. et al. (2013). Multimodal interaction with BCL-2 family proteins underlies the proapoptotic activity of PUMA BH3. *Chemistry and Biology* 20 (7): 888.

44 Stewart, M.L., Fire, E., Keating, A.E., and Walensky, L.D. (2010). The MCL-1 BH3 helix is an exclusive MCL-1 inhibitor and apoptosis sensitizer. *Nature Chemical Biology* 6 (8): 595.

45 Liao, H., Li, X., Zhao, L. et al. (2020). A PROTAC peptide induces durable beta-catenin degradation and suppresses Wnt-dependent intestinal cancer. *Cell Discovery* 6: 35.

46 Li, Y., Wu, M., Fu, Y. et al. (2024). Therapeutic stapled peptides: efficacy and molecular targets. *Pharmacological Research* 203: 107137.

47 Xia, S., Zhu, Y., Liu, M. et al. (2020). Fusion mechanism of 2019-nCoV and fusion inhibitors targeting HR1 domain in spike protein. *Cellular and Molecular Immunology* 17 (7): 765.

48 Xia, S., Yan, L., Xu, W. et al. (2019). A pan-coronavirus fusion inhibitor targeting the HR1 domain of human coronavirus spike. *Science Advances* 5: eaav4580.

49 Zheng, M., Cong, W., Peng, H. et al. (2021). Stapled peptides targeting SARS-CoV-2 spike protein HR1 inhibit the fusion of virus to its cell receptor. *Journal of Medicinal Chemistry* 64 (23): 17486.

50 Chen, B., Liu, C., Cong, W. et al. (2023). Cyclobutane-bearing restricted anchoring residues enabled geometry-specific hydrocarbon peptide stapling. *Chemical Science* 14 (41): 11499.

51 Bird, G.H., Patten, J.J., Zavadoski, W. et al. (2024). A stapled lipopeptide platform for preventing and treating highly pathogenic viruses of pandemic potential. *Nature Communications* 15 (1): 274.

52 Zhang, H., Penninger, J.M., Li, Y. et al. (2020). Angiotensin-converting enzyme 2 (ACE2) as a SARS-CoV-2 receptor: molecular mechanisms and potential therapeutic target. *Intensive Care Medicine* 46 (4): 586.

53 Maas, M.N., Hintzen, J.C.J., Löffler, P.M.G., and Mecinović, J. (2021). Targeting SARS-CoV-2 spike protein by stapled hACE2 peptides. *Chemical Communications* 57 (26): 3283.

54 Zhang, H., Zhao, Q., Bhattacharya, S. et al. (2008). A cell-penetrating helical peptide as a potential HIV-1 inhibitor. *Journal of Molecular Biology* 378 (3): 565.

55 Vinogradov, A.A., Choo, Z.-N., Totaro, K.A., and Pentelute, B.L. (2016). Macrocyclization of unprotected peptide isocyanates. *Organic Letters* 18 (6): 1226.

56 Liu, T., Cong, W., Ye, L. et al. (2020). Rational design of stapled peptides targeting phosphorylated GSK3beta for regulating osteoclast differentiation. *RSC Advances* 10 (13): 7758.

57 Cong, W., Shen, H., Liao, X. et al. (2023). Discovery of an orally effective double-stapled peptide for reducing ovariectomy-induced bone loss in mice. *Acta Pharmaceutica Sinica B* https://doi.org/10.1016/j.apsb.2023.05.004 10.1016/j.apsb.2023.05.004.

58 Lai, Y., Tuvim, M.J., Leitz, J. et al. (2022). Screening of hydrocarbon-stapled peptides for inhibition of calcium-triggered exocytosis. *Frontiers in Pharmacology* 13: 891041.

59 Schreyer, E., Obringer, C., Messaddeq, N. et al. (2022). PATAS, a first-in-class therapeutic peptide biologic, improves whole-body insulin resistance and associated comorbidities in vivo. *Diabetes* 71 (9): 2034.

60 Obrecht, D., Chevalier, E., Moehle, K., and Robinson, J.A. (2012). β-Hairpin protein epitope mimetic technology in drug discovery. *Drug Discovery Today: Technologies* 9 (1): e63.

61 Fasan, R., Dias, R.L., Moehle, K. et al. (2004). Using a beta-hairpin to mimic an alpha-helix: cyclic peptidomimetic inhibitors of the p53-HDM2 protein-protein interaction. *Angewandte Chemie (International Ed. in English)* 43 (16): 2109.

62 Makwana, K.M. and Mahalakshmi, R. (2015). Trp-Trp cross-linking: a structure-reactivity relationship in the formation and design of hyperstable peptide beta-hairpin and alpha-helix scaffolds. *Organic Letters* 17 (10): 2498.

63 Ganesh Kumar, M., Mali, S.M., Raja, K.M., and Gopi, H.N. (2015). Design of stable beta-hairpin mimetics through backbone disulfide bonds. *Organic Letters* 17 (2): 230.

64 Park, J.H. and Waters, M.L. (2013). Positional effects of click cyclization on beta-hairpin structure, stability, and function. *Organic & Biomolecular Chemistry* 11 (1): 69.

65 Wang, G., Drinkwater, N., Drew, D.R. et al. (2016). Structure-activity studies of beta-hairpin peptide inhibitors of the Plasmodium falciparum AMA1-RON2 interaction. *Journal of Molecular Biology* 428 (20): 3986.

66 Pace, J.R., Lampkin, B.J., Abakah, C. et al. (2021). Stapled beta-hairpins featuring 4-mercaptoproline. *Journal of the American Chemical Society* 143 (37): 15039.

67 Sawyer, N. and Arora, P.S. (2018). Hydrogen bond surrogate stabilization of beta-hairpins. *ACS Chemical Biology* 13 (8): 2027.

68 Ma, B., Liu, D., Zheng, M. et al. (2024). Development of a double-stapled peptide stabilizing both alpha-helix and beta-sheet structures for degrading transcription factor AR-V7. *JACS Au* 4 (2): 816.

69 Robinson, J.A. (2008). β-Hairpin peptidomimetics: design, structures and biological activities. *Accounts of Chemical Research* 41 (10): 1278.

70 Balboa, J.R., Essig, D.J., Ma, S. et al. (2023). Development of a potent cyclic peptide inhibitor of the nNOS/PSD-95 interaction. *Journal of Medicinal Chemistry* 66 (1): 976.

71 Walsh, C.T., Zydowsky, L.D., and McKeon, F.D. (1992). Cyclosporin A, the cyclophilin class of peptidylprolyl isomerases, and blockade of T cell signal transduction. *The Journal of Biological Chemistry* 267 (19): 13115.

72 Tsomaia, N. (2015). Peptide therapeutics: targeting the undruggable space. *European Journal of Medicinal Chemistry* 94: 459.

73 Krzyzanowski, A., Esser, L.M., Willaume, A. et al. (2022). Development of macrocyclic PRMT5-adaptor protein interaction inhibitors. *Journal of Medicinal Chemistry* 65 (22): 15300.

74 Fetse, J., Zhao, Z., Liu, H. et al. (2022). Discovery of cyclic peptide inhibitors targeting PD-L1 for cancer immunotherapy. *Journal of Medicinal Chemistry* 65 (18): 12002.

75 Adihou, H., Gopalakrishnan, R., Forster, T. et al. (2020). A protein tertiary structure mimetic modulator of the Hippo signalling pathway. *Nature Communications* 11 (1): 5425.

76 Diaz-Perlas, C., Ricken, B., Farrera-Soler, L. et al. (2023). High-affinity peptides developed against calprotectin and their application as synthetic ligands in diagnostic assays. *Nature Communications* 14 (1): 2774.

77 Ji, X., Nielsen, A.L., and Heinis, C. (2024). Cyclic peptides for drug development. *Angewandte Chemie (International Ed. in English)* 63 (3): e202308251.

78 Vazquez Torres, S., Leung, P.J.Y., Venkatesh, P. et al. (2024). De novo design of high-affinity binders of bioactive helical peptides. *Nature* 626 (7998): 435.

79 Ma, B., Liu, D., Wang, Z. et al. (2024). A top-down design approach for generating a peptide PROTAC drug targeting androgen receptor for androgenetic alopecia therapy. *Journal of Medicinal Chemistry* 67 (12): 10336.

80 Tucker, T.J., Embrey, M.W., Alleyne, C. et al. (2021). A series of novel, highly potent, and orally bioavailable next-generation tricyclic peptide PCSK9 inhibitors. *Journal of Medicinal Chemistry* 64 (22): 16770.

12

Small Molecule Inhibitors Targeting Protein–Protein Interactions in the BCL Protein

Wenhua Zhu[1,2], Yangbo He[1,2], Gang Chen[1,2], and Di Zhu[3]

[1]*Anhui University of Chinese Medicine, No. 350, Longzi Lake Road, Modian Street, Yaohai District, Hefei 230012, China*
[2]*Yangtze Delta Drug Advanced Research Institute and Yangtze Delta Pharmaceutical College, No. 100, Dongtinghu Road, Linjiang Town, Haimen District, Nantong 226133, China*
[3]*Fudan University, School of Basic Medical Sciences, No. 825 Chuansha Road, Zhangjiang Hi-Tech Park, Pudong New Area, Shanghai 201210, China*

12.1 Introduction

Protein–protein interactions (PPIs) are fundamental to cellular processes such as signal transduction, cell proliferation, and apoptosis [1, 2]. Recent research estimates that the human proteome encompasses approximately 650,000 PPIs, forming an intricate network known as the interactome, which is crucial for modulating various protein functions [3]. These interactions are central to the pathology of cancer, influencing tumor initiation, progression, invasion, and metastasis [4]. Targeting and modulating PPIs presents a promising therapeutic strategy, although the challenge of identifying small molecules that effectively disrupt these interactions is substantial due to the broad, flat, and variable nature of PPI interfaces [5, 6].

In PPI interfaces, amino acid residues contribute unequally to binding free energy, with certain residues—called “hotspots” [7]—often accounting for the majority of the binding energy. These hotspots typically form a highly dependent network within the interaction interface, referred to as the “hot zone.” This concept has significantly advanced the development of PPI inhibitors by providing a theoretical basis for the design of ligands that directly target these interaction interfaces. The rational design of PPI inhibitors is thus seen as a cutting-edge approach in drug discovery with considerable potential [8, 9].

PPI inhibitors, as a novel class of therapeutics, have been extensively studied and exhibit substantial therapeutic promise. By disrupting the formation or function of critical PPIs, these inhibitors can modulate the activity of signaling pathways and halt aberrant biological processes associated with diseases. Compared to traditional small-molecule drugs, PPI inhibitors offer greater specificity and fewer side effects due to their precise targeting of specific protein complexes [10].

Recent research has focused on PPI inhibitors targeting key interactions between proteins such as the BCL-2 family, BCL9/β-catenin, and BCL-6. These interactions

Targeting Protein–Protein Interactions for Drug Discovery, First Edition. Edited by Jian Zhang.

play pivotal roles in the development and progression of various diseases, including cancer, neurodegenerative disorders, and immune system dysregulation [10]. This review provides a comprehensive evaluation of inhibitors targeting BCL protein PPIs. It emphasizes strategies for the discovery and optimization of small-molecule drugs, focusing on specificity toward different BCL protein targets and their potential for cancer therapy. The goal is to guide medicinal chemists in designing effective modulators of BCL proteins to advance more effective cancer treatment strategies.

12.2 Inhibitors of BCL-2 Family Antiapoptotic Proteins

12.2.1 Members and Structure of BCL-2 Family Proteins

The B-cell lymphoma 2 (BCL-2) protein family plays a crucial role in the regulation of apoptosis through a complex network of protein interactions. Members of this family are categorized based on their function and the number of homologous domains they contain into antiapoptotic proteins, proapoptotic proteins, and those containing only the BCL-2 homology 3 (BH3) domain. Antiapoptotic proteins include Bcl-2, Bcl-xL, Mcl-1, Bcl-w, Bfl-1 (Bcl-2A1), and Bcl-b [11]. The first five of these possess four BH domains (BH1 through BH4) and a C-terminal transmembrane (TM) domain. They are primarily localized to the outer mitochondrial membrane or, in some cases, are found in the cytosol, where they interact with proapoptotic proteins, and are also present in the endoplasmic reticulum and other organelles (Figure 12.1a). In contrast, Bcl-b is structurally and functionally distinct from other antiapoptotic BCL-2 proteins, containing only BH1 and BH2 domains and a hydrophobic TM region [12, 13], but lacking the BH3 domain. Notably, Bcl-b has a characteristic BH1 sequence, "TWGR," and can bind to the BH3 domain of Bax (Figure 12.1a) [14]. Proapoptotic proteins, such as Bak, Bax, and Bok, are essential for the execution of apoptosis. These proteins generally possess three BH domains (BH1-3) and a C-terminal TM domain. They exist in a dormant state and are distributed in the cytosol or loosely associated with membranes. Upon receiving apoptotic signals, these proteins undergo conformational changes, translocate to the outer mitochondrial membrane, and form dimers that facilitate apoptosis. BH3-only proteins act as sensors in the apoptosis signaling pathway (Figure 12.1a). They include proteins with both BH3 and TM domains (e.g., Bim, Bmf, Hrk, Bik), as well as those containing only the BH3 domain (e.g., Bid, Bad, Noxa, Puma). These proteins play a critical role in regulating apoptosis by modulating the interactions between antiapoptotic and proapoptotic proteins [14, 15].

Structural biology research indicates that antiapoptotic proteins are typically composed of eight to nine α-helices, with their BH1 to BH3 domains forming a narrow, V-shaped groove that facilitates interactions with other proteins or molecules (Figure 12.1b,c) [16]. In contrast, proapoptotic proteins or BH3-only proteins possess a BH3 domain that adopts a tightly packed α-helical structure, which effectively binds to the V-shaped groove of antiapoptotic proteins, thereby activating proapoptotic signaling pathways and inducing apoptosis. Within the

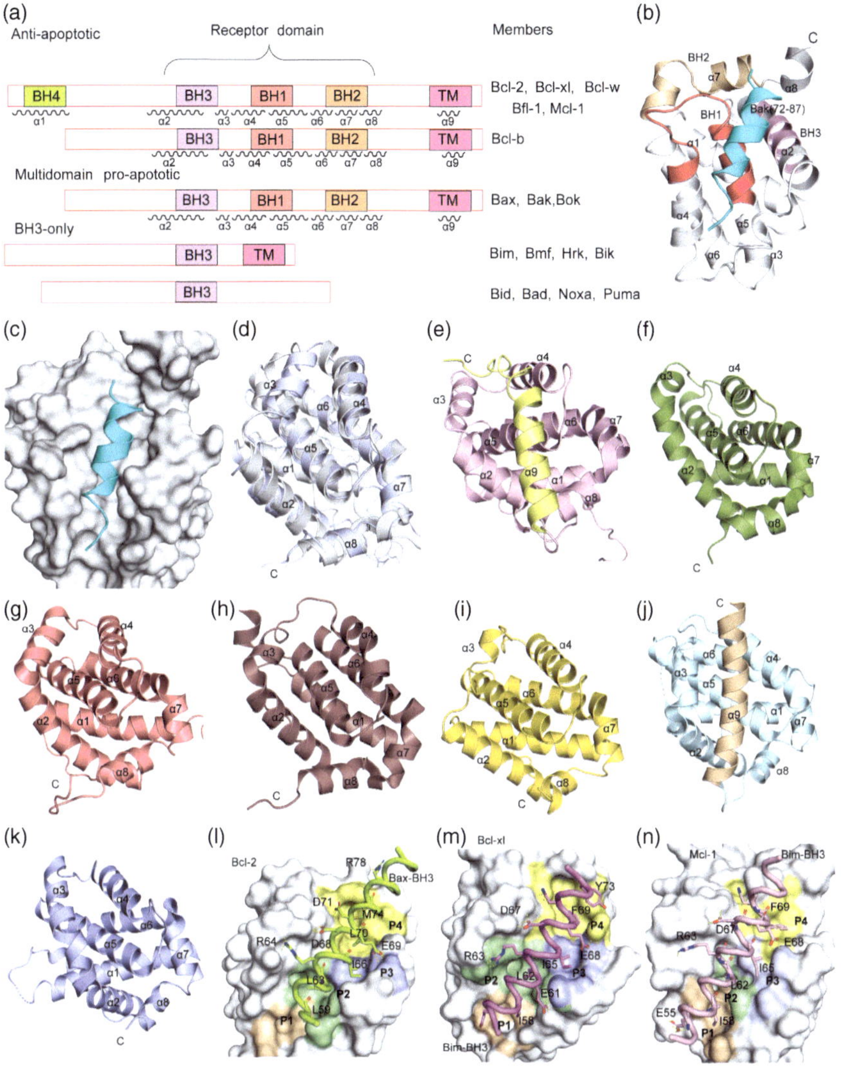

Figure 12.1 Structures of BCL-2 family proteins. (a) composition of BCL-2 family protein members; (b, c) crystal structure of Bcl-xL-Bak BH3 domain complex (PDB ID: 1BXL); (d) structural superposition of Bcl-2 and Bcl-xL (Bcl-2 PDB ID: 1G5M, Bcl-xL PDB ID: 1BXL); (e) crystal structure of Bcl-w (PDB ID: 1O0L); (f) crystal structure of Bfl-1 (PDB ID: 5W62); (g) crystal structure of Mcl-1 (PDB ID: 4LXD); (h) crystal structure of Bcl-b (PDB ID: 4B4S); (i) crystal structure of Bok (PDB ID: 6CKV); (j) crystal structure of Bax (PDB ID: 4S0O); (k) crystal structure of Bak (PDB ID: 2YV6). (l) crystal structure of Bcl-2-Bax BH3 domain complex (PDB ID: 2XA0); (m) crystal structure of Bcl-xL-Bim BH3 domain complex (PDB ID: 4QVF); (n) crystal structure of Mcl-1-Bim BH3 domain complex (PDB ID: 2NL9). Source: Wenhua Zhu, Yangbo He, Gang Chen, Di Zhu.

Bcl-2 family, Bcl-xL and Bcl-2 exhibit highly similar structures, centered around two hydrophobic helices (α5 and α6), with α3 and α4 helices forming a slender hydrophobic groove that mediates interactions between antiapoptotic and proapoptotic proteins (Figure 12.1d) [17, 18]. However, Bcl-w and Mcl-1, while similar in overall structure to Bcl-xL and Bcl-2, differ in that Bcl-w's C-terminal α9 helix folds in a manner more akin to the BH3 domain of proapoptotic proteins (Figure 12.1e,g) [19, 20]. This folding characteristic restricts Bcl-w's interaction with proapoptotic proteins such as Bid or Bak, demonstrating its unique strategy in protein interaction regulation (Figure 12.1e). In contrast, Bfl-1 features a distinctive structure with multiple amphipathic helices centered around the hydrophobic helices and an external hydrophobic groove (Figure 12.1f) [21]. Bcl-b differs from other BCL-2 family members by inserting an unstructured loop between α5 and α6 helices (Figure 12.1h) [12]. Proapoptotic proteins also typically consist of eight to nine α-helices. Notably, Bok protein exhibits significant structural variation in the region between α2 and α4 helices, with a distinctive hydrophobic groove characteristic (Figure 12.1i) [22]. Bax's α9 helix, part of its BH3 domain, interacts with other BCL-2 family members to regulate cellular survival and death decisions (Figure 12.1j) [23]. Bak's core structure is composed of two principal α-helices (α5 and α6), which are predominantly hydrophobic, with α3 and α4 helices on one side and α1 and α2 helices on the other, and its BH3 domain situated between α2 and α3 helices (Figure 12.1k) [24]. Collectively, the BCL-2 family proteins regulate cellular survival and apoptosis through complex domain interactions, highlighting their crucial role in cellular homeostasis.

12.2.2 Binding Sites and Key Interactions of BCL-2 Family Proteins

The BH3 domains of proapoptotic and BH3-only proteins exhibit amphipathic α-helical characteristics and are highly conserved. These domains bind to the hydrophobic grooves of antiapoptotic proteins through hydrophobic and electrostatic interactions, displaying similar hydrophobic and charge properties [16, 25, 26]. For instance, in the Bax BH3 domain, hydrophobic residues L59, L63, I66, L70, and M74 form hydrophobic interactions with corresponding residues in Bcl-2 (Figure 12.1l), with these residues being highly conserved among proapoptotic proteins and contributing to the formation of four pockets (P1-P4) in the hydrophobic groove (Figure 12.1l) [27]. Additionally, residues R64, D68, E69, D71, and R78 in the Bax BH3 domain engage in hydrogen bonding and salt bridge interactions with analogous residues in Bcl-2 (Figure 12.1l). In the Bim BH3 domain, residues I58, L62, I65, and F69 form hydrophobic interactions with corresponding residues in Bcl-xL, while residues E61, R63, D67, E68, and Y73 establish hydrogen bonds and salt bridges with equivalent residues in Bcl-xL (Figure 12.1m) [28]. Other proapoptotic proteins exhibit similar or analogous interaction patterns at corresponding positions. Mcl-1 and other antiapoptotic proteins possess conserved tertiary structures but differ in residue sequences [27]. Mcl-1 features a distinct hydrophobic groove that includes lysine and histidine, which contrasts with other antiapoptotic proteins by creating a more positively charged electrostatic environment. This unique feature facilitates the binding of negatively charged inhibitors. Notably, the critical binding hotspots for Mcl-1 are located within the P2 and P3 pockets (Figure 12.1n).

12.2.3 Antiapoptotic Proteins of the BCL-2 Family and Cancer

In malignant tumors, the high expression of BCL-2 family members often leads to enhanced antiapoptotic capabilities in tumor cells, thereby promoting tumor progression and drug resistance. Specifically, the aberrant expression of Bc-2, Bcl-xL, and Mcl-1 is notably significant in this process. BCL-2 is upregulated in various cancers, including prostate cancer, lung cancer, chronic lymphocytic leukemia, acute myeloid leukemia (AML), mixed lineage leukemia, and mantle cell lymphoma, and it promotes malignancy and chemoresistance through multiple mechanisms. Similarly, Bcl-xL exhibits excessive expression in colorectal cancer, T-cell lymphoma, multiple myeloma, and chronic lymphocytic leukemia [29–35]. Animal model studies indicate that Bcl-xL expression correlates closely with survival rates in colorectal cancer, and inhibition of Bcl-xL significantly suppresses tumor growth, enhances chemotherapeutic-induced tumor cell death, reduces cancer cell invasiveness, and increases sensitivity to radiotherapy, highlighting the therapeutic potential of targeting Bcl-xL [36, 37]. Additionally, other antiapoptotic proteins such as Mcl-1, BFL-1, and BCL-W also show varying degrees of overexpression in different tumors. Mcl-1 is highly expressed in hematologic malignancies and various solid tumors [38, 39], where its overexpression disrupts the balance between antiapoptotic and proapoptotic proteins, thus preventing apoptosis and leading to malignant proliferation [40]. BFL-1 is significantly upregulated in melanoma cell lines, and overexpression of BCL2A1 promotes growth in intracranial tumors [41]. In breast cancer specimens, BCL-W mRNA levels are significantly higher compared to adjacent normal tissues [42, 43], with BCL-W transcriptional levels elevated in the plasma of patients with metastatic disease compared to those with primary tumors [42].

12.2.4 BCL-2 Family Antiapoptotic Protein Inhibitors

With the gradual elucidation of the crystal structures of BCL-2 family proteins, such as Bcl-2, Bcl-xL, and Mcl-1, and their BH3 domain complexes [16, 27, 28], detailed insights into the ligand-binding pockets of Bcl-2 and Bcl-xL have emerged, providing a profound understanding for targeted drug design. However, the structural similarity between Bcl-2 and Bcl-xL presents challenges in drug design, particularly regarding selectivity and small-molecule development. Nevertheless, through the simulation of BH3 domains, researchers have successfully developed a series of targeted inhibitors, including both peptide and small-molecule inhibitors, collectively referred to as "BH3 mimetics" [44–48]. Furthermore, the interaction of Bax and Bak with Bcl-2 impedes the apoptotic process in tumor cells. Thus, modeling the α-helical structure within the BH3 domain is anticipated to be an effective strategy for targeting the Bcl-2 and Bax interaction, potentially offering new avenues for cancer therapy.

12.2.4.1 Selective Bcl-2/Bcl-xL Inhibitors

Compounds 1-8 Over the past two decades, the development of selective small-molecule inhibitors targeting Bcl-2 antiapoptotic proteins has become a significant area of research in cancer therapy. BH3 mimetics, as a promising drug strategy, achieve this by binding with high affinity to the hydrophobic groove of Bcl-2 antiapoptotic proteins, leading to the discovery and development of related inhibitors (Table 12.1). Dr. Huang and his team were the first to report the discovery of Bcl-2 small-molecule inhibitors. Using computational structure databases for screening, they identified the organic small molecule HA14-1(**1**), which effectively induced apoptosis in human AML (HL-60) cells overexpressing Bcl-2, despite an IC_{50} of 9 μM in binding assays [49]. However, HA14-1 was chemically unstable. Subsequently, two independent studies reported the discovery of novel small-molecule inhibitors of Bcl-2 or Bcl-xL. Dr. Yuan and his team identified two different classes of Bcl-xL small-molecule inhibitors through fluorescence polarization-based high-throughput screening (HTS), with compounds **2** and **3** being the most effective representatives [50]. Dr. Hockenbery and colleagues discovered that the antibiotic antimitotic agent A (compound **4**) exhibited potential as a Bcl-2/Bcl-xL small-molecule inhibitor, with an IC_{50} of 2.5 μM [51, 52]. These studies collectively indicate that small-molecule organic inhibitors binding to the BH3 domain of Bcl-2/Bcl-xL can effectively block their antiapoptotic functions, thereby inducing apoptosis in cells overexpressing Bcl-2/Bcl-xL. Inspired by these findings, the Enyedy team utilized structure-based computational screening methods to identify a class of Bcl-2 small-molecule inhibitors, with the most representative compound being **5** (FP IC_{50} = 10.4 μM) [53]. Based on these initial lead compounds, the team designed and synthesized new analogs, including compound **6** (YC137), which demonstrated for the first time that Bcl-2-specific inhibitors can selectively induce apoptosis in tumor cells overexpressing or dependent on this antiapoptotic protein, with no significant effects on various primary cells [54]. Mai Nguyen developed a hydrophobic compound **7** (obatoclax) through structure-based drug screening and optimization, effectively overcoming challenges associated with targeting Mcl-1 antiapoptotic proteins. This established a foundation for the combination of obatoclax with other Mcl-1-targeted anticancer agents. Although obatoclax has shown potential therapeutic advantages in combination studies, such as with PD-1 antibodies for liver cancer or enhancing the activity of anti-multiple myeloma agents, its lack of specificity and high cytotoxicity limit its further development in cancer therapy [55].

Targeting differences in residue binding within the hydrophobic groove represents an effective strategy for achieving Bcl-2 specificity. Although all antiapoptotic Bcl-2 family members interact with proapoptotic partners along their hydrophobic grooves, designing specific inhibitors can leverage sequence variability and subtle conformational differences between them. Inspired by this approach, the Iyer research group used 3-substituted indolinone compounds with known anticancer properties as structural scaffolds. Through computational docking, they identified the small molecule **8** (Disarib), which exhibits specificity for Bcl-2 with a dissociation constant (K_d) of 28 nM. Disarib primarily targets the BH1 domain while

Table 12.1 Selective Bcl-2/Bcl-xL inhibitors.

Compound	Structure	Targets	Affinity	Development status
1 HA14-1	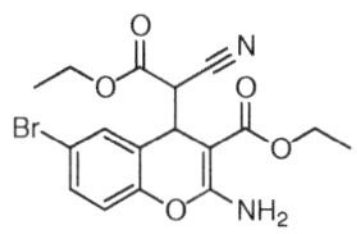	BCL-2	Bcl-2 FP $IC_{50} = 9\,\mu M$	Preclinical
2	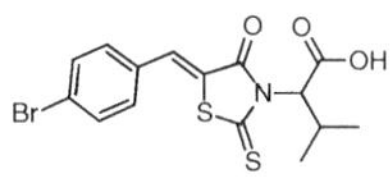	Bcl-xL	Bcl-xL FP $K_i = 2.4\,\mu M$	Preclinical
3	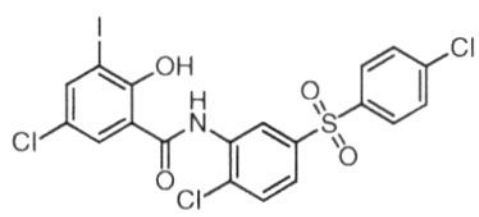	Bcl-xL	Bcl-xL FP $IC_{50} = 3.3\,\mu M$	Preclinical
4 Antimycin A3	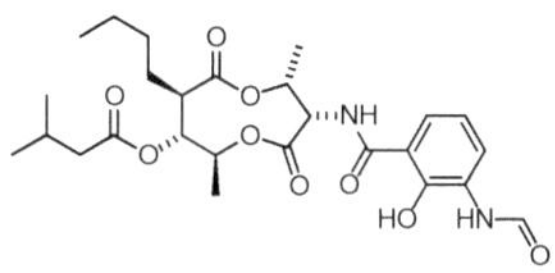	BCL-2 Bcl-xL	Bcl-2/Bcl-xL FP $IC_{50} = 3.3\,\mu M$	Preclinical
5	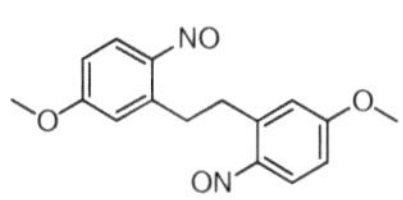	BCL-2 Bcl-xL	Bcl-2 FP $IC_{50} = 10.4\,\mu M$	Preclinical
6 YC137	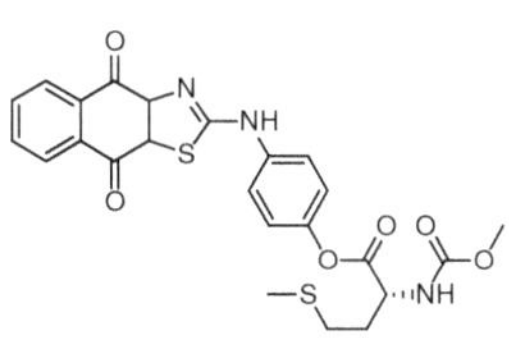	BCL-2 Bcl-xL	Bcl-2 FP $K_i = 1.3\,\mu M$ Bcl-xL FP $K_i > 100\,\mu M$	Preclinical
7 Obatoclax	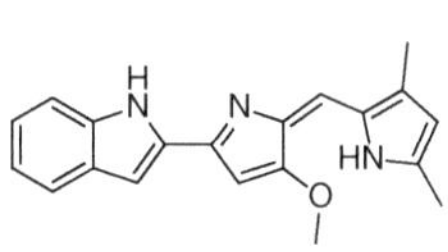	BCL-2 Bcl-xL	Bcl-xL $IC_{50} = 4.69\,\mu M$ Bcl-2 $IC_{50} = 1.11\,\mu M$ Mcl-1 $IC_{50} = 2.90$	Phase III clinical trial
8 Disarib	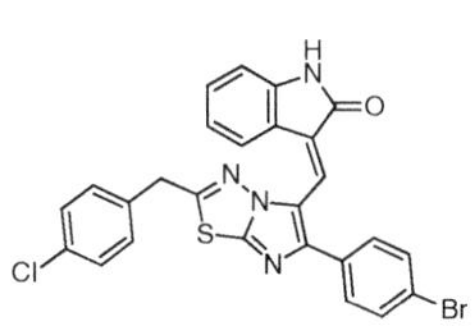	BCL-2	Bcl-2 $K_d = 28\,nM$	Preclinical

(Continued)

Table 12.1 (Continued)

Compound	Structure	Targets	Affinity	Development status
9	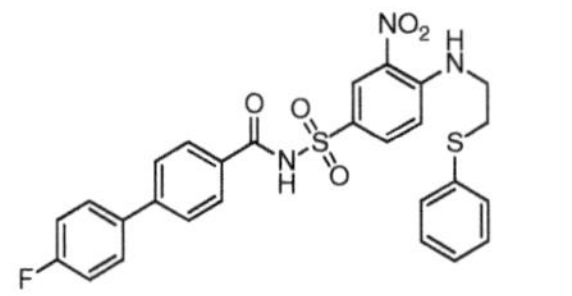	BCL-2 Bcl-xL	$K_i = 36\,nM$	Preclinical
10 ABT-737	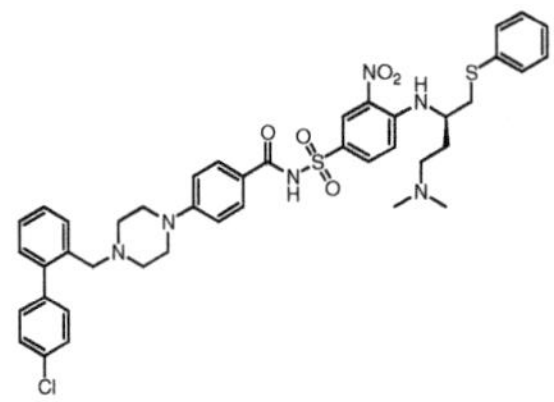	BCL-2 Bcl-xL	Bcl-xL $IC_{50} = 5\,nM$ Bcl-2 $IC_{50} = 7\,nM$ Bcl-w $IC_{50} = 66\,nM$ Mcl-1 $IC_{50} > 10{,}000\,nM$	Phase II clinical trial
11 ABT-263	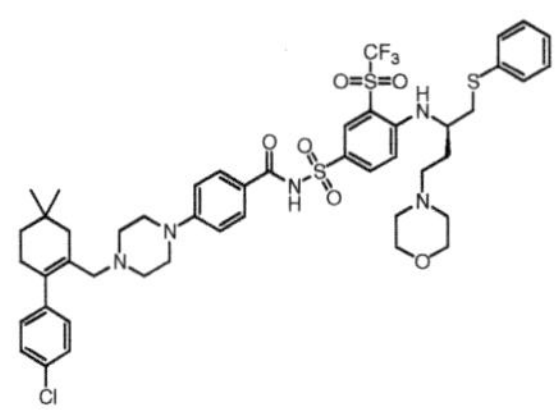	BCL-2/ Bcl-xL	Bcl-xL $K_i < 1\,nM$ Bcl-2 $K_i < 1\,nM$	Preclinical
12	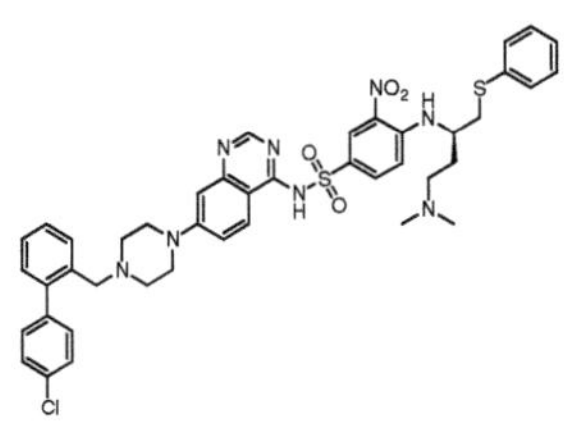	BCL-2 Bcl-xL	Bcl-xL $IC_{50} = 7\,nM$ Bcl-2 $IC_{50} = 8.7\,nM$ Bcl-w $IC_{50} = 440\,nM$ Mcl-1 $IC_{50} > 1000\,nM$	Preclinical
13 S44563	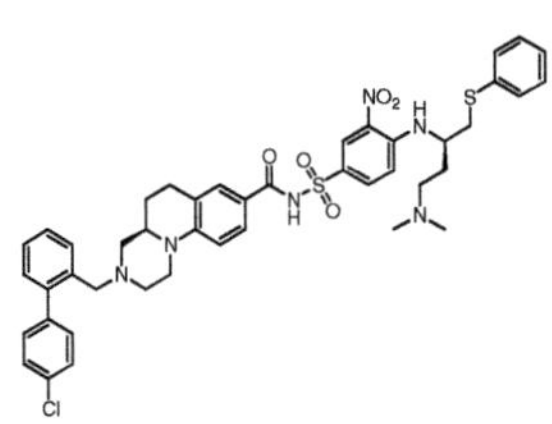	BCL-2/ Bcl-xL	Bcl-2 FP $K_i = 131\,nM$ Bcl-xL FP $K_i = 140\,nM$	Preclinical

Table 12.1 (Continued)

Compound	Structure	Targets	Affinity	Development status
14 ABT-199	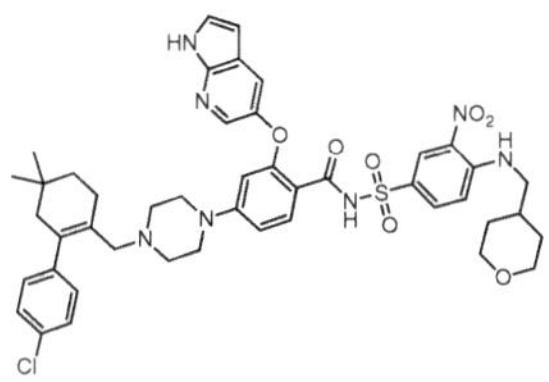	BCL-2 Bcl-xL	Bcl-xL $K_i = 48\,nM$ Bcl-2 $K_i < 0.01\,nM$ Bcl-w $K_i = 245\,nM$ Mcl-1 $K_i > 444\,nM$	Preclinical
15 APG2575 Lisaftoclax	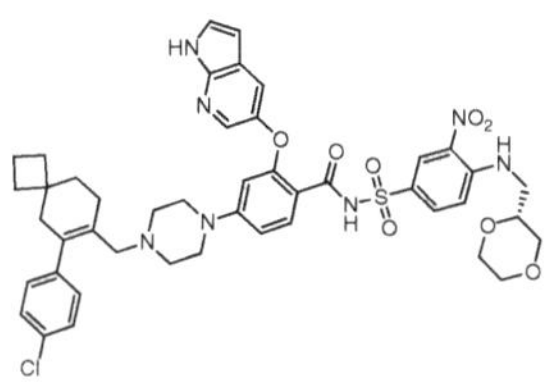	BCL-2 Bcl-xL	Bcl-xL $K_i < 0.5\,nM$ Bcl-2 $K_i < 0.1\,nM$	Phase III clinical trial
16 Sonrotoclax	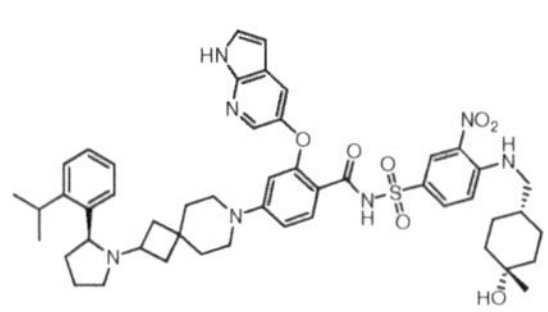	BCL-2 Bcl-xL	Bcl-2 $IC_{50} = 0.014\,nM$ Bcl-xL $IC_{50} = 28\,nM$	Phase III clinical trial
17	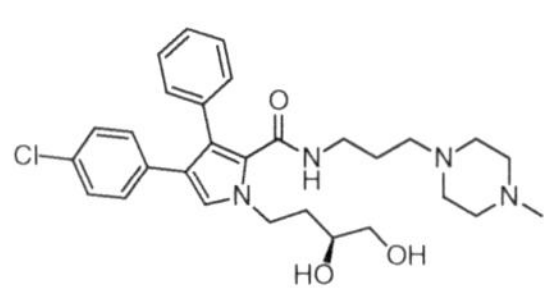	BCL-2 Bcl-xL	Bcl-2 FP $IC_{50} = 0.21\,\mu M$ Bcl-xL FP $IC_{50} = 0.45\,\mu M$	Preclinical
18	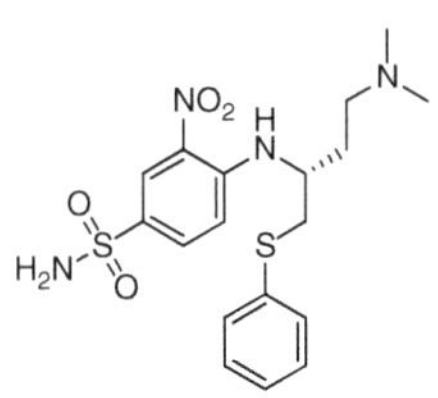	BCL-2 Bcl-xL	Bcl-2 FP $IC_{50} > 100\,\mu M$ Bcl-xL FP $IC_{50} = 0.24\,\mu M$	Preclinical

(Continued)

Table 12.1 (Continued)

Compound	Structure	Targets	Affinity	Development status
19	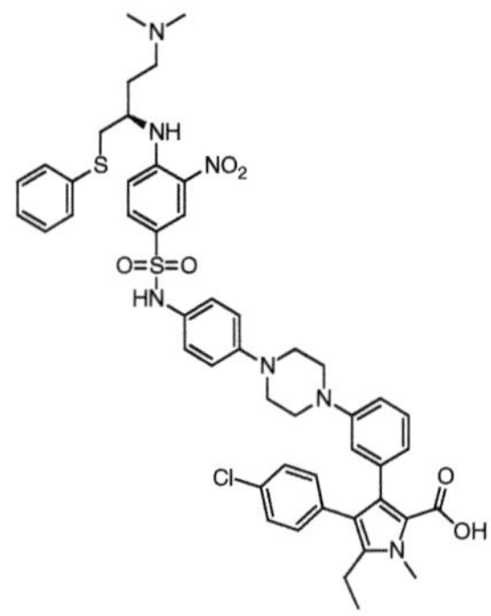	BCL-2 Bcl-xL	Bcl-2 FP $IC_{50} = 4.4\,nM$ Bcl-xL FP $IC_{50} = 8.9\,nM$	Preclinical
20 BM903	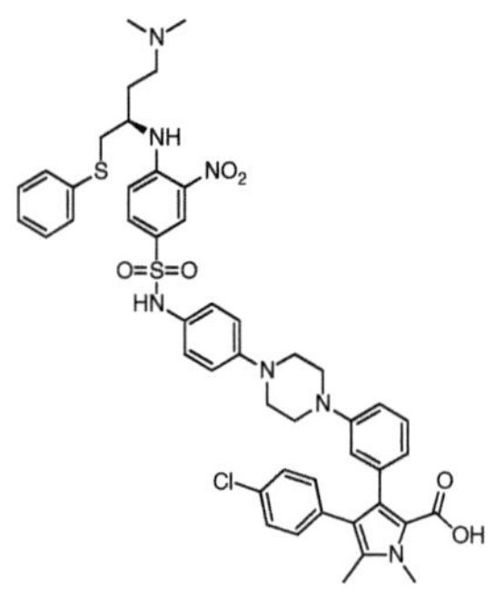	BCL-2 Bcl-xL	/	Preclinical
21 BM957	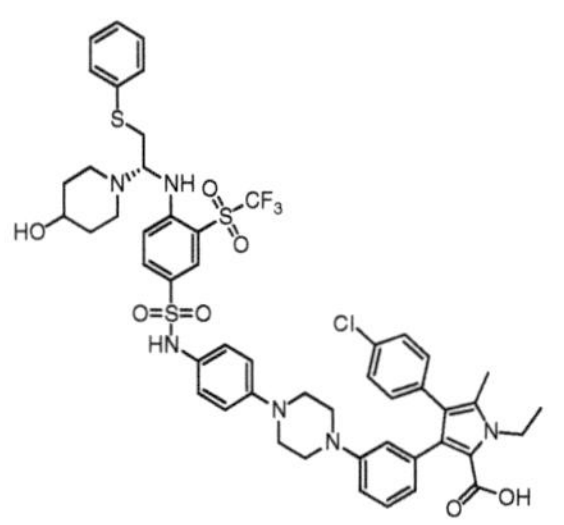	BCL-2 Bcl-xL	$K_i = < 1\,nM$ Bcl-2 FP $IC_{50} = 5.4\,nM$ Bcl-xL FP $IC_{50} = 6.0\,nM$	Preclinical
22 BM1074	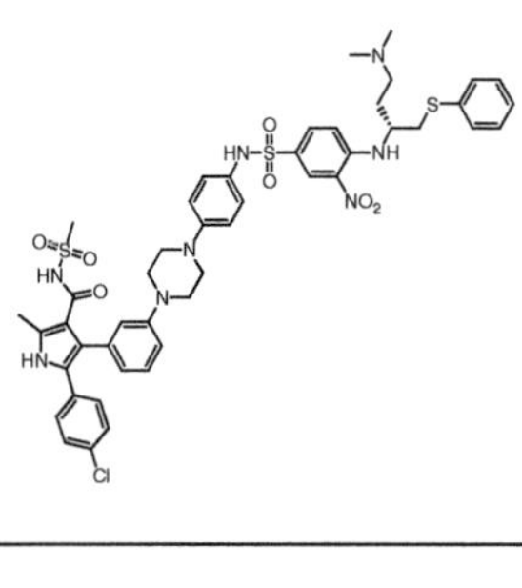	BCL-2 Bcl-xL	$K_i = < 1\,nM$ Bcl-2 FP $IC_{50} = 1.8\,nM$ Bcl-xL FP $IC_{50} = 6.9\,nM$	Preclinical

Table 12.1 (Continued)

Compound	Structure	Targets	Affinity	Development status
23 BM1197	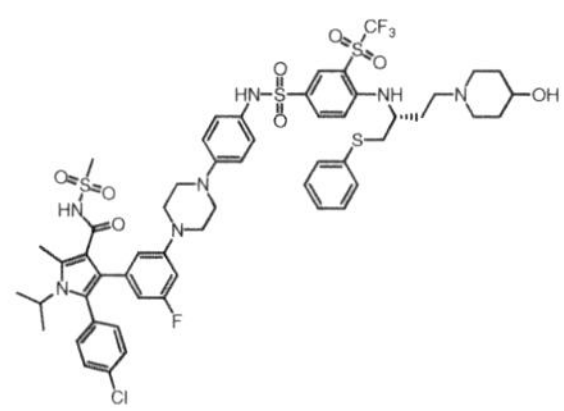	BCL-2 Bcl-xL	K_i = < 1 nM Bcl-2 FP IC_{50} = 3.5 nM Bcl-xL FP IC_{50} = 5.2 nM	Preclinical
24 BM-1252 APG-1252 Pelcitoclax	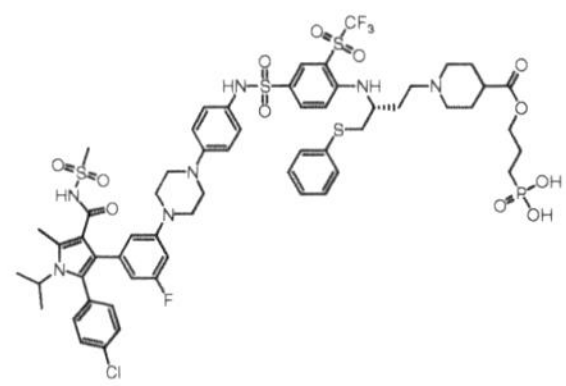	BCL-2 Bcl-xL	Bcl-2/Bcl-xL K_i = <1 nM	Phase II clinical trial
25 AZD4320	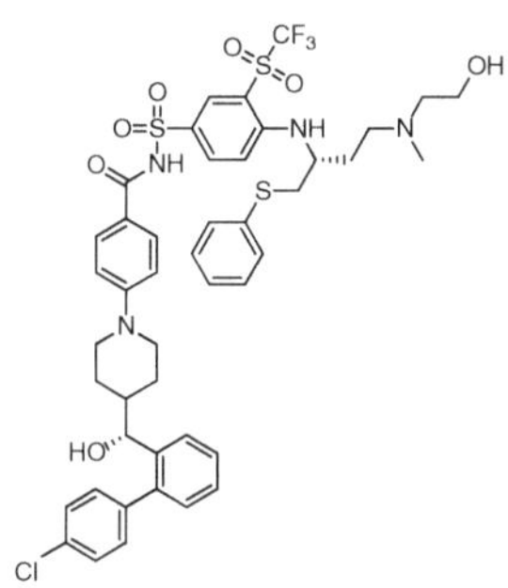	BCL-2 Bcl-xL	/	Preclinical
26 AZD0466	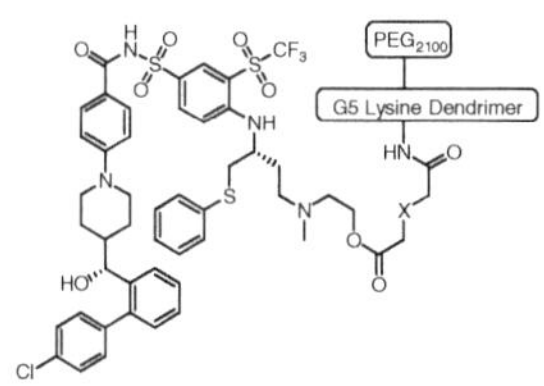	BCL-2 Bcl-xL	/	Preclinical
27	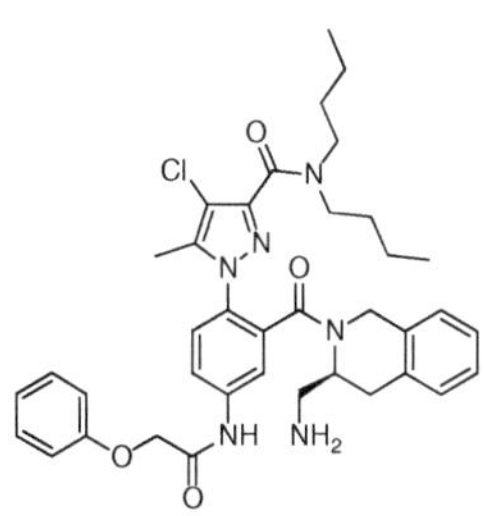	BCL-2 Bcl-xL	Bcl-2 IC_{50} = 0.03 μM Bcl-xL IC_{50} = 0.81 μM	Preclinical

(Continued)

Table 12.1 (Continued)

Compound	Structure	Targets	Affinity	Development status
28	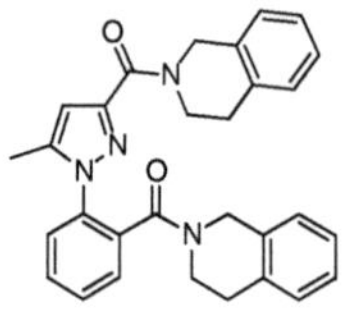	BCL-2 Bcl-xL	Bcl-2 IC_{50} = 1.5 μM Bcl-xL IC_{50} = 16.6 μM	Preclinical
29 S55746	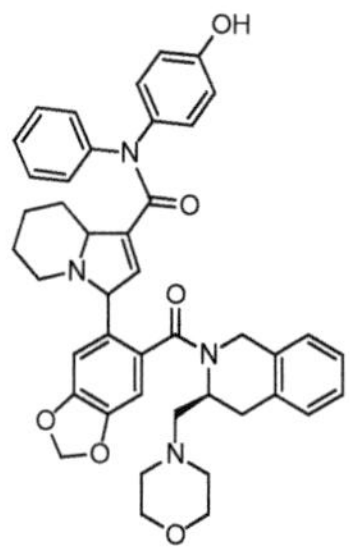	BCL-2	Bcl-2 FP K_i = 1.3 nM Bcl-xL FP K_i = 520 nM	Preclinical
30	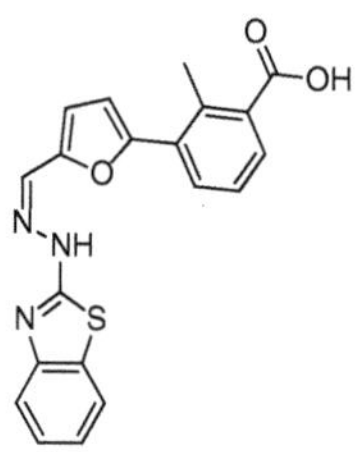	Bcl-xL	AlphaScreen Bcl-xL IC_{50} = 7.7 μM	Preclinical
31 WEHI-539	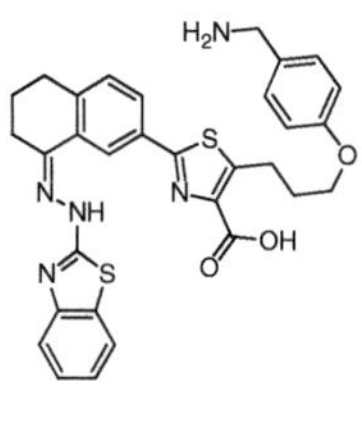	Bcl-xL	AlphaScreen Bcl-xL IC_{50} = 1.1 nM	Preclinical
32 A-1155463	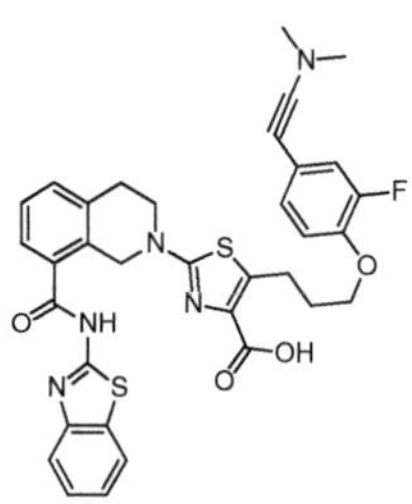	Bcl-xL	Bcl-2 K_i = 80 nM Bcl-xL K_i < 0.01 nM	Preclinical

Table 12.1 (Continued)

Compound	Structure	Targets	Affinity	Development status
33 A-1331852	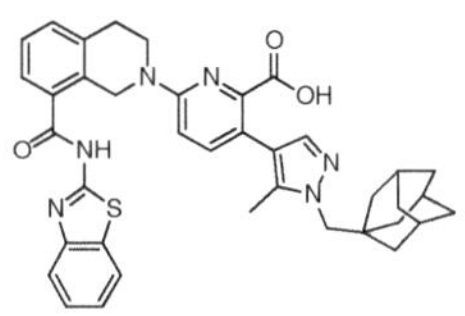	Bcl-xL	Bcl-2 $K_i = 6.1\,nM$ Bcl-xL $K_i < 0.01\,nM$	Preclinical
34 A-1293102	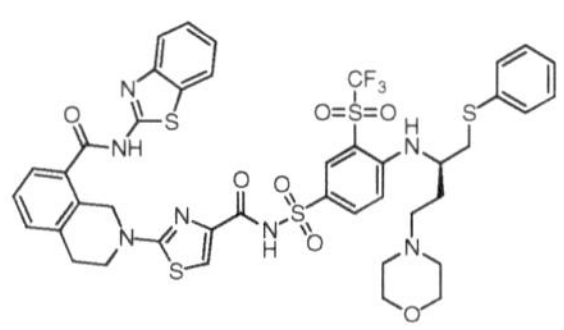	Bcl-xL	Bcl-xL $K_i = 0.43\,nM$ Bcl-2 $K_i = 193\,nM$ Mcl-1 $K_i > 3900\,nM$	Preclinical

sparing other antiapoptotic family members. In Bcl-2-driven tumor models, Disarib demonstrates significant tumor regression effects [56].

ABT-737 and Its Related Derivatives Understanding the biological functions of Bcl-2 proteins and their potential in anticancer therapy, particularly through advances in structural biology techniques, has been pivotal in the development of first-generation high-selectivity Bcl-2 inhibitors such as ABT-737(compound **10**) (Table 12.1) [57]. AbbVie, focusing on Bcl-2 protein inhibitors, employed NMR-based fragment screening to identify compounds capable of binding to the hydrophobic BH3-binding groove of Bcl-2. A representative compound, **9**, exhibits significant binding affinity for Bcl-xL ($K_i = 36 \pm 1.6\,nM$). By introducing a 2-dimethylaminoethyl group at the 1-amino position of the thioethylamino linker and replacing the fluorophenyl group with piperazine, further refinements led to ABT-737. ABT-737 mimics the BH3 domain of the bad protein and shows a high affinity for Bcl-xL, Bcl-2, and Bcl-w, but not Mcl-1 [57]. Despite its substantial activity against solid tumors in animal models, ABT-737 faced challenges related to pharmacokinetics and solubility in preclinical studies. Through targeted modification of critical sites, ABT-737-derived compound ABT-263 (compound **11**, navitoclax) was developed [58]. Phase I clinical trials demonstrated that ABT-263 is well-tolerated and safe in patients with non-small cell lung cancer (NSCLC) and lymphoid malignancies [59, 60]. However, dose-dependent thrombocytopenia resulting from Bcl-xL inhibition limited its efficacy in cancer cell suppression. Currently, ABT-263 is being evaluated in multiple clinical trials for its effectiveness in treating various malignancies.

A distinctive feature of these inhibitors is the inclusion of the acylsulfonamide structure, which may present key metabolic issues [61]. Addressing these, Sleebs employed a spatial replacement strategy, substituting the benzoyl sulfonamide moiety with a quinazoline sulfonamide scaffold to develop a new class of Bcl-2 family inhibitors. A representative compound, **12** (Table 12.1), shows nanomolar activity against Bcl-2 and Bcl-xL but is ineffective against Mcl-1 [62]. The quinazoline sulfonamide exhibits significantly lower inhibition of Bcl-w compared to Bcl-xL and Bcl-2. Additionally, this class of compounds demonstrates submicromolar activity in small-cell lung cancer cell lines (e.g., H146, H889, H1963). However, pharmacokinetic studies in rats reveal high clearance rates and poor oral bioavailability for these derivatives [62, 63].

Building on ABT-737, S44563(compound **13**) was developed through conformational constraints of piperazine and the phenyl ring and demonstrated excellent antitumor activity in patient-derived uveal melanoma xenograft models [64, 65]. Subsequently, Souers et al. optimized BH3 domain mimetics ABT-737 and ABT-263 to develop ABT-199 (compound **14**, venetoclax) [66]. ABT-199 shows sub-nanomolar selectivity for Bcl-2 and significantly reduces the impact on other Bcl-2 family members [67]. *In vitro*, it exhibits effective cytotoxicity and demonstrates significant antitumor efficacy in vivo. Crystal complex interaction analysis revealed two distinct conformations of ABT-199 bound to Bcl-2 [67], with the 4,4-dimethylcyclohexene ring (4DM) undergoing positional flipping in the P2 pocket of Bcl-2 and the benzoylamide (BA) acyl oxygen displaying two different conformations in the P4 pocket (Figure 12.2a,b). One of these conformations forms a hydrogen bond interaction with the N143 residue of Bcl-2. The specificity of ABT-199 is primarily determined by the hydrogen bond formed between D103 in the BH3 domain of Bcl-2 and the indole (AI) moiety (D103 is substituted by E103 in Bcl-xL). These structural features are crucial for understanding the precise binding mechanism of ABT-199 to its target Bcl-2. Notably, ABT-199's lesser selectivity for Bcl-xL allows it to achieve higher systemic drug concentrations without inducing severe thrombocytopenia, overcoming the limitations associated with ABT-263 [66]. Due to its significant clinical activity in various hematologic malignancies, particularly chronic lymphocytic leukemia, ABT-199 received FDA approval in 2016 and 2018 for monotherapy of chronic lymphocytic leukemia and in combination with hypomethylating agents or low-dose cytarabine for elderly patients with AML [68].

APG-2575 (**15**, Lisaftoclax), developed based on ABT-199, is a potent Bcl-2 inhibitor with modifications at site 1, replacing dimethyl with cyclobutyl, and at site 2, substituting tetrahydro-2H-pyran with 1,4-dioxane [69]. APG-2575 is currently under evaluation in Phase I/II clinical trials for CLL/SLL, chronic lymphocytic leukemia, non-Hodgkin lymphoma (NHL), AML, and breast cancer (NCT04215809, NCT03913949, NCT04964518, NCT04946864).

Despite the high rates of durable remission achieved with Venetoclax (ABT-199) therapy, relapse is widespread, indicating the development of resistance. The G101V mutation in BCL2, commonly observed in relapsed patients treated with venetoclax, is sufficient to confer resistance by disrupting compound binding. To address this issue, the Liu research group employed scaffold screening, co-crystal structure analysis of BCL2 inhibitors, and structure-based drug design to identify a series of BCL2 inhibitors containing 7-azabicyclo [3.5] nonane. Subsequent optimization of these inhibitors for potency, selectivity, pharmacokinetics, and drug metabolism led to the identification of Sonrotoclax (compound **16**). Sonrotoclax inhibits the

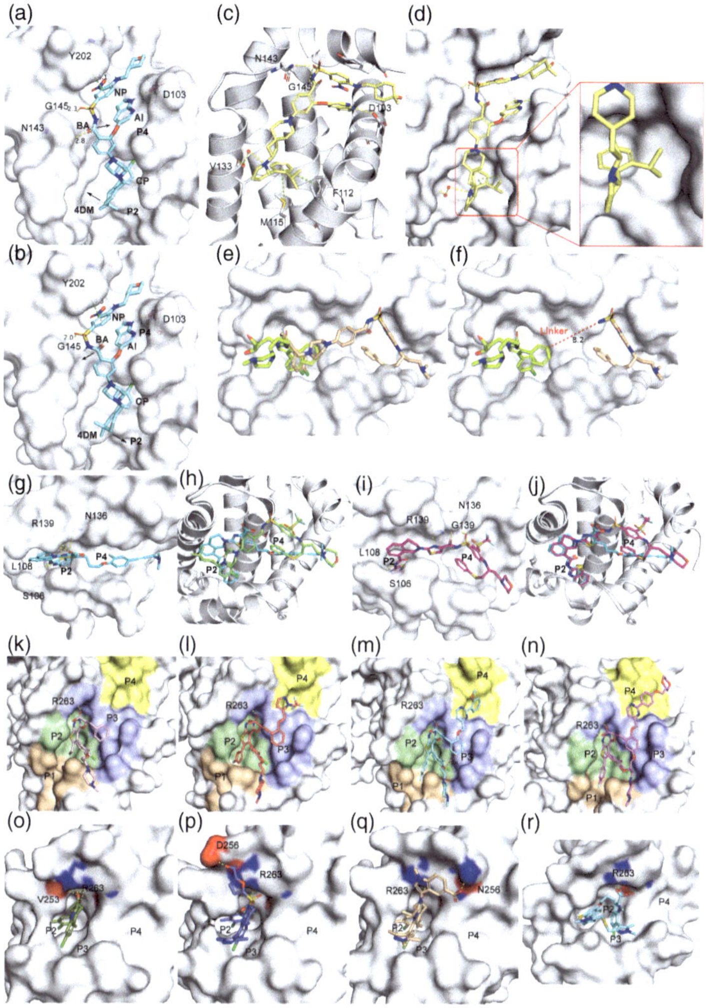

Figure 12.2 The interaction analysis of representative molecules with Bcl-2/Bcl-xL/Mcl-1. (a, b) two binding conformations of ABT-199 with Bcl-2 (PDB ID:6O0K); (c) interaction analysis of the Bcl-2/Sonrotoclax complex (PDB ID:8HOG); (d) close-up of the sub-pocket formed by the phenyl-2-isopropyl group of Sonrotoclax at the P2 edge of Bcl-2. (e) superposition of the crystal structure of 17 (limon green) onto the crystal structure of 10 (wheat) in complex with Bcl-xL (PDB ID:3SPF, 2YXJ); (f) measurement of the distance (8.2 Å) between 10 and 18. (g) interaction analysis of the A-1155463/Bcl-xL complex (PDB ID:4QVX); (h) superposition of the crystal structure of A-1155463 (cyan) onto the crystal structure of 11 (green) in complex with Bcl-xL (PDB ID:4QVX, 4QNQ); (i) interaction analysis of the A-1293102/Bcl-xL complex (PDB ID:7LH7); (j) superposition of the crystal structure of A-1155463 (cyan) onto the crystal structure of A-1293102 (magenta) in complex with Bcl-xL (PDB ID:4QVX, 4QNQ). (k) interaction analysis of the 50/Mcl-1 complex (PDB ID:6QYO); (l) interaction analysis of the S63845/Mcl-1 complex (PDB ID:5LOF); (m) interaction analysis of the S64513/Mcl-1 complex (PDB ID:6YBL); (n) interaction analysis of the ABBV-467/Mcl-1 complex (PDB ID:8EI1). (o) interaction analysis of the 54/Mcl-1 complex (PDB ID:4HW2); (p) interaction analysis of the 56/Mcl-1 complex (PDB ID:5FDR); (q) interaction analysis of the 58/Mcl-1 complex (PDB ID:6NE5). (r) interaction analysis of AZD5991 with Mcl-1 (PDB ID:6FS0). Source: Wenhua Zhu, Yangbo He, Gang Chen, Di Zhu.

Bcl-2-Bak interaction with an IC_{50} of 0.014 nM and exhibits a 2000-fold selectivity over Bcl-xL. Compared to Venetoclax, Sonrotoclax demonstrates a 14-fold increase in potency. It shows superior cytotoxic activity in various hematologic cancer cell lines and more profound inhibition of tumor growth in multiple hematologic malignancy models compared to Venetoclax. Notably, Sonrotoclax effectively inhibits the Venetoclax-resistant BCL2 G101V variant [70].

Structural analysis of Sonrotoclax binding indicates that its isopropyl group interacts at the entrance of the P2 pocket, while the connected phenyl ring forms beneficial edge π–π interactions with the F112 side chain, contributing to the enhanced binding affinity of Sonrotoclax (Figure 12.2c). Additionally, the isopropyl substitution induces the formation of a sub-pocket above the P2 site (Figure 12.2d), comprised of Y108, F112, D111, and M115, which provides extra hydrophobic interactions. The phenyl ring of Sonrotoclax interacts with the sulfur atom of M115 through a sulfur-π interaction, and the pyrrolidine forms a water-bridge hydrogen bond with the backbone carbonyl of V133. Such interactions are absent in the BCL2 co-crystal structure with Venetoclax. These interactions collectively enhance the binding affinity of Sonrotoclax compared to Venetoclax. Sonrotoclax is currently undergoing evaluation in multiple clinical trials [70].

BM903 and Its Related Derivatives Zhou et al. conducted a detailed structural analysis of the Bcl-xL-Bad BH3 peptide complex, elucidating the characteristics of the binding pocket. Through pharmacophore screening, they designed a novel small-molecule scaffold, compound **17**, targeting Bcl-xL (Table 12.1). Based on the interaction patterns of the crystal structures of **17** and **10** with the Bcl-xL complex (Figure 12.2e), large fragments (compound **18**) from **17** and **10** were used for structural assembly (Figure 12.2f), ultimately leading to the successful development of a new, highly efficient Bcl-2/Bcl-xL small molecule inhibitor **19** through structural optimization (Table 12.1). This inhibitor has a K_i value of 0.28 nM for binding to Bcl-2 and Bcl-xL [71]. In the H146 xenograft tumor model, compound **19** demonstrated significant in vivo antitumor activity and induced tumor regression, though it did not achieve sustained complete tumor eradication. To address this limitation, the research team optimized the nitro and soluble thiophenyl "tail" groups of a compound **19** analogue, compound **20** (BM-903) (Table 12.1), leading to the design of BM-957(compound **21**). Additionally, they optimized the pyrrole core structure, resulting in compound BM1074 (compound **22**), both of which exhibit sub-nanomolar affinity ($K_i < 1$ nM) for Bcl-2 and Bcl-xL and achieved complete and sustained tumor regression in the H146 xenograft model [72, 73].

Subsequently, by integrating the two optimization strategies, the team developed compound BM1197 (compound **23**) (Table 12.1), which demonstrated a sub-nanomolar affinity for Bcl-2 and Bcl-xL and effectively disrupted the interactions between proapoptotic and antiapoptotic Bcl-2 family proteins, thereby inducing BAX-bak dependent apoptosis in SCLC cell lines [74]. In SCLC xenograft models, BM-1197 achieved persistent tumor regression following daily or weekly intravenous administration. Recent studies indicate that BM-1197 exhibits potent antitumor activity against malignant lymphoma cells both in vitro and in vivo, with its mechanism of action involving induction of apoptosis via the endogenous apoptotic pathway, highlighting its potential for the treatment of malignant lymphomas [75].

To address the issue of platelet reduction associated with Bcl-xL inhibitors while maintaining robust antitumor efficacy, the research team developed a "masked" strategy by limiting the cellular permeability of the drug to reduce its accumulation in platelets, resulting in BM-1252 (compound **24**, APG-1252, Pelcitoclax) [76]. BM-1252 has completed clinical trials for small cell lung cancer and solid tumors (NCT03080311), and is undergoing Phase II trials for relapsed or refractory NHL, as well as recurrent ovarian and endometrial cancers (NCT05186012, NCT05691504, NCT04893759). APG-1252 is metabolized into a more potent derivative, APG-1252-12A, which exhibits sub-nanomolar affinity for Bcl-2 and Bcl-xL. APG-1252-12A targets mitochondria, inducing cytochrome c release, PARP cleavage, and caspase activation in HL-60 cells, thereby promoting caspase-dependent apoptosis and demonstrating significant antitumor efficacy in AML [77]. In vivo, both APG-1252 and APG-1252-12A achieved complete tumor regression in various human cancer animal models while avoiding the common off-target toxicities associated with Bcl-xL inhibitors. Furthermore, APG-1252-12A demonstrated excellent antitumor effects in gastric and liver cancers, both as a monotherapy and in combination with chemotherapy [78, 79].

AZD4320, AZD0466, Compound 27, and S55746 AZD4320 (compound **25**, Table 12.1) is a dual Bcl-2/Bcl-xL inhibitor developed by Ariscon, which has demonstrated the ability to induce complete and sustained tumor regression in the RS4; ALL model. To address thrombocytopenia, intravenous administration and planned management strategies were employed, resulting in high efficacy in preclinical models of AML. However, the development of AZD4320 was hindered by observed cardiovascular toxicity in preclinical studies [80]. To overcome this issue, Patterson et al. conjugated AZD4320 to a polyethylene glycolylated polylysine dendrimer and, through mathematical modeling, designed AZD0466 (compound **26**, Table 12.1). In human leukemia xenograft mouse models, AZD0466 exhibited significantly enhanced preclinical activity and improved tolerability compared to AZD4320 [81]. AZD0466 entered clinical trials in 2021 and 2022 as a monotherapy or in combination with other anticancer agents for the treatment of hematologic malignancies and NHL. However, the development of AZD0466 was halted due to strategic or safety concerns.

John Porter's team identified a moderately potent and selective Bcl-2 inhibitor, phenylpyrazole derivative **27** [82], through HTS. They then optimized a series of phenylpyrazole compounds with tetrahydroisoquinoline amide substitutions, showing high affinity for Bcl-2 and Bcl-xL, with better selectivity for Bcl-2. One representative compound, **28** (Table 12.1) [82], and its evolved analog S55746 (compound **29**, Table 12.1), a Bcl-2 selective BH3 mimetic, exhibited strong antitumor activity in Bcl-2 dependent lymphoma xenograft models while preserving platelets. Crystallographic analysis revealed S55746 occupies the S1/2/3 region, contrasting with ABT-199's larger binding surface area [66, 67, 83]. S55746, now an oral compound, is the second Bcl-2 selective inhibitor in clinical development, currently in phase I trials for hematologic malignancies (Clinical Trial IDs: NCT02920697, NCT02920541, NCT02603445).

WEHI-539, A-1155463, A-1331852, and A-1293102 Lessene identified a novel scaffold (compound **30**, Table 12.1) through high-throughput chemical screening and employed structure-guided drug design to discover a highly selective Bcl-xL small molecule inhibitor, WEHI-539 (compound **31**, Table 12.1). WEHI-539 binds very tightly to Bcl-xL, with its half-maximal inhibitory concentration (IC_{50}) and dissociation constant (K_d) values approaching or below 1 nM [84], and exhibits a slow dissociation rate. However, the utility of WEHI-539 as a tool molecule is limited by the presence of unstable and potentially toxic functional groups. Additionally, the poor physicochemical properties of this compound render its dosage in vivo impractical. Subsequent structure-based medicinal chemistry efforts led to the development of Bcl-xL selective inhibitors A-1155463 (compound **32**) and A-1331852 (compound **33**), with A-1331852 being the first truly effective, selective, and orally bioavailable Bcl-xL inhibitor [85–87]. Nevertheless, recent studies have shown that A-1331852 exhibits off-target effects by inhibiting MCL-1 transcription, leading to cell death in U937 and HL-60 cells, necessitating its use in combination with other BH3 mimetics to mitigate adverse effects [88]. The X-ray co-crystal structure of A-1155463 with Bcl-xL reveals that the alkyne amine group is positioned in the hydrophobic P4 pocket, which is critical for the tight binding of the pharmacophore. The benzothiazole group of A-1155463 is deeply embedded in the P2 pocket, forming a unique network of hydrogen bonds that drives the selectivity of the pharmacophore (Figure 12.2g). Overlaying the X-ray structures of navitoclax (compound **11**) and A-1155463 bound to Bcl-xL reveals that the thiazole carboxylate of A-1155463 directly points to the "bent-back" unit of navitoclax's P4 binding (Figure 12.2h). This view prompted the hypothesis that hybrid molecules could maintain the high selectivity of A-1155463 for Bcl-xL through P2 binding of the benzothiazole amide while achieving high efficacy similar to navitoclax, leading to the development of A-1293102(compound **34**). A-1293102 exhibits picomolar affinity for Bcl-xL and effectively and selectively induces cell death in Bcl-xL-dependent tumor cells [89]. X-ray crystallographic analysis shows a key hydrogen bond network between A-1293102 and Bcl-xL in the P2 binding pocket, with the fluorobenzene moiety of A-1155463 positioned similarly to the phenylthio group of A-1293102 in the P4 pocket (Figure 12.2i). Furthermore, the dimethyl and morpholine groups of A-1155463 and A-1293102 are exposed to the solvent and are closely aligned within the overlay, while the bent portions effectively occupy the P4 pocket in a manner similar to navitoclax (Figure 12.2j). A-1293102 represents one of the few unique structural series of selective Bcl-xL inhibitors, thus serving as a valuable tool for biological research and a leading compound for further optimization [89].

12.2.4.2 Selective Mcl-1 Inhibitor

Among the Bcl-2 family members, Mcl-1 is frequently amplified in human tumors and is often associated with tumor recurrence and chemoresistance, particularly with resistance to ABT-737, ABT-263, and ABT-199 [90], and has been identified as a potential resistance factor to these inhibitors [91–95]. Alanine scanning of Bcl-2 and Bcl-xL indicates that P4 binding is not crucial for Mcl-1 activity [96]. The conformational rigidity of the hydrophobic binding groove of Mcl-1 [97], combined with the exceptionally high affinity of its endogenous ligands, complicates the design of small molecules that can directly inhibit Mcl-1 and induce mechanism-based cell death. Significant reviews and viewpoints have previously discussed small heterocyclic molecules as Mcl-1 inhibitors [98–100]. A recent review has compiled structures of various Mcl-1 inhibitors reported before 2020, including those based on diverse heterocycles (such as indole, imidazole, thiophene, nicotinic acid, piperazine, triazine, thiazole, isoprenyl), oligomers (tert-butyl, quaternary pyridine), polyphenols, phenols, anilino benzoic acids, anthraquinones, macrocycles, natural products, and metal-based complexes [101]. These insights provide a comprehensive understanding of strategies targeting the Mcl-1 protein. Here, we primarily review the discovery and design strategies of some representative Mcl-1 inhibitors, identifying compounds with high affinity for Mcl-1 through natural product screening, small molecule library collection, and structure-based drug design.

TW-37, MIM-1, UMI-77, Compound 9, and TTX-810 Monique Verhaegen utilized computational modeling to develop a novel Bcl-2 homology domain 3 (BH3) mimetic, TW-37(compound **35**, Table 12.2), which represents the first rationally designed small molecule with high affinity for Mcl-1, Bcl-xL, and Bcl-2 [102, 103]. TW-37 demonstrates a significantly stronger binding affinity for Mcl-1 compared to Bcl-2 or Bcl-xL. Subsequently, Nicole A. Cohen identified MIM-1(compound **36**, Table 12.2) [104], a new small molecule inhibitor selectively targeting the BH3 binding groove of Mcl-1, through competitive screening using Mcl-1 peptide mimetics. MIM-1 shows at least a 10-fold higher binding affinity for Mcl-1 compared to Bcl-xL. However, additional studies indicate that despite MIM-1's BAK-dependent apoptotic activity, its efficacy may be limited and cell-type dependent, as it failed to induce apoptosis in two Mcl-1-dependent cell lines [105]. Abulwerdi identified 3-substituted (4-hydroxy-naphthyl) aryl sulfonamides against Mcl-1 using HTS. Following structure-activity relationship (SAR) optimization via nuclear magnetic resonance (NMR), compound UMI-77(compound **37**) was discovered, which selectively binds to Mcl-1 with a K_i of 490 nM and effectively inhibits pancreatic cancer growth both in vitro and in vivo [106]. Additionally, David J. Richard employed a combination of HTS and directed hit optimization to identify compound **38** (Table 12.2), a Mcl-1 inhibitor, which reduces affinity for Bcl-xL by over 100-fold. Notably, compound **38** exhibits micromolar activity against the Mcl-1-dependent H929 cell line [107]. TTX-810(compound **39**, Table 12.2), discovered collaboratively by the Broad Institute of Harvard University, MIT, and Bayer Pharmaceuticals, is an effective, selective, and rapidly cleared Mcl-1 inhibitor that shows effective and reversible binding to the BH3 binding groove of human Mcl-1 (K_d = 0.3 nM). The outstanding efficacy of TTX-810 was achieved through structure-guided design and comprehensive analysis across over 500 cancer cell lines. It inhibits Mcl-1 interaction with proapoptotic protein BAK or the BH3-only protein BIM in a dose- and time-dependent manner, leading to the activation of intrinsic apoptotic cascades and cancer cells [108].

Table 12.2 Selective Mcl-l inhibitors.

Compound	Structure	Targets	Affinity
41	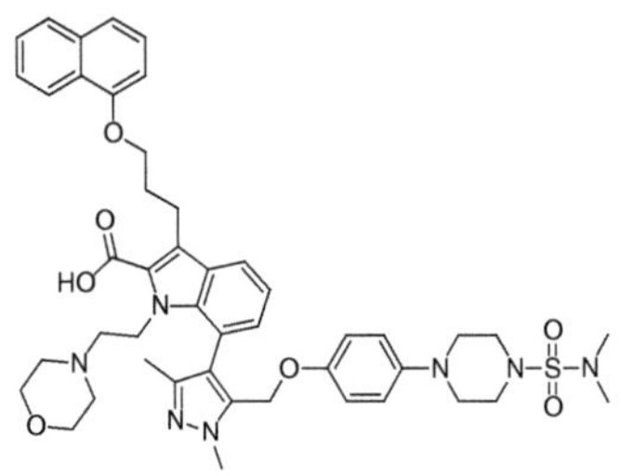	Mcl-1	A-1210477(41) Mcl-1 $K_i = 0.43$ nM
42	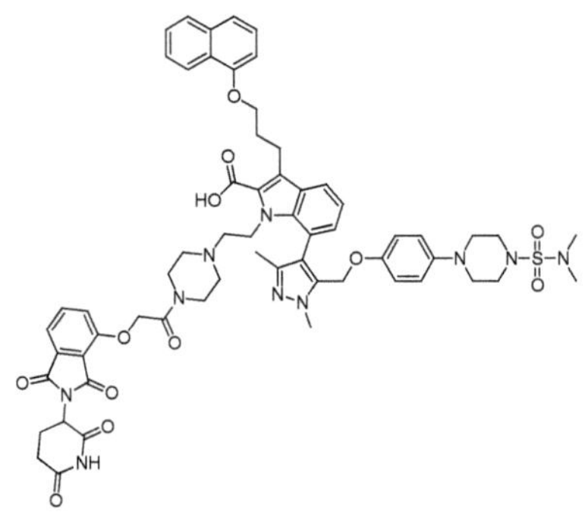	Mcl-1	dMcl-1-2(42) $K_d = 30$ nM
43	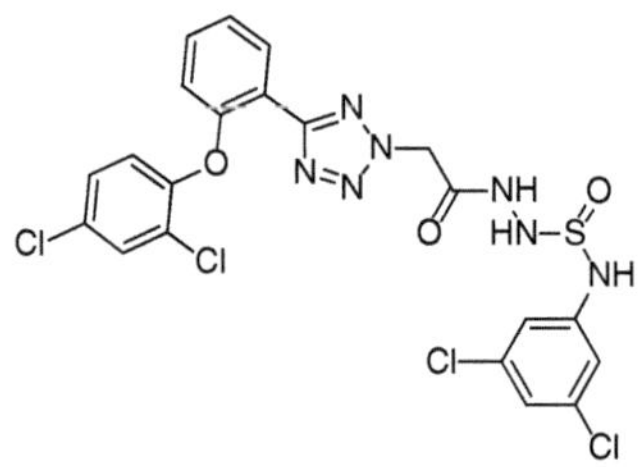	Mcl-1	Mcl-1 $K_i = 3.0$ μM
44	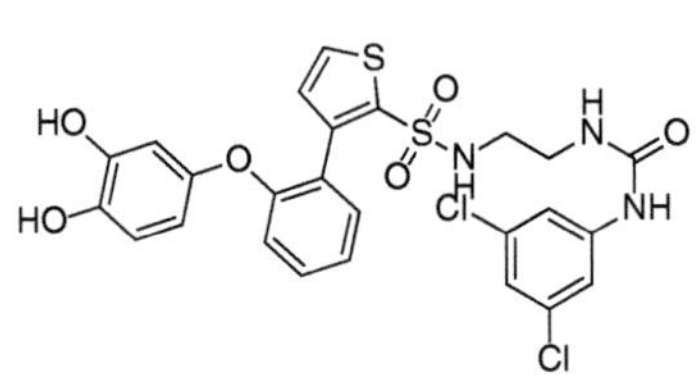	Mcl-1	Mcl-1 $K_i = 0.33$ M Mcl-1 $K_i = 0.83$ M
45 **FWJ-04**	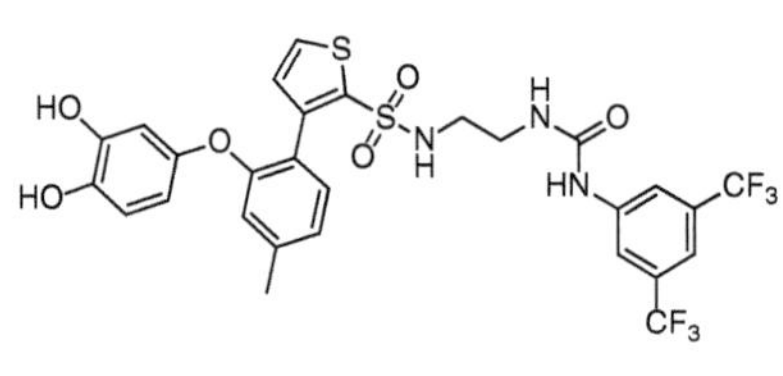	Mcl-1	Mcl-1 $K_i = 0.49$ μM no binding

Table 12.2 (Continued)

Compound	Structure	Targets	Affinity
46	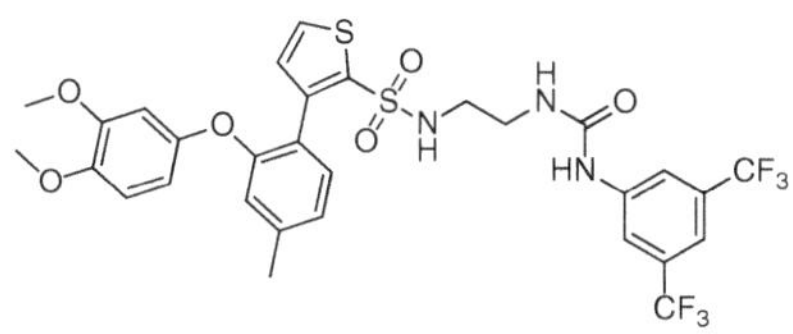	Mcl-1	Mcl-1 $K_i = 4.9\,\mu M$ no binding
47 **LC126**	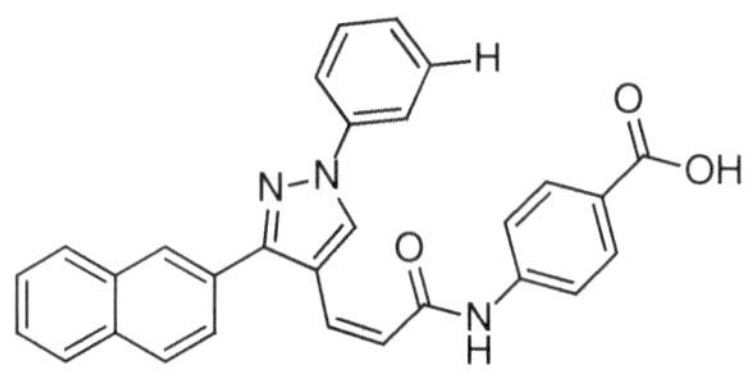	Mcl-1	Mcl-1 $K_i = 13\,\mu M$ Mcl-1 $K_i = 10\,\mu M$ Bcl-xL weak binding
48 **GQN-B37-E**	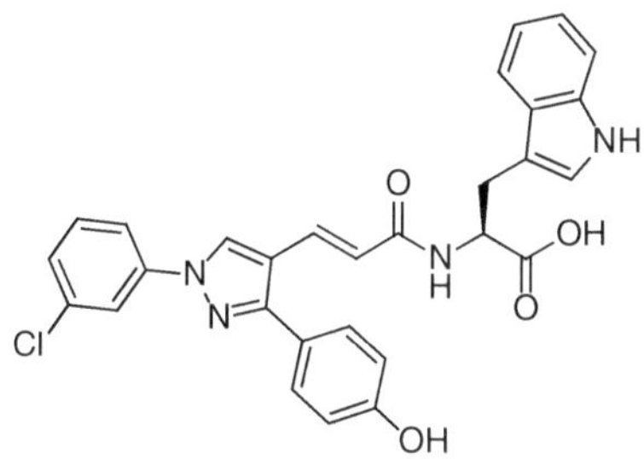	Mcl-1	Mcl-1 $K_i = 0.6\,\mu M$ Mcl-1 no binding Bcl-xL no binding
49	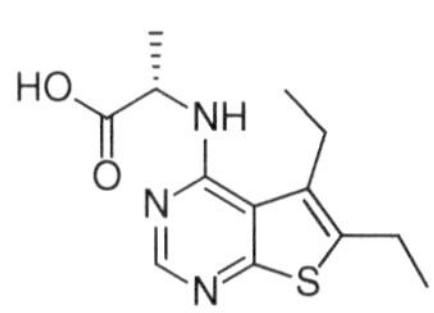	Mcl-1	Mcl-1 $K_i = 50\,\mu M$ Mcl-1 $K_i = 74\,\mu M$
50	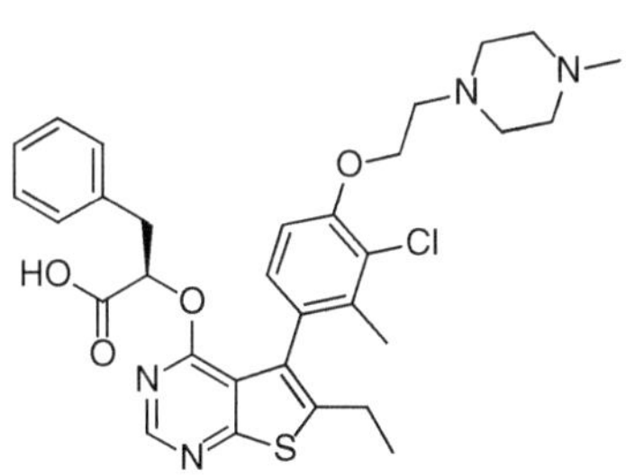	Mcl-1	Mcl-1 $K_i = 28\,nM$

(Continued)

Table 12.2 (Continued)

Compound	Structure	Targets	Affinity
51 **S63845**	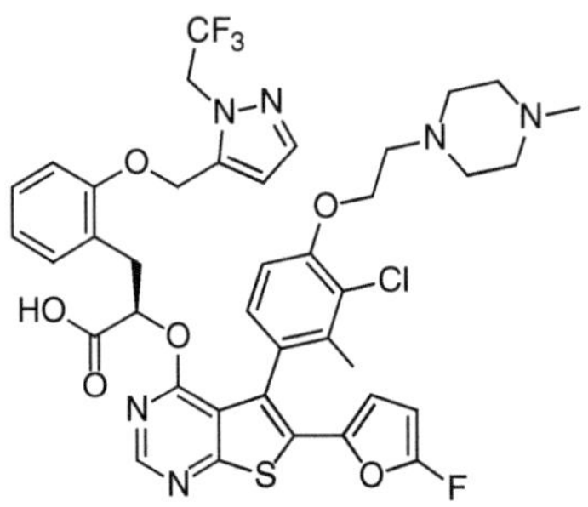	Mcl-1	Mcl-1 $K_i < 1.2$ nM Mcl-1 K_i >10,000 nM Bcl-xL K_i >10,000 nM
52 **S64315**	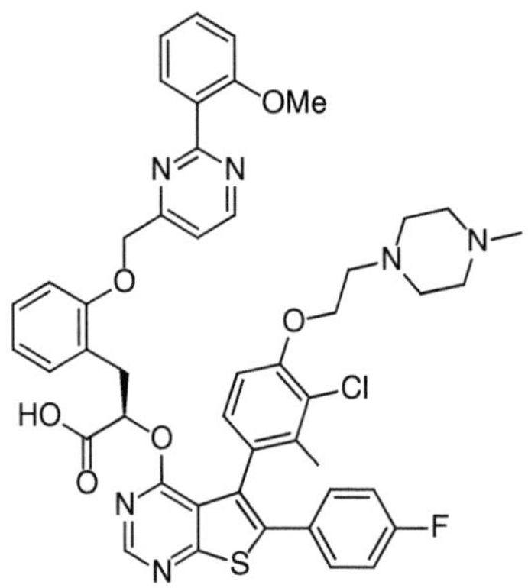	Mcl-1	Mcl-1 $K_i = 0.029$ nM
53 **ABBV-467**	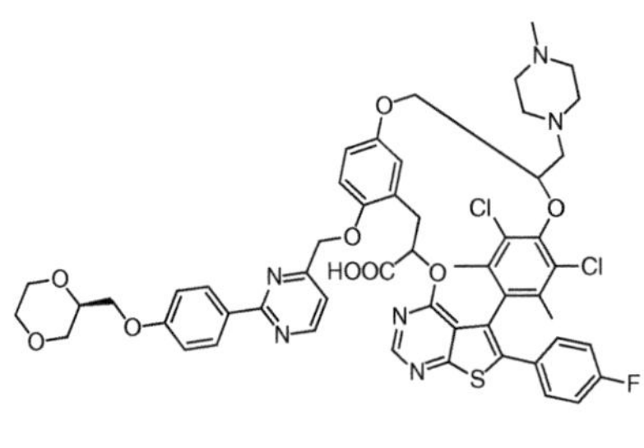	Mcl-1	Mcl-1 $K_i < 0.01$ nM
54	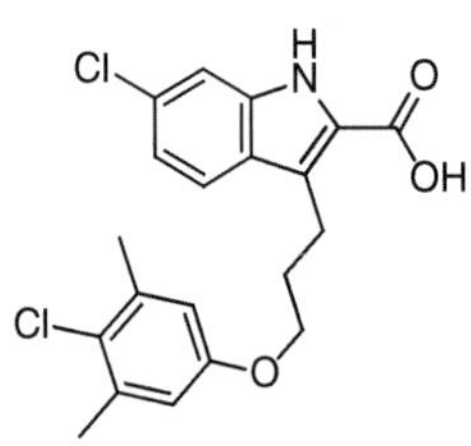	Mcl-1	Mcl-1 $K_i = 0.055$ μM Bcl-xL $K_i > 15$ μM Mcl-1 $K_i = 0.87$ μM

Table 12.2 (Continued)

Compound	Structure	Targets	Affinity
55	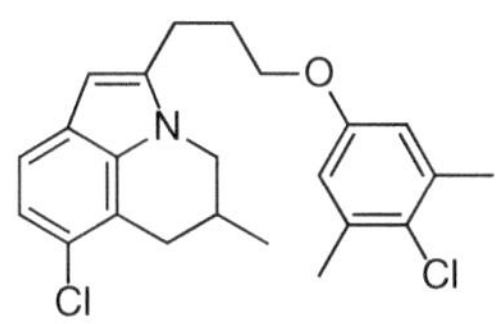	Mcl-1	Mcl-1 $K_i = 3$ nM Bcl-xL $K_i = 5200$ nM Mcl-1 $K_i = 770$ nM
56	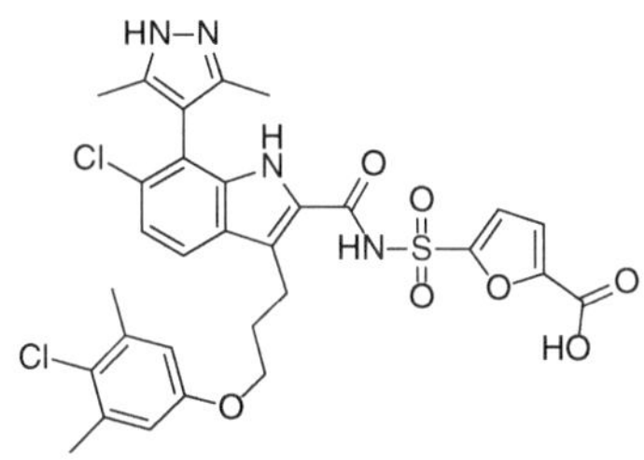	Mcl-1	Mcl-1 $K_i = 0.36$ nM
57	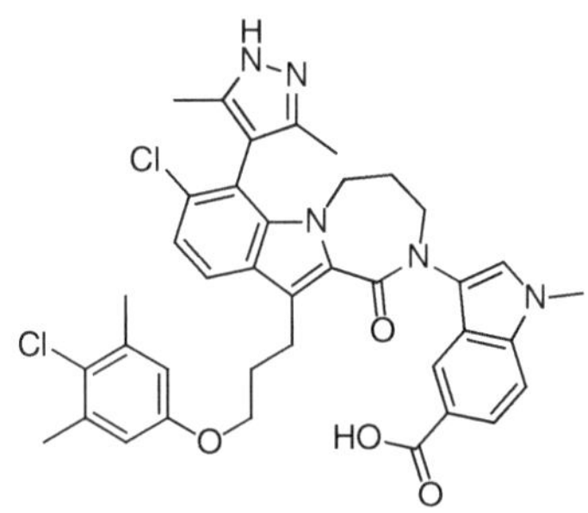	Mcl-1	Mcl-1 $K_i = 0.089$ nM
58	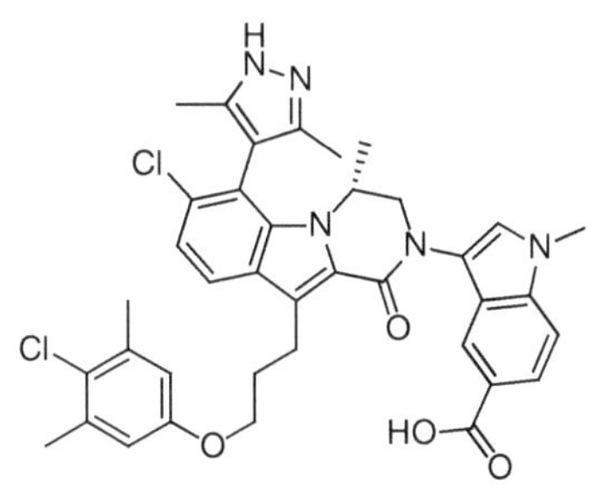	Mcl-1	Mcl-1 $K_i = 0.03$ nM

(Continued)

Table 12.2 (Continued)

Compound	Structure	Targets	Affinity
59	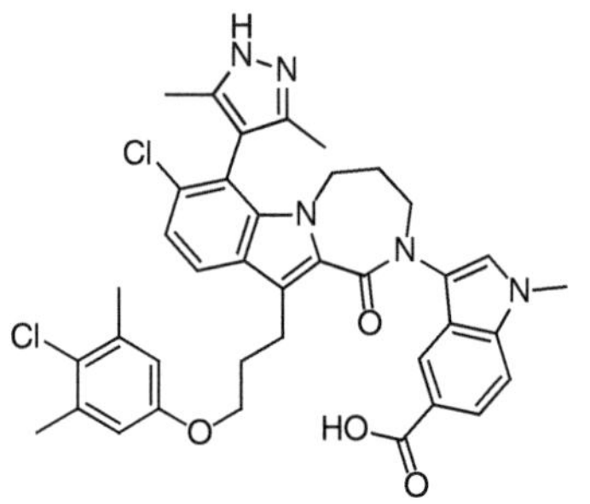	Mcl-1	Mcl-1 $K_i = 0.089$ nM
60	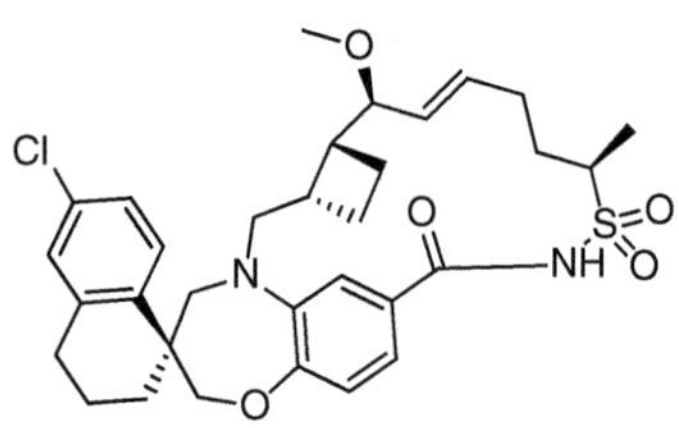	Mcl-1	60 Mcl-1 $K_i = 0.04$ nM
61 **AMG-8621**	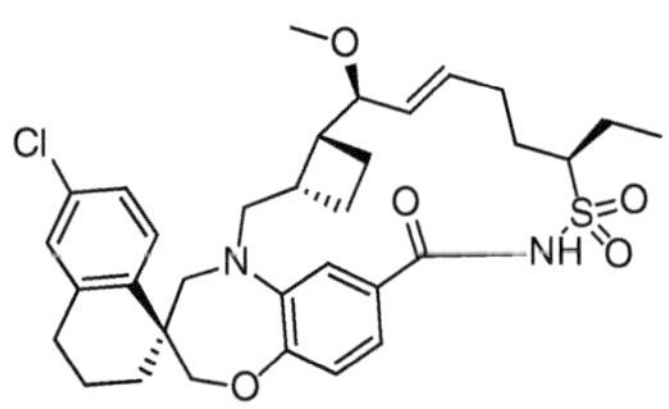	Mcl-1	Mcl-1 $K_i = 0.04$ nM Bcl-xL $K_i = 1900$ nM Mcl-1 $K_i = 2500$ nM
62 **AMG-176**	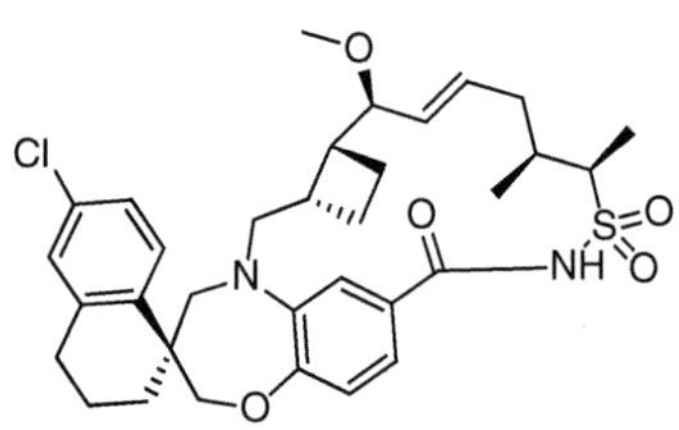	Mcl-1	Mcl-1 $K_i = 0.06$ nM Bcl-xL $K_i = 700$ nM Mcl-1 $K_i = 950$ nM

Table 12.2 (Continued)

Compound	Structure	Targets	Affinity
63 **AMG-397**	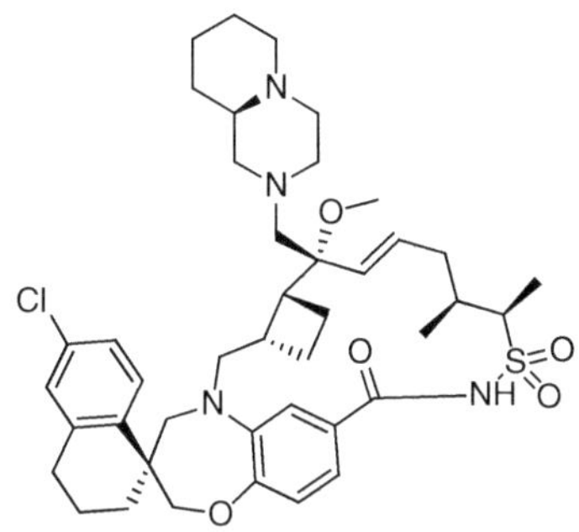	Mcl-1	Mcl-1 $K_i = 0.015\,nM$
64	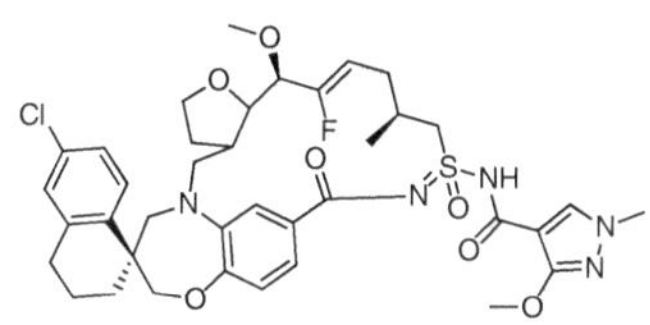	Mcl-1	Mcl-1 $K_i = 0.011\,nM$
65	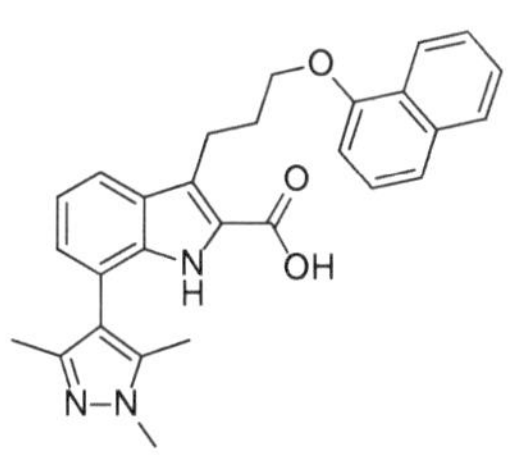	Mcl-1	Mcl-1 TR FRET $IC_{50} = 0.29\,\mu M$
66	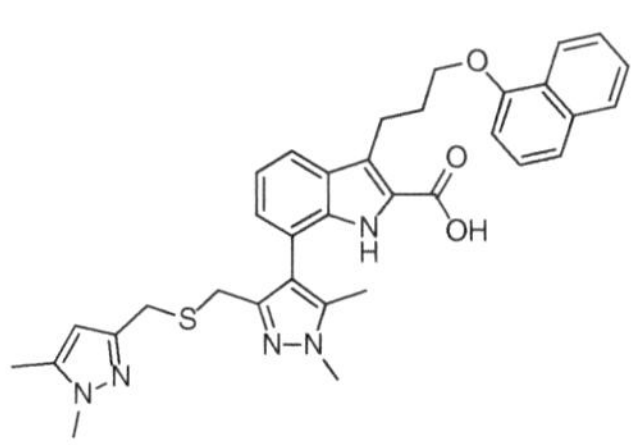	Mcl-1	Mcl-1 TR FRET $IC_{50} = 0.042\,\mu M$
67	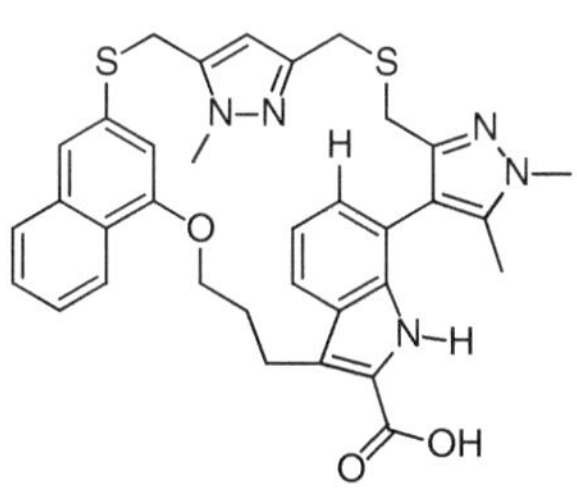	Mcl-1	Mcl-1 TR FRET $IC_{50} = 0.042\,\mu M$ $IC_{50} = 0.019\,\mu M$

(Continued)

Table 12.2 (Continued)

Compound	Structure	Targets	Affinity
68 **AZD5991**	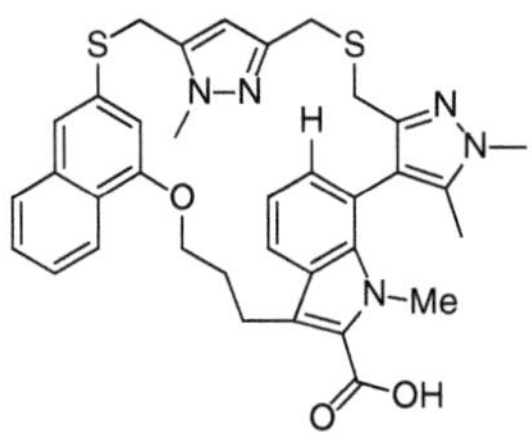	Mcl-1	Mcl-1 K_i = 200pM Mcl-1 K_i = 6.8 μM Bcl-xL K_i = 18 μM
69 **JNJ-4355**	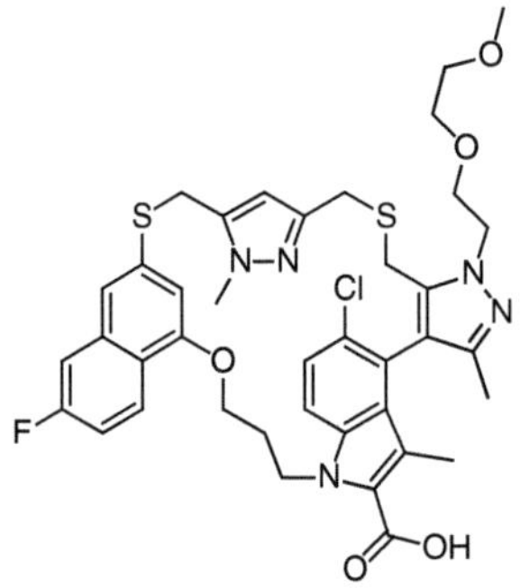	Mcl-1	Mcl-1 K_i = 0.015 nM
70 **JNJ-78394355**	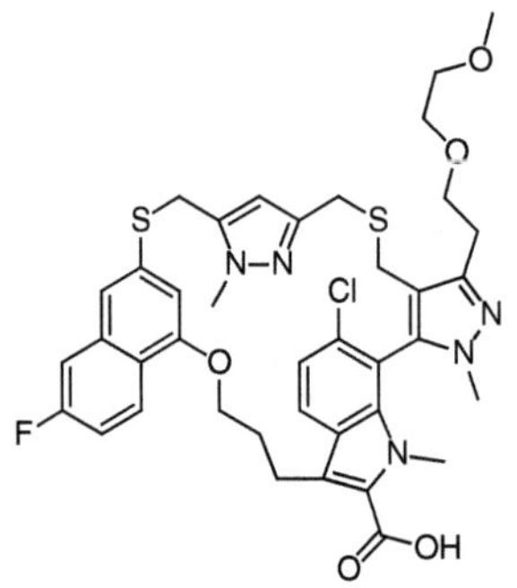	Mcl-1	Mcl-1 K_i = 0.018 nM
71	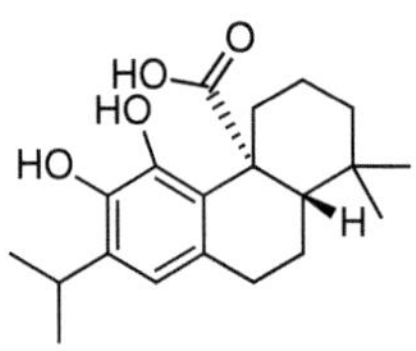	BCL9	K_i(ELISA) = 3.3 μM

Table 12.2 (Continued)

Compound	Structure	Targets	Affinity
72	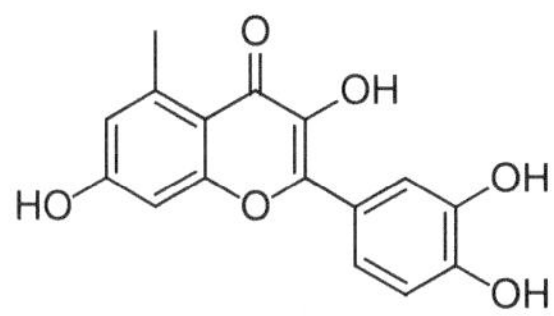	BCL9	IC_{50} (FP) = 10.10 μM IC_{50} (CT26) = 14.40 μM
73	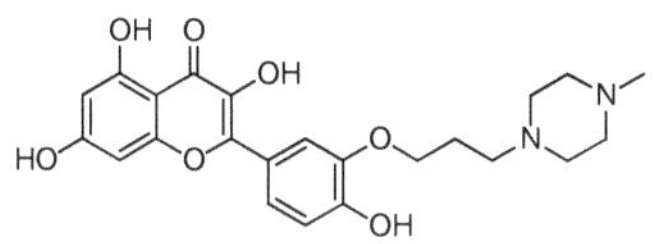	BCL9	IC_{50} (FP) = 2.78 μM IC_{50} (Axin) = 2.81 μM IC_{50} (CT26) = 3.21 μM
74	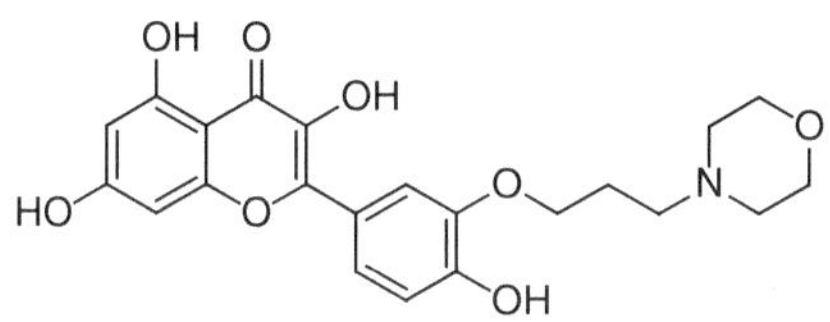	BCL9	IC_{50} (FP) =3.13 μM IC_{50} (Axin) =6.29 μM IC_{50} (CT26) =7.85 μM
75	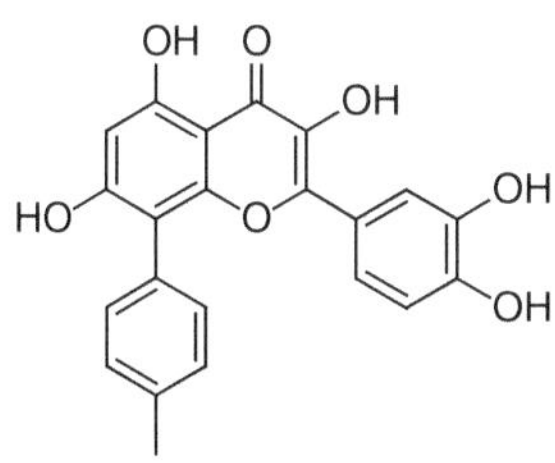	BCL9	IC_{50} (FP) = 2.25 μM SPR K_d = 1.44 μM
76 **CP-868388**	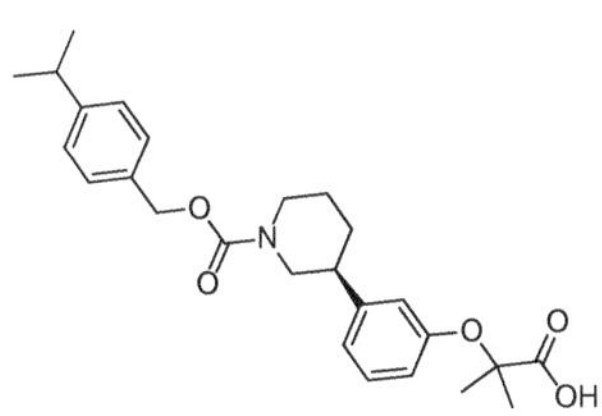	BCL9	K_i (ALPHA Screen) = 1.2 μM
77	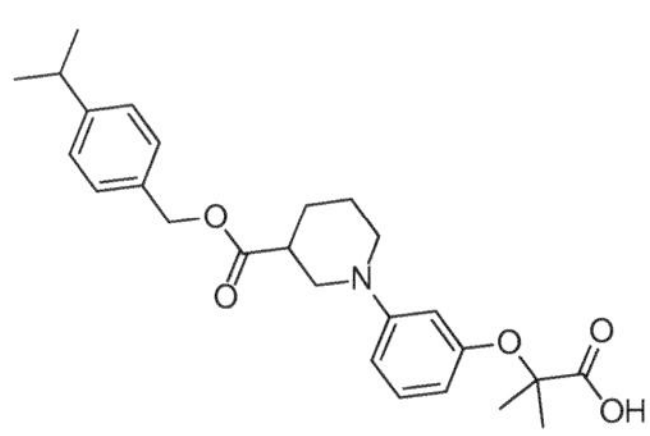	BCL9	K_i (ALPHA Screen) = 77 μM

(Continued)

Table 12.2 (Continued)

Compound	Structure	Targets	Affinity
78 **CPPAA-30**	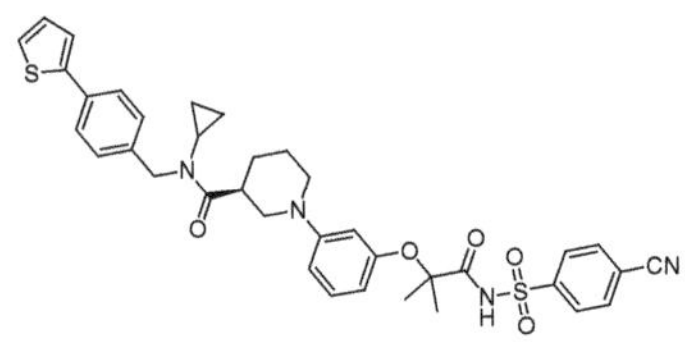	BCL9	K_i (ALPHA Screen) = 3.6 μM
79 **ZW4864**	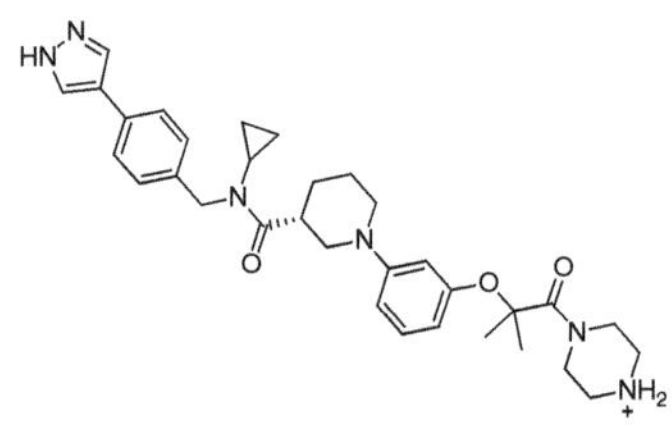	BCL9	K_i (ALPHA Screen) = 0.76 μM
80	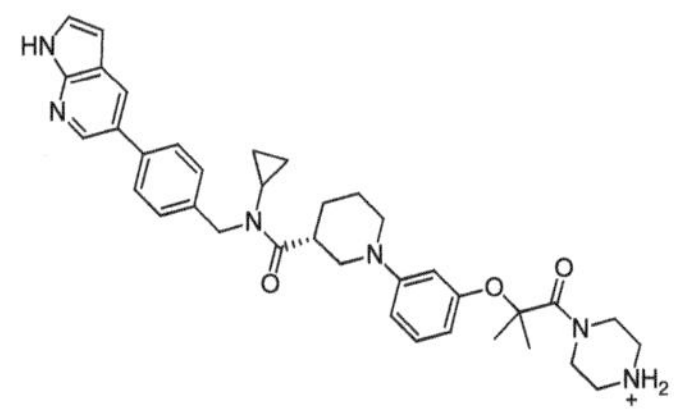	BCL9	K_i (ALPHA Screen) = 2.7 μM

A-1210477 and dMcl-1-2 Milan Bruncko, building on a previously identified indole-2-carboxylic acid compound **40** through HTS, designed A-1210477 (compound **41**, Table 12.2) using co-crystal structures and structure-based design. This inhibitor has sub-nanomolar binding affinity for Mcl-1 (K_i = 0.43 nM) and shows negligible affinity for Bcl-2, Bcl-xL, or Bcl-w (all >10 μM) [109]. Prior SAR studies indicated that the aryl piperazine substituent in A-1210477 still interacts effectively with the BH3 binding site. Inspired by this, Papatzimas and colleagues replaced the piperazine with a morpholine group to link with the cereblon (CRBN) ligand, and through coupling and *N*-alkylation of the indole, successfully synthesized dMcl-1-2(compound **42,** Table 12.2). Surface plasmon resonance (SPR) studies further confirmed that the dissociation constant (K_d) of A-1210477 is 19 nM, while that of dMcl-1-2 is 30 nM. These results validate that the indole substitution and piperazine part serve as suitable linker positions, effectively occupying the BH3 binding site without affecting Mcl-1 binding. This chemical probe technique provides a solid foundation for future optimization and demonstrates the potential of designing heterobifunctional small molecules, particularly Mcl-1-targeted protein degraders (PROTACs), to induce successful protein degradation, making these PROTACs powerful tools in the development of novel antiapoptotic drugs targeting Bcl-2 family proteins [110].

FWJ-D4, FWJ-D5, and GQN-B37-E Chengwen Yang's team conducted virtual screening based on docking to identify a hit compound, BCL-VS-156(compound **43**, Table 12.2), featuring a phenyl tetrazole and hydrazine thioformamide moiety that targets the BH3 binding pocket on Mcl-1 [111]. They optimized this lead compound into a series of compounds with a 3-phenyl thiophene-2-sulfonamide core, which exhibited micromolar binding affinity for both Mcl-1 and Bcl-2. Further structure optimization through scaffold hopping yielded a new derivative, YCW-E11(compound **44**, Table 12.2), with sub-micromolar binding affinity. YCW-E11 effectively induced apoptosis in HL-60 cells via the mitochondrial pathway [111]. Building on structural analysis and molecular modeling, the team improved the selectivity of these compounds for Mcl-1. Notably, representative compounds FWJ-D4(compound **45**, Table 12.2) and its precursor FWJ-D5(compound **46**, Table 12.2) induced mitochondrial apoptosis in various cancer cell lines, with FWJ-D5 demonstrating significant tumor suppression in xenograft models [112]. Recently, the laboratory developed another class of small molecule Mcl-1 inhibitors through virtual screening targeting Mcl-1, discovering LC126(compound **47**, Table 12.2). Structural optimization led to several compounds with selective binding to Mcl-1 at sub-micromolar concentrations. Cellular studies indicated that these compounds disrupted the interaction between BIM and Mcl-1, primarily inducing mitochondrial apoptosis in leukemia cell lines. The representative compound GQN-B37-E(compound **48**, Table 12.2) selectively bound to Mcl-1 (K_i = 0.6 μM), and its precursor GQN-B37-Me was particularly effective against leukemia cells H929 and MV-4-11, inducing caspase-dependent apoptosis at micromolar concentrations [113].

S63845, S64513, and ABBV-467 Zoltan Szlávik identified the thienyl and pyrimidine amino acid derivative **49** (Table 12.2) through fragment-based screening using NMR, which exhibited considerable affinity for both Mcl-1 and Bcl-2 targets (K_i values of 50 and 74 μM, respectively). Through fragment growing and X-ray crystallographic structure-guided optimization, compound **50** (Table 12.2) was obtained. This compound demonstrated high binding affinity for Mcl-1 and induced targeted cell killing in the H929 cancer cell line via apoptosis activation [114]. Co-crystal complex analysis revealed that compound 18a binds to the hydrophobic P2 and P3 pockets of Mcl-1 and forms hydrogen bonds with R263. Notably, the ethyl group did not fully occupy the P2 pocket, while the P4 pocket remained completely exposed (Figure 12.2k). This suggests that further structural optimization should focus on filling the P2 and P4 pockets. The final optimized compounds, S63845(compound **51**, Table 12.2) and S64315(compound **52**, Table 12.2), effectively occupy the hydrophobic P2 and P4 pockets of Mcl-1 (Figure 12.2l). S63845 demonstrated significant sensitivity in hematological malignancies and solid tumor cell lines and achieved notable tumor regression in AMO1 and MV4-11 xenograft models. However, administration of S63845 was associated with adverse effects such as anemia, somnolence, weight loss, and hemolysis [115]. S64315 also showed significant tumor regression in AMO-1 xenograft studies [116] and is currently undergoing clinical trials in AML phase I/II trials, myelodysplastic syndrome phase I trials, and multiple myeloma (MM), lymphoma, and diffuse large B-cell lymphoma (DLBCL) phase I trials (NCT04629443, NCT02979366, NCT02992483, NCT04702425). Additionally, the phase I clinical trial combining VOB560 (a prodrug of S55746) with S64315 is ongoing, targeting Bcl-2 and Mcl-1 in NHL, MM, or AML. Analysis of the S64513/Mcl-1 crystal complex binding mode revealed that S64513 interacts extensively with the hydrophobic BH3 binding groove of Mcl-1 and forms hydrogen bonds between its carboxyl group and the side chain of R263 in Mcl-1 (Figure 12.2m). Researchers at AbbVie noted that although S64315 is structurally linear, its compact and folded binding conformation suggests that macrocyclic constraints could orient the molecule into a highly active conformation. The carboxyl-connected phenyl ring occupies the P3 pocket, with an oxygen atom extending into the solvent region at a distance of 3.8 Å, providing a suitable conformational restriction site (Figure 12.2m). Extensive research led to the substitution of the nitrogen-containing linker with a carbon ring, incorporation of a symmetrical hexasubstituted phenyl core, and extension of a methoxy side chain to occupy the P4 pocket, alongside the introduction of a polar dioxane ring to occupy a broader solvent region (Figure 12.2n), ultimately yielding the highly effective and selective Mcl-1 inhibitor ABBV-467(compound **53**, Table 12.2). ABBV-467 exhibited potent Mcl-1 inhibitory activity and high selectivity toward other antiapoptotic proteins. Moreover, ABBV-467 induced rapid, mechanism-based apoptosis in Mcl-1-dependent tumor cell lines. In vivo, ABBV-467, administered as a monotherapy or in intermittent dosing in combination with venetoclax, effectively inhibited the growth of xenografts derived from human hematologic malignancies [117]. ABBV-467 entered clinical trials in November 2019, with intravenous administration in patients with relapsed/refractory MM (NCT04178902). However, in 2021, the recruitment for this trial was terminated due to "strategic considerations."

12.2.4.3 Compounds 54-59

The team led by Stephen W. Fesik utilized fragment-based screening methods to identify several structurally diverse small molecules, which were subsequently developed into highly effective Mcl-1 inhibitors through structure-based design. These inhibitors demonstrated superior selectivity for Mcl-1 over Bcl-xL and Bcl-2. NMR-derived models indicated that the hydrophobic regions of Hit 1 and Hit 2 bind adjacent to the large pocket. Based on these findings, structural concatenation of these fragments led to the identification of a series of potent Mcl-1 inhibitors, with the most promising being the optimized 2-indolecarboxylic acid derivative **54** (Table 12.2), which binds Mcl-1 with a K_i value of 55 nM [118]. The structure of compound **54**/Mcl-1 complex revealed that the Hit 2 fragment binds deeply within the hydrophobic P2 pocket (Figure 12.2o). Encouraged by this observation, the team incorporated the Hit 2 fragment into a Hit 3 molecule featuring a tricyclic indole-2-carboxylic acid core structure and further optimized it based on structural data to obtain compound **55**, which exhibited high Mcl-1 efficacy and selectivity [119]. To develop more effective Mcl-1 inhibitors, the team continued to extend the 2-indolecarboxylic acid series to occupy the entire BH3 binding interface. This approach rapidly led to the development of new, effective Mcl-1 inhibitors capable of accessing an additional pocket (P3) within the BH3 peptide binding groove. The most representative molecule, **56**, achieved a binding affinity of 0.36 nM with Mcl-1, demonstrating that filling the additional pocket in the BH3 groove can significantly enhance affinity (Figure 12.2p) [120]. Subsequent modifications guided by X-ray structural data resulted in a new tricyclic indole-diazabenzodiazepinone derivative **57**, which showed picomolar binding affinity and targeted antiproliferative activity in Mcl-1-sensitive cells [121]. Despite the sub-nanomolar binding affinity, achieving in vivo efficacy required further improvement in cellular potency. Compound **57** exhibited high clearance rates, low exposure, and a short half-life in rats, which are undesirable pharmacokinetic characteristics. To address these issues, the team optimized the three-carbon linker of the diazepinone, leading to the development of compound **58**, which demonstrated targeted activity across various cancer models. In addition to showing cellular efficacy in multiple Mcl-1-sensitive cell lines, compound **58** exhibited favorable pharmacokinetic properties in mouse xenograft cancer models and demonstrated synergistic tumor growth inhibition when combined with standard chemotherapeutic agents such as doxorubicin in HCC-1187 and BT-20 triple-negative breast cancer xenograft models. Structural interaction analysis of compound **58**/Mcl-1 complex indicated that the dimethylchlorobenzyl ether is more deeply integrated into the P2 hydrophobic pocket compared to leucine in BH3 members of the Bcl-2 family (Figure 12.2q) [122]. Furthermore, compound **58** does not fully occupy the P3 pocket, with only the methyl of the trimethylpyrazole pointing toward this pocket, and does not reach the P4 pocket. However, compound **58** creates a pseudo-hydrophobic pocket around residue R263 through interactions with the indole head group, which significantly improves affinity by sequestering this hydrophilic residue from solvent. Additionally, the indole head group positions the carboxylate to form favorable hydrogen bonding with Asn 260, further stabilizing the complex. Recently, the Fesik team further optimized this series of inhibitors to identify compounds suitable for clinical development, improving cellular potency, pharmaceutical properties, and pharmacokinetics. This led to the design and synthesis of a new Mcl-1 inhibitor, compound **59**, which demonstrates higher potency and improved pharmacokinetic characteristics, notably a 20-fold increase in mouse IV clearance compared to previously disclosed inhibitors. Moreover, compound **59** achieved tumor regression in hematologic and lung cancer xenograft models and showed enhanced efficacy when combined with chemotherapeutic agents such as docetaxel or topotecan [123].

AMG-176 and AMG-397 Sean Caenepeel utilized HTS and structure-based design to optimize a series of helical-cyclic molecules, employing conformational constraints as a guiding principle. Through central ring expansion, carboxylic acid bioisosteric replacement, and the introduction of trans-olefins to impose further conformational constraints, compounds **60** (Table 12.3) and **61** (AMG-8621, Table 12.3) were developed, exhibiting appropriate potency and selectivity. However, pharmacokinetic studies revealed that compounds **60** and **61** had short half-lives and poor oral bioavailability. Simple methylation modifications led to the development of AMG 176(compound **62**, Table 12.3), which displayed improved pharmacokinetic properties, including enhanced oral bioavailability [124]. AMG 176 is the first selective oral MCL-1 inhibitor to enter human clinical trials and has been advanced in trials for relapsed or refractory AML and relapsed or refractory multiple myeloma. Subsequently, a combination of structure-guided and ligand-based design approaches was employed to support more active conformations. The introduction of the polar (R)-piperidinylpiperazine side chain led to the development of the second-generation acylsulfonamide AMG 397(compound **63**, Table 12.3), which demonstrated further improvements in potency and pharmacokinetic properties. AMG 397 exhibited picomolar affinity (K_i = 15 pM) for Mcl-1 and selectively competed with proapoptotic Bcl-2 family members, such as BIM, for binding to the BH3 binding groove of Mcl-1 [125]. In cellular assays, AMG 397 effectively disrupted the interaction between Mcl-1 and BIM. The second-generation acylsulfonamide AMG 397 with the polar (R)-piperidinylpiperazine side chain showed enhanced pharmacokinetic properties that supported oral administration. However, clinical development of AMG 397 was halted in February 2020 in favor of intravenous AMG-176. Despite AMG-176's slow absorption and very long half-life ($t_{1/2} > 12$ hours) leading to extended target coverage time and increased risk of myocardial exposure and mitochondrial toxicity, exploratory medicinal chemistry strategies based on AMG-176 focused on enhancing inhibitor efficacy, optimizing physicochemical and pharmacokinetic properties, and minimizing cardiac toxicity risk. This effort revealed the macrocyclic 2-allyltetrahydrofuran derivative **64** (allyl-THFs, Table 12.3), which inhibits the interaction between Mcl-1 and the BH3 domain of proapoptotic proteins. Compound **64** exhibited high biochemical and cellular activity and balanced in vitro ADME characteristics, including moderate lipophilicity, high aqueous solubility, and low turnover in liver microsomes. Additionally, given that compound **64** showed minimal cardiac ion channel liability potential upon oral administration in mice and had optimized short half-life pharmacokinetics, it is unlikely to produce off-target cardiac toxic effects, thereby further complicating the safety profile of targeted therapy [126].

AZD5991, JNJ-4355, and JNJ-78934355 Adriana E. Tron initiated a series of studies on indole-2-carboxylic acids (compounds **64** and **65**) [118], applying structure-based drug design to explore novel binding modes with known Mcl-1 inhibitors' scaffolds, which led to the identification of an induced fit pocket in Mcl-1. By investigating protein dynamics and the conformations of crystallized inhibitors, a series of macrocyclic inhibitors targeting Mcl-1 were designed and synthesized. Specifically, hit compound **66** (Table 12.3) was synthesized as a by-product and demonstrated significant in vitro activity (IC_{50} = 0.042 μM). These compounds adopt a unique "U"-shaped conformation, providing favorable spatial sites for macrocyclization. The introduction of a nonpolar flexible linker connecting pyrazole 5-methyl and naphthyl further enhanced the 3-carbon atom, resulting in compound **67** (Table 12.3) with high affinity. Subsequently, the incorporation of 6-Cl and 1-N-Me substituents on the indole moiety contributed to molecular conformation stability, ultimately leading to the development of AZD5991(compound **68**, Table 12.3) [127]. AZD5991 exhibits high affinity and selectivity for Mcl-1, with X-ray crystallography (PDB ID: 6FS0) revealing that AZD5991 primarily interacts with the P2 and P3 pockets (Figure 12.2r). The hydrophobic naphthyl group penetrates into the P2 pocket, while the indole core predominantly interacts with the P3 residues.

Table 12.3 β-catenin/BCL9 inhibitors.

Compound	Structure	Targets	Affinity
71	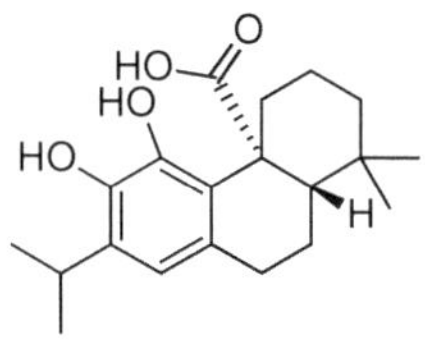	BCL9	K_i (ELISA) = 3.3 μM
72	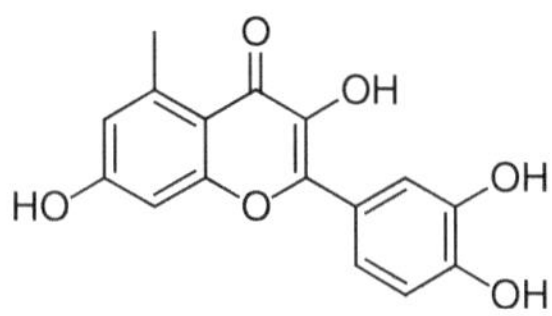	BCL9	IC_{50} (FP) = 10.10 μM IC_{50} (CT26) = 14.40 μM
73	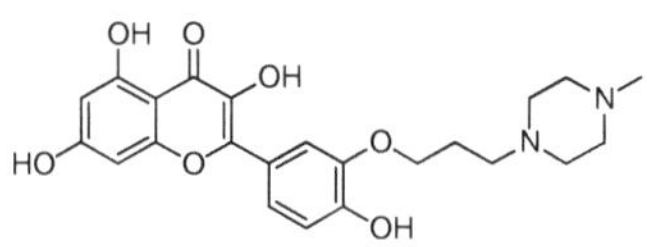	BCL9	IC_{50} (FP) = 2.78 μM IC_{50} (Axin) = 2.81 μM IC_{50} (CT26) = 3.21 μM
74	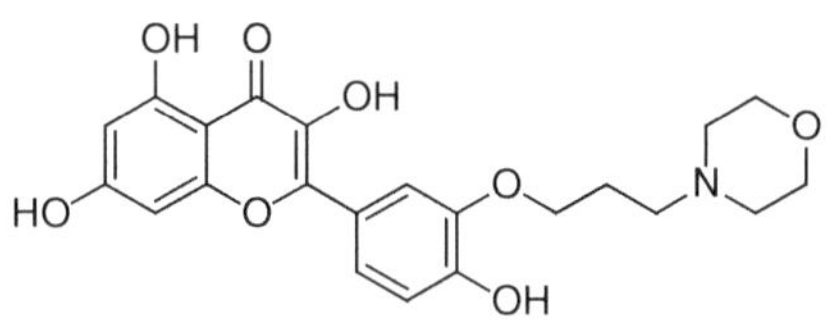	BCL9	IC_{50} (FP) =3.13 μM IC_{50} (Axin) =6.29 μM IC_{50} (CT26) =7.85 μM
75	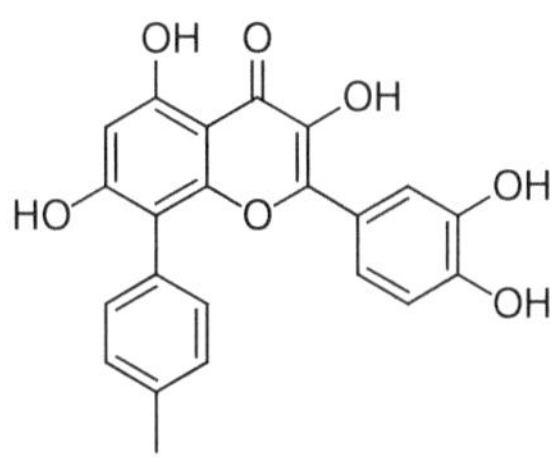	BCL9	IC_{50} (FP) = 2.25 μM SPR K_d = 1.44 μM
76 **CP-868388**	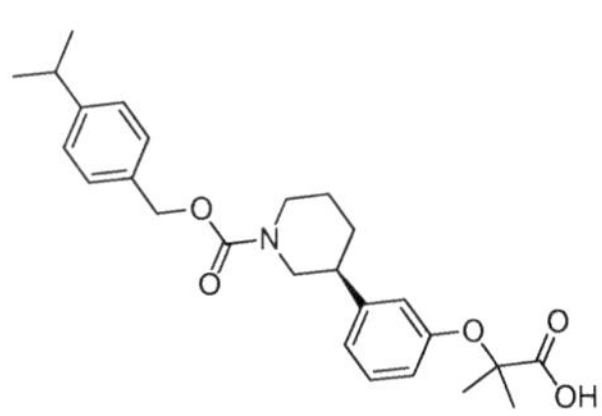	BCL9	K_i (ALPHA Screen) = 1.2 μM

(Continued)

Table 12.3 (Continued)

Compound	Structure	Targets	Affinity
77	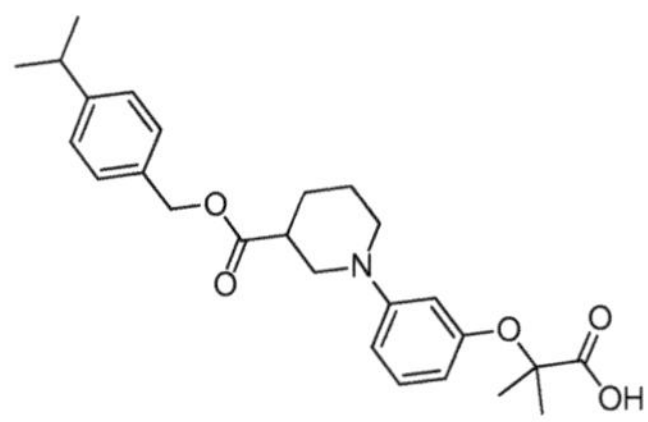	BCL9	K_i (ALPHA Screen) = 77 μM
78 **CPPAA-30**	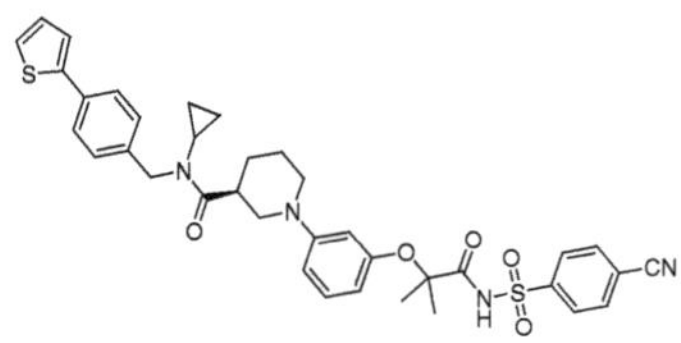	BCL9	K_i (ALPHA Screen) = 3.6 μM
79 **ZW4864**	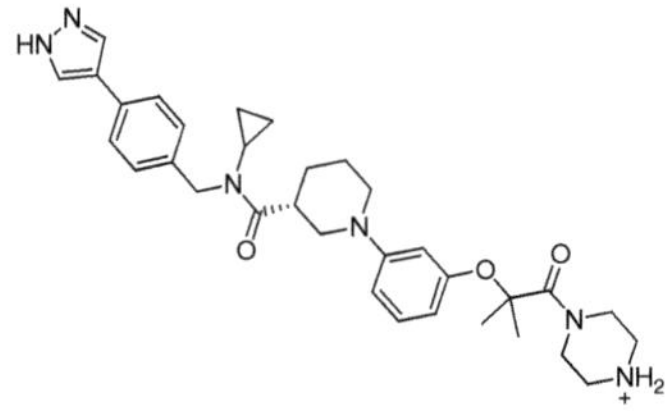	BCL9	K_i (ALPHA Screen) = 0.76 μM
80	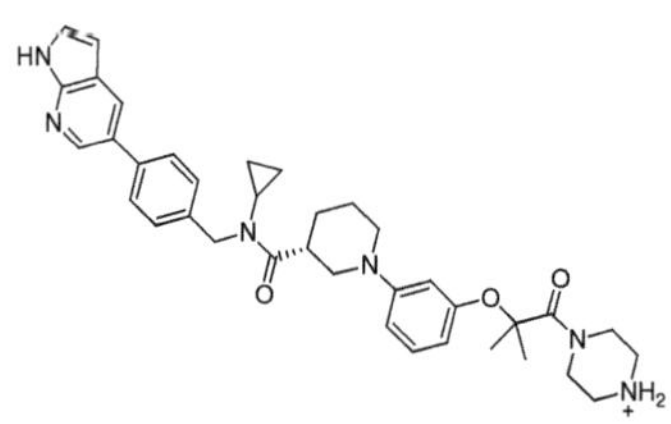	BCL9	K_i (ALPHA Screen) = 2.7 μM
81	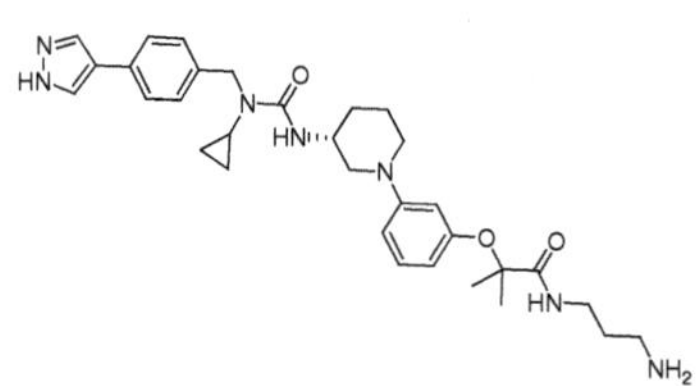	BCL-9	K_d (SPR) = 0.13 μM FP IC_{50} = 2.59 μM

Table 12.3 (Continued)

Compound	Structure	Targets	Affinity
82	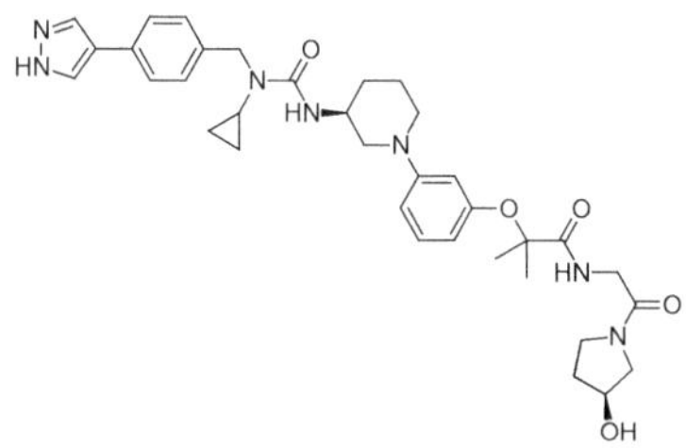	BCL-9	K_d (SPR) = 0.63 μM FP IC_{50} = 1.61 μM
83	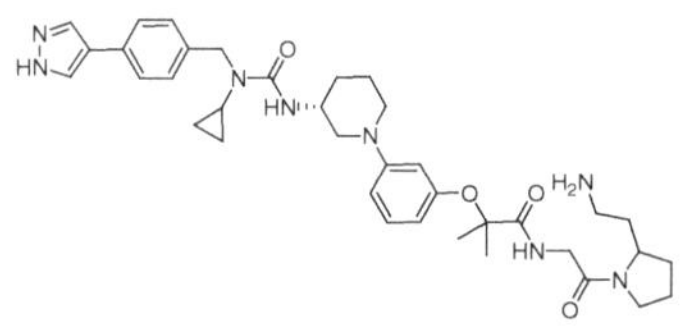	BCL-9	K_d (SPR) = 0.082 μM FP IC_{50} = 3.92 μM
84	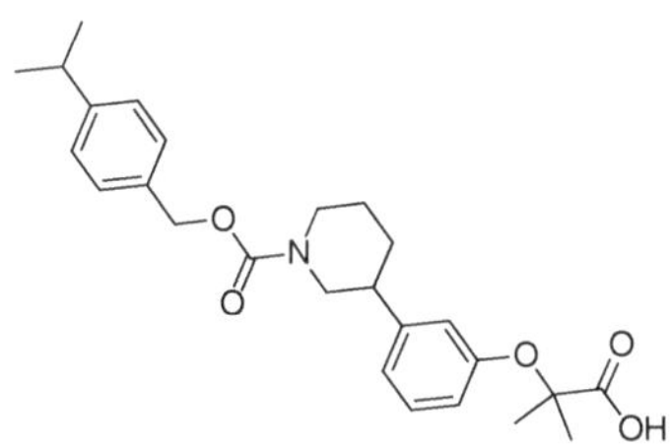	BCL-9	IC_{50} (FP) = 5.39 μM IC_{50} (Axin) = 14.56 μM
85	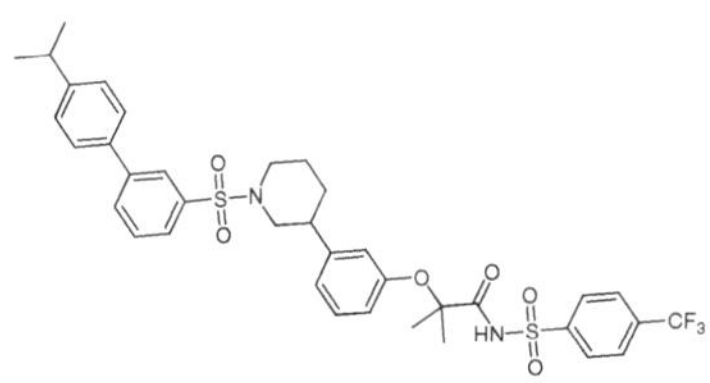	BCL-9	IC_{50} (FP) = 0.72 μM IC_{50} (Axin) = 0.82 μM
86	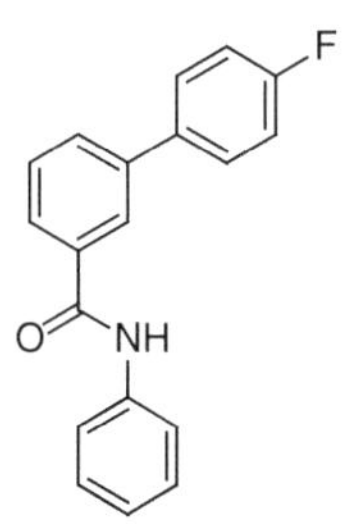	BCL-9	/

(Continued)

Table 12.3 (Continued)

Compound	Structure	Targets	Affinity
87 **PNPB-22**	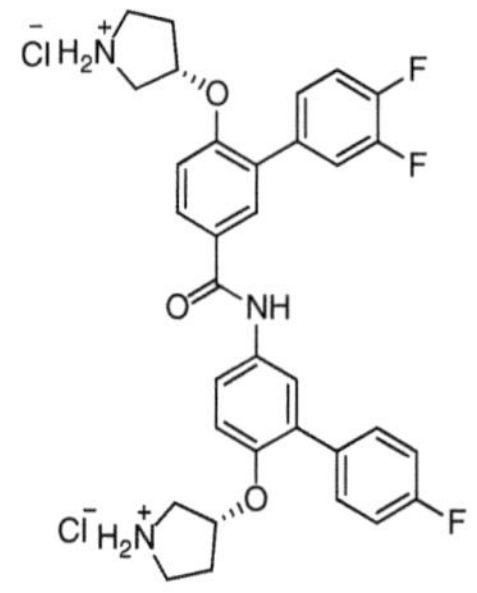	BCL-9	K_i (Alpha Screen) = 2.1 μM K_d (TIC) = 0.33 μM
88	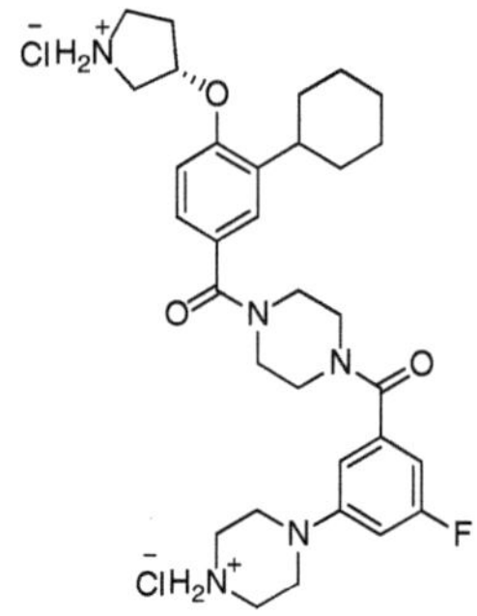	BCL-9	K_i (Alpha Screen) = 5.2 μM
89 **PNPB-29**	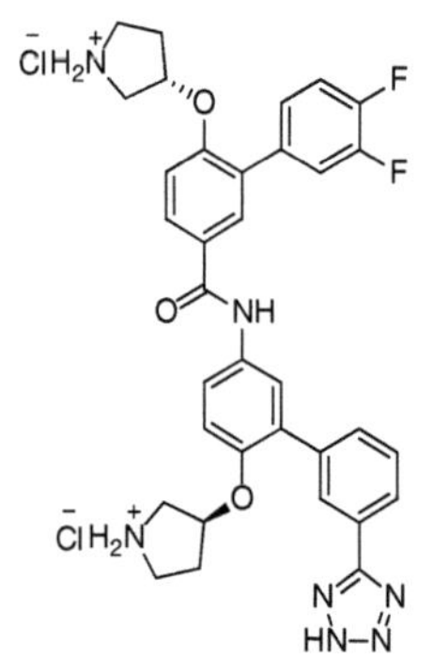	BCL-9	K_i (Alpha Screen) = 0.47 μM K_d (TIC) = 0.33 μM

Table 12.3 (Continued)

Compound	Structure	Targets	Affinity
90 **Z13138**	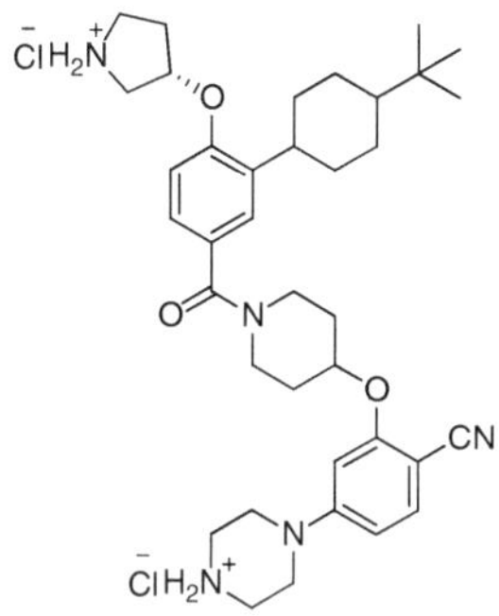	BCL-9	K_i (Alpha Screen) = 0.96 μM
91 **L814-1428**	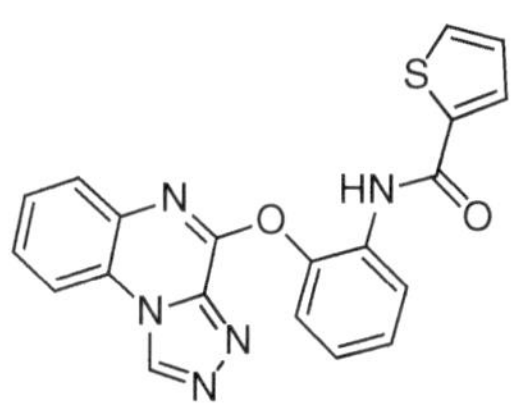	BCL-9	IC_{50} (Alpha Screen) = 7.4 μM IC_{50} (TIC) = 2.2 μM
92 **E722-2648**	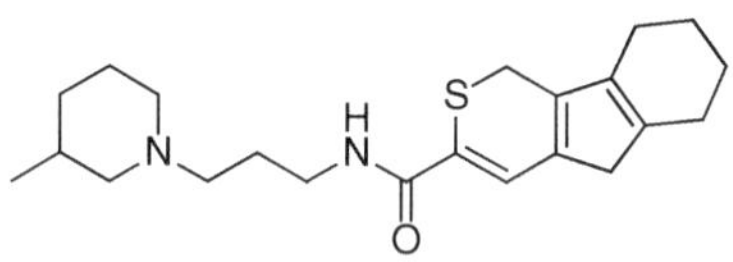	BCL-9	(AlphaScreen) IC_{50} = 9.0 μM (TIC) IC_{50} = 1.0 μM

Additionally, a strong salt bridge interaction is observed between the carboxyl group adjacent to the indole ring and Arg263. AZD5991 has entered Phase I/II clinical trials to assess its efficacy in five hematologic malignancies. However, in 2021, the trial was paused for "further evaluation of safety-related information." To improve the physicochemical properties of AZD5991, Frederik J. Rombouts' team further optimized it, resulting in JNJ-4355(compound **69**, Table 12.3). This compound significantly increased the equilibrium solubility (3.14 mM in pH 7 buffer) and reduced protein binding (99.93%) [128]. Furthermore, JNJ-4355 demonstrated promising in vitro potency in cancer cell lines and patient-derived AML samples (cell-killing AC50 ranging from 0.29 to 75 nM in 25/27 evaluable samples). In vivo studies confirmed the disruption of Mcl-1:BAK complexes in MOLM13 (AML) xenografts in mice. The efficacy of JNJ-4355 was also demonstrated in MOLP8 (multiple myeloma) xenografts, resulting in complete tumor regression following a single intravenous dose. Subsequent exploration of the impact of pyrazole and indole isomerism on the overall properties and potency of the molecules revealed that incorporating fluorine into the naphthyl portion was essential for enhancing cellular potency. Additionally, introducing various side chains on the pyrazole significantly improved solubility. This exploration ultimately led to the development of compound JNJ-78394355(**70**, Table 12.3), which exhibits remarkable cellular potency and desirable properties [129].

12.3 Inhibitors of β-catenin/BCL9

12.3.1 Physiological Functions of BCL9

B-cell lymphoma 9 (BCL9) is a protein encoded by the BCL9 gene located on chromosome 1. In 1998, Willis et al. first identified the overexpression of BCL9 in B-cell acute lymphoblastic leukemia cell lines, associated with the *t*(1;14) (q21;q32) translocation [130]. BCL9 is a protein consisting of 1394 amino acid residues and serves as a crucial transcriptional co-activator in the Wnt signaling pathway. It contains three conserved regions: the Pygopus binding domain (HD1), the β-catenin binding domain (HD2), and a typical nuclear localization signal peptide (HD3) [131]. Increasing evidence suggests the role of BCL9 in various diseases, particularly its significance in cancer [132]. Initially identified as a protein interacting with the transcription factor β-catenin, BCL9 plays a key role in regulating β-catenin-dependent gene expression, affecting cell proliferation, differentiation, and survival [132]. As a member of the BCL9 family, BCL9's unique function is its interaction with β-catenin. The Wnt signaling pathway, essential for embryonic development and stem cell maintenance, relies on the translocation of β-catenin into the nucleus and its binding to TCF/LEF1 transcription factors to activate target genes. BCL9 enhances this process by stabilizing β-catenin and promoting its interaction with TCF/LEF1, thereby facilitating the expression of Wnt target genes [133]. In cancer biology, the role of BCL9 has garnered significant attention, with high expression levels detected in multiple myeloma, colorectal cancer, breast cancer, lung cancer, liver cancer, and ovarian cancer. Aberrant activation of the Wnt signaling pathway is often associated with mutations in key components such as β-catenin or APC, which are closely linked to the development of various cancers [134]. In the canonical Wnt pathway, in the absence of Wnt ligands, cytoplasmic β-catenin is phosphorylated and degraded by the destruction complex composed of Axin, adenomatous polyposis coli (APC), and glycogen synthase kinase 3 [135]. BCL9 stabilizes β-catenin, promotes its binding to TCF/LEF1 transcription factors, and forms complexes with other related factors, effectively activating the Wnt signaling pathway [136].

12.3.2 β-catenin/BCL9 PPI

The BCL9 protein primarily consists of two domains: the N-terminal cysteine-rich domain (Cys-rich domain), which is rich in cysteine and can bind to the regulatory region of the β-catenin protein, thereby inhibiting the degradation of β-catenin and leading to its accumulation in the cytoplasm; and the C-terminal basic domain (B-domain), which binds to the transcriptional activation domain of β-catenin, promoting the interaction between β-catenin and TCF/LEF1 transcription factors. Additionally, BCL9 binds to β-catenin through its Armadillo repeat 1 (R1). Therefore, inhibiting the formation of the β-catenin/BCL9 complex may not affect the interaction between β-catenin and its protein partners (such as Axin and APC), significantly reducing side effects [137]. Molecules that disrupt the BCL9/β-catenin PPI can effectively inhibit Wnt/β-catenin signaling, making the BCL9/β-catenin complex a novel target for drug development. The buried surface area of the β-catenin/BCL9 interaction interface is approximately 1450 $Å^2$, with a binding constant (K_d) of about 0.50 μM. This interaction is mediated by a helical segment of about 25 amino acid residues of BCL9 and involves two key hotspots [138]. The D162A mutation in β-catenin reduces its binding affinity to BCL9 by fourfold, while the D164A mutation reduces the interaction by 25-fold. These β-catenin residues form salt bridges with H358 and R359 of BCL9. Additionally, alanine mutations of H358 or R359 in BCL9 significantly reduce its binding affinity to β-catenin. Therefore, hotspot 1 includes β-catenin residues D162, E163, and D164, which form an acidic protrusion and interact with H358 and R359 of BCL9 (Figure 12.3a) [139]. However, no druggable pocket has been observed in this hotspot region. Mutations L366A and I369A in BCL9 disrupt the hydrophobic interactions at the C-terminal end of the BCL9 helix, leading to the loss of β-catenin binding [140]. Similarly, L366K and L373 mutations in BCL9 prevent β-catenin binding. Kawamoto et al. reported similar results and further found that mutating F374 in BCL9 to 2-naphthylalanine (2-Nal) increased its binding affinity to β-catenin by approximately threefold. These BCL9 residues interact with the hydrophobic residues L156, L159, and L178 of β-catenin. Additionally, the L156A and L159A double mutation in β-catenin abolishes its interaction with BCL9. This observation was confirmed and extended in 2015 to include the L156S and L159S as well as the L156S and L178S double mutations, which also abolished binding to BCL9. Hotspot 2 is a hydrophobic pocket in β-catenin, including residues A149, A152, L156, L159, L160, V167, A171, M174, and L178, which interact with residues L366, I369, L373, and F374 of BCL9. Due to the binding of BCL9 residues L366, I369, L373, and F374 in a broad and slightly shallow pocket of β-catenin, hotspot 2 has been extensively studied in the design and optimization of β-catenin/BCL9 interaction inhibitors (Figure 12.3b,c) [140–142].

12.3.3 Targeting β-catenin/BCL9 Small Molecule Inhibitors

12.3.3.1 Natural Products and Their Derivatives

In 2012, Roche et al. discovered a novel β-catenin/BCL9 inhibitor, Natural Product **71** (Carnosic Acid, Table 12.3) [143], through an ELISA screen of 47,500 compounds from the LOPAC/PhytoPure/MRCT libraries. This compound specifically binds to HD2-ARD with an inhibition constant (K_i) of 3.3 μM. NMR and crystallographic analyses demonstrated that compound **71** interacts directly with the first four ARD regions of β-catenin, with binding dependent on the adjacent unstable α-helix at the BCL9 binding site. Natural Product **71** promotes the degradation of active β-catenin, which correlates with its inhibition of TOPFlash luciferase activity and reduction in

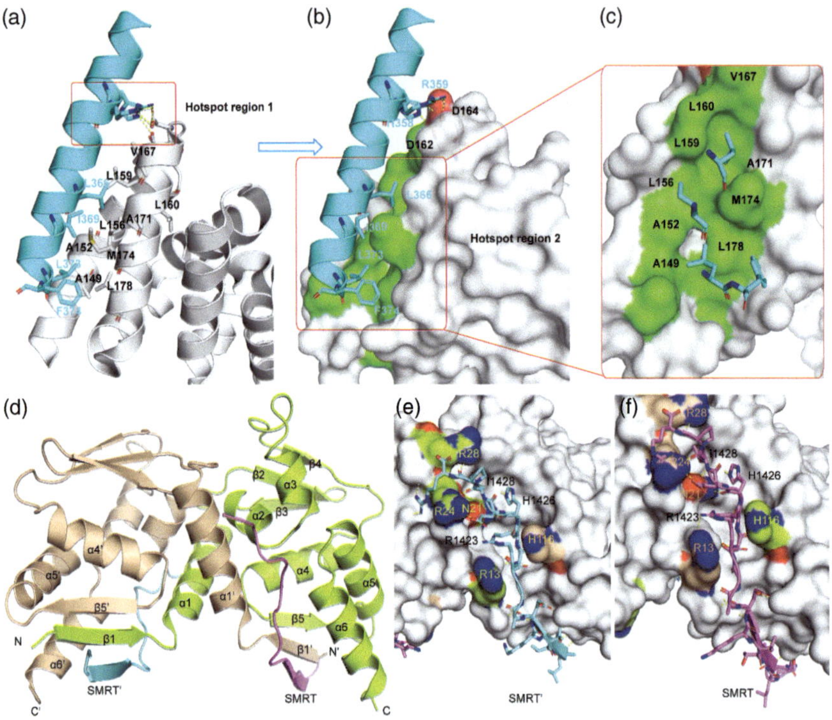

Figure 12.3 β-Catenin/BCL9 binding sites and interaction characteristics, as well as structural characteristics of the BCL6 BTB/POZ domain. (a) interaction between β-catenin and BCL9, with yellow dashed lines indicating hydrogen bonds; (b). contact map showing the hydrophobic interface between BCL9 and β-catenin hotspot region 2; (c). schematic representation of β-catenin hotspot region 2. (d) structure of the BCL6 BTB domain/SMRT-BBD complex; (e, f) "Bottom" of the complex, viewed along the molecular pseudo-twofold axis. Source: Wenhua Zhu, Yangbo He, Gang Chen, Di Zhu.

the expression of Wnt target genes such as Axin2 and B9L in HeLa, HCT116, and SW480 cells [144]. However, due to the presence of the catechol moiety – a known PAINS substructure – **71** exhibits various biological activities, and further research is yet to be reported.

Quercetin (compound **72**, FP IC_{50} = 10.10 μM) demonstrates notable inhibitory effects on the β-catenin/BCL9 interaction, suggesting that quercetin and its derivatives could serve as promising scaffolds for developing more effective β-catenin/BCL9 PPI inhibitors. In a study conducted by Shen et al., a series of novel compounds were designed and synthesized with various substituents at the 5-OH, 7-OH, 8-OH, and 3′-OH positions, based on quercetin. These compounds were assessed for their inhibitory activity against β-catenin/BCL9 PPI using competitive fluorescence polarization (FP) assays. Among them, two 3′-OH substituted quercetin derivatives, **73** (IC_{50} = 2.78 ± 0.39 μM) and **74** (IC_{50} = 3.13 ± 0.33 μM), exhibited the most potent inhibition. Further evaluation was performed to assess the effects of compounds **73** and **74** on the expression of the Wnt/β-catenin pathway-specific target gene Axin2 in CT26 cells, using quantitative real-time PCR (qPCR). Axin2, a downstream gene in the Wnt/β-catenin signaling pathway, serves as a key indicator of pathway activity [142]. The results revealed that compound **73** showed the strongest inhibition of Axin2 expression (IC_{50} = 2.81 ± 0.01 μM), while compound **74** demonstrated relatively weaker inhibition (IC_{50} = 6.29 ± 0.71 μM). Additionally, the impact of compound **73** on Wnt/β-catenin signaling hyperactivation in CT26 cancer cells was evaluated using the CCK-8 assay. Compound **73** exhibited significant growth inhibition with an IC_{50} value of 3.21 ± 0.04 μM, whereas compound 74 displayed a lower efficacy with an IC_{50} value of 7.85 ± 0.01 μM. Both compounds were more effective than quercetin (compound **72**, IC_{50} = 14.40 ± 0.04 μM). These findings indicate that compound 72 possesses the strongest activity, both at the protein level and in the cellular context [145].

Subsequently, the research team identified a novel 8-substituted quercetin derivative, **75**, which demonstrates significant inhibitory activity against the β-catenin/BCL9 PPI. Competitive fluorescence polarization (FP) assays revealed that compound 5 has an IC_{50} of 2.25 μM, while SPR analysis determined a dissociation constant (K_d) of 1.44 μM. Furthermore, compound **75** selectively inhibits the proliferation of colorectal cancer (CRC) cells and effectively suppresses Wnt signaling activation, concomitantly downregulating the expression of associated oncogenic Wnt target genes. In vivo studies show that compound **75** exhibits substantial anti-CRC activity, achieving a tumor growth inhibition rate (TGI) of 75.99% and modulating the tumor immune microenvironment [146]. These findings offer promising lead compounds and strategies for developing β-catenin/BCL9 PPI inhibitors based on natural product structures.

12.3.3.2 Phenyl-piperidine Derivatives

The Ji laboratory employed a highly selective ALPHAScreen assay to evaluate inhibitors of the β-catenin/BCL9 interaction. Screening 207 natural products and LOPAC Pfizer compounds led to the identification of **76 (CP-868388**, Table 12.3**)**, which demonstrated a K_i value of 1.2 μM [147]. This compound is notably 44 times more selective for the β-catenin/BCL9 interaction compared to β-catenin/E-cadherin, without affecting transactivation in Wnt/β-catenin signaling. Due to the hydrolytic instability of the carbamate group in **76** and challenges in synthesizing the chiral carbon, a scaffold hopping strategy was employed to develop derivative **77**(Table 12.3) [148]. Derivative **77**, incorporating a phenylpiperidine moiety, exhibited a K_i value of 71 μM, which is sixfold less effective than **76** [148]. Further optimization involved replacing the isopropyl group on the benzyl ring and using a bioisosteric replacement for the carboxyl group, resulting in **78 (CPPAA-30)**. This compound, particularly the (S)-isomer, showed enhanced efficacy with a K_i value of 3.6 μM. Compound **78** selectively disrupts the β-catenin/BCL9 interaction while sparing β-catenin/E-cadherin, inhibiting Wnt signaling in SW480 cells with an IC_{50} of 6.6 μM, and downregulating Wnt target genes Axin2 and cyclin D1. It significantly inhibits the growth of SW480, MDA-MB-231, and MDA-MB-468 cells, with IC_{50} values of 16, 14, and 18 μM, respectively. Further modifications, including substituting the isopropyl group on the benzyl ring with a 1*H*-pyrazole heterocycle and adding a piperazine moiety, yielded compound **79 (ZW4864**, Table 12.3**). It** demonstrated a K_i value of 0.76 μM and featured the phenylpiperidine scaffold. Direct binding assays confirmed its selective disruption of the β-catenin/BCL9 interaction [149]. **It also** showed moderate liver microsomal stability in humans ($T_{1/2}$ = 38.4 min) and mice ($T_{1/2}$ = 9.6 min), inhibited tumor growth in a TNBC PDX model at 90 mg/kg, and exhibited 83% oral bioavailability in mice. This has driven further development of small-molecule β-catenin/BCL9 inhibitors. Replacement of the 1*H*-pyrazole ring with other heterocyclic rings containing hydrogen-bond donors led to the creation of compound **80**. Although **80** showed improved cellular activity against MDA-MB-468 cells (IC_{50} = 5.0 μM) compared to compound **79**, its efficacy as a β-catenin/BCL9 interaction inhibitor (K_i = 2.7 μM) is about 3.5 times lower. Like compound **79**, **80** directly binds to β-catenin, disrupts the β-catenin/BCL9 interaction, and downregulates Wnt target gene expression [150].

Zhu et al. enhanced molecular interactions by substituting urea groups for amide groups, resulting in the development of a new structure based on ZW4864 and its analogs. They identified **81**, **82**, and **83**, with notable target specificity and efficacy. Compound **82**, in particular, exhibited a K_d value of 82 nM, marking it as one of the highest-affinity compounds for β-catenin reported to date. In vitro inhibition studies revealed that **81** (K_d = 0.13 μM) and **82** (K_d = 0.63 μM) are the most promising candidates for further evaluation, demonstrating significant antitumor effects in the CT26 mouse model. These compounds not only inhibited the expression of various cancer-related genes but also enhanced the activation of dendritic cells (DCs) and T cells. This effect increased the maturation of cDC1 cells, improved antigen presentation efficiency, and stimulated CD8+ T cells through co-stimulatory signals, establishing a direct positive feedback loop. This mechanism optimized the antitumor immune response and offered new insights into targeting BCL9/BCL9L for immune enhancement. Additionally, compound **81** showed moderate liver microsomal stability ($T_{1/2}$ = 31.5 min) and metabolic bioavailability (MF = 40%), significantly outperforming ZW4864 ($T_{1/2}$ = 2.37 min; MF = 4.77%). Although compound **83** displayed good inhibitory activity in FP and Axin2 assays, it had poor liver microsomal stability ($T_{1/2}$ = 3.74 min; MF = 7.33%) [151].

In their study, Zhang et al. designed compounds based on **CP-86388 (76)** and utilized molecular docking to predict its binding mode with β-catenin. Their results revealed that the 4-isopropylphenyl group effectively occupies the hydrophobic pocket of β-catenin (comprising residues L159, L160, V167, and A171), thereby enhancing binding affinity (Figure 12.4a). Crucially, the binding site for the BCL9 hotspot residues was mimicked by the 4-isopropylphenyl, piperidinyl, phenyl, and carboxyl groups (Figure 12.4b). Additionally, the carboxyl group of **76** forms a key hydrogen bond with K181, facilitating the introduction of large substituents. The enantiomer of **76**, **84**, was found to inhibit the β-catenin/BCL9 interaction with an IC_{50} of 5.39 μM. However, qPCR analysis indicated that compound **84** displayed weaker inhibitory activity against Axin2 expression in HCT116 cells, with an IC_{50} of 14.56 μM. Further modifications, including the incorporation of biphenyl, sulfonyl, and carboxylated *N*-acyl sulfonamide groups, led to the development of **85**, which exhibited the most potent inhibition of the β-catenin/BCL9 interaction, with an IC_{50} of 0.72 μM. Compound **85** also demonstrated significant inhibition of Axin2 expression, with an IC_{50} of 0.82 μM. Additionally, **85** showed favorable microsomal

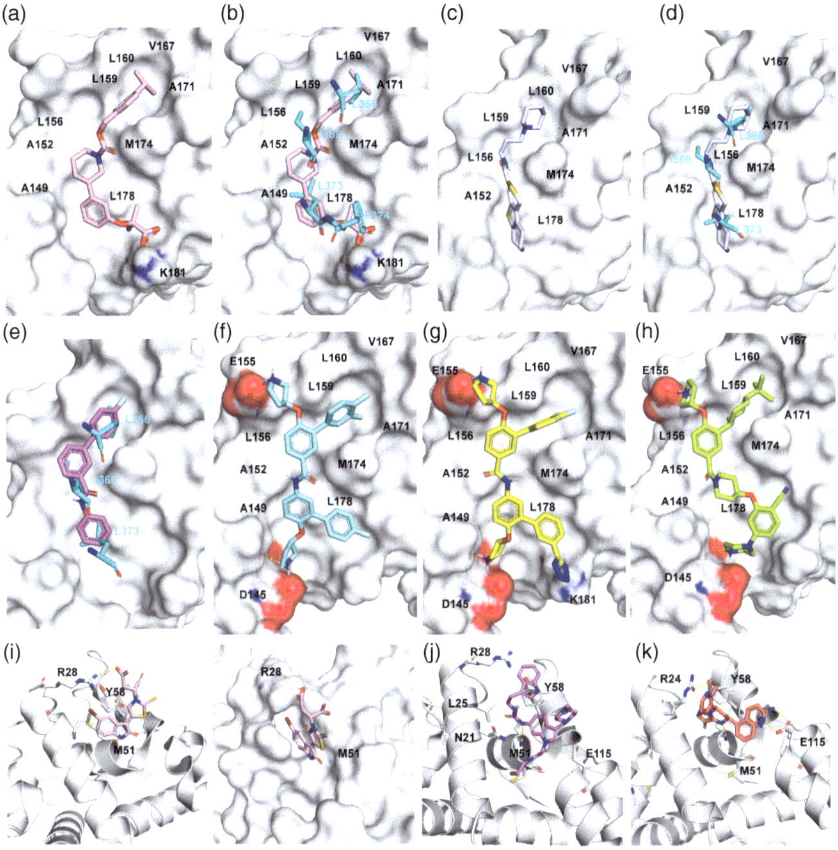

Figure 12.4 The interaction analysis of representative molecules with β-catenin/BCL6. (a) predicted binding mode of 76 with β-catenin; (b) BCL9 key residues overlaid on the predicted binding mode of 76 with β-catenin (PDB ID:2GL7). (c) predicted binding mode of 92 with β-catenin; (d) BCL9 key residues overlaid on the predicted binding mode of 92 with β-catenin (PDB ID: 3SL9). (e) BCL9 key residues overlaid on the predicted binding mode of 86 with β-catenin; (f) predicted binding mode of 87 with β-catenin; (g) Predicted binding mode of 89 with β-catenin; (h) predicted binding mode of 90 with β-catenin. (i) Interaction analysis of 79-6 with BCL-6 (PDB ID:3LBZ). (j) the interaction analysis of 109/GSK137 with BCL-6 (PDB ID:6CL3); (k) the interaction analysis of 109/GSK137 with BCL-6 (PDB ID:7DBE). Source: Wenhua Zhu, Yangbo He, Gang Chen, Di Zhu.

stability in mouse models, and CCK8 assays confirmed its effective suppression of Wnt-dependent CRC cell proliferation. qPCR analysis further validated that **85** significantly inhibited Axin2 expression in CT26 (IC_{50} = 0.66 μM), HCT116 (IC_{50} = 0.82 μM), and SW480 (IC_{50} = 5.15 μM) cells [152].

12.3.3.3 3-(4-fluorophenyl)-*N*-phenylbenzamide Derivatives

Residues L366, I369, and L373 of BCL9 are crucial hotspots for β-catenin-partner interactions, deeply embedded within β-catenin's hydrophobic pocket. This pocket has been targeted to design the general fragment **86** (Table 12.3), which selectively mimics the binding modes of L366 (*i*), I369 ($i + 3$), and L373 ($i + 7$) in the α-helix (Figure 12.4e). However, due to its relatively simple structure, this fragment exhibited weak binding affinity and selectivity. The introduction of two positively charged pyrrolidine groups and a 4-fluorophenyl moiety led to the development of compound **87 (PNPB-22)**, which showed a substantial enhancement in binding affinity, with an ALPHAScreen-derived K_i of 2.1 μM (Table 12.3). Compound **87** effectively inhibited the β-catenin/BCL9 interaction in ITC measurements, exhibiting a K_d value of 0.33 μM, and demonstrated 125-fold selectivity over the β-catenin/E-cadherin interaction. Molecular docking studies, supported by SAR and site-directed mutagenesis data, revealed that the two positively charged pyrrolidine groups form salt bridges with β-catenin residues D145 and E155 (Figure 12.4f), while the 3, 4-difluorophenyl group occupies the hydrophobic pocket formed by β-catenin residues L159, L160, V167, and A171 (Figure 12.4f). Cellular assays showed that **87** has good cell permeability and effectively inhibits the proliferation of Wnt-dependent colorectal cancer and triple-negative breast cancer cells, as well as suppresses transcriptional activation associated with the Wnt/β-catenin pathway [142]. Thus, compound **87** represents a significant advancement in the development of small-molecule β-catenin/BCL9 inhibitors.

Mutagenesis and crystallographic studies have identified two critical hotspot regions in the β-catenin/BCL9 interaction. These findings have propelled the SAR studies and optimization of compound **87**. By introducing a tetrazole group to form an additional hydrogen bond with K181, the first nanomolar small-molecule β-catenin/BCL9 inhibitor, **88 (PNPB-29)**, was successfully synthesized. This compound exhibited a K_i value of 0.47 μM in ALPHAScreen assays [153]. It demonstrated 1900-fold selectivity over β-catenin/E-cadherin and effectively down-regulated Wnt target gene expression in SW480 cells in a dose-dependent manner. However, compound **88** showed IC_{50} values ranging from 41 to 73 μM in HCT116, SW480, and MDA-MB-231 cancer cells, indicating that further optimization is needed to enhance its anticancer efficacy.

To further evaluate the potential of β-catenin pocket binding with ligands, the research team replaced the two phenyl rings in compound **87** with a piperazine ring and a cyclohexyl group, and introduced a piperazine-1-yl group, resulting in the highly promising inhibitor, **89** (Table 12.3), which exhibited a K_i value of 5.2 μM in the ALPHAScreen assay [154]. Molecular docking studies revealed that the pyrrolidine-3-yl and piperazine-1-yl groups formed salt bridges with E155 and D145 in β-catenin (Figure 12.4g), respectively, while the cyclohexyl group occupied the hydrophobic binding pocket of BCL9 L366. This approach was validated through SAR studies and site-directed mutagenesis. Compound **89** demonstrated 98-fold higher selectivity for β-catenin/BCL9 compared to β-catenin/E-cadherin and displayed moderate efficacy in inhibiting TOPFlash luciferase activity and cell viability in SW480, HCT116, and A549 cells, with IC_{50} values ranging from 22 to 300 μM. However, its inhibitory activity against full-length β-catenin was notably lower ($IC_{50} > 50$ μM). Subsequently, the team optimized the 1,4-diphenylcarbonylpiperazine scaffold to a 1-phenylcarbonyl-4-phenoxy-piperidine scaffold, which alleviated conformational strain and adjusted physicochemical properties. By incorporating a tert-butyl group into the cyclohexyl moiety and substituting the 4-cyanophenyl group with a

5-fluorophenyl group, they developed compound 20 (ZL3138), which demonstrated enhanced activity with a K_i value of 0.96 μM in the ALPHAScreen assay [155]. Docking studies showed that in compound **90**, the pyrrolidine-3-yl and piperazine-1-yl groups formed salt bridges with D145 and E155 in β-catenin (Figure 12.4h), while the 1-phenylcarbonyl-4-phenoxy-piperidine scaffold and tert-butyl cyclohexyl group bound to β-catenin's hydrophobic pocket (Figure 12.4h). This binding mode was further supported by SAR and site-directed mutagenesis studies. Biotin-tagged protein pull-down assays and co-immunoprecipitation (co-IP) experiments confirmed that compound **90** directly interacts with β-catenin, disrupting the β-catenin/BCL9 interaction. Additional cellular assays indicated that compound **90** selectively inhibits TOPFlash luciferase activity and suppresses the growth of SW480, HCT116, MDA-MB-231, and MDA-MB-468 cells by downregulating Wnt target genes Axin2, BCL9L, and MMP7 (IC_{50} values of 3.9, 6.2, 9.7, and 4.1 μM, respectively), though its in vivo efficacy remains unreported (Figure 12.4i).

12.3.3.4 Other β-catenin/BCL9 Small Molecule Inhibitors

Recently, Tanton performed a two-round screening of 320,000 small molecules using the ALPhaScreen assay, identifying 16 active compounds [156]. The binding thermodynamics of these compounds were validated through ITC analysis, and their cellular inhibition of the β-catenin/BCL9 interaction was confirmed via co-immunoprecipitation (co-IP) experiments. Notably, compound **91 (L814-1428)** and **92 (E722-2648)** emerged as effective β-catenin/BCL9 inhibitors (Table 12.3). In the ALPHAScreen assay, the IC_{50} values for β-catenin/BCL9 interaction were 7.4 and 9.0 μM for **91** and **92**, respectively, while ITC measurements indicated IC_{50} values of 1.0 and 2.2 μM. Both compounds significantly downregulated the expression of Wnt target genes Axin2 and CD4 in CRC cell lines, inhibited the Wnt signaling pathway, and reduced the proliferation of Colo320 and HCT116 cells. In vivo studies showed that compound **92** demonstrated inhibitory effects in HCT116 NOD scid-γ and NCr nude mouse models, with intraperitoneal administration at doses of 4 and 3 mg/kg, respectively, without notable effects on animal body weight. Molecular docking calculations revealed that compound **92** precisely binds to the BCL9-binding pocket in the first armadillo repeat of β-catenin (Figure 12.4c), occupying the central part of the BCL9 cleft (Especially residues S366, I369, and L373, Figure D) and interacting primarily through van der Waals forces with the hydrophobic surface (including residues A152-M174).

12.4 Targeting BCL-6 Small Molecule Inhibitors

12.4.1 Biological Functions of BCL6

The BCL6 protein represents a highly promising target for the treatment of various lymphomas, including non-Hodgkin's lymphoma, follicular lymphoma, and diffuse large B-cell lymphoma (DLBCL) [157, 158]. BCL6 suppresses genes that are involved in differentiation, inflammation, apoptosis, and cell cycle regulation in germinal center (GC) B cells [159]. It achieves this either by directly binding to the DNA sequence 5′-TTCCTAGAA-3′ (the BCL6 binding site) or by indirectly inhibiting the transcriptional activity of other factors [160]. Additionally, BCL6 autoregulates its own transcriptional expression and indirectly upregulates certain genes essential for the GC response, such as *AICDA*, by inhibiting the expression of microRNAs like miR155. As expected, knocking down BCL6 in lymphoma cells results in tumor arrest [161]. However, unconditional knockout of *BCL6* can result in severe, potentially fatal systemic inflammatory diseases [162]. BCL6 deficiency in macrophages results in accelerated atherosclerosis in murine models [163]. Experiments have shown that BCL6-deficient mice die rapidly within a few weeks and develop a massive inflammatory infiltrate in all their organs [164].

12.4.2 Structural Characteristics of the BCL6 BTB/POZ Domain

BCL6 is an oncogene that belongs to the BTB/POZ (bric-a-brac, tram track, broad complex/poxvirus zinc finger) family and is located on chromosome 3q27 [165–167]. The BCL6 protein is composed of 706 amino acids and spans a total length of 24 kb [168, 169]. Structurally and functionally, the protein can be divided into three distinct regions: the N-terminal BTB/POZ domain (amino acid residues 5-129), the C-terminal C_2H_2 zinc finger domain (amino acid residues 518-681), and an intermediate region containing three PEST domains (RD2) [160, 168]. Among these, Zinc Fingers are involved in macrophage-driven inflammatory responses by binding to the promoters of BCL6-regulated genes. The RD2 region facilitates B-cell trafficking. The BTB/POZ domain functions through dimerization and PPIs [170].

X-ray crystallography reveals that the BCL6 BTB structural domain forms a stable homodimer, functioning exclusively as such in solution [171]. BCL6 exerts transcriptional repression by either directly recruiting coactivators or indirectly inhibiting transcriptional regulation via class I and II histone deacetylases (HDACs) that are recruited by the BTB structural domain [172, 173]. The BCL6 BTB structural domain interacts in a mutually exclusive manner with three co-repressor proteins: SMRT (silencing mediator of retinoid or thyroid hormone receptors), N-COR (nuclear receptor co-repressor), and B-COR (BCL6 co-repressor) [174–176]. These co-repressor proteins bind to the lateral groove (LG), which is the exposed surface at the interface of the two chains in the BCL6 BTB homodimer. SMRT, N-COR, and B-COR recognize the lateral groove (LG) through their respective BCL6 BTB-binding domains (BBDs). Although the BBD of B-COR is distinct from those of SMRT and N-COR, all three co-repressor proteins bind to the same region of the BCL6 BTB structural domain [171, 177].

The BTB domain is considered crucial for the BCL6 protein to interact with other factors, with its side pocket serving as a key binding site for various proteins (Figure 12.3d). This domain consists of a symmetric dimer (see Fig. 1B, PDB: 1R28, 1R29), and the interaction between the BTB subunits of BCL6 is primarily mediated by α1, α2, α6, β1, and β5. The bottom of the dimer features a large hydrophobic pocket mainly formed by β1 and α6′, β1′ and/or α6, while the top pocket is shaped by α3 and β4 (Figure 12.3d). Research has shown that mouse N-CoR (residues 1351-1616) and human SMRT (residues 1414-1498) can directly bind to the BCL6

BTB domain [178]. SMRT binds to the α1′, β1′, α2, α3, and α6 chains of the dimer (Figure 12.3d). The deletion of SMRT residues 1414-1420 completely eliminates complex formation. The middle region of SMRT, Ser-1424, and Ile-1425, interacts significantly with the pocket, which is enveloped by the BTB dimer's Arg-13 (α1′) and His-116 (α6), with SMRT's Ile-1425 forming hydrophobic interactions with BTB domain's Val-18′ (α1′) and Cys-53 (α2). Mutating residues 1427-1430 to alanine disrupts the complex, indicating that this segment is crucial for complex stability [171]. These findings provide a valuable reference for peptide and small-molecule drug design. Researchers tested interactions using full-length BCL6 with mutations and found that the mutant proteins were unaffected by the presence of SMRT, demonstrating that the BTB side pocket is essential for the cooperative repression by full-length BCL6 [171]. To investigate key sites, they mutated H116 to A and N21 to K in the BCL6 BTB domain and found that, compared to the wild type, the mutant protein domain's transcriptional repression was significantly reduced, although other sites still provided compensatory repression [173].

Analyzing the structure of the BCL6 protein is an important step in understanding how to interfere with its stability and activity. It has been demonstrated that the effective targeting region of BCL6 for PPIs with three co-deterrent proteins (SMRT, N-COR, B-COR) is the LG. To target this region, BCL6 small-molecule inhibitors have been demonstrated to disrupt the interaction between BCL6 and BBD in vitro [167, 179]. BTB inhibitors have shown significant potential in treating diseases driven by BCL6. This approach not only offers good targeting specificity but also effectively avoids the side effects associated with BCL6 inactivation. Structural analysis of the BCL6-BBD complex reveals that the groove region associated with SMRT residues 1423-1428 exhibits high complexity and density of intermolecular contacts between the protein and the BBD peptide (Figure 12.3e,f). Alanine scanning mutagenesis confirmed that all six BBD residues are essential for the stability of the complex. Consequently, the binding pocket centered on this region has been utilized for the development and design of inhibitors.

12.4.3 BCL6-targeted Diseases

BCL6 is expressed in both Germinal Center B-cell (GCB-type) and Activated B-cell (ABC-type) diffuse Large B-cell Lymphomas (DLBCL) [180]. Knocking down BCL6 using shRNA has an inhibitory effect on both ABC-DLBCL and GCB-DLBCL cells. Existing research has shown that inhibiting the aberrant expression of BCL6 has therapeutic effects on various cancers. The primary disease targeted by BCL6 inhibitors is DLBCL, and studies have found that BCL6-targeted drugs inhibit multiple DLBCL cell lines, including GCB-type DLBCL, ABC-type DLBCL, and BCR-type DLBCL [180–183]. Additionally, BCL6 is involved in other diseases such as follicular lymphoma, chronic myeloid leukemia, breast cancer, and NSCLC [184–188]. Experimental evidence also suggests that inhibiting BCL6 can alleviate chronic cardiac transplant rejection by regulating Tfh cell expansion and the humoral response [189].

12.4.4 BCL6 Inhibitor

12.4.4.1 79-6, FX1 and AP-4-287

Air Melnick et al. employed a pocket-based analysis of the BCL6 BTB domain to identify 79-6(compound **93,** Table 12.4) as a potent inhibitor via HTS [157]. Co-crystallization of 79-6 with the BCL6 BTB domain (PDB: 3LBZ) revealed that 79-6 binds symmetrically within two hydrophobic pockets of the BCL6 BTB domain [157]. The indoline-2-one ring of 79-6 interacts with Tyr-58, while the bromine atom on this ring forms a hydrophobic interaction with Leu-25. The indole ring further stabilizes 79-6 within the binding site (fig) [157]. Importantly, 79-6 exhibits high specificity by selectively binding to the BTB domain of BCL6 and not to other BTB domains. It disrupts the co-repressor interactions of N-CoR and SMRT with BCL6, resulting in the upregulation of the art gene, without affecting non-BCL6-dependent BTB repression pathways, thereby highlighting its specificity [157].

Due to the lower affinity of the first-generation BCL6 inhibitor 79-6 compared to endogenous co-repressors of BCL6, researchers developed a series of molecules based on 79-6 using ligand competition saturation techniques (Monte Carlo-SILCS, MC-SILCS). Among these, FX1(compound **94,** Table 12.4) exhibited the highest activity. The structural differences between FX1 and 79-6, while keeping the core scaffold unchanged, primarily include the removal of the carboxyl group from the carboxylic acid tail's γ-carbon and the substitution of the bromine atom on the indoline-2-one ring with a chlorine atom [158]. Treatment with FX1 led to the upregulation of 1027 genes in three DLBCL cell lines (SUDHL-4, SUDHL-6, Farage) [158]. Gene enrichment analysis of genes upregulated by BCL6 knockdown and FX1 treatment revealed a high degree of overlap between the two sets [158]. The combination of FX1 with doxorubicin resulted in a differential reduction in the GI_{50} of doxorubicin during GCB-DLBCL treatment. Additionally, FX1 induces the expression of p53, p21, and p27, thereby modulating downstream pathways related to cell death and antiproliferation [158].

Due to the limitations in water solubility and pharmacokinetics of FX1, Luis J. Montaner and colleagues designed the prodrug AP-4-287(compound **95,** Table 12.4) based on FX1. This molecule exhibits improved water solubility, with a solubility of 218 ng/ml in PBS buffer, representing approximately a 150-fold increase compared to FX1 [190]. After administration, AP-4-287 shows lower side effects and retains stability and a half-life in human plasma similar to FX1 [190].

12.4.4.2 BI3812 and TMX-2164

Yusuke Kamada employed a biophysics-driven fragment-based approach combined with SPR fragment screening to successfully identify fragment **96** (SPR K_d = 1200 μM, Table 12.4). This fragment was subsequently optimized into a more effective inhibitor, compound **97**, which demonstrated a significantly improved binding affinity (SPR K_d = 0.078 μM) and cellular activity [191]. Nina Kerres utilized fluorescence polarization (FP) analysis and HTS to identify a pyrimidine diamine-containing compound, **98** (Table 12.4), which exhibited a weak binding affinity to BCL6. Further structural optimization led to the development

Table 12.4 BCL6 inhibitors.

Compound	Structure	Targets	Affinity
93 79-6	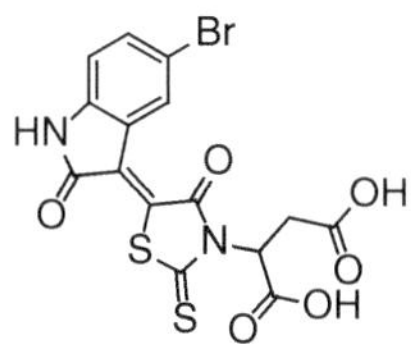	BCL-6	$K_d = 138\,\mu M$
94 FX-1	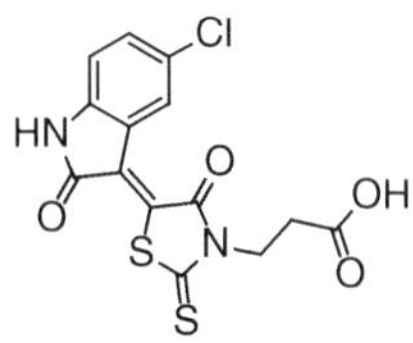	BCL-6	$K_d = 7\,\mu M$
95 AP-4-287	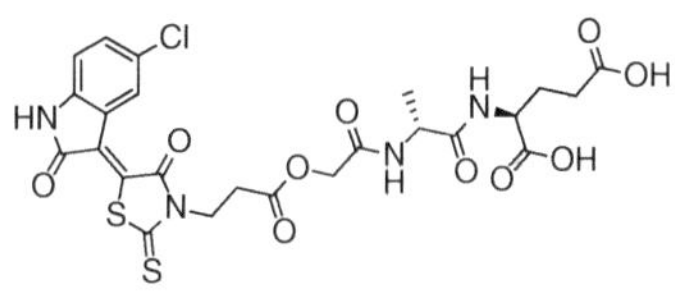	BCL-6	/
96	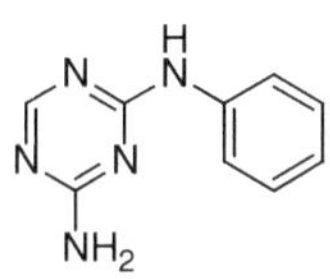	BCL-6	SPR $K_d = 1200\,\mu M$
97	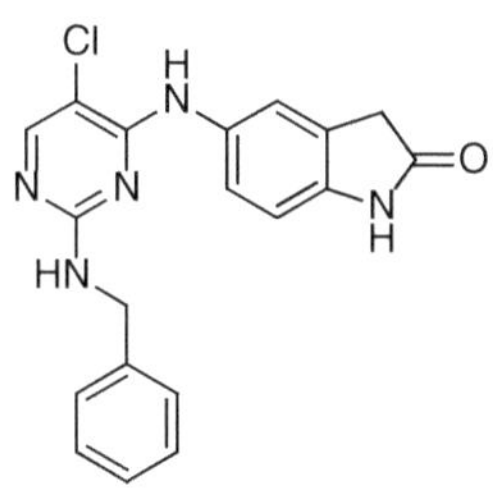	BCL-6	SPR $K_d = 0.078\,\mu M$
98	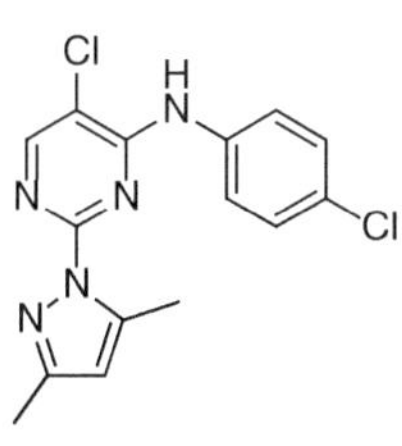	BCL-6	$K_d = 20\,\mu M$ FP = $18\,\mu M$

(Continued)

Table 12.4 (Continued)

Compound	Structure	Targets	Affinity
99 BI3812	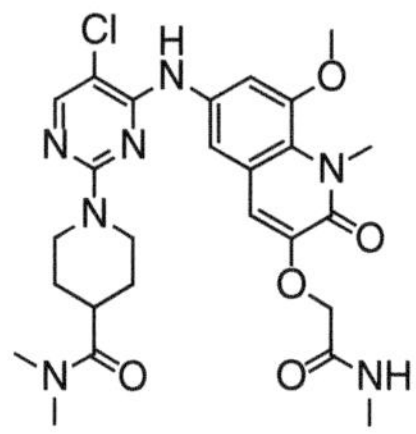	BCL-6	/
100 TMX-2164	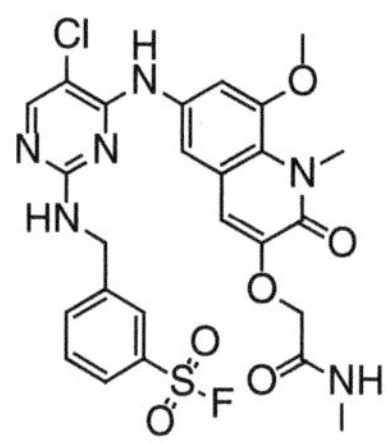	BCL-6	$IC_{50} = 152\,nM$
101 TMX-2177	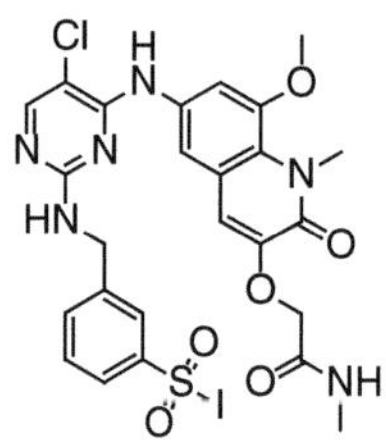	BCL-6	\
102	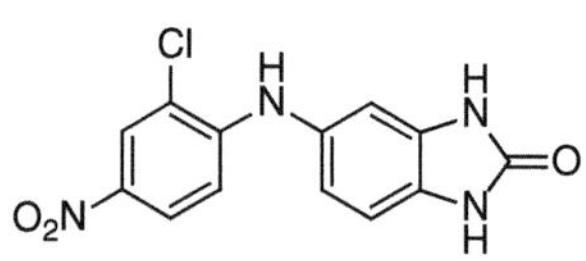	BCL-6	ELISA $IC_{50} = 4.5\,\mu M$
103	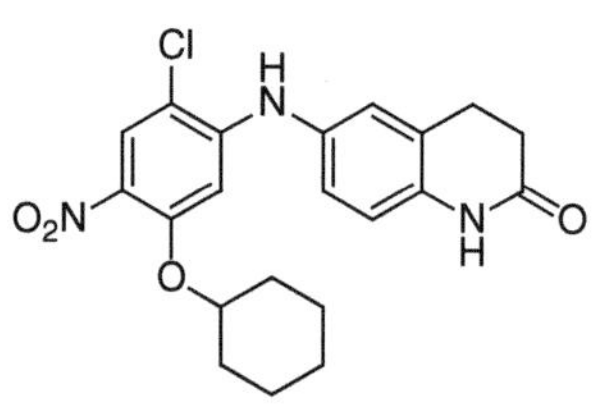	BCL-6	$IC_{50} = 0.1\,\mu M$

Table 12.4 (Continued)

Compound	Structure	Targets	Affinity
104 BCL6-i	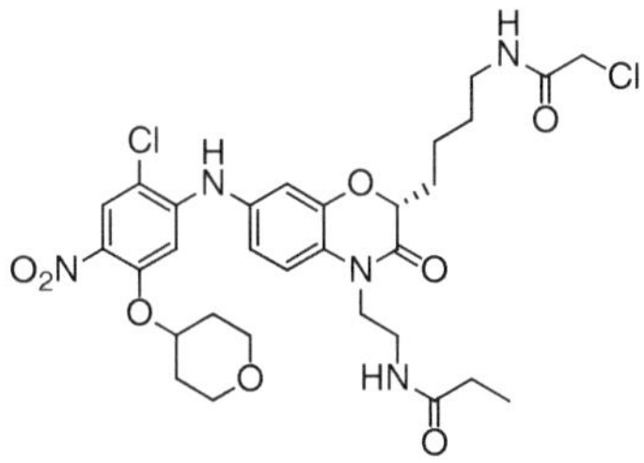	BCL-6	Kinact/KI a (M-1s −1) = (1.9 + 0.6) × 104
105	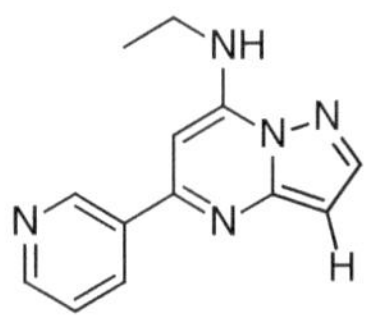	BCL-6	$K_d = 689\,\mu M$
106	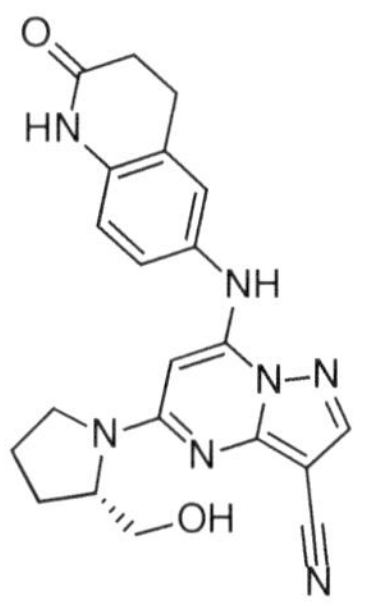	BCL-6	BCL6 SPR $K_d = 0.5\,\mu M$ BCL6 FRET $IC_{50} = 0.37\,\mu M$
107	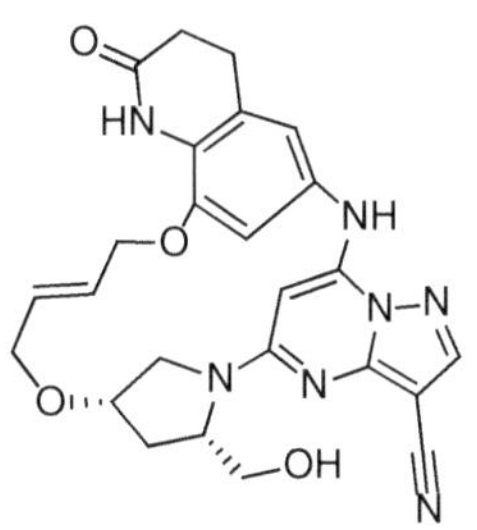	BCL-6	BCL6 SPR $K_d = 0.00065\,\mu M$ BCL6 FRET $IC_{50} = 0.0029\,\mu M$
108	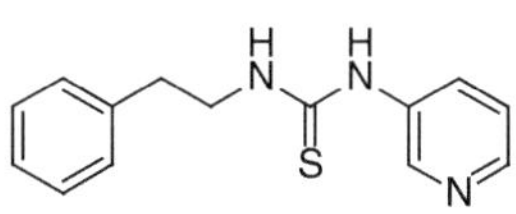	BCL-6	LE = −1.79

(Continued)

Table 12.4 (Continued)

Compound	Structure	Targets	Affinity
109	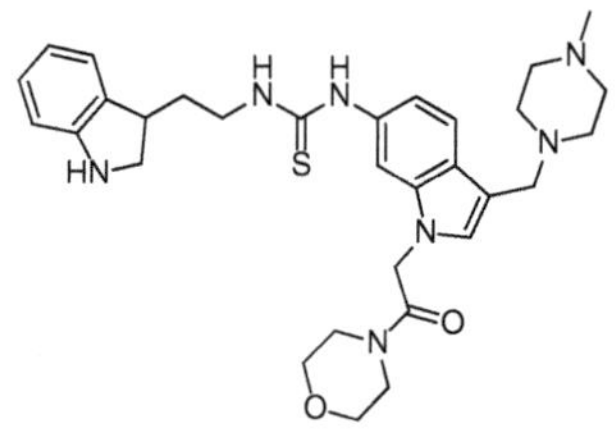	BCL-6	BCL6 K_d = 36 μM ITC K_d = 36 μM
110	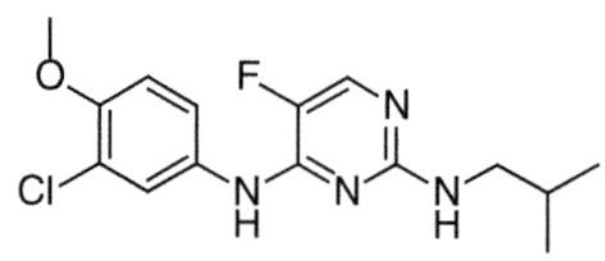	BCL-6	HTRF:IC = 19.4 μM
111	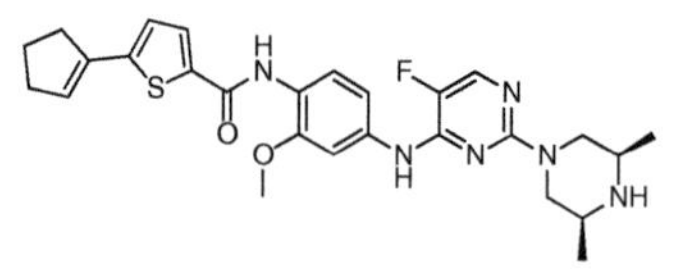	BCL-6	HTRF IC_{50} = 0.47 μM BLI K_d = 366 nM
112 WK500B	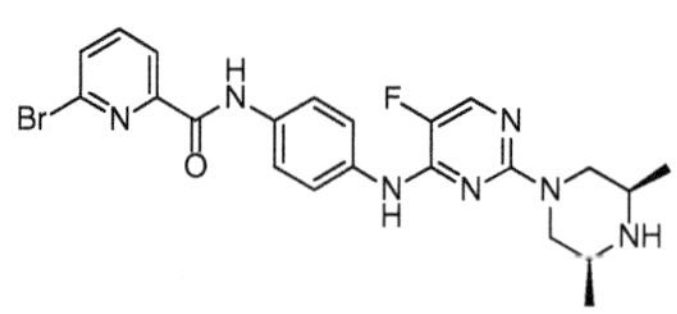	BCL-6	IC_{50} = 1.37 μM SPR K_d = 1.61 μM
113 GSK137	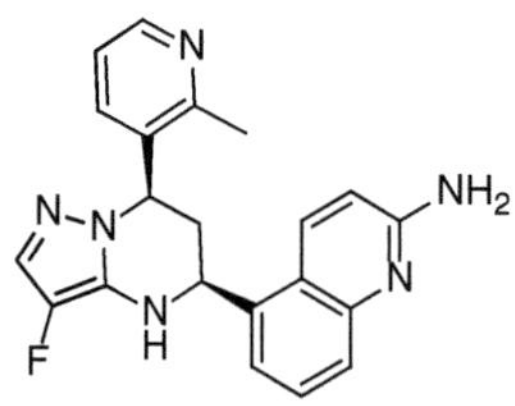	BCL-6	TR-FRET pIC_{50} = 8
114 CCT369347	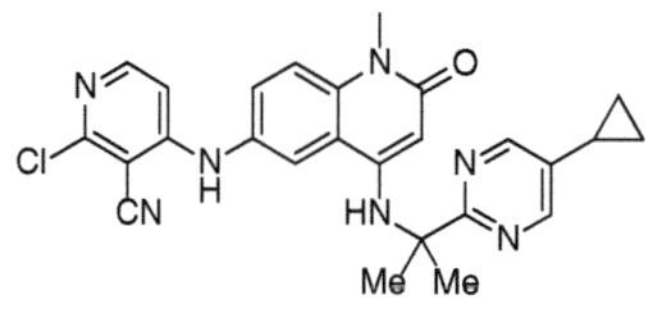	BCL-6	TR-FRET IC_{50} = 0.026 μM NanoBRET IC_{50} = 0.57 μM Solubility (NMR) = 29 μM

Table 12.4 (Continued)

Compound	Structure	Targets	Affinity
115	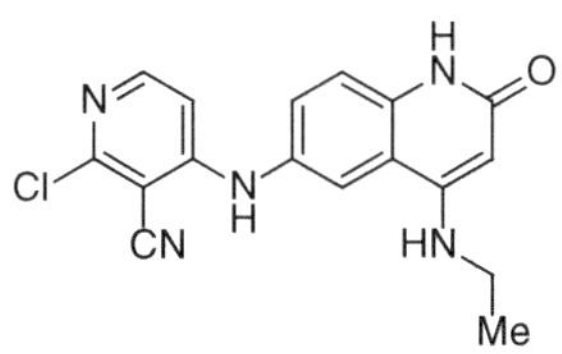	BCL-6	BCL6 TR-FRET $IC_{50} = 1300\,nM$
116 CCT372064	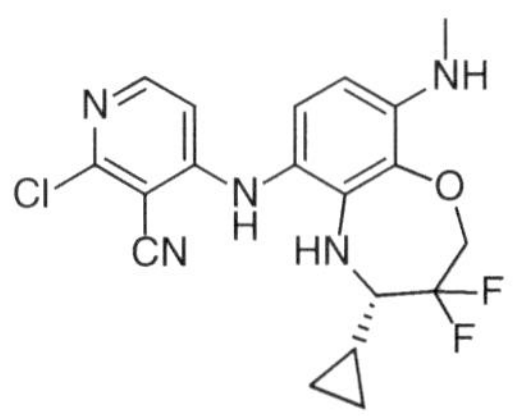	BCL-6	BCL6 TR-FRET $IC_{50} = 0.0039\,\mu M$ NanoBRET $IC_{50} = 0.012\,\mu M$
117	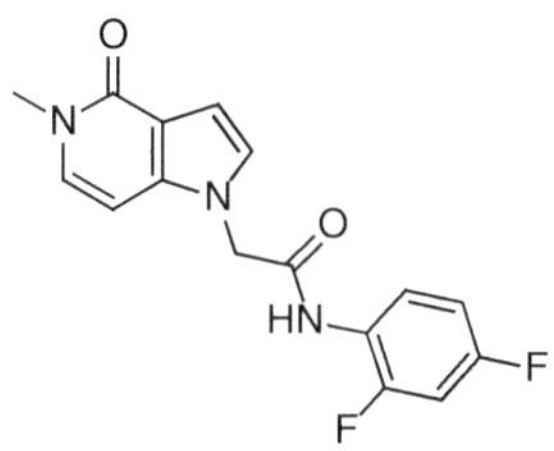	BCL-6	SPR $K_d = 282\,\mu M$
118 OICR12694	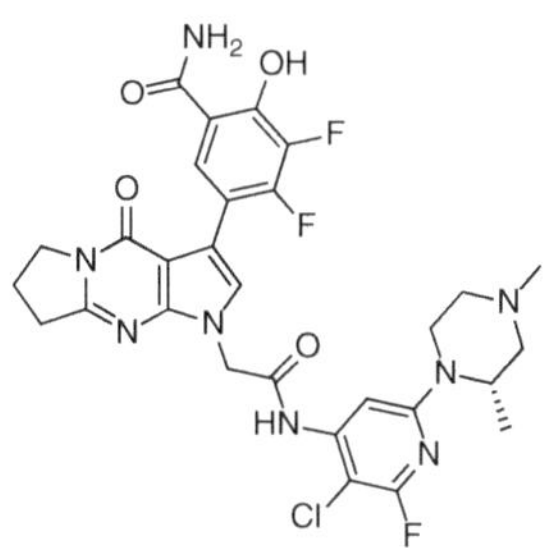	BCL-6	SPR $K_d = 0.005\,\mu M$

of compound **99** (BI3812), which displayed nanomolar-level inhibition of the BCL6-co-repressor interaction, with an ULight IC_{50} value of <3 nM [192]. Based on modifications of compounds **97** and **99**, irreversible BCL6 inhibitor **100** (TMX-2164, Table 12.4) and reversible inhibitor **101**(TMX-2177) were designed. Compound **100** is an electrophilic sulfonyl fluoride covalent inhibitor, and mass spectrometry confirmed its covalent binding to Tyr58 of BCL6. Compound **100** demonstrated approximately twice the binding affinity to BCL6 compared to the structurally similar reversible inhibitor **101** [193].

12.4.4.3 BCL6-i and Compound 107

Through HTS, diphenylamine derivatives (compound **102**, Table 12.4) were identified as BCL6 inhibitors, and subsequent structural optimization yielded compound **103** (Table 12.4). This compound has an IC_{50} of 0.1 μM for inhibiting the BCL6-corepressor (BCoR) interaction, approximately 40 times more effective than its lead compound **102** [194]. Further optimization led to the development of compound **104**(BCL6-i, Table 12.4) BCL-i is an irreversible inhibitor targeting BCL6, specifically interacting with Cys53 in the BTB domain of BCL6 [195].

Compound **106** (Table 12.4) is a pyrazolo[1,5-a] pyrimidine series BCL6 inhibitor optimized from compound **105** (Table 12.4). Xiahui Zhu and colleagues found that in non-macrocyclic compounds, the flexibility of the lactam and pyrrolidine rings significantly affects binding affinity [196]. Introducing a macrocycle (compound **107**, Table 12.4) to constrain these rings into the bioactive conformation greatly improves affinity, with differences in affinity greater than 100-fold before and after cyclization [196]. Thus, incorporating a bicyclic lactam group to enhance interactions and using macrocycles to stabilize conformation is a preferred strategy for designing BCL6 inhibitors [196].

12.4.4.4 Compounds 109, 111, WK500B, and GSK137

Chen Used NMR-based screening of a library of fragment-like small molecules and identified a thiourea compound **108** (7cc5, Table 12.4) that binds to the BCL6BTB. From this hit, the application of computer-aided drug design (CADD), medicinal chemistry, NMR spectroscopy, and X-ray crystallography has yielded an inhibitor **109** (Table 12.4), which demonstrated over 100-fold improved potency for the BCL6BTB. This compound mimics the binding pattern of the corepressor SMRT. Structurally, the thiourea group in 18 forms hydrophobic interactions with Asn21, Leu25, Met51, Ala52, and Tyr58. Additionally, Met51 engages in a hydrogen bond with the carbonyl group of 18. The piperazine moiety establishes an electrostatic interaction with Glu115, while the morpholine group optimally occupies the HDCH site (Figure 12.4j). This detailed binding interaction provides insight into the design of effective BCL6 inhibitors [197].

Compound **111** (Table 12.4), an *N*-phenyl-4-pyrimidinamine derivative, was optimized from a hit compound **110** identified through HTS [198]. Experimental studies indicate that **111** strongly inhibits DLBCL in both in vivo and in vitro settings, suppresses germinal center (GC) formation, and reduces the proportion of Tfh cells [198].

WK500B(compound **112**, Table 12.4), identified through luciferase reporter gene assays and homogeneous time-resolved fluorescence (HTRF) screening, exhibits superior oral bioavailability compared to other BCL6 inhibitors [199]. Although its IC_{50} is about 20 times higher than FX1, WK500B effectively inhibits tumor growth in DLBCL mouse models and shows strong pharmacokinetic properties [199]. Molecular docking studies reveal that WK500B interacts with the BCL6 BTB domain through its carbonyl oxygen and pyridine nitrogen, potentially forming interactions with Arg24, while His14 forms a hydrogen bond with the piperazine ring of WK500B [199]. This disruption of the BCL6 BTB-SMRT complex modulates

BCL6 target gene expression, highlighting WK500B's potential as an effective therapeutic agent [199].

GSK137, a potent small-molecule BCL6 inhibitor with in vivo activity, suppresses antibody responses in mice. This is the primary literature source for the compound GSK137. (compound **113**, Table 12.4) is a novel, effective, and cell-permeable small molecule inhibitor that binds to the BCL6 BTB-POZ domain. It exhibits an in vitro pIC_{50} value of 8 for binding to the BCL6 BTB-POZ domain and a cellular pIC_{50} of 7.3. The compound has good solubility (128 μg/ml) and high permeability (86 nM/s). Despite the limited impact on cell viability and proliferation in four BCL6-expressing B cell lymphoma cell lines, GSK137 induces a slight accumulation of cells in the G1 phase. Pharmacokinetic studies show that it effectively targets the intended domain and, when administered orally, can inhibit immunoglobulin G responses, reducing the number of germinal centers and germinal center B cells in mice. This indicates that GSK137 has potential in vivo activity, capable of inhibiting T-dependent immune responses. X-ray crystallography of the complex shows that GSK137 binds within the two lateral grooves of the BTB-POZ domain dimer through key interactions. The aminoquinoline group is partially exposed on the surface, forming double hydrogen bonds with the Glu115 side chain and interacting with the small loop structure (Cys53 to Gly55). The pyrazole-tetrahydropyridine core is located deep within the binding pocket, interacting with the Tyr58 aromatic side chain through π–π stacking and forming hydrogen bonds with the Met51 backbone carbonyl and the Arg24 side chain. The hydrogen bond between Arg24 and the Tyr58 hydroxyl further stabilizes the complex. In addition, the methylpyridine group, similar to the aminoquinoline group, is exposed to the solvent and stabilized by π–π stacking with the R24 guanidine group (Figure 12.4k).

12.4.4.5 CCT369347, CCT372064, and OICR12694

Swen Hoelder and colleagues, in their study of the quinolone scaffold, identified an unoccupied sub-pocket below the scaffold within the BCL6 BTB domain [200]. X-ray crystallography revealed that this sub-pocket is occupied by five water molecules, which are tightly bound to BCL6. Researchers discovered that the compound CCT369347 can displace three of these water molecules (positions 1, 3, and 5) from the pocket [200]. Additionally, the pyridine ring of CCT369347 interacts with the remaining adjacent water molecules, thus enhancing the binding affinity of the compound for the BCL6 BTB domain [200].

CCT372064 was developed by fusing an additional ring to the previously identified quinolone molecule 115 [201]. This added ring better fills the HDCH pocket, resulting in a 300-fold increase in activity compared to the lead molecule **115**. Crystal structure analysis reveals that CCT372064, which contains electronegative fluorine atoms, interacts closely with the protein surface [201]. The molecule also demonstrates significant cellular activity and an improved NanoBRET IC_{50} value, showing approximately a 50-fold enhancement compared to CCT369347 [201].

Compound 118 (OICR12694, Table 12.4) is a novel BCL6 inhibitor that was developed through a process of virtual screening and structural optimization. Initially, starting from a virtual screening hit with a binding affinity of 282 μM, a series of highly effective BCL6-BTB inhibitors were identified through structure-based design and SAR studies. Ultimately, the candidate compound OICR12694 was selected due to its low nanomolar DLBCL cell proliferation inhibition activity and favorable oral pharmacokinetic properties. Regarding biological activity, OICR12694 demonstrated potent growth inhibition in BCL6-dependent cell lines such as Karpas-422 across multiple in vitro assays. It also exhibited high selectivity for BCL6-BTB compared to other BTB family members. Furthermore, OICR12694 showed good safety in a series of in vitro toxicity tests and exhibited excellent oral bioavailability in both mice and dogs. Based on these characteristics, OICR12694 is considered one of the most effective oral bioavailable BCL6-BTB inhibitors to date, with significant preclinical potential, particularly for the treatment of DLBCL and other BCL6-driven cancers [202].

12.5 BCL-3 Inhibitors

12.5.1 Structural Characteristics of BCL-3

B-cell CLL/lymphoma 3 (BCL-3) is a member of the non-classical inhibitor of NF-κB (IκB) family. The BCL-3 gene is located at 19q13 and was first identified near the breakpoint of the translocation *t*(14;19) (q32;q13.1) in patients with B-cell chronic lymphocytic leukemia [203–205]. Unlike other IκB family members, BCL-3 is a nuclear protein primarily localized in the cell nucleus [206]. Human BCL-3 is a 446-residue protein with a molecular weight of approximately 47 kDa, featuring a central domain composed of seven ankyrin repeat sequences (ANK) (Figure 12.5a). Each ankyrin repeat consists of approximately 30 amino acids and includes an initial β-hairpin followed by a helix-turn-helix motif oriented nearly perpendicular to the β-hairpin. The overall shape of the ankyrin repeat domain resembles a "cup-shaped hand," with the β-hairpin forming the fingers, the α1 helix forming the palm, and the α2 helix forming the back of the hand [207, 208]. Additionally, the BCL-3 protein contains an N-terminal domain rich in proline and a C-terminal transcriptional activation domain rich in serine/proline (Figure 12.5b) [209].

12.5.2 Biological Functions of BCL-3

Studies have shown that BCL-3 regulates the subcellular localization and DNA-binding activity of NF-κB. BCL-3 binds to the p52 or p50 homodimers of the NF-κB protein family, causing these dimers to dissociate from the DNA κB binding sites [210–212]. Once BCL-3 dissociates, the p52 or p50 homodimers can rebind to the DNA κB sites but lack the transcriptional activation domains, thereby competitively inhibiting the transcriptional activation of NF-κB/RelA dimers [210, 213]. BCL-3 is currently recognized for its diverse physiological functions, including regulation of cell differentiation, adhesion, proliferation, apoptosis, cell cycle, and DNA damage response [214–219]. In humans, BCL-3 is implicated in the pathogenesis of various diseases such as psoriasis, chronic kidney disease, fibrosis, and atherosclerosis [220–222]. Recent research has increasingly focused on

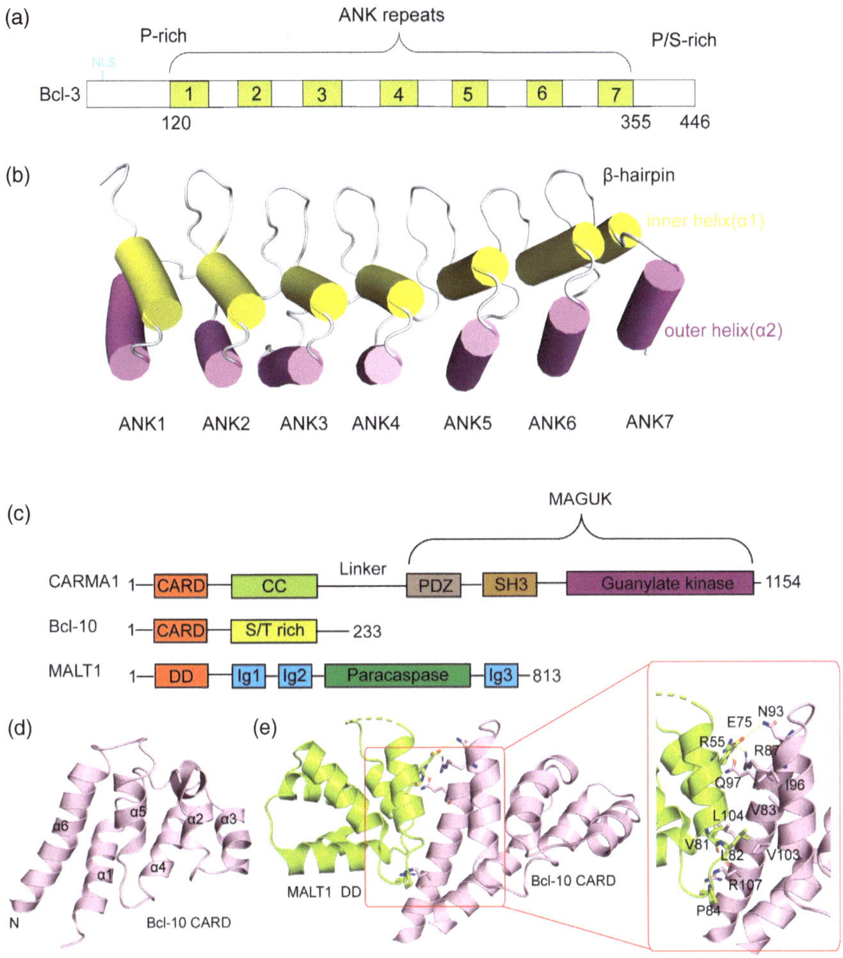

Figure 12.5 Structural characteristics of BCL-3 and BCL-10. (a) domain organization of BCL-3; (b) ribbon diagram of the BCL-3 ARD (PDB ID: 6GK2). (c) domain organization of CARMA1, BCL10, MALT1; (d) the crystal structure of BCL10 CARD domain (PDB ID: 6GK2); (e) the BCL10 and MALT1 interaction site I (BM-I) (PDB ID: 6GK2), The key residues of BCL10 and MALT1 are represented as pink and lemon sticks, respectively. Source: Wenhua Zhu, Yangbo He, Gang Chen, Di Zhu.

the role of BCL-3 in inflammation and adaptive immune responses. Additionally, BCL-3's impact on tumor cell proliferation and apoptosis in the development of malignancies has garnered significant attention. BCL-3 is often considered an oncogene, with its phosphorylation linked to tumorigenesis [213]. Elevated expression of BCL-3 protein has been observed in a range of malignancies, including diffuse large B-cell lymphoma, gastric cancer, cervical cancer, hepatocellular carcinoma, colorectal cancer, and ovarian cancer [223–231]. Notably, mice with constitutive BCL-3 deficiencies exhibit only mild immune defects [232, 233], and BCL-3's role in inducing immune checkpoint PD-L1 expression further underscores its potential as a novel therapeutic target [234].

The anchor protein repeat domain of BCL-3 and its regulatory protein partner p50 exhibit a crucial PPI. Mutants of p50 and BCL-3 generated via site-directed mutagenesis (ANKM123) have confirmed the importance of the BCL-3/p50 interaction in driving BCL-3 functionality. When these mutants are transfected into HEK293 cells, they exhibit a loss of p50 binding capability and are unable to effectively mediate NF-κB signaling [235, 236]. Although targeting the NF-κB signaling pathway has been extensively explored as a therapeutic strategy for various pathological conditions, including cancer, no selective NF-κB pathway inhibitors have been approved for clinical use to date, primarily due to potential toxicity associated with global inhibition of NF-κB signaling [237]. BCL-3 regulates transcription by interacting with NF-κB family proteins p50 and p52. Targeting the NF-κB pathway co-activator BCL-3 and its interacting partners holds potential for inhibiting NF-κB signaling. Despite the direct association of BCL-3 with tumor cell metastasis, its constitutive deletion in mouse models demonstrates good tolerance, making BCL-3 a promising therapeutic target [232].

12.5.3 BCL-3/P50 Inhibitors

To date, only one small molecule BCL-3/p50 inhibitor has been reported. The team led by Jitka Soukupová employed virtual drug design and screening methods to model the interaction between BCL-3 and NF-κB1 (p50), identifying a unique target on BCL-3. The interactions include residues from p50 (Lys 275, Asp 297, Ser 299, Pro 300, Thr 301, Asp 302, Val 303, His 304, and Arg 305) and residues from BCL-3's anchor protein repeats 6 (Met 298, Tyr 299, Ser 300, Gly 301, Ser 302, Ser 303, His 306, and Ser 307) and unique anchor protein repeat 7 (Asn 331, Cys 332, His 333, Asn 334, Asp 335, Val 340, and Arg 342). The screening results identified a compound, JS6, which fits well into the BCL-3 binding pocket in a manner highly similar to p50 residues. JS6 demonstrated significant intracellular BCL-3 inhibition activity, reducing BCL-3-NF-κB1 binding, tumor colony formation, and cancer cell migration in vitro while showing tumor stasis and anti-metastatic activity in vivo without notable systemic toxicity. Although these findings might reflect the broader effects of the NF-κB signaling pathway, further investigation is needed to ascertain whether BCL-3 inhibition primarily involves the classical or non-classical NF-κB pathway or if BCL-3 has alternative functions. Overall, these results underscore the effectiveness of computational screening in identifying novel intracellular PPI inhibitors and confirming BCL-3 as a promising anti-metastatic target [238].

12.6 BCL-10 Inhibitors

12.6.1 Biological Functions of BCL-10

The B-cell CLL-lymphoma 10 (BCL-10) gene was identified in 1999 by Willis et al. from the MALT1 (mucosa-associated lymphoid tissue lymphoma translocation gene 1) *t*(1;14)(p22;q32) chromosomal translocation breakpoint 1p22 clone. This gene encodes a cytoplasmic protein consisting of 233 amino acid residues with an approximate molecular weight of 32 kDa, and is widely expressed across various human tissues [239]. BCL-10 homologs include c-E10, Ciper, CLAP, or Carmen. BCL-10 significantly induces NF-κB activation through the oligomerization of its CARD (caspase recruitment domain) structure. BCL-10 can interact via CARD-CARD interactions with scaffold proteins containing CARD domains, such as CARMA1 (CARD11), CARMA2 (CARD14), CARMA3 (CARD10), and CARD9, forming the CARMA-BCL-10-MALT1 (CBM) complex (Figure 12.5c) [240, 241]. Within the organism, the CBM complex acts as a crucial regulator of the immune system, mediating NF-κB and mitogen-activated protein kinase (MAPK) signaling pathways. NF-κB signaling plays a pivotal role in both innate and adaptive immune regulation [242]. In lymphocytes, the CBM complex modulates T-cell receptor (TCR) and B-cell receptor (BCR) signaling, thereby regulating lymphocyte activation and adaptive immunity [243–245]. Additionally, the CBM complex participates in various signaling pathways in immune cells, such as BAFF and TLR4 signaling pathways in B cells [246, 247], and NK1.1, Ly49H, NKG2D, and Ly49D signaling pathways in natural killer (NK) cells [248, 249].

12.6.2 Structural Characteristics of BCL-10

BCL-10 serves as a core adaptor protein in the CBM (CARD-BCL-10-MALT1) signaling complex. The N-terminal of BCL-10 contains a caspase recruitment domain (CARD), which mediates interactions with CARD domains of CARD9, CARD10, CARD11, CARD14, and the CARD domain of BCL-10 itself [239, 250–252]. The C-terminal of BCL-10 is rich in serine/threonine residues, which are characteristically phosphorylated and further facilitate binding with MALT1 [253]. The core domain of BCL-10 is the CARD, a highly conserved protein structural domain typically composed of six α-helices. Specific amino acid residues between the α2 and α3 helices form hydrogen bonds, stabilizing the overall protein conformation. Hydrophobic amino acids between the α1 and α4 helices interact through hydrophobic interactions, maintaining the stability of the CARD domain. Additionally, charged amino acid residues between the α5 and α6 helices interact through salt bridges, further enhancing structural stability (Figure 12.5d). The CARD-CARD interactions of BCL-10 can be categorized into three types: Type I inter-chain contacts, Type II inter-chain contacts, and Type III intra-chain contacts. These interactions collectively maintain the functional and structural integrity of the BCL-10 protein [254]. In the interaction between BCL-10 and MALT1, BCL-10 associates with the death domain (DD) of MALT1 via its CARD domain. Specifically, the BCL-10-MALT1 interaction site I (BM-I) indicates that the core interaction region of MALT1 binds to the C-terminal portion of BCL-10, particularly the α6 helix of BCL-10. Within BM-I, two critical interaction residues of MALT1, V81 and L82, are located in the α4 helix. The corresponding interface on BCL-10 is comprised of a hydrophobic region formed by residues L104 and V103 in the α6 helix, along with V83 and I96 in the α5 helix. The stability of this hydrophobic core is further reinforced by multiple hydrogen bonds and salt bridges, thereby enhancing the overall stability of the BCL-10-MALT1 interaction (Figure 12.5e) [255].

12.6.3 BCL-10 and Cancer

Gene abnormalities in BCL-10, such as IGH/BCL-10 fusion, result in BCL-10 overexpression and activation, which in turn leads to abnormal proliferation and antiapoptotic behavior of B cells, ultimately contributing to lymphomagenesis. Following DNA damage, BCL-10 can assemble the CBM complex, inhibiting apoptosis and promoting cell survival, thereby contributing to the development of tumor resistance [256]. BCL-10 plays a crucial role in various tumor types, including ovarian cancer, head and neck squamous cell carcinoma, cervical cancer, breast cancer, and clear cell renal carcinoma. Abnormal expression of BCL-10 is frequently associated with enhanced cancer cell growth and invasiveness [253, 257–260]. BCL-10 not only regulates apoptosis but is also closely related to the progression of MALT lymphoma [261]. Studies both in vivo and in vitro have demonstrated that, in breast cancer, the CBM complex is activated by HER2, leading to the activation of the NF-κB signaling pathway, which promotes tumor malignancy by increasing tumor migration, invasion, and resistance to apoptosis [262]. Research by Sung-Hsin Kuo and colleagues has shown that BCL-10 accelerates the cell cycle and promotes cervical cancer cell growth by modulating the NF-κB-dependent expression of Cyclin D1 [263]. Additionally, BCL-10 is involved in the ubiquitination of histones at DNA damage sites, providing a platform for the assembly of repair factors such as BRCA1, RAP, and MRN, and facilitating homologous recombination repair of double-strand breaks, thus supporting the survival of damaged cells [264]. Extensive research indicates that BCL-10 promotes tumor cell proliferation, invasion, and angiogenesis, thereby driving tumor progression in various cancers, including ovarian cancer, NSCLC, head and neck squamous cell carcinoma, hepatocellular carcinoma, clear cell renal carcinoma, and colorectal cancer [264–270]. Research indicates that disrupting the CARMA1-BCL-10-MALT1 (CBM) signaling complex leads to the production of IFN-γ by most tumor-infiltrating Treg cells, subsequently resulting in delayed tumor growth. In the pool of autoreactive Tregs, partial disruption of the CBM complex and induction of IFN-γ secretion do not cause systemic autoimmunity but are sufficient to initiate a successful immune checkpoint therapy in the tumor microenvironment. Therefore, targeting the CBM complex could potentially lay the groundwork for successful immune checkpoint therapy. However, to date, no small-molecule inhibitors specifically targeting the CBM complex have been developed [271, 272].

12.7 Summary

Over the past two decades, drug development targeting PPIs has undergone significant advancements, particularly in the realm of BCL proteins, enhancing the prospects for cancer treatment. X-ray crystallography and NMR spectroscopy have played crucial roles in elucidating the interactions between small molecules and BCL proteins. These techniques have enabled researchers to gain detailed insights into how small molecules bind to target proteins, facilitating the design of more effective inhibitors. However, traditional high-throughput virtual screening methods have faced numerous challenges in addressing the complex and dynamic nature of PPI interfaces, limiting the accuracy of predicting and optimizing small-molecule binding. To overcome these challenges, researchers have turned to fragment-based drug discovery strategies. This approach utilizes low-affinity fragments as starting points, providing a foundation for designing high-affinity inhibitors through preliminary binding with target proteins. X-ray crystallography, NMR spectroscopy, and molecular docking have been instrumental in this process, offering detailed structural information about the binding of low-affinity fragments to target proteins. Such data, combined with chemical modifications and structural optimization of the fragments, have led to the development of several compounds with notable biological activity.

To further enhance the success rate of BCL protein drug development, researchers have employed innovative strategies. For instance, the "constrained macrocyclic conformation" method improves binding affinity by stabilizing the spatial structure of fragments; the "guided optimization" technique enhances specificity by optimizing fragment binding modes; the "scaffold hopping" strategy broadens binding possibilities by modifying fragment scaffolds; and the "fragment merging" approach generates higher-affinity compounds by combining multiple fragments. These strategies have not only achieved success in BCL protein drug development but also provided valuable insights and experiences for developing other challenging drug targets. In summary, the integration of X-ray crystallography, NMR spectroscopy, and innovative drug development strategies has led to significant progress in developing PPI inhibitors. These methods and approaches have not only advanced BCL protein drug discovery but also offer important guidance for the development of other drug targets.

References

1 Ideker, T. and Sharan, R. (2008). Protein networks in disease. *Genome Research* 18 (4): 644–652.

2 Luck, K., Kim, D.K., Lambourne, L. et al. (2020). A reference map of the human binary protein interactome. *Nature* 580 (7803): 402–408.

3 Vidal, M., Cusick, M.E., and Barabási, A.L. (2011). Interactome networks and human disease. *Cell* 144 (6): 986–998.

4 Zeng, X.Y., Xie, H., Yuan, J. et al. (2019). M2-like tumor-associated macrophages-secreted EGF promotes epithelial ovarian cancer metastasis via activating EGFR-ERK signaling and suppressing lncRNA LIMT expression. *Cancer Biology & Therapy* 20 (7): 956–966.

5 Ni, D., Lu, S., and Zhang, J. (2019). Emerging roles of allosteric modulators in the regulation of protein-protein interactions (PPIs): a new paradigm for PPI drug discovery. *Medicinal Research Reviews* 39 (6): 2314–2342.

6 Ran, X. and Gestwicki, J.E. (2018). Inhibitors of protein-protein interactions (PPIs): an analysis of scaffold choices and buried surface area. *Current Opinion in Chemical Biology* 44: 75–86.

7 Cowman, S.J., Fuja, D.G., Liu, X.D. et al. (2020). Macrophage HIF-1α is an independent prognostic indicator in kidney cancer. *Clinical Cancer Research* 26 (18): 4970–4982.

8 Arkin, M.R. and Wells, J.A. (2004). Small-molecule inhibitors of protein-protein interactions: progressing towards the dream. *Nature Reviews. Drug Discovery* 3 (4): 301–317.

9 London, N., Miller, R.M., Krishnan, S. et al. (2014). Covalent docking of large libraries for the discovery of chemical probes. *Nature Chemical Biology* 10 (12): 1066–1072.

10 Wu, D., Li, Y., Zheng, L. et al. (2023). Small molecules targeting protein-protein interactions for cancer therapy. *Acta Pharmaceutica Sinica B* 13 (10): 4060–4088.

11 Youle, R.J. and Strasser, A. (2008). The BCL-2 protein family: opposing activities that mediate cell death. *Nature Reviews. Molecular Cell Biology* 9 (1): 47–59.

12 Rautureau, G.J., Yabal, M., Yang, H. et al. (2012). The restricted binding repertoire of Bcl-B leaves Bim as the universal BH3-only prosurvival Bcl-2 protein antagonist. *Cell Death & Disease* 3 (12): e443.

13 Rautureau, G.J., Day, C.L., and Hinds, M.G. (2010). The structure of Boo/Diva reveals a divergent Bcl-2 protein. *Proteins* 78 (9): 2181–2186.

14 Zhai, D., Ke, N., Zhang, H. et al. (2003). Characterization of the anti-apoptotic mechanism of Bcl-B. *The Biochemical Journal* 376 (Pt 1): 229–236.

15 Huang, D.C. and Strasser, A. (2000). BH3-only proteins-essential initiators of apoptotic cell death. *Cell* 103 (6): 839–842.

16 Sattler, M., Liang, H., Nettesheim, D. et al. (1997). Structure of Bcl-xL-Bak peptide complex: recognition between regulators of apoptosis. *Science* 275 (5302): 983–986.

17 Muchmore, S.W., Sattler, M., Liang, H. et al. (1996). X-ray and NMR structure of human Bcl-xL, an inhibitor of programmed cell death. *Nature* 381 (6580): 335–341.

18 Petros, A.M., Medek, A., Nettesheim, D.G. et al. (2001). Solution structure of the antiapoptotic protein bcl-2. *Proceedings of the National Academy of Sciences of the United States of America* 98 (6): 3012–3017.

19 Day, C.L., Chen, L., Richardson, S.J. et al. (2005). Solution structure of prosurvival Mcl-1 and characterization of its binding by proapoptotic BH3-only ligands. *The Journal of Biological Chemistry* 280 (6): 4738–4744.

20 Hinds, M.G., Lackmann, M., Skea, G.L. et al. (2003). The structure of Bcl-w reveals a role for the C-terminal residues in modulating biological activity. *The EMBO Journal* 22 (7): 1497–1507.

21 Harvey, E.P., Seo, H.S., Guerra, R.M. et al. (2018). Crystal structures of anti-apoptotic BFL-1 and its complex with a covalent stapled peptide inhibitor. *Structure* 26 (1): 153–60.e4.

22 Zheng, J.H., Grace, C.R., Guibao, C.D. et al. (2018). Intrinsic instability of BOK enables membrane permeabilization in apoptosis. *Cell Reports* 23 (7): 2083–94.e6.

23 Garner, T.P., Reyna, D.E., Priyadarshi, A. et al. (2016). An autoinhibited dimeric form of BAX regulates the BAX activation pathway. *Molecular Cell* 63 (3): 485–497.

24 Wang, H., Takemoto, C., Akasaka, R. et al. (2009). Novel dimerization mode of the human Bcl-2 family protein Bak, a mitochondrial apoptosis regulator. *Journal of Structural Biology* 166 (1): 32–37.

25 Fire, E., Gullá, S.V., Grant, R.A. et al. (2010). Mcl-1-Bim complexes accommodate surprising point mutations via minor structural changes. *Protein Science* 19 (3): 507–519.

26 Liu, X., Dai, S., Zhu, Y. et al. (2003). The structure of a Bcl-xL/Bim fragment complex: implications for Bim function. *Immunity* 19 (3): 341–352.

27 Ku, B., Liang, C., Jung, J.U. et al. (2011). Evidence that inhibition of BAX activation by BCL-2 involves its tight and preferential interaction with the BH3 domain of BAX. *Cell Research* 21 (4): 627–641.

28 Rajan, S., Choi, M., Baek, K. et al. (2015). Bh3 induced conformational changes in Bcl-xl revealed by crystal structure and comparative analysis. *Proteins* 83 (7): 1262–1272.

29 Kim, J.H., Lee, H., Shin, E.A. et al. (2017). Implications of Bcl-2 and its interplay with other molecules and signaling pathways in prostate cancer progression. *Expert Opinion on Therapeutic Targets* 21 (9): 911–920.

30 Ke, W., Zhao, X., and Lu, Z. (2021). Foeniculum vulgare seed extract induces apoptosis in lung cancer cells partly through the down-regulation of Bcl-2. *Biomedicine & Pharmacotherapy* 135: 111213.

31 Gross, A. and Katz, S.G. (2017). Non-apoptotic functions of BCL-2 family proteins. *Cell Death and Differentiation* 24 (8): 1348–1358.

32 Khan, N. and Kahl, B. (2018). Targeting BCL-2 in hematologic malignancies. *Targeted Oncology* 13 (3): 257–267.

33 Adams, J.M. and Cory, S. (2007). The Bcl-2 apoptotic switch in cancer development and therapy. *Oncogene* 26 (9): 1324–1337.

34 Delbridge, A.R. and Strasser, A. (2015). The BCL-2 protein family, BH3-mimetics and cancer therapy. *Cell Death and Differentiation* 22 (7): 1071–1080.

35 Yoshino, T., Shiina, H., Urakami, S. et al. (2006). Bcl-2 expression as a predictive marker of hormone-refractory prostate cancer treated with taxane-based chemotherapy. *Clinical Cancer Research* 12 (20 Pt 1): 6116–6124.

36 Ramesh, P., Lannagan, T.R.M., Jackstadt, R. et al. (2021). BCL-XL is crucial for progression through the adenoma-to-carcinoma sequence of colorectal cancer. *Cell Death and Differentiation* 28 (12): 3282–3296.

37 Yang, J., Sun, M., Zhang, A. et al. (2011). Adenovirus-mediated siRNA targeting Bcl-xL inhibits proliferation, reduces invasion and enhances radiosensitivity of human colorectal cancer cells. *World Journal of Surgical Oncology* 9: 117.

38 D'aguanno, S. and Del Bufalo, D. (2020). Inhibition of anti-apoptotic Bcl-2 proteins in preclinical and clinical studies: current overview in cancer. *Cells* 9 (5): 1287.

39 Wei, A.H., Roberts, A.W., Spencer, A. et al. (2020). Targeting MCL-1 in hematologic malignancies: rationale and progress. *Blood Reviews* 44: 100672.

40 Moujalled, D.M., Pomilio, G., Ghiurau, C. et al. (2019). Combining BH3-mimetics to target both BCL-2 and MCL1 has potent activity in pre-clinical models of acute myeloid leukemia. *Leukemia* 33 (4): 905–917.

41 Cruz-Muñoz, W., Jaramillo, M.L., Man, S. et al. (2012). Roles for endothelin receptor B and BCL2A1 in spontaneous CNS metastasis of melanoma. *Cancer Research* 72 (19): 4909–4919.

42 Kim, E.S., Choi, J.Y., Hwang, S.J. et al. (2019). Hypermethylation of miR-205-5p by IR governs aggressiveness and metastasis via regulating Bcl-w and Src. *Molecular Therapy - Nucleic Acids* 14: 450–464.

43 Ding, W., Ren, J., Ren, H. et al. (2017). Long noncoding RNA HOTAIR modulates MiR-206-mediated Bcl-w Signaling to facilitate cell proliferation in breast cancer. *Scientific Reports* 7 (1): 17261.

44 Ebrahim, A.S., Sabbagh, H., Liddane, A. et al. (2016). Hematologic malignancies: newer strategies to counter the BCL-2 protein. *Journal of Cancer Research and Clinical Oncology* 142 (9): 2013–2022.

45 Wang, S., Yang, D., and Lippman, M.E. (2003). Targeting Bcl-2 and Bcl-xL with nonpeptidic small-molecule antagonists. *Seminars in Oncology* 30 (5 Suppl 16): 133–142.

46 Gardner, C.R. (2004). Anticancer drug development based on modulation of the Bcl-2 family core apoptosis mechanism. *Expert Review of Anticancer Therapy* 4 (6): 1157–1177.

47 Ciardiello, F. and Tortora, G. (2002). Inhibition of bcl-2 as cancer therapy. *Annals of Oncology* 13 (4): 501–502.

48 Nakashima, T., Miura, M., and Hara, M. (2000). Tetrocarcin a inhibits mitochondrial functions of Bcl-2 and suppresses its anti-apoptotic activity. *Cancer Research* 60 (5): 1229–1235.

49 Wang, J.L., Liu, D., Zhang, Z.J. et al. (2000). Structure-based discovery of an organic compound that binds Bcl-2 protein and induces apoptosis of tumor cells. *Proceedings of the National Academy of Sciences of the United States of America* 97 (13): 7124–7129.

50 Degterev, A., Lugovskoy, A., Cardone, M. et al. (2001). Identification of small-molecule inhibitors of interaction between the BH3 domain and Bcl-xL. *Nature Cell Biology* 3 (2): 173–182.

51 Tzung, S.P., Kim, K.M., Basañez, G. et al. (2001). Antimycin a mimics a cell-death-inducing Bcl-2 homology domain 3. *Nature Cell Biology* 3 (2): 183–191.

52 Zhai, D., Jin, C., Satterthwait, A.C. et al. (2006). Comparison of chemical inhibitors of antiapoptotic Bcl-2-family proteins. *Cell Death and Differentiation* 13 (8): 1419–1421.

53 Enyedy, J. I, Ling, Y., Nacro, K. et al. (2001). Discovery of small-molecule inhibitors of Bcl-2 through structure-based computer screening. *Journal of Medicinal Chemistry* 44 (25): 4313–4324.

54 Real, P.J., Cao, Y., Wang, R. et al. (2004). Breast cancer cells can evade apoptosis-mediated selective killing by a novel small molecule inhibitor of Bcl-2. *Cancer Research* 64 (21): 7947–7953.

55 Nguyen, M., Marcellus, R.C., Roulston, A. et al. (2007). Small molecule obatoclax (GX15-070) antagonizes MCL-1 and overcomes MCL-1-mediated resistance to apoptosis. *Proceedings of the National Academy of Sciences of the United States of America* 104 (49): 19512–19517.

56 Vartak, S.V., Iyer, D., Santhoshkumar, T.R. et al. (2017). Novel BCL2 inhibitor, Disarib induces apoptosis by disruption of BCL2-BAK interaction. *Biochemical Pharmacology* 131: 16–28.

57 Oltersdorf, T., Elmore, S.W., Shoemaker, A.R. et al. (2005). An inhibitor of Bcl-2 family proteins induces regression of solid tumours. *Nature* 435 (7042): 677–681.

58 Tse, C., Shoemaker, A.R., Adickes, J. et al. (2008). ABT-263: a potent and orally bioavailable Bcl-2 family inhibitor. *Cancer Research* 68 (9): 3421–3428.

59 Wilson, W.H., O'Connor, O.A., Czuczman, M.S. et al. (2010). Navitoclax, a targeted high-affinity inhibitor of BCL-2, in lymphoid malignancies: a phase 1 dose-escalation study of safety, pharmacokinetics, pharmacodynamics, and antitumour activity. *The Lancet Oncology* 11 (12): 1149–1159.

60 Gandhi, L., Camidge, D.R., Ribeiro de Oliveira, M. et al. (2011). Phase I study of Navitoclax (ABT-263), a novel Bcl-2 family inhibitor, in patients with small-cell lung cancer and other solid tumors. *Journal of Clinical Oncology* 29 (7): 909–916.

61 Huang, S., Connolly, P.J., Lin, R. et al. (2006). Synthesis and evaluation of *N*-acyl sulfonamides as potential prodrugs of cyclin-dependent kinase inhibitor JNJ-7706621. *Bioorganic & Medicinal Chemistry Letters* 16 (14): 3639–3641.

62 Sleebs, B.E., Czabotar, P.E., Fairbrother, W.J. et al. (2011). Quinazoline sulfonamides as dual binders of the proteins B-cell lymphoma 2 and B-cell lymphoma extra long with potent proapoptotic cell-based activity. *Journal of Medicinal Chemistry* 54 (6): 1914–1926.

63 Touré, B.B., Miller-Moslin, K., Yusuff, N. et al. (2013). The role of the acidity of *N*-heteroaryl sulfonamides as inhibitors of bcl-2 family protein-protein interactions. *ACS Medicinal Chemistry Letters* 4 (2): 186–190.

64 Loriot, Y., Mordant, P., Dugue, D. et al. (2014). Radiosensitization by a novel Bcl-2 and Bcl-xL inhibitor S44563 in small-cell lung cancer. *Cell Death & Disease* 5 (9): e1423.

65 Némati, F., De Montrion, C., Lang, G. et al. (2014). Targeting Bcl-2/Bcl-xL induces antitumor activity in uveal melanoma patient-derived xenografts. *PLoS One* 9 (1): e80836.

66 Souers, A.J., Leverson, J.D., Boghaert, E.R. et al. (2013). ABT-199, a potent and selective BCL-2 inhibitor, achieves antitumor activity while sparing platelets. *Nature Medicine* 19 (2): 202–208.

67 Birkinshaw, R.W., Gong, J.N., Luo, C.S. et al. (2019). Structures of BCL-2 in complex with venetoclax reveal the molecular basis of resistance mutations. *Nature Communications* 10 (1): 2385.

68 Hafezi, S. and Rahmani, M. (2021). Targeting BCL-2 in cancer: advances, challenges, and perspectives. *Cancers (Basel)* 13 (6): 1292.

69 Luo, Q., Pan, W., Zhou, S. et al. (2020). A novel BCL-2 inhibitor APG-2575 exerts synthetic lethality with BTK or MDM2-p53 inhibitor in diffuse large B-cell lymphoma. *Oncology Research* 28 (4): 331–344.

70 Liu, J., Li, S., Wang, Q. et al. (2024). Sonrotoclax overcomes BCL2 G101V mutation-induced venetoclax resistance in preclinical models of hematologic malignancy. *Blood* 143 (18): 1825–1836.

71 Zhou, H., Chen, J., Meagher, J.L. et al. (2012). Design of Bcl-2 and Bcl-xL inhibitors with subnanomolar binding affinities based upon a new scaffold. *Journal of Medicinal Chemistry* 55 (10): 4664–4682.

72 Chen, J., Zhou, H., Aguilar, A. et al. (2012). Structure-based discovery of BM-957 as a potent small-molecule inhibitor of Bcl-2 and Bcl-xL capable of achieving complete tumor regression. *Journal of Medicinal Chemistry* 55 (19): 8502–8514.

73 Aguilar, A., Zhou, H., Chen, J. et al. (2013). A potent and highly efficacious Bcl-2/Bcl-xL inhibitor. *Journal of Medicinal Chemistry* 56 (7): 3048–3067.

74 Bai, L., Chen, J., Mceachern, D. et al. (2014). BM-1197: a novel and specific Bcl-2/Bcl-xL inhibitor inducing complete and long-lasting tumor regression in vivo. *PLoS One* 9 (6): e99404.

75 Sun, Y.L., Jiang, W.Q., Luo, Q.Y. et al. (2019). A novel Bcl-2 inhibitor, BM-1197, induces apoptosis in malignant lymphoma cells through the endogenous apoptotic pathway. *BMC Cancer* 20 (1): 1.

76 Bai, L., Chen, J., Liu, L. et al. (2014). 338 BM-1252 (APG-1252): a potent dual specific Bcl-2/Bcl-xL inhibitor that achieves complete tumor regression with minimal platelet toxicity. *European Journal of Cancer* 50: 109–110.

77 Wang, J., Yang, D., Luo, Q. et al. (2017). APG-1252-12A induces mitochondria-dependent apoptosis through inhibiting the antiapoptotic proteins Bcl-2/Bcl-xL in HL-60 cells. *International Journal of Oncology* 51 (2): 563–572.

78 Di, T., Luo, Q.Y., Song, J.T. et al. (2024). APG-1252 combined with Cabozantinib inhibits hepatocellular carcinoma by suppressing MEK/ERK and CREB/Bcl-xL pathways. *International Immunopharmacology* 139: 112615.

79 Yi, H., Qiu, M.Z., Yuan, L. et al. (2020). Bcl-2/Bcl-xL inhibitor APG-1252-M1 is a promising therapeutic strategy for gastric carcinoma. *Cancer Medicine* 9 (12): 4197–4206.

80 Balachander, S.B., Criscione, S.W., Byth, K.F. et al. (2020). AZD4320, a dual inhibitor of Bcl-2 and Bcl-x(L), induces tumor regression in hematologic cancer models without dose-limiting thrombocytopenia. *Clinical Cancer Research* 26 (24): 6535–6549.

81 Patterson, C.M., Balachander, S.B., Grant, I. et al. (2021). Design and optimisation of dendrimer-conjugated Bcl-2/x(L) inhibitor, AZD0466, with improved therapeutic index for cancer therapy. *Communications Biology* 4 (1): 112.

82 Porter, J., Payne, A., de Candole, B. et al. (2009). Tetrahydroisoquinoline amide substituted phenyl pyrazoles as selective Bcl-2 inhibitors. *Bioorganic & Medicinal Chemistry Letters* 19 (1): 230–233.

83 Casara, P., Davidson, J., Claperon, A. et al. (2018). S55746 is a novel orally active BCL-2 selective and potent inhibitor that impairs hematological tumor growth. *Oncotarget* 9 (28): 20075–20088.

84 Lessene, G., Czabotar, P.E., Sleebs, B.E. et al. (2013). Structure-guided design of a selective BCL-X(L) inhibitor. *Nature Chemical Biology* 9 (6): 390–397.

85 Tao, Z.F., Hasvold, L., Wang, L. et al. (2014). Discovery of a potent and selective BCL-XL inhibitor with in vivo activity. *ACS Medicinal Chemistry Letters* 5 (10): 1088–1093.

86 Koehler, M.F., Bergeron, P., Choo, E.F. et al. (2014). Structure-guided rescaffolding of selective antagonists of BCL-XL. *ACS Medicinal Chemistry Letters* 5 (6): 662–667.

87 Wang, L., Doherty, G.A., Judd, A.S. et al. (2020). Discovery of A-1331852, a first-in-class, potent, and orally-bioavailable BCL-X(L) inhibitor. *ACS Medicinal Chemistry Letters* 11 (10): 1829–1836.

88 Chiou, J.T., Wu, Y.Y., Lee, Y.C. et al. (2023). BCL2L1 inhibitor A-1331852 inhibits MCL1 transcription and triggers apoptosis in acute myeloid leukemia cells. *Biochemical Pharmacology* 215: 115738.

89 Tao, Z.F., Wang, X., Chen, J. et al. (2021). Structure-based design of A-1293102, a potent and selective BCL-X(L) inhibitor. *ACS Medicinal Chemistry Letters* 12 (6): 1011–1016.

90 Bose, P. and Grant, S. (2013). Mcl-1 as a therapeutic target in acute myelogenous Leukemia (AML). *Leukemia Research Reports* 2 (1): 12–14.

91 Konopleva, M., Contractor, R., Tsao, T. et al. (2006). Mechanisms of apoptosis sensitivity and resistance to the BH3 mimetic ABT-737 in acute myeloid leukemia. *Cancer Cell* 10 (5): 375–388.

92 Van Delft, M.F., Wei, A.H., Mason, K.D. et al. (2006). The BH3 mimetic ABT-737 targets selective Bcl-2 proteins and efficiently induces apoptosis via Bak/Bax if Mcl-1 is neutralized. *Cancer Cell* 10 (5): 389–399.

93 Tahir, S.K., Yang, X., Anderson, M.G. et al. (2007). Influence of Bcl-2 family members on the cellular response of small-cell lung cancer cell lines to ABT-737. *Cancer Research* 67 (3): 1176–1183.

94 Gores, G.J. and Kaufmann, S.H. (2012). Selectively targeting Mcl-1 for the treatment of acute myelogenous leukemia and solid tumors. *Genes & Development* 26 (4): 305–311.

95 Zhang, H., Guttikonda, S., Roberts, L. et al. (2011). Mcl-1 is critical for survival in a subgroup of non-small-cell lung cancer cell lines. *Oncogene* 30 (16): 1963–1968.

96 Lee, E.F., Czabotar, P.E., van Delft, M.F. et al. (2008). A novel BH3 ligand that selectively targets Mcl-1 reveals that apoptosis can proceed without Mcl-1 degradation. *The Journal of Cell Biology* 180 (2): 341–355.

97 Lee, E.F., Czabotar, P.E., Yang, H. et al. (2009). Conformational changes in Bcl-2 pro-survival proteins determine their capacity to bind ligands. *The Journal of Biological Chemistry* 284 (44): 30508–30517.

98 Chen, L. and Fletcher, S. (2017). Mcl-1 inhibitors: a patent review. *Expert Opinion on Therapeutic Patents* 27 (2): 163–178.

99 Wan, Y., Dai, N., Tang, Z. et al. (2018). Small-molecule Mcl-1 inhibitors: emerging anti-tumor agents. *European Journal of Medicinal Chemistry* 146: 471–482.

100 Hird, A.W. and Tron, A.E. (2019). Recent advances in the development of Mcl-1 inhibitors for cancer therapy. *Pharmacology & Therapeutics* 198: 59–67.

101 Negi, A. and Murphy, P.V. (2021). Development of Mcl-1 inhibitors for cancer therapy. *European Journal of Medicinal Chemistry* 210: 113038.

102 Verhaegen, M., Bauer, J.A., Martín de la Vega, C. et al. (2006). A novel BH3 mimetic reveals a mitogen-activated protein kinase-dependent mechanism of melanoma cell death controlled by p53 and reactive oxygen species. *Cancer Research* 66 (23): 11348–11359.

103 Mohammad, R.M., Goustin, A.S., Aboukameel, A. et al. (2007). Preclinical studies of TW-37, a new nonpeptidic small-molecule inhibitor of Bcl-2, in diffuse large cell lymphoma xenograft model reveal drug action on both Bcl-2 and Mcl-1. *Clinical Cancer Research* 13 (7): 2226–2235.

104 Cohen, N.A., Stewart, M.L., Gavathiotis, E. et al. (2012). A competitive stapled peptide screen identifies a selective small molecule that overcomes MCL-1-dependent leukemia cell survival. *Chemistry & Biology* 19 (9): 1175–1186.

105 Varadarajan, S., Vogler, M., Butterworth, M. et al. (2013). Evaluation and critical assessment of putative MCL-1 inhibitors. *Cell Death and Differentiation* 20 (11): 1475–1484.

106 Abulwerdi, F., Liao, C., Liu, M. et al. (2014). A novel small-molecule inhibitor of mcl-1 blocks pancreatic cancer growth in vitro and in vivo. *Molecular Cancer Therapeutics* 13 (3): 565–575.

107 Richard, D.J., Lena, R., Bannister, T. et al. (2013). Hydroxyquinoline-derived compounds and analoguing of selective Mcl-1 inhibitors using a functional biomarker. *Bioorganic & Medicinal Chemistry* 21 (21): 6642–6649.

108 Rauh, U., Kaushik, V., Pilari, S. et al. (2023). Abstract LB_A01: rapidly cleared MCL1 inhibitor TTX-810 showing robust efficacy in preclinical tumor models with no effects on cardiomyocytes in vitro and in vivo. *Molecular Cancer Therapeutics* 22 (12_Supplement): LB_A01-LB_A.

109 Bruncko, M., Wang, L., Sheppard, G.S. et al. (2015). Structure-guided design of a series of MCL-1 inhibitors with high affinity and selectivity. *Journal of Medicinal Chemistry* 58 (5): 2180–2194.

110 Papatzimas, J.W., Gorobets, E., Maity, R. et al. (2019). From inhibition to degradation: targeting the antiapoptotic protein myeloid cell leukemia 1 (MCL1). *Journal of Medicinal Chemistry* 62 (11): 5522–5540.

111 Yang, C., Chen, S., Zhou, M. et al. (2014). Development of 3-phenyl-*N*-(2-[3-phenylureido]ethyl)-thiophene-2-sulfonamide compounds as inhibitors of anti-apoptotic Bcl-2 family proteins. *ChemMedChem* 9 (7): 1436–1452.

112 Li, Y., Fan, W., Gong, Q. et al. (2021). Structure-based optimization of 3-phenyl-*N*-(2-[3-phenylureido]ethyl)thiophene-2-sulfonamide derivatives as selective Mcl-1 inhibitors. *Journal of Medicinal Chemistry* 64 (14): 10260–10285.

113 Gong, Q., Li, C., Wang, H. et al. (2024). Discovery of phenylpyrazole derivatives as a new class of selective inhibitors of MCL-1 with antitumor activity. *ACS Omega* 9 (25): 27369–27396.

114 Szlávik, Z., Ondi, L., Csékei, M. et al. (2019). Structure-guided discovery of a selective Mcl-1 inhibitor with cellular activity. *Journal of Medicinal Chemistry* 62 (15): 6913–6924.

115 Kotschy, A., Szlavik, Z., Murray, J. et al. (2016). The MCL1 inhibitor S63845 is tolerable and effective in diverse cancer models. *Nature* 538 (7626): 477–482.

116 Szlavik, Z., Csekei, M., Paczal, A. et al. (2020). Discovery of S64315, a potent and selective Mcl-1 inhibitor. *Journal of Medicinal Chemistry* 63 (22): 13762–13795.

117 Yuda, J., Will, C., Phillips, D.C. et al. (2023). Selective MCL-1 inhibitor ABBV-467 is efficacious in tumor models but is associated with cardiac troponin increases in patients. *Communications Medicine* 3 (1): 154.

118 Friberg, A., Vigil, D., Zhao, B. et al. (2013). Discovery of potent myeloid cell leukemia 1 (Mcl-1) inhibitors using fragment-based methods and structure-based design. *Journal of Medicinal Chemistry* 56 (1): 15–30.

119 Burke, J.P., Bian, Z., Shaw, S. et al. (2015). Discovery of tricyclic indoles that potently inhibit Mcl-1 using fragment-based methods and structure-based design. *Journal of Medicinal Chemistry* 58 (9): 3794–3805.

120 Pelz, N.F., Bian, Z., Zhao, B. et al. (2016). Discovery of 2-indole-acylsulfonamide myeloid cell leukemia 1 (Mcl-1) inhibitors using fragment-based methods. *Journal of Medicinal Chemistry* 59 (5): 2054–2066.

121 Shaw, S., Bian, Z., Zhao, B. et al. (2018). Optimization of potent and selective tricyclic indole diazepinone myeloid cell leukemia-1 inhibitors using structure-based design. *Journal of Medicinal Chemistry* 61 (6): 2410–2421.

122 Lee, T., Christov, P.P., Shaw, S. et al. (2019). Discovery of potent myeloid cell leukemia-1 (Mcl-1) inhibitors that demonstrate in vivo activity in mouse xenograft models of human cancer. *Journal of Medicinal Chemistry* 62 (8): 3971–3988.

123 Tarr, J.C., Salovich, J.M., Aichinger, M. et al. (2024). Discovery of a myeloid cell leukemia 1 (Mcl-1) inhibitor that demonstrates potent in vivo activities in mouse models of hematological and solid tumors. *Journal of Medicinal Chemistry* 67: 14370–14393.

124 Caenepeel, S., Brown, S.P., Belmontes, B. et al. (2018). AMG 176, a selective MCL1 inhibitor, is effective in hematologic cancer models alone and in combination with established therapies. *Cancer Discovery* 8 (12): 1582–1597.

125 Caenepeel, S., Karen, R., Belmontes, B. et al. (2020). Abstract 6218: discovery and preclinical evaluation of AMG 397, a potent, selective and orally bioavailable MCL1 inhibitor. *Cancer Research* 80 (16_Supplement): 6218.

126 Romanov-Michailidis, F., Hsiao, C.C., Urner, L.M. et al. (2023). Discovery of an oral, beyond-rule-of-five Mcl-1 protein-protein interaction modulator with the potential of treating hematological malignancies. *Journal of Medicinal Chemistry* 66 (9): 6122–6148.

127 Tron, A.E., Belmonte, M.A., Adam, A. et al. (2018). Discovery of Mcl-1-specific inhibitor AZD5991 and preclinical activity in multiple myeloma and acute myeloid leukemia. *Nature Communications* 9 (1): 5341.

128 Rombouts, F.J., William, L., Velter, I. et al. (2022). Abstract 2133: in pursuit of MCL-1 inhibitors with improved therapeutic window for the treatment of hematological malignancies: discovery of JNJ-4355. *Cancer Research* 82 (12_Supplement): 2133.

129 Demin, S., Peschiulli, A., Velter, A. I, et al. (2023). Macrocyclic carbon-linked pyrazoles as novel inhibitors of MCL-1. *ACS Medicinal Chemistry Letters* 14 (7): 955–961.

130 Willis, T.G., Zalcberg, R. I, Coignet, L.J. et al. (1998). Molecular cloning of translocation *t*(1;14)(q21;q32) defines a novel gene (BCL9) at chromosome 1q21. *Blood* 91 (6): 1873–1881.

131 Wang, J., Zheng, M., Zhu, L. et al. (2019). Low BCL9 expression inhibited ovarian epithelial malignant tumor progression by decreasing proliferation, migration, and increasing apoptosis to cancer cells. *Cancer Cell International* 19: 330.

132 Wu, M., Dong, H., Xu, C. et al. (2024). The Wnt-dependent and Wnt-independent functions of BCL9 in development, tumorigenesis, and immunity: implications in therapeutic opportunities. *Genes & Diseases* 11 (2): 701–710.

133 Liu, J., Xiao, Q., Xiao, J. et al. (2022). Wnt/β-catenin signalling: function, biological mechanisms, and therapeutic opportunities. *Signal Transduction and Targeted Therapy* 7 (1): 3.

134 Zhan, T., Rindtorff, N., and Boutros, M. (2017). Wnt signaling in cancer. *Oncogene* 36 (11): 1461–1473.

135 Anastas, J.N. and Moon, R.T. (2013). WNT signalling pathways as therapeutic targets in cancer. *Nature Reviews. Cancer* 13 (1): 11–26.

136 Wang, Z., Li, Z., and Ji, H. (2021). Direct targeting of β-catenin in the Wnt signaling pathway: current progress and perspectives. *Medicinal Research Reviews* 41 (4): 2109–2129.

137 Sampietro, J., Dahlberg, C.L., Cho, U.S. et al. (2006). Crystal structure of a beta-catenin/BCL9/Tcf4 complex. *Molecular Cell* 24 (2): 293–300.

138 Valenta, T., Hausmann, G., and Basler, K. (2012). The many faces and functions of β-catenin. *The EMBO Journal* 31 (12): 2714–2736.

139 Hoffmans, R. and Basler, K. (2004). Identification and in vivo role of the armadillo-legless interaction. *Development* 131 (17): 4393–4400.

140 De La Roche, M., Worm, J., and Bienz, M. (2008). The function of BCL9 in Wnt/beta-catenin signaling and colorectal cancer cells. *BMC Cancer* 8: 199.

141 Kawamoto, S.A., Thompson, A.D., Coleska, A. et al. (2009). Analysis of the interaction of BCL9 with beta-catenin and development of fluorescence polarization and surface plasmon resonance binding assays for this interaction. *Biochemistry* 48 (40): 9534–9541.

142 Hoggard, L.R., Zhang, Y., Zhang, M. et al. (2015). Rational design of selective small-molecule inhibitors for β-catenin/B-cell lymphoma 9 protein-protein interactions. *Journal of the American Chemical Society* 137 (38): 12249–12260.

143 De La Roche, M., Rutherford, T.J., Gupta, D. et al. (2012). An intrinsically labile α-helix abutting the BCL9-binding site of β-catenin is required for its inhibition by carnosic acid. *Nature Communications* 3: 680.

144 Schlauderer, F., Lammens, K., Nagel, D. et al. (2013). Structural analysis of phenothiazine derivatives as allosteric inhibitors of the MALT1 paracaspase. *Angewandte Chemie (International Ed. in English)* 52 (39): 10384–10387.

145 Shen, L.A., Peng, X., Bao, Y. et al. (2023). Design, synthesis and biological evaluation of quercetin derivatives as novel β-catenin/B-cell lymphoma 9 protein-protein interaction inhibitors. *European Journal of Medicinal Chemistry* 247: 115075.

146 Peng, X., Shen, L.A., Bao, Y. et al. (2024). Design, synthesis, and biological evaluation of novel 8-substituted quercetin derivatives targeting the β-catenin/B-cell lymphoma 9 interaction. *Bioorganic & Medicinal Chemistry Letters* 98: 129591.

147 Zhang, M., Wisniewski, J.A., and Ji, H. (2015). AlphaScreen selectivity assay for β-catenin/B-cell lymphoma 9 inhibitors. *Analytical Biochemistry* 469: 43–53.

148 Wang, Z., Zhang, M., Luo, W. et al. (2021). Discovery of 2-(3-(3-carbamoylpiperidin-1-yl)phenoxy)acetic acid derivatives as novel small-molecule inhibitors of the β-catenin/B-cell lymphoma 9 protein-protein interaction. *Journal of Medicinal Chemistry* 64 (9): 5886–5904.

149 Wang, Z., Zhang, M., Quereda, V. et al. (2021). Discovery of an orally bioavailable small-molecule inhibitor for the β-catenin/B-cell lymphoma 9 protein-protein interaction. *Journal of Medicinal Chemistry* 64 (16): 12109–12131.

150 Wang, Z., Zhang, M., Thompson, H.M. et al. (2022). New ZW4864 derivatives as small-molecule inhibitors for the β-catenin/BCL9 protein-protein interaction. *ACS Medicinal Chemistry Letters* 13 (5): 865–870.

151 Zhu, W., Liu, C., Xi, K. et al. (2024). Discovery of novel 1-phenylpiperidine urea-containing derivatives inhibiting β-catenin/BCL9 interaction and exerting antitumor efficacy through the activation of antigen presentation of cDC1 cells. *Journal of Medicinal Chemistry* 67 (15): 12485–12520.

152 Zhang, H., Liu, C., Chen, Q. et al. (2023). Discovery of novel 3-phenylpiperidine derivatives targeting the β-catenin/B-cell lymphoma 9 interaction as a single agent and in combination with the anti-PD-1 antibody for the treatment of colorectal cancer. *Journal of Medicinal Chemistry* 66 (2): 1349–1379.

153 Zhang, M., Wang, Z., Zhang, Y. et al. (2018). Structure-based optimization of small-molecule inhibitors for the β-catenin/B-cell lymphoma 9 protein-protein interaction. *Journal of Medicinal Chemistry* 61 (7): 2989–3007.
154 Wisniewski, J.A., Yin, J., Teuscher, K.B. et al. (2016). Structure-based design of 1,4-dibenzoylpiperazines as β-catenin/B-cell lymphoma 9 protein-protein interaction inhibitors. *ACS Medicinal Chemistry Letters* 7 (5): 508–513.
155 Li, Z., Zhang, M., Teuscher, K.B. et al. (2021). Discovery of 1-benzoyl 4-phenoxypiperidines as small-molecule inhibitors of the β-catenin/B-cell lymphoma 9 protein-protein interaction. *Journal of Medicinal Chemistry* 64 (15): 11195–11218.
156 Tanton, H., Sewastianik, T., Seo, H.S. et al. (2022). A novel β-catenin/BCL9 complex inhibitor blocks oncogenic Wnt signaling and disrupts cholesterol homeostasis in colorectal cancer. *Science Advances* 8 (17): eabm3108.
157 Cerchietti, L.C., Ghetu, A.F., Zhu, X. et al. (2010). A small-molecule inhibitor of BCL6 kills DLBCL cells in vitro and in vivo. *Cancer Cell* 17 (4): 400–411.
158 Cardenas, M.G., Yu, W., Beguelin, W. et al. (2016). Rationally designed BCL6 inhibitors target activated B cell diffuse large B cell lymphoma. *Journal of Clinical Investigation* 126 (9): 3351–3362.
159 Ranuncolo, S.M., Polo, J.M., and Melnick, A. (2008). BCL6 represses CHEK1 and suppresses DNA damage pathways in normal and malignant B-cells. *Blood Cells, Molecules & Diseases* 41 (1): 95–99.
160 Hatzi, K. and Melnick, A. (2014). Breaking bad in the germinal center: how deregulation of BCL6 contributes to lymphomagenesis. *Trends in Molecular Medicine* 20 (6): 343–352.
161 Schlager, S., Salomon, C., Olt, S. et al. (2020). Inducible knock-out of BCL6 in lymphoma cells results in tumor stasis. *Oncotarget* 11 (9): 875–890.
162 Toney, L.M., Cattoretti, G., Graf, J.A. et al. (2000). BCL-6 regulates chemokine gene transcription in macrophages. *Nature Immunology* 1 (3): 214–220.
163 Barish, G.D., Yu, R.T., Karunasiri, M.S. et al. (2012). The Bcl6-SMRT/NCoR cistrome represses inflammation to attenuate atherosclerosis. *Cell Metabolism* 15 (4): 554–562.
164 Ye, B.H., Cattoretti, G., Shen, Q. et al. (1997). The BCL-6 proto-oncogene controls germinal-centre formation and Th2-type inflammation. *Nature Genetics* 16 (2): 161–170.
165 Ye, B.H., Lista, F., Lo Coco, F. et al. (1993). Alterations of a zinc finger-encoding gene, BCL-6, in diffuse large-cell lymphoma. *Science* 262 (5134): 747–750.
166 Tahara, K., Takizawa, M., Yamane, A. et al. (2017). Overexpression of B-cell lymphoma 6 alters gene expression profile in a myeloma cell line and is associated with decreased DNA damage response. *Cancer Science* 108 (8): 1556–1564.
167 Cardenas, M.G., Oswald, E., Yu, W. et al. (2017). The expanding role of the BCL6 oncoprotein as a cancer therapeutic target. *Clinical Cancer Research* 23 (4): 885–893.
168 Basso, K. and Dalla-Favera, R. (2012). Roles of BCL6 in normal and transformed germinal center B cells. *Immunological Reviews* 247 (1): 172–183.

169 Onizuka, T., Moriyama, M., Yamochi, T. et al. (1995). BCL-6 gene product, a 92- to 98-kD nuclear phosphoprotein, is highly expressed in germinal center B cells and their neoplastic counterparts. *Blood* 86 (1): 28–37.

170 Gu, H., He, J., Li, Y. et al. (2022). B-cell lymphoma 6 inhibitors: current advances and prospects of drug development for diffuse large B-cell lymphomas. *Journal of Medicinal Chemistry* 65 (23): 15559–15583.

171 Ahmad, K.F., Melnick, A., Lax, S. et al. (2003). Mechanism of SMRT corepressor recruitment by the BCL6 BTB domain. *Molecular Cell* 12 (6): 1551–1564.

172 Wong, C.W. and Privalsky, M.L. (1998). Components of the SMRT corepressor complex exhibit distinctive interactions with the POZ domain oncoproteins PLZF, PLZF-RARalpha, and BCL-6. *The Journal of Biological Chemistry* 273 (42): 27695–27702.

173 Lemercier, C., Brocard, M.P., Puvion-Dutilleul, F. et al. (2002). Class II histone deacetylases are directly recruited by BCL6 transcriptional repressor. *The Journal of Biological Chemistry* 277 (24): 22045–22052.

174 Dhordain, P., Albagli, O., Lin, R.J. et al. (1997). Corepressor SMRT binds the BTB/POZ repressing domain of the LAZ3/BCL6 oncoprotein. *Proceedings of the National Academy of Sciences of the United States of America* 94 (20): 10762–10767.

175 Huynh, K.D. and Bardwell, V.J. (1998). The BCL-6 POZ domain and other POZ domains interact with the co-repressors N-CoR and SMRT. *Oncogene* 17 (19): 2473–2484.

176 Huynh, K.D., Fischle, W., Verdin, E. et al. (2000). BCoR, a novel corepressor involved in BCL-6 repression. *Genes & Development* 14 (14): 1810–1823.

177 Ghetu, A.F., Corcoran, C.M., Cerchietti, L. et al. (2008). Structure of a BCOR corepressor peptide in complex with the BCL6 BTB domain dimer. *Molecular Cell* 29 (3): 384–391.

178 Melnick, A., Carlile, G., Ahmad, K.F. et al. (2002). Critical residues within the BTB domain of PLZF and Bcl-6 modulate interaction with corepressors. *Molecular and Cellular Biology* 22 (6): 1804–1818.

179 Zacharchenko, T., Kalverda, A.P., and Wright, S.C. (2022). Structural basis of Apt48 inhibition of the BCL6 BTB domain. *Structure* 30 (3): 396–407.e3.

180 Cerchietti, L.C., Yang, S.N., Shaknovich, R. et al. (2009). A peptomimetic inhibitor of BCL6 with potent antilymphoma effects in vitro and in vivo. *Blood* 113 (15): 3397–3405.

181 Béguelin, W., Teater, M., Gearhart, M.D. et al. (2016). EZH2 and BCL6 cooperate to assemble CBX8-BCOR complex to repress bivalent promoters, mediate germinal Center formation and lymphomagenesis. *Cancer Cell* 30 (2): 197–213.

182 Wells, J.A. and Mcclendon, C.L. (2007). Reaching for high-hanging fruit in drug discovery at protein–protein interfaces. *Nature* 450 (7172): 1001–1009.

183 Fujita, N., Jaye, D.L., Geigerman, C. et al. (2004). MTA3 and the Mi-2/NuRD complex regulate cell fate during B lymphocyte differentiation. *Cell* 119 (1): 75–86.

184 Wu, M., Xie, J., Xing, Y. et al. (2024). Selectively targeting BCL6 using a small molecule inhibitor is a potential therapeutic strategy for ovarian cancer. *International Journal of Biological Sciences* 20 (2): 486–501.

185 Duy, C., Hurtz, C., Shojaee, S. et al. (2011). BCL6 enables Ph+ acute lymphoblastic leukaemia cells to survive BCR-ABL1 kinase inhibition. *Nature* 473 (7347): 384–388.

186 Alizadeh, A.A., Eisen, M.B., Davis, R.E. et al. (2000). Distinct types of diffuse large B-cell lymphoma identified by gene expression profiling. *Nature* 403 (6769): 503–511.

187 Walker, S.R., Liu, S., Xiang, M. et al. (2015). The transcriptional modulator BCL6 as a molecular target for breast cancer therapy. *Oncogene* 34 (9): 1073–1082.

188 Hideshima, T., Mitsiades, C., Ikeda, H. et al. (2010). A proto-oncogene BCL6 is up-regulated in the bone marrow microenvironment in multiple myeloma cells. *Blood* 115 (18): 3772–3775.

189 Xia, Y., Jin, S., and Wu, Y. (2023). Small-molecule BCL6 inhibitor protects chronic cardiac transplant rejection and inhibits T follicular helper cell expansion and humoral response. *Frontiers in Pharmacology* 14.

190 Cai, Y., Poli, A.N.R., Vadrevu, S. et al. (2021). BCL6 BTB-specific inhibitor reversely represses T-cell activation, Tfh cells differentiation, and germinal center reaction in vivo. *European Journal of Immunology* 51 (10): 2441–2451.

191 Kamada, Y., Sakai, N., Sogabe, S. et al. (2017). Discovery of a B-cell lymphoma 6 protein-protein interaction inhibitor by a biophysics-driven fragment-based approach. *Journal of Medicinal Chemistry* 60 (10): 4358–4368.

192 Kerres, N., Steurer, S., Schlager, S. et al. (2017). Chemically induced degradation of the oncogenic transcription factor BCL6. *Cell Reports* 20 (12): 2860–2875.

193 Teng, M., Ficarro, S.B., Yoon, H. et al. (2020). Rationally designed covalent BCL6 inhibitor that targets a tyrosine residue in the homodimer Interface. *ACS Medicinal Chemistry Letters* 11 (6): 1269–1273.

194 Yasui, T., Yamamoto, T., Sakai, N. et al. (2017). Discovery of a novel B-cell lymphoma 6 (BCL6)–corepressor interaction inhibitor by utilizing structure-based drug design. *Bioorganic & Medicinal Chemistry* 25 (17): 4876–4886.

195 Sameshima, T., Yamamoto, T., Sano, O. et al. (2018). Discovery of an irreversible and cell-active BCL6 inhibitor selectively targeting Cys53 located at the protein–protein interaction interface. *Biochemistry* 57 (8): 1369–1379.

196 Mccoull, W., Abrams, R.D., Anderson, E. et al. (2017). Discovery of pyrazolo[1,5-a]pyrimidine B-cell lymphoma 6 (BCL6) binders and optimization to high affinity macrocyclic inhibitors. *Journal of Medicinal Chemistry* 60 (10): 4386–4402.

197 Cheng, H., Linhares, B.M., Yu, W. et al. (2018). Identification of thiourea-based inhibitors of the B-cell lymphoma 6 BTB domain via NMR-based fragment screening and computer-aided drug design. *Journal of Medicinal Chemistry* 61 (17): 7573–7588.

198 Guo, W., Xing, Y., Zhang, Q. et al. (2020). Synthesis and biological evaluation of B-cell lymphoma 6 inhibitors of *N*-phenyl-4-pyrimidinamine derivatives bearing potent activities against tumor growth. *Journal of Medicinal Chemistry* 63 (2): 676–695.

199 Xing, Y., Guo, W., Wu, M. et al. (2022). An orally available small molecule BCL6 inhibitor effectively suppresses diffuse large B cell lymphoma cells growth in vitro and in vivo. *Cancer Letters* 529: 100–111.

200 Lloyd, M.G., Huckvale, R., Cheung, K.-M.J. et al. (2021). Into deep water: optimizing BCL6 inhibitors by growing into a solvated pocket. *Journal of Medicinal Chemistry* 64 (23): 17079–17097.

201 Davis, O.A., Cheung, K.-M.J., Brennan, A. et al. (2022). Optimizing shape complementarity enables the discovery of potent tricyclic BCL6 inhibitors. *Journal of Medicinal Chemistry* 65 (12): 8169–8190.

202 Mamai, A., Chau, A.M., Wilson, B.J. et al. (2023). Discovery of OICR12694: a novel, potent, selective, and orally bioavailable BCL6 BTB inhibitor. *ACS Medicinal Chemistry Letters* 14 (2): 199–210.

203 Ohno, H., Takimoto, G., and Mckeithan, T.W. (1990). The candidate proto-oncogene bcl-3 is related to genes implicated in cell lineage determination and cell cycle control. *Cell* 60 (6): 991–997.

204 Podgornik, H., Pretnar, J., Skopec, B. et al. (2014). Concurrent rearrangements of BCL2, BCL3, and BCL11A genes in atypical chronic lymphocytic leukemia. *Hematology* 19 (1): 45–48.

205 Alpatov, R., Carstens, B., Harding, K. et al. (2015). Rare double-hit with two translocations involving IGH both, with BCL2 and BCL3, in a monoclonal B-cell lymphoma/leukemia. *Molecular Cytogenetics* 8: 101.

206 Puccio, I., Khan, S., Butt, A. et al. (2018). Immunohistochemical assessment of survivin and Bcl3 expression as potential biomarkers for NF-κB activation in the Barrett metaplasia-dysplasia-adenocarcinoma sequence. *International Journal of Experimental Pathology* 99 (1): 10–14.

207 Jacobs, M.D. and Harrison, S.C. (1998). Structure of an IkappaBalpha/NF-kappaB complex. *Cell* 95 (6): 749–758.

208 Michel, F., Soler-Lopez, M., Petosa, C. et al. (2001). Crystal structure of the ankyrin repeat domain of Bcl-3: a unique member of the IkappaB protein family. *The EMBO Journal* 20 (22): 6180–6190.

209 Herrington, F.D. and Nibbs, R.J. (2016). Regulation of the adaptive immune response by the IκB family protein Bcl-3. *Cells* 5 (2): 14.

210 Bours, V., Franzoso, G., Azarenko, V. et al. (1993). The oncoprotein Bcl-3 directly transactivates through kappa B motifs via association with DNA-binding p50B homodimers. *Cell* 72 (5): 729–739.

211 Fujita, T., Nolan, G.P., Liou, H.C. et al. (1993). The candidate proto-oncogene bcl-3 encodes a transcriptional coactivator that activates through NF-kappa B p50 homodimers. *Genes & Development* 7 (7b): 1354–1363.

212 Collins, P.E., Grassia, G., Colleran, A. et al. (2015). Mapping the interaction of B cell leukemia 3 (BCL-3) and nuclear factor κB (NF-κB) p50 identifies a BCL-3-mimetic anti-inflammatory peptide. *The Journal of Biological Chemistry* 290 (25): 15687–15696.

213 Wang, V.Y., Li, Y., Kim, D. et al. (2017). Bcl3 phosphorylation by Akt, Erk2, and IKK is required for its transcriptional activity. *Molecular Cell* 67 (3): 484–97.e5.

214 Clarkson, R.W., Heeley, J.L., Chapman, R. et al. (2000). NF-kappaB inhibits apoptosis in murine mammary epithelia. *The Journal of Biological Chemistry* 275 (17): 12737–12742.

215 Chen, C.Y., Lee, D.S., Yan, Y.T. et al. (2015). Bcl3 bridges LIF-STAT3 to Oct4 signaling in the maintenance of naïve pluripotency. *Stem Cells* 33 (12): 3468–3480.

216 Lace, B., Kempa, I., Klovins, J. et al. (2012). BCL3 gene role in facial morphology. *Birth Defects Research. Part A, Clinical and Molecular Teratology* 94 (11): 918–924.

217 Mitsui, T., Ishida, M., Izawa, M. et al. (2011). Inhibition of Bcl3 gene expression mediates the anti-proliferative action of estrogen in pituitary lactotrophs in primary culture. *Molecular and Cellular Endocrinology* 345 (1–2): 68–78.

218 Ren, X., Song, W., Liu, W. et al. (2013). Rhomboid domain containing 1 inhibits cell apoptosis by upregulating AP-1 activity and its downstream target Bcl-3. *FEBS Letters* 587 (12): 1793–1798.

219 García, I., Cosío, G., Lizárraga, F. et al. (2013). Bcl-3 regulates UVB-induced apoptosis. *Human Cell* 26 (2): 47–55.

220 Tohyama, M., Shirakata, Y., Hanakawa, Y. et al. (2018). Bcl-3 induced by IL-22 via STAT3 activation acts as a potentiator of psoriasis-related gene expression in epidermal keratinocytes. *European Journal of Immunology* 48 (1): 168–179.

221 Chen, R., Wang, L., Liu, S. et al. (2017). Bcl-3 is a novel biomarker of renal fibrosis in chronic kidney disease. *Oncotarget* 8 (57): 97206–97216.

222 Marchetti, G., Girelli, D., Zerbinati, C. et al. (2015). An integrated genomic-transcriptomic approach supports a role for the proto-oncogene BCL3 in atherosclerosis. *Thrombosis and Haemostasis* 113 (3): 655–663.

223 Wang, H., Xiong, M., Hu, Y. et al. (2016). MicroRNA-19b inhibits proliferation of gastric cancer cells by targeting B-cell CLL/lymphoma 3. *Oncology Reports* 36 (4): 2079–2086.

224 Ibrahim, H.A., Amen, F., Reid, A.G. et al. (2011). BCL3 rearrangement, amplification and expression in diffuse large B-cell lymphoma. *European Journal of Haematology* 87 (6): 480–485.

225 Maldonado, V., Espinosa, M., Pruefer, F. et al. (2010). Gene regulation by BCL3 in a cervical cancer cell line. *Folia Biologica* 56 (4): 183–193.

226 Tu, K., Liu, Z., Yao, B. et al. (2016). BCL-3 promotes the tumor growth of hepatocellular carcinoma by regulating cell proliferation and the cell cycle through cyclin D1. *Oncology Reports* 35 (4): 2382–2390.

227 Urban, B.C., Collard, T.J., Eagle, C.J. et al. (2016). BCL-3 expression promotes colorectal tumorigenesis through activation of AKT signalling. *Gut* 65 (7): 1151–1164.

228 Legge, D.N., Shephard, A.P., Collard, T.J. et al. (2019). BCL-3 promotes a cancer stem cell phenotype by enhancing β-catenin signalling in colorectal tumour cells. *Disease Models & Mechanisms* 12 (3).

229 Legge, D.N., Chambers, A.C., Parker, C.T. et al. (2020). The role of B-cell lymphoma-3 (BCL-3) in enabling the hallmarks of cancer: implications for the treatment of colorectal carcinogenesis. *Carcinogenesis* 41 (3): 249–256.

230 Gaire, B., Padmanabhan, S., Zou, Y. et al. (2023). IFNγ induces Bcl3 expression by JAK1/STAT1/p65 signaling, resulting in increased IL-8 expression in ovarian cancer cells. *FEBS Open Bio* 13 (8): 1495–1506.

231 Seaton, G., Smith, H., Brancale, A. et al. (2024). Multifaceted roles for BCL3 in cancer: a proto-oncogene comes of age. *Molecular Cancer* 23 (1): 7.

232 Schwarz, E.M., Krimpenfort, P., Berns, A. et al. (1997). Immunological defects in mice with a targeted disruption in Bcl-3. *Genes & Development* 11 (2): 187–197.

233 Wakefield, A., Soukupova, J., Montagne, A. et al. (2013). Bcl3 selectively promotes metastasis of ERBB2-driven mammary tumors. *Cancer Research* 73 (2): 745–755.

234 Zou, Y., Uddin, M.M., Padmanabhan, S. et al. (2018). The proto-oncogene Bcl3 induces immune checkpoint PD-L1 expression, mediating proliferation of ovarian cancer cells. *The Journal of Biological Chemistry* 293 (40): 15483–15496.

235 Collins, P.E., Kiely, P.A., and Carmody, R.J. (2014). Inhibition of transcription by B cell leukemia 3 (Bcl-3) protein requires interaction with nuclear factor κB (NF-κB) p50. *The Journal of Biological Chemistry* 289 (10): 7059–7067.

236 Keutgens, A., Shostak, K., Close, P. et al. (2010). The repressing function of the oncoprotein BCL-3 requires CtBP, while its polyubiquitination and degradation involve the E3 ligase TBLR1. *Molecular and Cellular Biology* 30 (16): 4006–4021.

237 Cartwright, T., Perkins, N.D., and Wilson, C.L. (2016). NFKB1: a suppressor of inflammation, ageing and cancer. *The FEBS Journal* 283 (10): 1812–1822.

238 Soukupová, J., Bordoni, C., Turnham, D.J. et al. (2021). The discovery of a novel antimetastatic Bcl3 inhibitor. *Molecular Cancer Therapeutics* 20 (5): 775–786.

239 Willis, T.G., Jadayel, D.M., Du, M.Q. et al. (1999). Bcl10 is involved in *t*(1;14)(p22;q32) of MALT B cell lymphoma and mutated in multiple tumor types. *Cell* 96 (1): 35–45.

240 Ruland, J. and Hartjes, L. (2019). CARD-BCL-10-MALT1 signalling in protective and pathological immunity. *Nature Reviews. Immunology* 19 (2): 118–134.

241 Lu, H.Y., Bauman, B.M., Arjunaraja, S. et al. (2018). The CBM-opathies-A rapidly expanding spectrum of human inborn errors of immunity caused by mutations in the CARD11-BCL10-MALT1 complex. *Frontiers in Immunology* 9: 2078.

242 Alsaidalani, A.A., García-Solís, B., Bukhari, E. et al. (2023). Inherited human BCL10 deficiencies. *Journal of Clinical Immunology* 44 (1): 13.

243 Ruland, J., Duncan, G.S., Elia, A. et al. (2001). Bcl10 is a positive regulator of antigen receptor-induced activation of NF-kappaB and neural tube closure. *Cell* 104 (1): 33–42.

244 Ruland, J., Duncan, G.S., Wakeham, A. et al. (2003). Differential requirement for Malt1 in T and B cell antigen receptor signaling. *Immunity* 19 (5): 749–758.

245 Wang, D., You, Y., Case, S.M. et al. (2002). A requirement for CARMA1 in TCR-induced NF-kappa B activation. *Nature Immunology* 3 (9): 830–835.

246 Fischer, K.D., Tedford, K., and Wirth, T. (2004). New roles for Bcl10 in B-cell development and LPS response. *Trends in Immunology* 25 (3): 113–116.

247 Yu, M., Chen, Y., He, Y. et al. (2012). Critical role of B cell lymphoma 10 in BAFF-regulated NF-κB activation and survival of anergic B cells. *Journal of Immunology* 189 (11): 5185–5193.

248 Gross, O., Grupp, C., Steinberg, C. et al. (2008). Multiple ITAM-coupled NK-cell receptors engage the Bcl10/Malt1 complex via Carma1 for NF-kappaB and MAPK activation to selectively control cytokine production. *Blood* 112 (6): 2421–2428.

249 Malarkannan, S., Regunathan, J., Chu, H. et al. (2007). Bcl10 plays a divergent role in NK cell-mediated cytotoxicity and cytokine generation. *Journal of Immunology* 179 (6): 3752–3762.

250 Bertin, J., Guo, Y., Wang, L. et al. (2000). CARD9 is a novel caspase recruitment domain-containing protein that interacts with BCL10/CLAP and activates NF-kappa B. *The Journal of Biological Chemistry* 275 (52): 41082–41086.

251 Wang, L., Guo, Y., Huang, W.J. et al. (2001). Card10 is a novel caspase recruitment domain/membrane-associated guanylate kinase family member that interacts with BCL10 and activates NF-kappa B. *The Journal of Biological Chemistry* 276 (24): 21405–21409.

252 Bertin, J., Wang, L., Guo, Y. et al. (2001). CARD11 and CARD14 are novel caspase recruitment domain (CARD)/membrane-associated guanylate kinase (MAGUK) family members that interact with BCL10 and activate NF-kappa B. *The Journal of Biological Chemistry* 276 (15): 11877–11882.

253 An, J., Liu, H., Magyar, C.E. et al. (2013). Hyperactivated JNK is a therapeutic target in pVHL-deficient renal cell carcinoma. *Cancer Research* 73 (4): 1374–1385.

254 David, L., Li, Y., Ma, J. et al. (2018). Assembly mechanism of the CARMA1-BCL10-MALT1-TRAF6 signalosome. *Proceedings of the National Academy of Sciences of the United States of America* 115 (7): 1499–1504.

255 Schlauderer, F., Seeholzer, T., Desfosses, A. et al. (2018). Molecular architecture and regulation of BCL10-MALT1 filaments. *Nature Communications* 9 (1): 4041.

256 Zhang, S., Pan, D., Jia, X.M. et al. (2017). The CARMA3-BCL10-MALT1 (CBM) complex contributes to DNA damage-induced NF-κB activation and cell survival. *Protein & Cell* 8 (11): 856–860.

257 Mahanivong, C., Chen, H.M., Yee, S.W. et al. (2008). Protein kinase C alpha-CARMA3 signaling axis links Ras to NF-kappa B for lysophosphatidic acid-induced urokinase plasminogen activator expression in ovarian cancer cells. *Oncogene* 27 (9): 1273–1280.

258 Rehman, A.O. and Wang, C.Y. (2009). CXCL12/SDF-1 alpha activates NF-kappaB and promotes oral cancer invasion through the Carma3/Bcl10/Malt1 complex. *International Journal of Oral Science* 1 (3): 105–118.

259 Kuo, S.H., Chou, C.H., Cheng, A.L. et al. (2012). Expression of BCL10 in cervical cancer has a role in the regulation of cell growth through the activation of NF-κB-dependent cyclin D1 signaling. *Gynecologic Oncology* 126 (2): 245–251.
260 Jiang, T., Grabiner, B., Zhu, Y. et al. (2011). CARMA3 is crucial for EGFR-induced activation of NF-κB and tumor progression. *Cancer Research* 71 (6): 2183–2192.
261 O'Connor, M.J. (2015). Targeting the DNA damage response in cancer. *Molecular Cell* 60 (4): 547–560.
262 Shrivastav, M., De Haro, L.P., and Nickoloff, J.A. (2008). Regulation of DNA double-strand break repair pathway choice. *Cell Research* 18 (1): 134–147.
263 Lieber, M.R. (2010). The mechanism of double-strand DNA break repair by the nonhomologous DNA end-joining pathway. *Annual Review of Biochemistry* 79: 181–211.
264 Farmer, H., Mccabe, N., Lord, C.J. et al. (2005). Targeting the DNA repair defect in BRCA mutant cells as a therapeutic strategy. *Nature* 434 (7035): 917–921.
265 D'Amours, D., Desnoyers, S., D'Silva, I. et al. (1999). Poly(ADP-ribosyl)ation reactions in the regulation of nuclear functions. *The Biochemical Journal* 342 (Pt 2): 249–268.
266 Tutt, A., Robson, M., Garber, J.E. et al. (2010). Oral poly(ADP-ribose) polymerase inhibitor olaparib in patients with BRCA1 or BRCA2 mutations and advanced breast cancer: a proof-of-concept trial. *Lancet* 376 (9737): 235–244.
267 Shah, A.P., Patel, C.N., Sureja, D.K. et al. (2018). A review on DNA repair inhibition by PARP inhibitors in cancer therapy. *Folia Medica* 60 (1): 39–47.
268 Conde, C., Mark, M., Oliver, F.J. et al. (2001). Loss of poly(ADP-ribose) polymerase-1 causes increased tumour latency in p53-deficient mice. *The EMBO Journal* 20 (13): 3535–3543.
269 Tong, W.M., Yang, Y.G., Cao, W.H. et al. (2007). Poly(ADP-ribose) polymerase-1 plays a role in suppressing mammary tumourigenesis in mice. *Oncogene* 26 (26): 3857–3867.
270 Audeh, M.W., Carmichael, J., Penson, R.T. et al. (2010). Oral poly(ADP-ribose) polymerase inhibitor olaparib in patients with BRCA1 or BRCA2 mutations and recurrent ovarian cancer: a proof-of-concept trial. *Lancet* 376 (9737): 245–251.
271 Bao, W., Sun, C., Sun, X. et al. (2020). Targeting BCL10 by small peptides for the treatment of B cell lymphoma. *Theranostics* 10 (25): 11622–11636.
272 Di Pilato, M., Kim, E.Y., Cadilha, B.L. et al. (2019). Targeting the CBM complex causes T(reg) cells to prime tumours for immune checkpoint therapy. *Nature* 570 (7759): 112–116.

13

Molecular Glues as Activators for PPI

Xiangbing Qi

National Institute of Biological Science, Chemistry Center, 7 Science Park Road ZGC Life Science Park, Beijing 102206, People's Republic of China

13.1 Introduction

13.1.1 Significance of PPI in Biology and Disease

Human cells contain approximately 20,000 proteins, which participate in over 130,000–650,000 types of protein–protein interactions (PPIs) [1, 2]. The vast number of PPIs serves as the fundamental component for signaling pathways, governing numerous physiological processes within the body. Consequently, the modulation of PPI emerges as a pivotal target for therapeutic interventions [3]. Disruptions in PPI often lead to aberrant signaling cascades, serving as pathological triggers for various diseases, such as cancer and neurodegenerative diseases [1, 2, 4].

In the context of cancer, the activation of oncogenes and the inactivation of tumor suppressor genes drive cancer development and progression [5]. The tumor suppressor p53 (encoded by the *TP53* gene) plays a central role in regulating apoptosis, DNA repair, and the cell cycle progression, serving as an essential defense mechanism against cancer [6–8]. MDM2 is a proto-oncoprotein that binds to the transcriptional activation domain of p53, thereby promoting its degradation [5, 9–11]. In human cancers, frequent mutations or deletions of *TP53,* or suppression of wild-type p53 function by heightened MDM2 levels contribute to the downregulation of p53-mediated tumor-suppressive pathways [5, 12, 13]. Consequently, disrupting the PPI between MDM2 and p53 holds promise for reinstating p53's inherent tumor-suppressive capacities in tumors with wild-type p53.

Neurodegenerative diseases are often characterized by the pathological accumulation of neurotoxic proteins, such as α-synuclein in Parkinson's disease (PD), amyloid-β (Aβ) and tau in Alzheimer's disease (AD), and mutant huntingtin (mHTT) in Huntington's disease (HD) [4, 14–16]. Tau, a microtubule-associated protein, is a key pathological protein in AD [17–19]. In the brains of AD patients, tau aggregation can exacerbate the toxicity of amyloid-β (Aβ) [20–23]. Consequently, targeting tau protein to an E3 ligase to enhance its ubiquitination and subsequent degradation is a promising therapeutic strategy for treating AD [24].

Targeting Protein–Protein Interactions for Drug Discovery, First Edition. Edited by Jian Zhang.

13.1.2 Opportunities and Challenges for PPI Drug Discovery

The manipulation of PPIs with small molecules has the potential to influence the interplay between two proteins, providing enhanced specificity compared to conventional small molecules that target individual proteins [25, 26]. Therefore, small-molecule PPI modulators offer unique advantages for diseases that pose challenges for conventional treatment modalities [27]. Recently, several PPI modulators have advanced to clinical trials, with the most advanced ones gaining regulatory approval.

Despite encouraging progress, developing PPI modulation for clinical applications remains a formidable challenge for several reasons: (i) typical PPI interfaces are broad and flat in nature, complicating identification of ligandable pockets akin to those found in traditional druggable targets [28]; (ii) the protein conformations delineated in crystal structures of protein complexes may not faithfully represent the dynamic PPIs in physiological settings [29]; (iii) PPIs typically lack natural or endogenous ligands, resulting in a lack of molecular frameworks for modulating these interactions [30]; and (iv) the assessment of PPI activity is challenging due to the scarcity of appropriate technologies, alongside the need for more efficient strategies for screening compound libraries [31]. These challenges need to be overcome to advance PPI modulation into clinically relevant interventions.

Despite these challenges, several drugs targeting PPIs have received FDA approval, and many others are undergoing clinical evaluation (Table 13.1) [22, 25, 28–38]. Notable examples include venetoclax, a BCL-2 inhibitor developed by AbbVie for the treatment of chronic lymphocytic leukemia (CLL), which induces apoptosis by disrupting the BCL-2/BAX interaction [39]. RO-5045337, an orally bioavailable and selective p53/MDM2 inhibitor, has demonstrated significant anticancer efficacy [40, 41]. These examples highlight the therapeutic potential of small molecule PPI modulators in treating cancer.

13.1.3 Four Classes of PPI Modulators

The PPI interface, albeit broad and extensive, often relies on the interaction of a small number of crucial amino acid residues for mutual recognition. These amino acid residues are referred to as "hotspots" [42–45]. Based on the binding site on proteins, small-molecule PPI modulators can be divided into orthosteric regulators and allosteric regulators (Figure 13.1). Orthosteric regulators engage the PPI binding interface of two proteins, directly interacting with the critical hotspot residues (see Figure 13.1a,b). In contrast, allosteric regulators target the non-PPI binding site of one of the interacting proteins, inducing changes in the conformation of the hotspot residues. In terms of their biological effects, small-molecule PPI modulators can be further classified into PPI stabilizers and PPI inhibitors (see Figure 13.1c,d). Orthosteric PPI stabilizers are often referred to as molecular glues (MGs), which will be the main focus of this chapter.

Table 13.1 Representative small-molecule PPI modulators approved by FDA or in clinical trials.

Drug	PPI	Pathology	Research institution	Stages
Venetoclax	BCL2/BAX	Chronic lymphocytic leukemia	AbbVie	Approved (2016)
Maraviroc	CCR5/gp120	HIV/AIDS	Pfizer	Approved (2007)
Tirofiban	FGG/ITGA2B/3	Cardiovascular disease	Abbott	Approved (1998)
Tacrolimus	FKBP12/CNA/B	Immuno-suppression	Astellas Pharma	Approved (1994)
Sirolimus	FKBP12/mTOR	Immuno-suppression	Pfizer	Approved (1999)
Temsirolimus	FKBP12/mTOR	Immuno-suppression	Accord Healthcare	Approved (2018)
Tafamidis	TTR tetramer	Cardiomyopathy	FoldRX	Approved (2019)
Eribulin	α-tubulin/β-tubulin	Cancer	Eisai	Approved (2010)
Ixabepilone	α-tubulin/β-tubulin	Cancer	R-Pharm-US	Approved (2007)
Paclitaxel	α-tubulin/β-tubulin	Cancer	Accord Healthcare	Approved (2018)
Docetaxel	α-tubulin/β-tubulin	Caner	Hospira	Approved (2016)
Cabazitaxel	α-tubulin/β-tubulin	Cancer	Sanofi	Approved (2010)
Lifitegrast	LFA-1/ICAM-1	Xerophthalmia	Lifelong Vision Foundation	Phase IV
RVX-208	BD/histone	Type II diabetes, coronary artery disease	Resverlogix	Phase III
GSK525762	BD/histone	Cancer	GSK	Phase II
Pevonedistat	NEDD8/APPBP1	Cancer	Takeda	Phase II
Navitoclax	BCL2/BAX	Cancer	AbbVie	Phase II
LCL-161	XIAP/caspase 9	Cancer	Novartis	Phase II
TL32711	XIAP/caspase 9	Cancer	TetraLogic	Phase II

(Continued)

Table 13.1 (Continued)

Drug	PPI	Pathology	Research institution	Stages
PRI-724	β-catenin/CBP	Cirrhosis, myeloid leukemia	Prism Pharma	Phase II
ASTX-660	XIAP/caspase 9	Cancer	Astex, Otsuka	Phase I/II
AMG-232	p53/MDM2	Cancer	Amgen	Phase I/II
DS-3032b	p53/MDM2	Cancer	Daiichi Sankyo	Phase I/II
CGM097	p53/MDM2	p53 wild-type solid tumor	Novartis	Phase I
RO-5045337	p53/MDM2	Cancer	Roche	Phase I
SAR-405838	p53/MDM2	Tumor	Sanofi	Phase I
Idasanutlin	p53/MDM2	Cancer	Roche	Phase I
GDC-0917	XIAP/caspase 9	Solid tumor	Genentech	Phase I
CA-170	PD-1/PD-L1	Cancer	Curis	Phase I

Ref: https://go.drugbank.com, https://clinicaltrials.gov (Last updated: 2021).

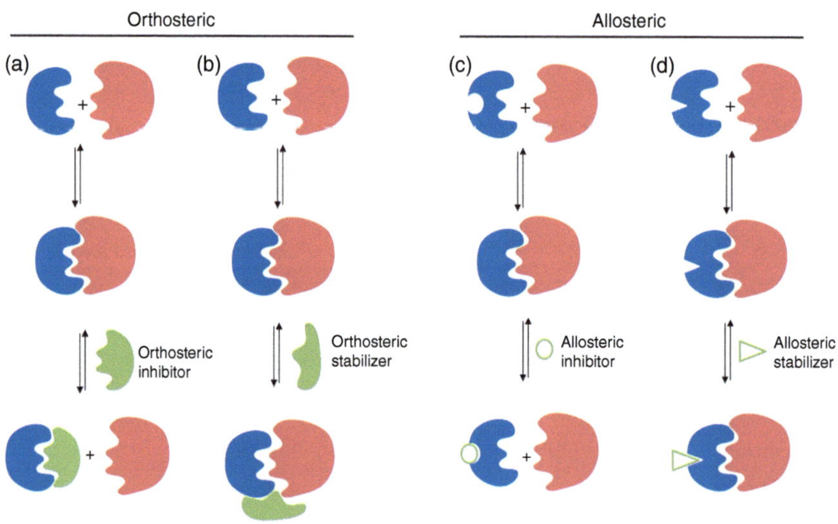

Figure 13.1 Four different classes of small-molecule PPI modulators. Regulating mechanism of small-molecule PPI modulators including (a) orthosteric inhibitor, (b) orthosteric stabilizer, (c) allosteric inhibitor, and (d) allosteric stabilizer. (This figure was created with BioRender.com.) Source: Xiangbing Qi.

13.2 Molecular Glues as Orthosteric PPI Stabilizers/Activators

13.2.1 Design Principles of MG

The term "molecular glue " has been used since the 1980s to describe proteins that bridge the interaction of other proteins [46–48]. However, its application to small molecules gained wider recognition only after it was employed to elucidate the mechanism of action of auxin, the first reported MG degrader [49]. Recently, Ning Zheng et al. [46] provided a formal definition of MGs as a class of compounds that enhance preexisting weak PPIs. Importantly, the formal definition of MGs significantly differs from its broader use as a term for chemical inducers of proximity (CIPs). Although both MGs and bifunctional compounds like PROteolysis TArgeting Chimeras (PROTACs) fall under the general category of CIPs, they are distinguished by their thermodynamic and kinetic properties [44]. PROTACs link two proteins of interest (POIs) without exhibiting cooperative effects, while MGs induce PPI in a cooperative manner [50]. (Figure 13.2a–c) To achieve this, molecular glues must bind to PPI surfaces of two proteins, thereby enhancing and stabilizing the interaction between them. In contrast, PROTACs are heterobifunctional molecules composed of a POI warhead and an E3 ligase ligand connected by a linker (Figure 13.2d). At elevated concentrations, bifunctional degraders can form binary complexes that impair target degradation, an effect known as the Hook effect (Figure 13.2e). Due to their cooperative nature, MGs are resistant to the Hook effect.

13.2.2 Classification of MGs

MGs can elicit a variety of downstream effects, including degradation, stabilization, and inhibition, each tailored to achieve specific biological outcomes for disease modulation (Figure 13.3). In the following section, we provide examples illustrating these three distinct categories of MGs and describe the methodologies employed in their discovery.

13.2.2.1 MG Degraders (MGDs)

MGsthat promote targeted protein degradation work by promoting the interaction between a POI with an E3 ubiquitin ligase. This interaction facilitates the E3 ligase's catalysis of ubiquitin transfer to substrate proteins, marking them for proteasomal degradation. Notable examples include auxin, immunomodulatory drugs (IMiDs), and indisulam.

Auxin The first reported MGDs trace back to the plant hormone auxin [49]. In 2007, Tan et al. [49] elucidated the structural basis of auxin's action, revealing that this natural endogenous MG facilitates PPIs between the transcriptional repressor AUX/IAA and the substrate TIR1 of the E3 ubiquitin ligase (Figure 13.4). This interaction induced the degradation of AUX/IAA through the SCF^{TIR1} complex, thereby regulating plant growth and development [49, 51–54].

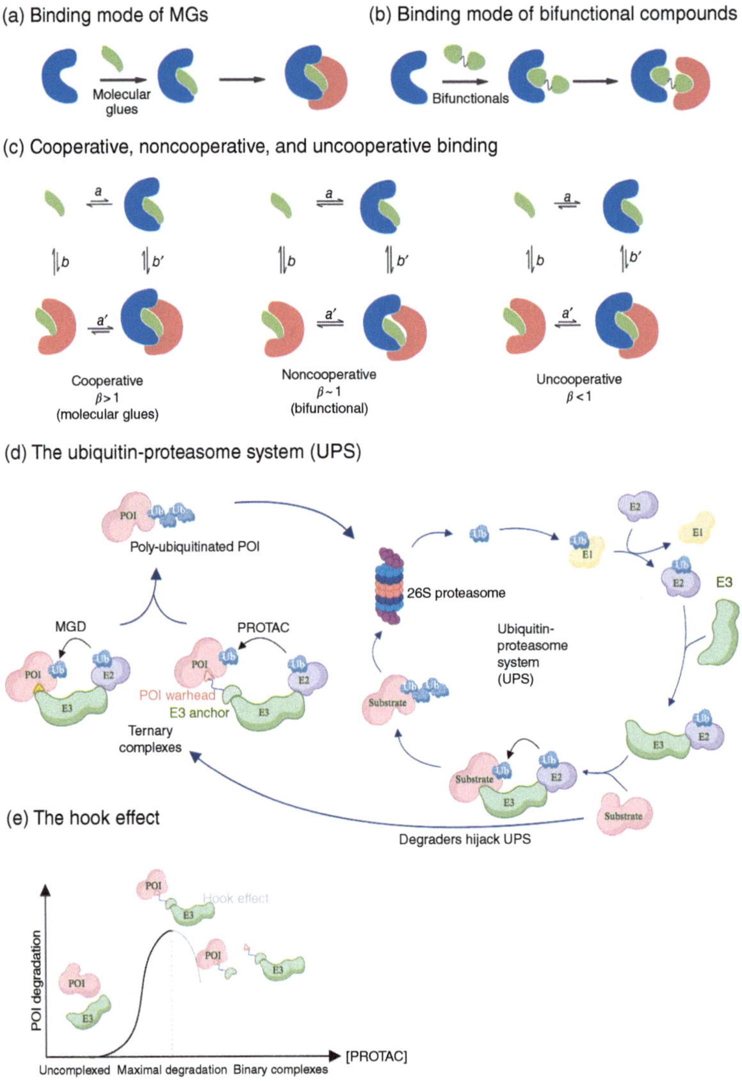

Figure 13.2 Binding modes of two different chemical inducers of proximity (CIPs). (a) Molecular glues (MGs) (green) and (b) bifunctional compound (green). (c) A schematic illustrating the difference between cooperative, noncooperative, and uncooperative binding; MGs and bifunctional compounds (green) can be distinguished quantitatively by the cooperative factor, α, in their induced ternary complexes with two protein targets (red and blue). α is calculated by $K_D^{binary}/K_D^{ternary}$, denoted as k_1/k_1' or k_2/k_2'. MGs exhibit high cooperativity with $\alpha > 1$ (cooperative) and bifunctional compounds typically exhibit $\alpha \sim 1$ (noncooperative). (d) The ubiquitin-proteasome system (UPS) operates through a cascade involving three enzymes that sequentially conjugate ubiquitin molecules to target proteins, thereby tagging them for degradation by the proteasome. Degraders leverage this system by bringing a target protein (POI) into proximity with an E3 ligase, which promotes the ubiquitination and subsequent degradation of the target protein. (e) The hook effect. (This figure was generated using BioRender.com.) Source: Xiangbing Qi.

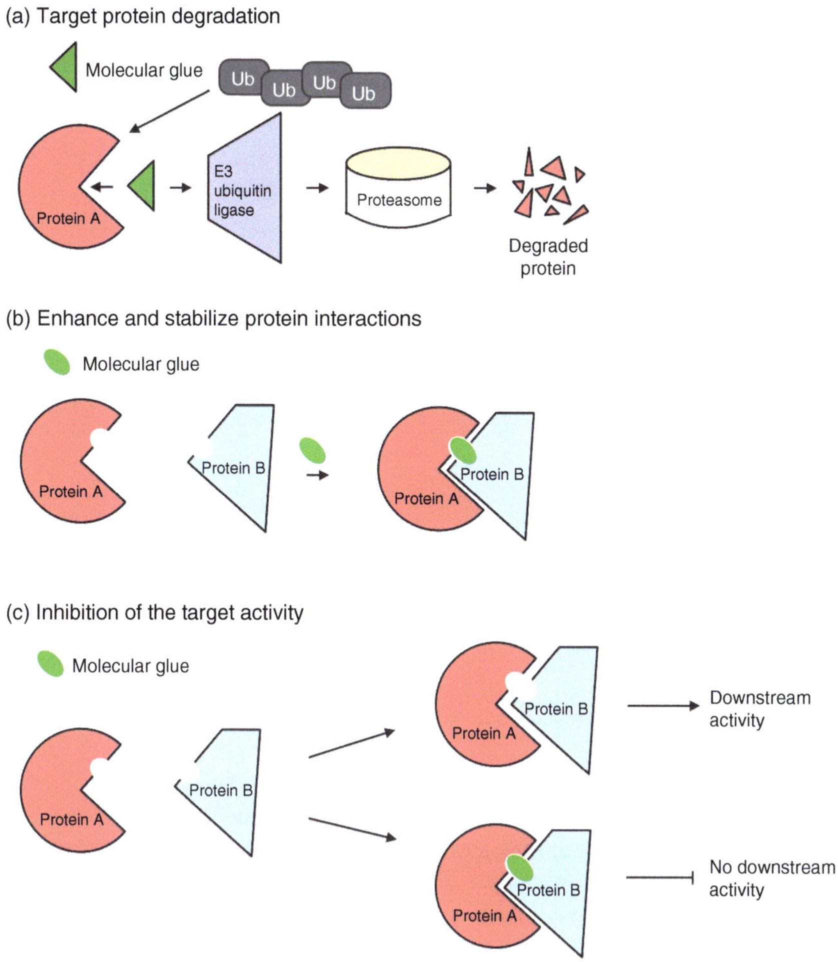

Figure 13.3 Three categories of MGs based on downstream biological effects, including (a) target protein degradation, (b) stabilization of the target of interest, and (c) inhibition of the target activity. (This figure was created with BioRender.com.) Source: Xiangbing Qi.

Immunomodulatory Drugs (IMiDs) Initially introduced in Europe in 1954, thalidomide garnered widespread use due to its potent sedative and antiemetic effects. However, following the emergence of teratogenic defects, commonly referred to as "seal limb" deformities, thalidomide was withdrawn from the market in 1961 [55]. Thalidomide's therapeutic potential was later recognized in the treatment of erythema nodosum leprosum (ENL), leading to its reapproval by the FDA in 1998 [56]. Today, analogs of thalidomide known as IMiDs are extensively employed in the treatment of multiple myeloma [56–58].

Over the past decade, work from many labs collectively revealed that IMiDs function as MGs between cereblon (CRBN), an adaptor protein within the DDB1-CUL4-RBX1 ubiquitin ligase complex (CRL4) and a variety of neo-substrate. Crystallographic analysis revealed that IMiDs bind to a shallow, hydrophobic pocket

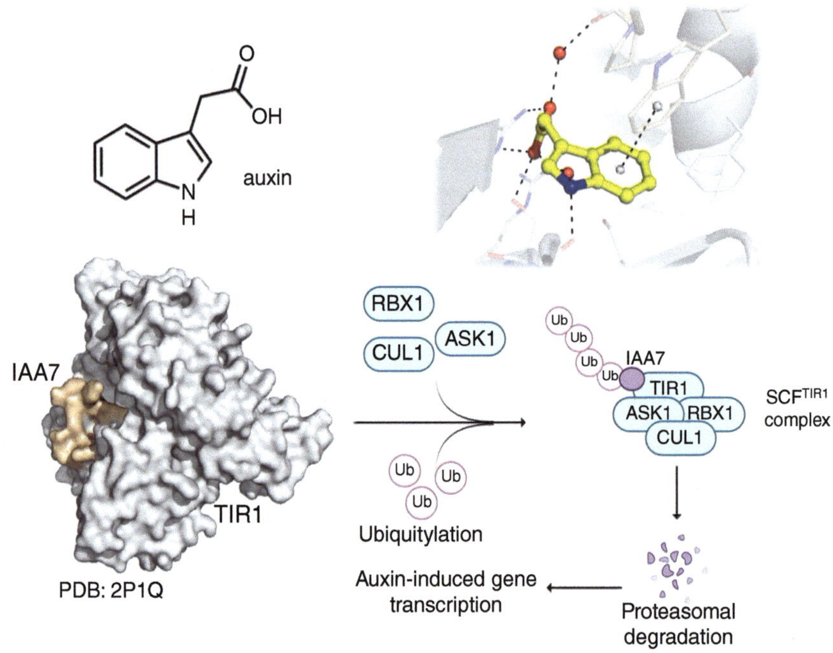

Figure 13.4 Mechanism of action of auxin. (This figure was created with pymol (version:1.7.2.1) and BioRender.com.) Source: Xiangbing Qi.

on the CRBN surface, interacting with three tryptophan residues (Trp380, Trp386, and Trp400) [59, 60]. Furthermore, these studies demonstrated that IMiDs alter the surface of CRBN, resulting in the recruitment of neo substrates of CRBN, such as CK1α and IKZF1/3.

Optimization of CRBN modulators has led to the development of a new generation of IMiDs/CRBN E3 ligase modulators (CELMoDs) [61–65]. For example, in 2016, Chamberlain et al. [66] screened a library of thalidomide derivatives, identifying CC-885 as a novel CRBN modulator with potent antiproliferative effects by degrading GSPT1 (Figure 13.5). In 2020, two CRBN E3 ligase modulators (CELMoDs), CC-3060 and CC-647, were identified as degraders of the transcription factor ZBTB1 [67]. Subsequent research by Watson et al. [68] in 2022 demonstrated that IMiD/CELMoD compounds could induce distinct conformational states in CRBN. Their study revealed that pomalidomide, iberdomide (CC-220) and mezigdomide (CC-92480) each induced unique conformations of CRBN, which directly correlated with their potency. This work underscored the significance of allosteric regulation and conformational control in CRBN MGDs, thereby facilitating the rational design of MGDs with improved neo-substrate selectivity and efficacy. Enhanced understanding of IMiDs' molecular mechanisms has led to the development of novel degraders with better efficacy, specificity, and drug-like properties. Notably, new analogues were developed and assessed for GSPT1 degradation activity. Structure–activity relationship (SAR) analysis identified CC-90009, which is currently in phase 1 clinical trials, as a potent and selective GSPT1 degrader

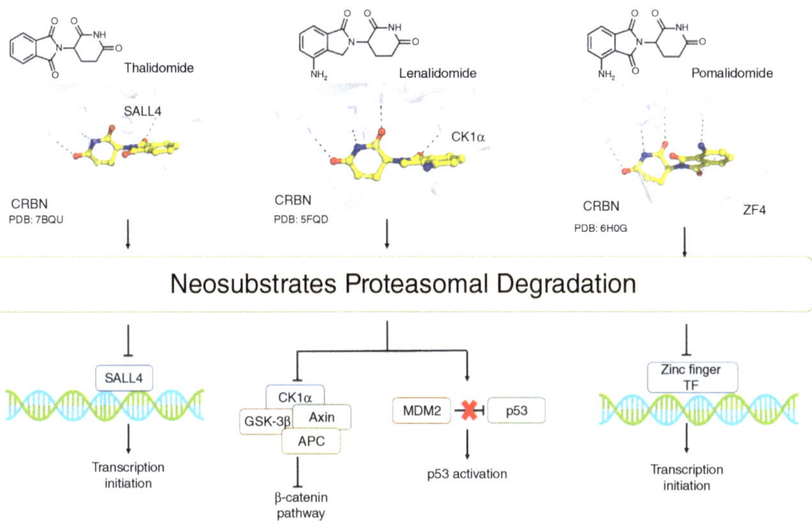

Figure 13.5 Mechanism of action of immunomodulatory drugs (IMiDs). (This figure was created with pymol (version:1.7.2.1) and BioRender.com.) Source: Xiangbing Qi.

[69, 70]. CC-90009 has demonstrated significant potential in treating acute myeloid leukemia, showing favorable pharmacokinetic, pharmacodynamic, and safety profiles [69].

RBM39 Degraders Sulfonamides were identified as another class of MGDs targeting the splicing factor RBM39 by recruiting the CRL4^{DCAF15} E3 ligase [71]. Notable examples of sulfonamide-based degraders include tasisulam, E7820, chloroquinoxaline sulfonamide, and indisulam (Figure 13.6) [72]. Indisulam, an aryl sulfonamide compound first synthesized by Eisai Pharmaceuticals in 1992, was initially developed as an antibacterial agent similar to sulfonamides [73]. However, by 1999, Owa et al. [74] demonstrated its significant anticancer properties. Subsequent studies revealed that indisulam could inhibit cancer cell proliferation [75]. Yoshino et al. [74] proposed that indisulam could selectively target critical proteins and pathways during the G1 phase of the cell cycle, leading to G1 phase arrest and cell death [76]. In 2017, Han et al. [71] elucidated the specific mechanism of action of indisulam and its analogues, showing that they acted as primary growth factors for the RBM39-DCAF15 complex. DCAF15 is a substrate receptor of the CUL4-DDB1-RBX1 E3 ubiquitin ligase complex (Figure 13.6). Subsequent structural biology studies revealed the structural basis of these classes of MG compounds.

Cyclin K Degraders In 2020, (*R*)-CR8, a cyclin-dependent kinase (CDK) inhibitor, was identified as a MGD that could facilitate the formation of a complex between CDK12/cyclin K and DDB1 (Figure 13.9, compound 1) [76]. To elucidate the mechanism of (R)-CR8, Słabicki et al. [76] determined the crystal structure of the complex (Figure 13.9). The binding of (*R*)-CR8 significantly enhanced the interaction between the kinase active domain of CDK12 and the BPA and

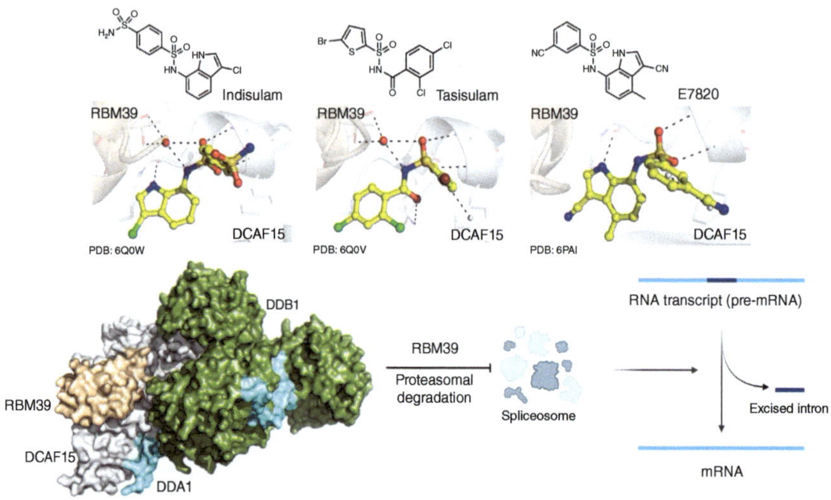

Figure 13.6 Mechanism of action of indisulam, tasisulam, and E7820. (This figure was created with pymol (version:1.7.2.1) and BioRender.com.) Source: Xiangbing Qi.

BPC domains of DDB1, increasing the binding affinity by a factor of 500–1000 (Figure 13.7). This enhanced interaction converts CDK12 into a substrate receptor of the CUL4-DDB1-RBX1 E3 ubiquitin ligase complex, leading to the ubiquitination of cyclin K (Figure 13.8). Structural analysis revealed that the phenylpyridine ring of (R)-CR8 engaged in hydrophobic interactions with I909 and cationic-π interactions with R928 and R947 of DDB1. Modifications to the pyridine ring, such as its removal or substitution of benzylpyridine with phenylpyridine, resulted in reduced compound activity, underscoring the importance of the benzylpyridine structure for efficacy [77]. Structure-guided mutational analyses and TR-FRET assays demonstrated that, in the presence of (*R*)-CR8, CDK12 functioned as a substrate receptor facilitated by the MG, positioning cyclin K in a location typically occupied by CRL4 substrates.

Concurrent to the CR8 work, Han et al. identified a new type of MG (Figure 13.9, compound 2) that functions as a cyclin K degrader [77]. Initially identified as a cytotoxic agent through high-throughput screening (HTS), HQ461's role as a MG was further characterized using chemical genetics and biochemical reconstitution approaches. HQ461 binds to the ATP-binding pocket of CDK12, modifying CDK12's surface to enhance its interaction with DDB1 (Figure 13.8). Through biochemical assays and chemical cross-linking mass spectrometry techniques, HQ461 was confirmed to be a MG. Following SAR optimization of HQ461, a more potent and bioavailable lead compound, HQ005 (Figure 13.9, compound 3), was developed [77].

Around the same time, Mayor-Ruiz et al. [78] employed scalable chemical profiling to screen a library of approximately 2000 cytotoxic small molecules, aiming to identify new MGs. This screening led to the identification of three compounds, dCeMM2, dCeMM3, and dCeMM4 (Figure 13.9, compounds 4–6) Further orthogonal experiments confirmed that these compounds could mediate PPIs between

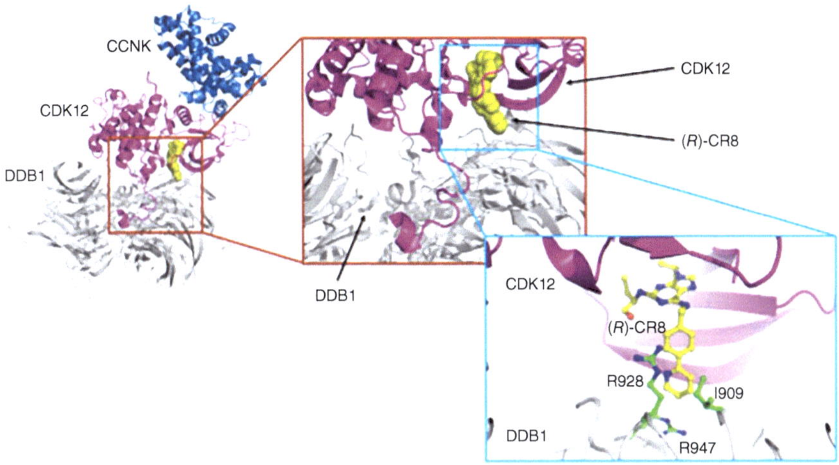

Figure 13.7 The crystal structure of DDB1 (white), CDK12 (magenta), CCNK (blue), and ***(R)*-CR8** (yellow). PDB: 6TD3 (This figure was created with pymol (version:1.7.2.1) and BioRender.com.) Source: Xiangbing Qi.

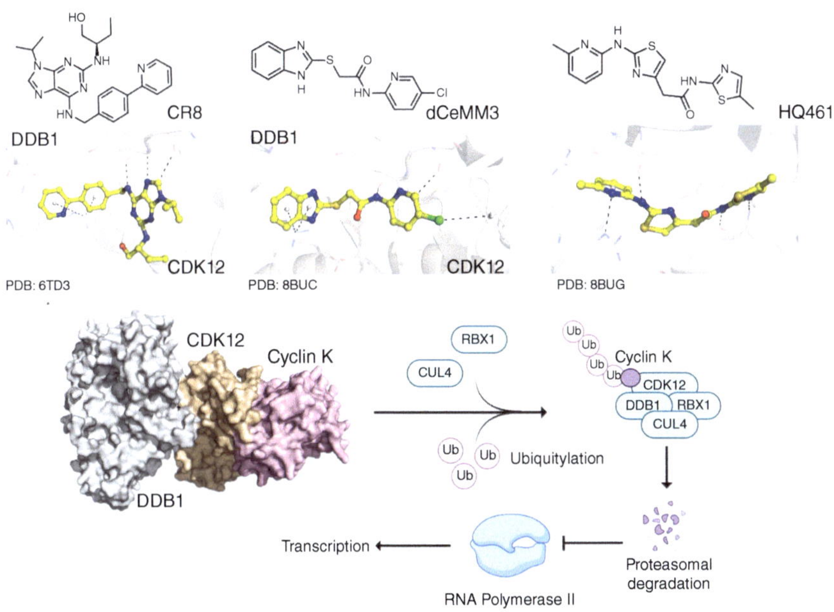

Figure 13.8 Mechanism of action of (*R*)-CR8, dCeMM3, and HQ461. (This figure was created with pymol (version:1.7.2.1) and BioRender.com.) Source: Xiangbing Qi.

CDK12 and DDB1, facilitating ubiquitination and subsequent degradation of the substrate CCNK [78].

In 2021, NCT02 (Figure 13.9, compound 7) was identified through HTS of an 80,000-compound library for its activity against patient-derived metastatic colorectal cancer spheroids [79]. NCT02 also functions as a MG that induces the

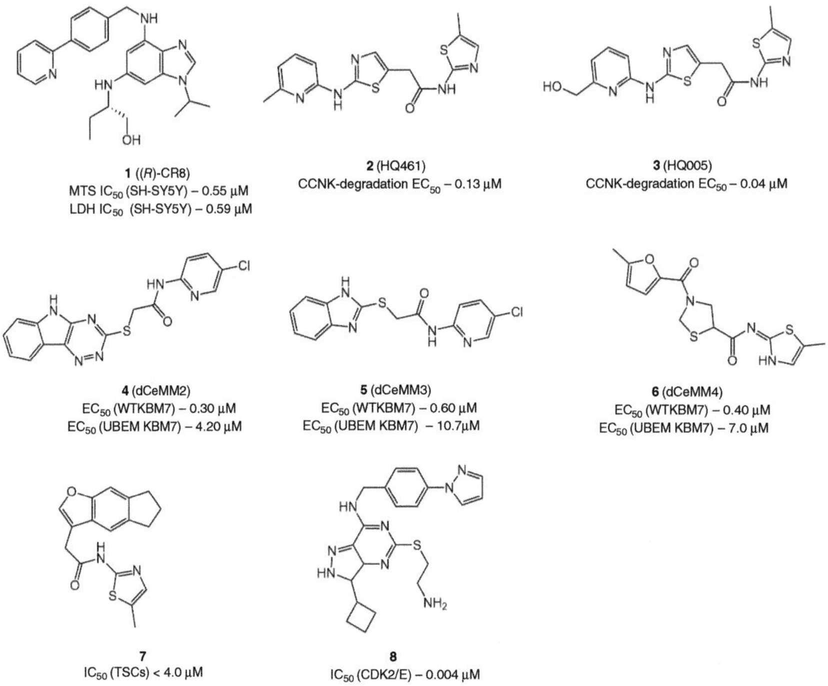

Figure 13.9 Molecular Glues that target CDK12-DDB1. (This figure was created with ChemDraw (version: 23.0.1).) Source: Xiangbing Qi.

ubiquitination of cyclin K. In 2022, Jarda et al. [80] reported that compound 8 (Figure 13.9), a CDK inhibitor based on a 3,5,7-substituted pyrazole[4,3-d] pyrimidine scaffold, effectively induced cyclin K degradation. Additionally, compound 8 was found to exhibit a binding mechanism similar to that of (*R*)-CR8.

13.2.2.2 MG Stabilizers

PDE3A–SLFN12 PDE3A, a member of the cyclic nucleotide phosphodiesterase (PDE) family, is pivotal in regulating intracellular levels of cAMP and cGMP [81]. While PDE3A is expressed in various human cells, it is particularly crucial for modulating the physiological and biochemical functions of cardiovascular cells [82]. Currently, two PDE3A inhibitors, cilostazol and milrinone, are approved for the treatment of cardiovascular diseases [83]. Recent studies have also indicated a connection between PDE3A expression and chemotherapy resistance in non–small-cell lung cancer (NSCLC) cells, suggesting a potential role for PDE3A in cancer resistance [84].

Electron microscopy studies revealed that MGs could enhance the proximity between the catalytic domain of PDE3A and the C-terminus of SLFN12, thereby promoting the formation of a PPI [85]. MGs specifically bound to a hydrophobic cavity at the interface of these two proteins (Figure 13.10a,b). Despite the diverse structural characteristics of PDE3A–SLFN12 MGs identified to date, they exhibited similar binding mechanisms: (i) formation of hydrogen bonds with residues H961

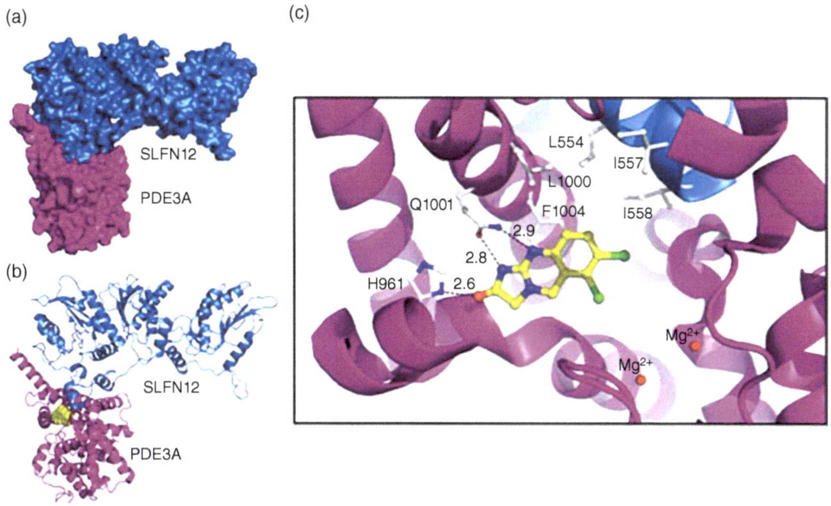

Figure 13.10 (a) The surface model of PDE3A (magenta) and SLFN12 (blue); (b) The atomic model of PDE3A–SLFN12 and **anagrelide** (yellow), one of their MGs; (c) The binding mode of **anagrelide** and PDE3A–SLFN12. The Mg^{2+} ions in PDE3A catalytic domain are shown in red spheres, PDB: 7EG0 (This figure was created with pymol (version:1.7.2.1) and BioRender.com.) Source: Xiangbing Qi.

and Q1001 of PDE3A; (ii) π–π stacking interactions with residue F1004 of PDE3A; and (iii) hydrophobic interactions with residues L1000 of PDE3A and L554, I557, and I558 of SLFN12 (Figure 13.10c).

In 2016, Meyerson and Greulich et al. utilized dynamic light scattering (DLS) to demonstrate that the compound DNMDP (Figure 13.11, compound 9) interacted with PDE3A, promoted the formation of a complex between PDE3A and SLFN12, and effectively inhibited cancer cell proliferation [86]. Biological assays revealed that the catalytic activity pocket of PDE3A was crucial for DNMDP's efficacy, underscoring its potential as an anticancer agent. This contrasts with traditional PDE3A inhibitors, which do not exhibit comparable anticancer effects, suggesting that DNMDP likely functions as a MG [87]. In 2021, the research team confirmed that SLFN12 acted as an RNase and that the PDE3A-DNMDP-SLFN12 complex enhanced RNase activity, thereby contributing to DNMDP's anticancer effect (Figure 13.12) [88]. They also reported the cryo-EM structure of the DNMDP-PDE3A-SLFN12 complex, providing direct evidence of DNMDP's role as an MG [89]. Structural optimization of DNMDP led to the development of BRD9500, which exhibited enhanced activity (Figure 13.11, compound 10) [90].

Anagrelide (Figure 13.11, compound 11), identified by Meanwell et al. through DLS screening of FDA-approved drugs, has been recognized as a MG for PDE3A–SLFN12. It enhanced the effects of cytokines such as IFN-α, thereby suppressing cancer cell proliferation by modulating the expression of various tumor suppressor genes [91]. In 2021, Wang et al. used cryo-electron microscopy to elucidate the ternary complex structure of anagrelide with PDE3A and SLFN12 (Figure 13.10c) [85]. A hydrophobic cavity was observed to be close to the 7-Cl

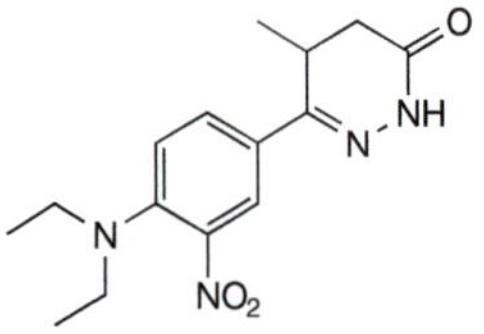

9 (DNMDP)
IC_{50} (SK-MEL-3) = 12 nM

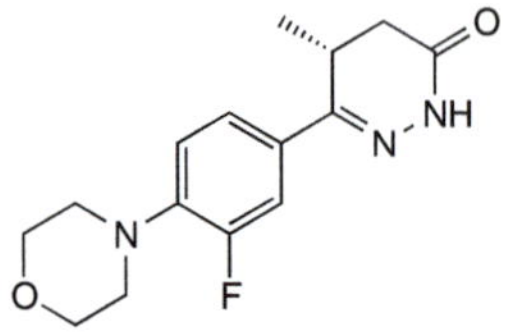

10 (BRD9500)
IC_{50} (SK-MEL-3) = 12 nM

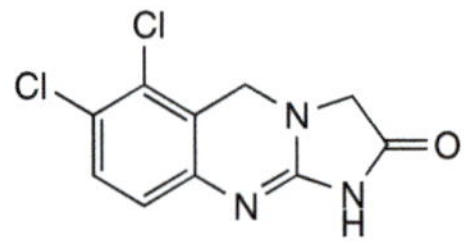

11 (Anagrelide)
IC_{50} (HeLa) = 6.67 nM

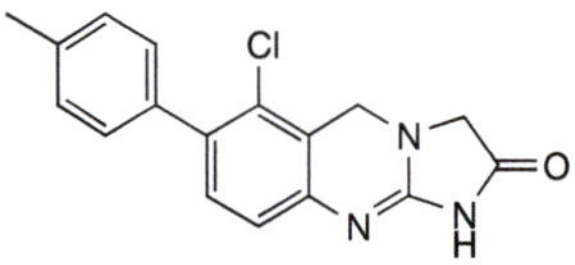

12 (A6)
IC_{50} (HeLa) = 0.30 nM

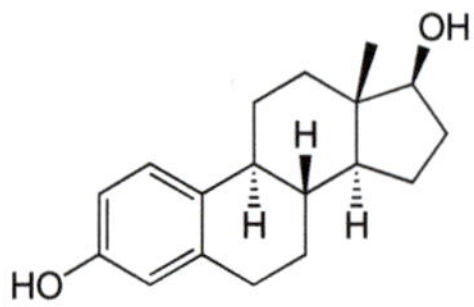

13 (E2)
IC_{50} (HeLa) = 1.5 μM

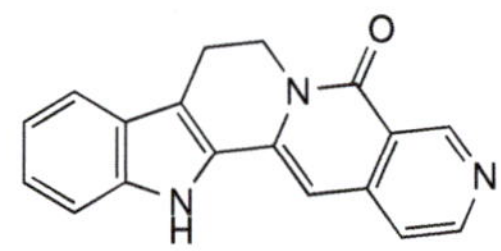

14 (Nauclefine)
IC_{50} (HeLa) = 5.2 nM

Figure 13.11 Molecular Glues that target PDE3A–SLFN12. (This figure was created with ChemDraw (version: 23.0.1).) Source: Xiangbing Qi.

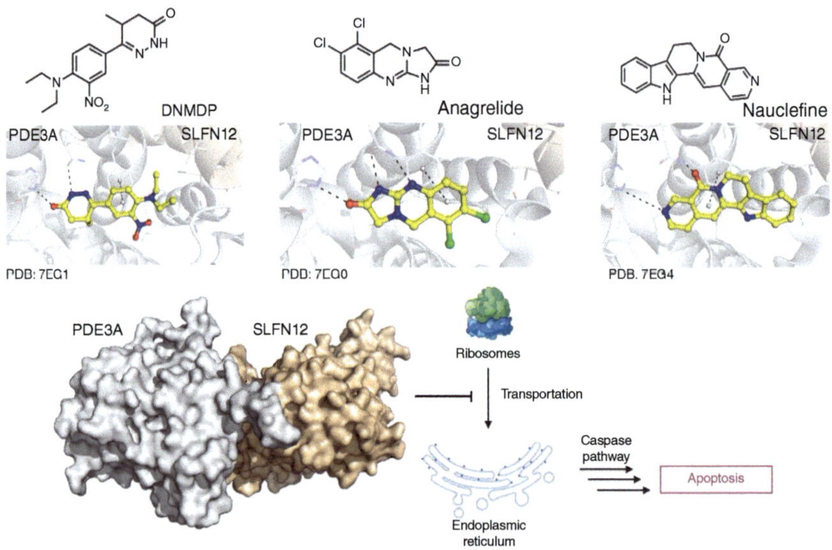

Figure 13.12 Mechanism of action of DNMDP, anagrelide, and nauclefine. (This figure was created with pymol (version:1.7.2.1) and BioRender.com.) Source: Xiangbing Qi.

position of anagrelide, which aligned precisely with the position occupied by the N, N-diethyl group in DNMDP (compound 9). A variety of hydrophobic groups was introduced to the phenyl ring of anagrelide, leading to the identification of compound A6 (Figure 13.11, compound 12) [85], which exhibited the highest cell activity among the reported PDE3A–SLFN12 MG compounds, with an IC_{50} value of 0.30 nmol/l.

In 2019, Wang et al. identified 17-β-estradiol (E2) (Figure 13.11, compound 13) as a regulator of the PDE3A–SLFN12 interaction [92]. Their studies revealed that the E2-PDE3A–SLFN12 complex bound to the ribosome, disrupting the translation of intracellular membrane proteins. This interaction led to a reduction in BCL-2 and MCL-1 levels, while failing to inhibit BAX and BAK effectively. The insufficient inhibition of BAX and BAK resulted in their oligomerization, forming pores in the mitochondrial outer membrane (Figure 13.12). This pore formation facilitated the release of cytochrome C into the cytoplasm, activating caspase-9, which in turn activated caspase-3, culminating in apoptosis and cell death [92]. In subsequent research, Qi et al. identified nauclefine (Figure 13.11, compound 14) as a potent MG for PDE3A–SLFN12, demonstrating remarkable anti-cancer activity with an IC_{50} value of less than 10 nmol/l. It's efficacy was confirmed in mouse models [89].

13.2.2.3 MG Inhibitors

Cyclosporin A/Rapamycin/FK506 Several classes of natural products function as non-degrader MGs [52]. Prominent examples include cyclosporin A, rapamycin, and FK506 (Figure 13.13). Cyclosporin A, an immunosuppressive agent, binds to both cyclophilin and calcineurin, a Ca^{2+}/calmodulin-dependent protein phosphatase. This interaction forms a ternary complex that inhibits calcineurin's phosphatase activity, subsequently modulating interleukin-2 activation [93]. Rapamycin, a macrolide antibiotic first synthesized in 1975 from *Streptomyces*, exhibits notable anti-tumor and immunosuppressive effects [94, 95]. It binds FK506-binding protein 12 (FKBP12) and FKBP-rapamycin-associated protein (FRAP), facilitating the recruitment of mTOR and enhancing its interaction with the FKBP-rapamycin binding domain (FRB) of mTOR. This binding influences downstream intracellular signaling pathways, as evidenced by *in vitro* and *in vivo* studies (Figure 13.9) [96]. Similarly, FK506, another macrolide antibiotic, promotes PPIs between FKBP and various downstream proteins, including calcineurin, thereby impacting intracellular signal transduction (Figure 13.15) [96–98].

MEK-KSP Recently, an unexpected MG interaction was discovered involving the MEK inhibitor trametinib and a naturally occurring PPI. In 2020, Khan et al. [99] elucidated the crystal structures of MEK in complex with the pseudokinase KSR (kinase suppressor of RAS) in the presence of MEK inhibitors. Trametinib, an FDA-approved drug, directly interacts with KSR at the MEK interface, forming a ternary MG-type complex (Figure 13.14) [99]. Trametinib acts as a kinase inhibitor by binding to an allosteric site on MEK, adjacent to the ATP-binding site, without competing with ATP. Its 3-substituted phenyl acetamide group extends into a sub-pocket, forming interactions at the KSR interface. This unique phenyl acetamide moiety differentiates trametinib from other MEK inhibitors by residing in a cavity formed between MEK and KSR, engaging both proteins. Trametinib disrupts the RAF–MEK complex due to steric clashes between the phenyl acetamide group and the pre-helix αG loop in BRAF. Building on this structural insight, a derivative named trametiglue was developed to investigate whether modifications at the KSRMEK and RAF-MEK interfaces could overcome adaptive resistance to MEK inhibitors. Trametiglue, incorporating a sulfamide group instead of the acetamide group, formed two additional hydrogen bonds at the KSR-MEK interface compared to trametinib. Coimmunoprecipitation experiments revealed that trametiglue significantly increased BRAF capture, indicating an enhanced interaction between BRAF and MEK. This approach underscores the potential for designing novel MEK inhibitors by targeting binding interfaces within regulatory complexes of the MAPK pathway with MG-type compounds.

KRAS-CYPA KRAS is a small GTPase that toggles between an inactive state bound to guanosine diphosphate (GDP) and an active state bound to guanosine triphosphate (GTP) [100]. In its active form, KRAS engages various effector proteins to regulate cellular growth and proliferation [101]. Oncogenic mutations in KRAS lead to dysregulated signaling pathways and unchecked cell proliferation [102]. Despite its critical role in cancer, KRAS has traditionally been considered "undruggable" due to the lack of suitable binding sites for small molecule inhibitors [103–106].

In 2023, Christopher J. Schulze et al. [103] demonstrated that the CYPA: drug: KRAS-G12C complex – where glycine-12 is mutated to cysteine – can effectively deactivate oncogenic signaling and induce tumor regression in various human cancer cell lines. Their research exploited the chaperone CYPA due to its advantageous electrostatic surface charge complementarity with residues on the KRAS effector binding interface. This characteristic is not shared by FKBP12. The team developed tricomplex inhibitors (Figure 13.15), including a tool ligand featuring a minimal CYPA-binding motif derived from sanglifehrin A, which binds CYPA with high affinity, coupled with a promiscuous cysteine-reactive warhead. Through SAR optimization, compound RMC-4998 was developed (Figure 13.16), demonstrating superior efficacy compared to existing inactive state-selective $KRAS^{G12C}$ inhibitors [103]. Notably, RMC-4998 specifically targets the active or GTP-bound state of KRAS-G12C; neither RMC-4998 nor CYPA could bind KRAS-G12C independently, and no complex formation was observed with GDP-bound KRAS-G12C, underscoring its selectivity for the active state.

Subsequent optimization of RMC-4998 led to the development of RMC-6291, which is currently undergoing phase I/IB clinical trials (NCT05462717). RMC-6291 specifically targets the active state of KRAS-G12C and exhibits closely related chemical structures and nearly identical kinetic constants compared to RMC-4998. Additionally, a reversible tricomplex inhibitor, RMC-6236 (RASMULT1), developed using this approach, is also in phase I/IB clinical trials (NCT05379985) (Figure 13.17) [103].

13.3 Methods for MG Discovery

Recently, a diverse range of methods has been developed to advance the field of MG discovery. This section will summarize the prominent methods used in the discovery of MGs and their relevant applications.

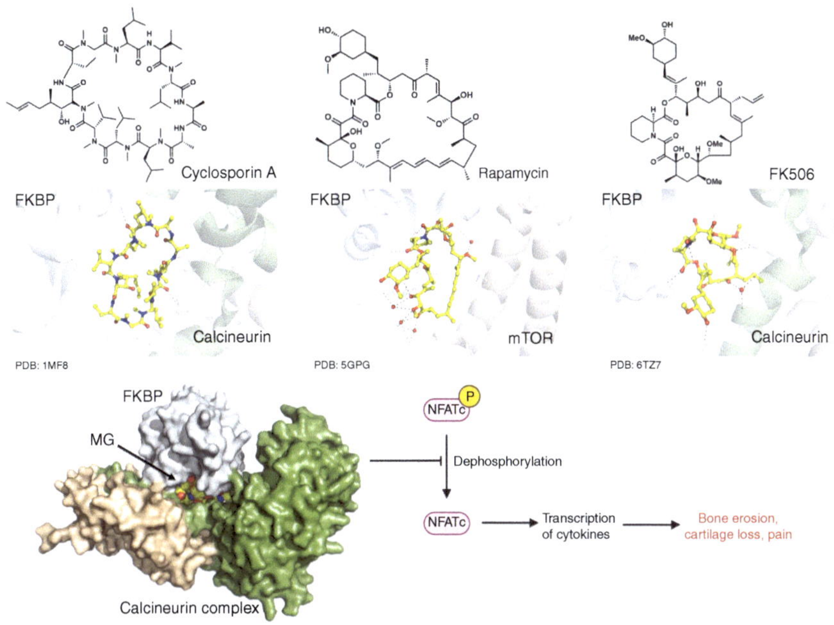

Figure 13.13 Mechanism of action of FK506, cyclosporin A, and rapamycin. (This figure was created with pymol (version:1.7.2.1) and BioRender.com.) Source: Xiangbing Qi.

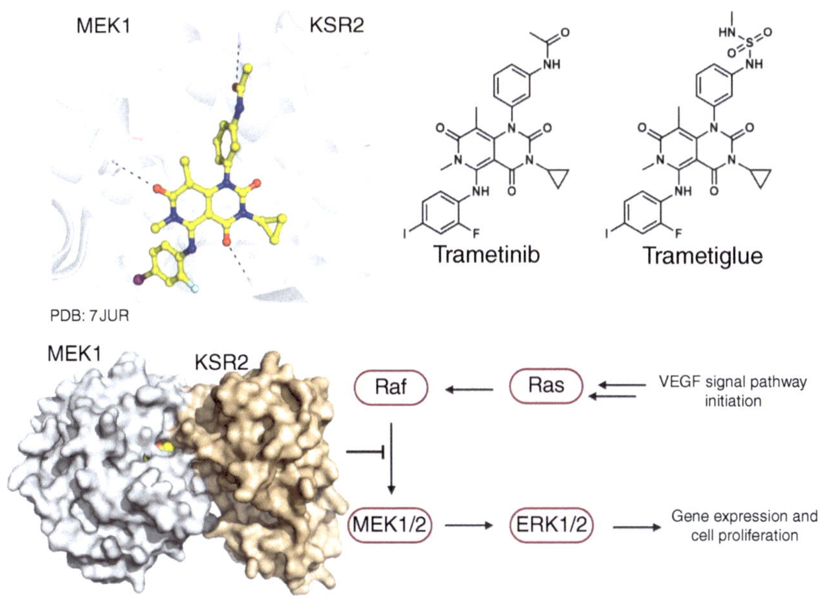

Figure 13.14 Mechanism of action of trametinib. (This figure was created with pymol (version:1.7.2.1) and BioRender.com.) Source: Xiangbing Qi.

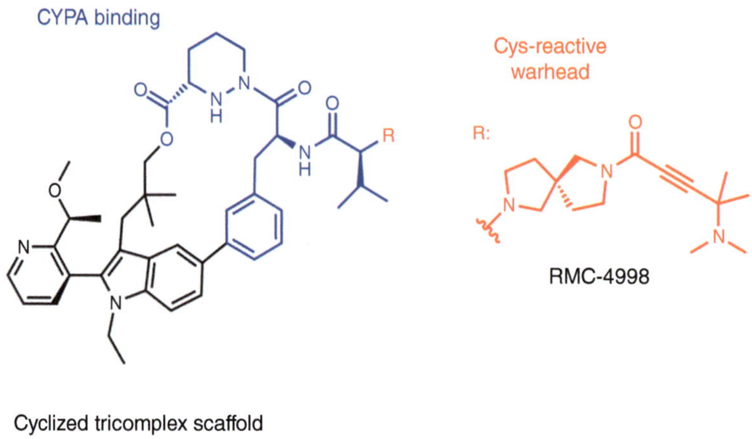

Figure 13.15 The chemical structure of the macrocyclized scaffold for subsequent compounds, with the sanglifehrin-derived CYPA-binding moiety shown in blue, and the chemical structure of RMC-4998. Source: Xiangbing Qi.

13.3.1 Modifications of E3 Ligands: Discovery of New CRBN Ligands

CRBN is the most extensively utilized E3 ligases in targeted protein degradation [107]. It has been effectively recruited by MGDs such as IMiDs [55, 62–65, 69]. Within the Ikaros family of transcription factors, Ikaros (IKZF1) and Aiolos (IKZF3) are susceptible to degradation by canonical IMiDs. This selective degradation is facilitated by a glutamine residue in their second zinc-finger domain, which is critical for IMiD-induced degradation. In contrast, Helios (IKZF2), which contains histidine residues, does not undergo degradation by IMiDs.

In 2021, Wang et al. [104] described their structure-guided development of CRBN-dependent molecular glue degraders specifically targeting Helios (IKZF2). Their objective was to reengineer CRBN to direct its degradation activity toward Helios. Initially, Wang et al. synthesized the IMiD derivative ALV-02-146-03 and confirmed its ability to bind to CRBN. They then used a TR-FRET-based CRBN: Helios dimerization assay to screen a focused library of analogues. Through iterative optimization, they developed two notable compounds, ALV1 and ALV2. ALV1 induces CRBN-dependent degradation of Helios while also destabilizing Ikaros, whereas ALV2 demonstrates relative selectivity for Helios. Eos (IKZF4), which also has a histidine residue at the same position as Helios, is similarly destabilized by both ALV1 and ALV2. The degradation of Helios by these molecular glues alleviates the anergic phenotype (T-cell unresponsiveness) and reduces the suppressive activity of Treg cells. Additionally, the Helios-targeted molecular glue degrader DKY709, developed by Novartis, is currently undergoing Phase I clinical trials for solid tumors, both as a monotherapy and in combination with spartalizumab, an anti–PD-1 antibody. Similar structure-guided approaches could be applied to develop MGDs for other targets. By reprogramming E3 ligase substrate specificity, it is possible to target a wider range of previously intractable proteins.

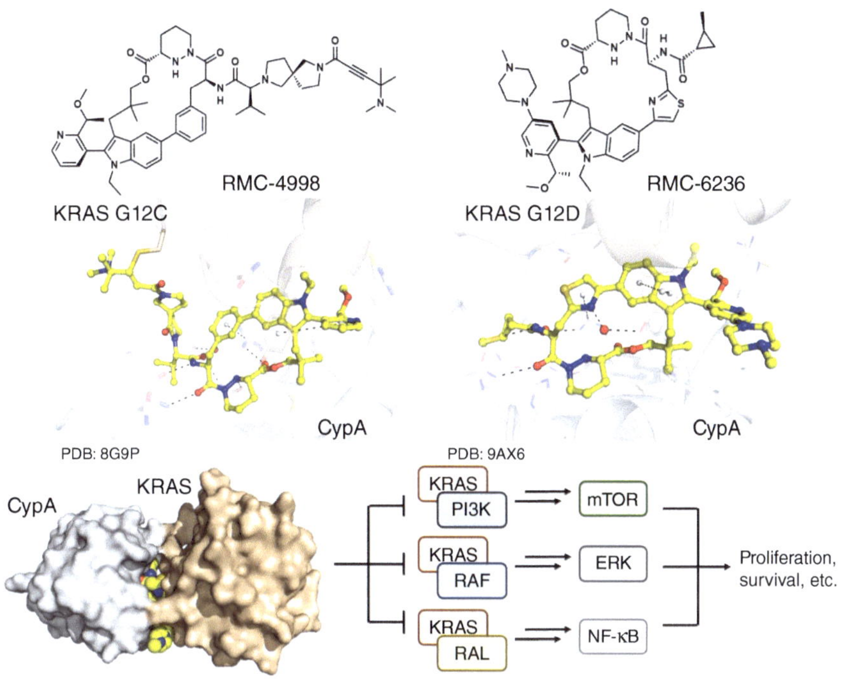

Figure 13.16 Mechanism of action of RMC-4998 and RMC-6236. (This figure was created with pymol (version:1.7.2.1) and BioRender.com.) Source: Xiangbing Qi.

13.3.2 Phenotype-Based Screening

13.3.2.1 Scalable Chemical Profiling

Mayor-Ruiz et al. [78] employed a phenotypic cell-based approach utilizing hyponeddylated models engineered by their team. Cullin-RING ligases (CRLs) are activated through reversible NEDD8 conjugation to the cullin component. Neddylation enhances CRLs' ubiquitination activity by promoting the assembly of the ubiquitin ligation complex and stabilizes CRLs by preventing the exchange of substrate receptors through CAND1. UBE2M, a crucial E2 enzyme in the neddylation pathway, is essential for maintaining CRL function; its dysfunction can result in resistance to degraders by inactivating a significant proportion of CRL complexes.

Utilizing UBE2M-mutant cancer cells, Mayor-Ruiz et al. [78] developed a scalable drug screening strategy based on the differential viability of wild-type (WT) versus UBE2M-mutant cell models. They initially discovered a novel sulfonamide scaffold, dCeMM1, which effectively recruits $CRL4^{DCAF15}$ to selectively degrade RBM39, thus validating their chemical profiling approach. Furthermore, they identified three small molecule drugs, dCeMM2, dCeMM3, and dCeMM4, which require functional CRLs to exert their toxic effects.

Through comprehensive intersection of quantitative proteomics and functional genomic strategies, these compounds were confirmed MGDs of cyclin K. Recently,

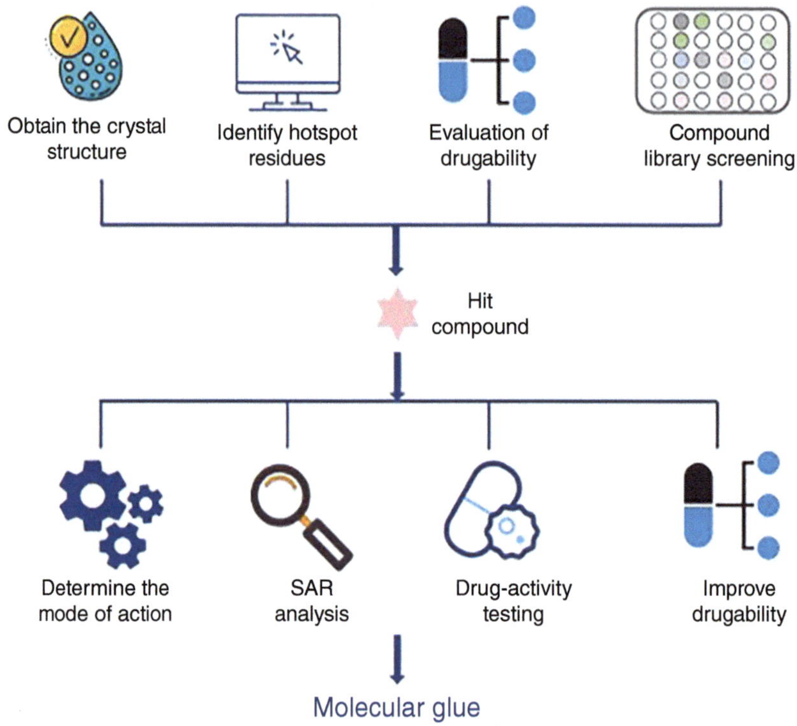

Figure 13.17 The basic process of rational design of MGs. (This figure was created with Microsoft PowerPoint.) Source: Xiangbing Qi.

Nomura et al. performed screening of a 750-member covalent fragment library using WT versus siUBE2M transfections in HAP1 cells. The fragment EN450 was identified to exhibit differential effects in WT versus siUBE2M cells and was found to act as a MG by covalently binding to the E2 enzyme UBE2D, thereby inducing the degradation of NFKB1 [105].

13.3.2.2 Proteomics-Based Screening

Proteomics-based screening is increasingly used to identify MGs by evaluating their selectivity and specificity toward protein targets [106]. Recent advancements in proteomic methodologies, including chemoproteomics, phosphoproteomics, and quantitative MS-based proteomics (such as SILAC, TMT, and label-free quantitative proteomics), have provided powerful tools for MG discovery [106, 108–110].

Traditional ligand identification methods typically involve screening recombinant proteins for small molecules that interact with specific structural regions or biochemical activities, employing techniques like surface plasmon resonance (SPR) and fluorescence polarization [107, 111]. These approaches assume that recombinant proteins accurately represent their physiological structure and function. Chemoproteomics, however, offers a more comprehensive strategy by characterizing ligand–protein interactions within intact cells, providing insights into off-target binding on a proteome-wide scale [112]. By combining drug-affinity enrichment with mass spectrometry (MS), chemoproteomics can uncover actionable binding sites within natural protein environments, facilitating the optimization of effective small-molecule ligands [113]. This approach enhances our understanding of small-molecule protein-binding mode, selectivity, and mechanisms.

Current methods for detecting POIs and E3 ligases often rely on biochemical techniques such as immunoblotting, which necessitate the development and validation of specific antibodies. In contrast, proteomic analysis provides an unbiased approach to discovering all relevant players and pathways involved in TPD [113]. MS-based proteomics allows for the identification and quantification of thousands of proteins from complex biological samples in a single experiment [114]. The sensitivity, speed, and precision of MS enable rapid profiling of protein abundance [115]. Advances such as data-independent acquisition (DIA) have further improved the accuracy and throughput of proteome analysis [116, 117]. Quantitative proteomics of both diseased and healthy samples can provide insights into protein abundance, localization, modifications, and interactions, aiding in the selection of targets and E3 ligases for TPD [118, 119]. Proteomic approaches are valuable for analyzing protein degradation, studying degrader–target interactions, and investigating downstream signaling pathways. Techniques such as thermal protein stability profiling, drug affinity-responsive target stability (DARTS), and proximity labeling can be employed to gain deeper insights into these processes [120–123].

13.3.3 Target-Based Screening

13.3.3.1 HTS for PPI Stabilizers

In 2019, Simonetta et al. [3] reported advancements in the design of small molecules that reestablish the interaction between CRL1β-TrCP and mutant β-catenin. Typically, CRL1βTrCP targets β-catenin for ubiquitination by recognizing a phospho-degron, including Ser33 and Ser37. However, mutations in β-catenin can stabilize it, leading to aberrant oncogenic transcription [38]. Focusing on Ser37, a critical mutation site, Simonetta's team utilized a phospho-degron peptide with weak E3 binding in a high-throughput biochemical screen based on fluorescence polarization [124]. This approach, applied to a library of 350,000 small compounds, identified several molecules such as NRX-1532 that enhanced the binding and CRL1β-TrCP–mediated ubiquitination of mutant β-catenin. Optimization efforts, including targeted molecular modifications at the β-catenin: β-TrCP interface, resulted in derivatives NRX-252114 and NRX-252262 with up to 10,000-fold increased potency. Notably, these compounds function by leveraging the high cooperativity of the ternary complex rather than binding to the E3 ligase alone, facilitating induced dimerization. This research illustrates the potential for designing MGDs that restore disrupted interactions between E3 ligases and their substrates. It provides a proof of concept for the development of future MGDs aimed at specific E3 ligases and substrates.

13.3.3.2 Rational Design: Discovery of KRAS-CYPA Molecular Glue

Cyclosporin A (CypA) has been exploited in designing MG inhibitors for challenging drug targets like KRAS [103]. Based on structure-guided design, researchers have developed a MG that specifically targets the active state of KRAS. This glue recruits CypA to KRAS, obstructing other oncogenic interactions and inhibiting KRAS-driven cell proliferation. Similarly, molecular glues have been engineered to stabilize the interaction between CRAF and its native inhibitor protein, 14–3–3 [125, 126]. These molecular glues lock the CRAF/14–3–3 complex in an inhibitory conformation, even in the absence of the phosphorylation typically required for complex stabilization.

13.4 Conclusions and Outlook

PPIs are essential for regulating a multitude of biological processes in both physiological and pathological states, positioning them as attractive targets for novel clinical therapies. MGs, which modulate these interactions, have transitioned from serendipitous discoveries to rational design. Several MGs are currently in clinical trials or already on the market, as discussed in this chapter. To further advance the development and application of MGs, several key directions could be highlighted:

1) Methods to identify pairs of proteins suitable for MG discovery: MGs typically enhance weak PPIs, which are difficult to capture. Therefore, techniques capable of detecting these weak PPIs are needed to identify suitable protein pairs for MG discovery.
2) There is a need for more Development of refined and HTS methods: There is a need to develope new assays that better capture the dynamics of PPIs and their modulation by MGs.
3) Structural insights into MG design principles: Greater emphasis on structural biology is required to elucidate the precise binding modes and conformational changes induced by MGs. This will aid in the rational design of more effective compounds.
4) Addressing the challenges in MG optimization: Further optimization of MGs to enhance specificity, efficacy, and pharmacokinetic properties is crucial. This includes fine-tuning the interaction interfaces and minimizing off-target effects.

Addressing these areas will be critical in advancing the field of MGs and their therapeutic applications.

Contributors

Drs. Xiangbing Qi and Ting Han conceived of the study; Linlin Fan, Ting Han and Xiangbing Qi prepared the manuscript; all authors discussed the results and commented on the manuscript.

References

1 Rapposelli, S., Gaudio, E., Bertozzi, F., and Gul, S. (2021). Editorial: Protein–protein interactions: drug discovery for the future. *Frontiers in Chemistry* 9: 811190.

2 Domostegui, A., Nieto-Barrado, L., Perez-Lopez, C., and Mayor-Ruiz, C. (2022). Chasing molecular glue degraders: screening approaches. *Chemical Society Reviews* 51 (13): 5498–5517.

3 Simonetta, K.R., Taygerly, J., Boyle, K. et al. (2019). Prospective discovery of small molecule enhancers of an E3 ligase-substrate interaction. *Nature Communications* 10 (1): 1402.

4 Peng, C., Trojanowski, J.Q., and Lee, V.M.Y. (2020). Protein transmission in neurodegenerative disease. *Nature Reviews Neurology* 16 (4): 199–212.

5 Nag, S. and Qin, J. (2013). The MDM2-p53 pathway revisited. *Journal of Biomedical Research* 27 (4): 254–271.

6 Gottlieb, T.M. and Oren, M. (1996). p53 in growth control and neoplasia. *Biochimica et Biophysica Acta (BBA) - Reviews on Cancer* 1287 (2–3): 77–102.

7 Levine, A.J. (1997). p53, the cellular gatekeeper for growth and division. *Cell* 88 (3): 323–331.

8 Bálint, É. and Vousden, K.H. (2001). Activation and activities of the p53 tumour suppressor protein. *British Journal of Cancer* 85 (12): 1813–1823.

9 Haupt, Y., Maya, R., Kazaz, A., and Oren, M. (1997). Mdm2 promotes the rapid degradation of p53. *Nature* 387 (6630): 296–299.

10 Kussie, P.H., Gorina, S., Marechal, V. et al. (1996). Structure of the MDM2 oncoprotein bound to the p53 tumor suppressor transactivation domain. *Science* 274 (5289): 948–953.

11 Wade, M., Li, Y.C., and Wahl, G.M. (2013). MDM2, MDMX and p53 in oncogenesis and cancer therapy. *Nature Reviews Cancer* 13 (2): 83–96.

12 Hainaut, P. and Hollstein, M. (1999). p53 and human cancer: the first ten thousand mutations. *Advances in Cancer Research* 77: 81–137.

13 Shangary, S. and Wang, S. (2009). Small-molecule inhibitors of the MDM2-p53 protein-protein interaction to reactivate p53 function: a novel approach for cancer therapy. *Annual Review of Pharmacology and Toxicology* 49 (1): 223–241.

14 Wong, Y.C. and Krainc, D. (2017). α-Synuclein toxicity in neurodegeneration: mechanism and therapeutic strategies. *Nature Medicine* 23 (2): 1–13.

15 Lee, S.J.C., Nam, E., Lee, H.J. et al. (2017). Towards an understanding of amyloid-β oligomers: characterization, toxicity mechanisms, and inhibitors. *Chemical Society Reviews* 46 (2): 310–323.

16 Tabrizi, S.J., Flower, M.D., Ross, C.A., and Wild, E.J. (2020). Huntington disease: new insights into molecular pathogenesis and therapeutic opportunities. *Nature Reviews Neurology* 16 (10): 529–546.

17 Rajalingam, K. and Dikic, I. (2016). Expanding the ubiquitin code. *Cell* 164 (5): 1074–1074.e1.

18 Kumar, D., Ambasta, R.K., and Kumar, P. (2020). Ubiquitin biology in neurodegenerative disorders: from impairment to therapeutic strategies. *Ageing Research Reviews* 61: 101078.

19 Wang, G.P., Khatoon, S., Iqbal, K., and Grundke-Iqbal, I. (1991). Brain ubiquitin is markedly elevated in Alzheimer disease. *Brain Research* 566 (1–2): 146–151.

20 Neumann, M., Sampathu, D.M., Kwong, L.K. et al. (2006). Ubiquitinated TDP-43 in frontotemporal lobar degeneration and amyotrophic lateral sclerosis. *Science* 314 (5796): 130–133.

21 Iqbal, K. and Grundke-Iqbal, I. (1991). Ubiquitination and abnormal phosphorylation of paired helical filaments in Alzheimer's disease. *Molecular Neurobiology* 5 (2–4): 399–410.

22 Abreha, M.H., Dammer, E.B., Ping, L. et al. (2018). Quantitative analysis of the brain ubiquitylome in Alzheimer's disease. *Proteomics* 18 (20): e1800108.

23 Hedna, R., Kovacic, H., Pagano, A. et al. (2022). Tau protein as therapeutic target for cancer? Focus on glioblastoma. *Cancers* 14 (21): 5386.

24 Chu, T.T., Gao, N., Li, Q.Q. et al. (2016). Specific knockdown of endogenous tau protein by peptide-directed ubiquitin-proteasome degradation. *Cell Chemical Biology* 23 (4): 453–461.

25 Lu, H., Zhou, Q., He, J. et al. (2020). Recent advances in the development of protein–protein interactions modulators: mechanisms and clinical trials. *Signal Transduction and Targeted Therapy* 5 (1): 213.

26 Skwarczynska, M. and Ottmann, C. (2015). Protein–protein interactions as drug targets. *Future Medicinal Chemistry* 7 (16): 2195–2219.

27 Conte, L.L., Chothia, C., and Janin, J. (1999). The atomic structure of protein-protein recognition sites. *Journal of Molecular Biology* 285 (5): 2177–2198.

28 DeLano, W.L., Ultsch, M.H., de Vos, A.M., and Wells, J.A. (2000). Convergent solutions to binding at a protein-protein interface. *Science* 287 (5456): 1279–1283.

29 Arkin, M.R. and Wells, J.A. (2004). Small-molecule inhibitors of protein–protein interactions: progressing towards the dream. *Nature Reviews Drug Discovery* 3 (4): 301–317.

30 Shin, W.H., Kumazawa, K., Imai, K. et al. (2020). Current challenges and opportunities in designing protein–protein interaction targeted drugs. *Advances and Applications in Bioinformatics and Chemistry* 13: 11–25.

31 Ni, D., Liu, N., and Sheng, C. (2019). Allosteric modulators of protein–protein interactions (PPIs). *Advances in Experimental Medicine and Biology* 1163: 313–334.

32 Tse, C., Shoemaker, A.R., Adickes, J. et al. (2008). ABT-263: a potent and orally bioavailable Bcl-2 family inhibitor. *Cancer Research* 68 (9): 3421–3428.

33 Pemmaraju, N., Garcia, J.S., Potluri, J. et al. (2022). Addition of navitoclax to ongoing ruxolitinib treatment in patients with myelofibrosis (REFINE): a post-hoc analysis of molecular biomarkers in a phase 2 study. *The Lancet Haematology* 9 (6): e434–e444.

34 Buckley, D.L., Gustafson, J.L., Van Molle, I. et al. (2012). Small-molecule inhibitors of the interaction between the E3 ligase VHL and HIF1α. *Angewandte Chemie International Edition* 51 (46): 11463–11467.

35 Buckley, D.L., Van Molle, I., Gareiss, P.C. et al. (2012). Targeting the von Hippel–Lindau E3 ubiquitin ligase using small molecules to disrupt

the VHL/HIF-1α interaction. *Journal of the American Chemical Society* 134 (10): 4465–4468.
36 Yap, J.L., Wang, H., Hu, A. et al. (2013). Pharmacophore identification of c-Myc inhibitor 10074-G5. *Bioorganic & Medicinal Chemistry Letters* 23 (1): 370–374.
37 Yin, H., Lee, G., Sedey, K.A. et al. (2005). Terphenyl-based Bak BH3 α-helical proteomimetics as low-molecular-weight antagonists of Bcl-x_L. *Journal of the American Chemical Society* 127 (29): 10191–10196.
38 Chen, L. (2005). p53-Helix mimetics antagonize p53/MDM2 interaction and activate p53. *Molecular Cancer Therapeutics* 4 (6): 1019–1025.
39 Byalik, T.E. (2018). Venetoclax in treatment of chronic lymphocytic leukemia (literature review). *Oncohematology* 13 (2): 32–38.
40 Vassilev, L.T. (2004). In vivo activation of the p53 pathway by small-molecule antagonists of MDM2. *Science* 303 (5659): 844–848.
41 Tovar, C., Rosinski, J., Filipovic, Z. et al. (2006). Small-molecule MDM2 antagonists reveal aberrant p53 signaling in cancer: Implications for therapy. *Proceedings of the National Academy of Sciences* 103 (6): 1888–1893.
42 Iancu-Rubin, C., Mosoyan, G., Glenn, K. et al. (2014). Activation of p53 by the MDM2 inhibitor RG7112 impairs thrombopoiesis. *Experimental Hematology* 42 (2): 137–145.e5.
43 Kwon, D.H., Eom, G.H., Ko, J.H. et al. (2016). MDM2 E3 ligase-mediated ubiquitination and degradation of HDAC1 in vascular calcification. *Nature Communications* 7 (1): 10492.
44 Ran, X. and Gestwicki, J.E. (2018). Inhibitors of protein–protein interactions (PPIs): an analysis of scaffold choices and buried surface area. *Current Opinion in Chemical Biology* 44 (44): 75–86.
45 Martino, E., Chiarugi, S., Margheriti, F., and Garau, G. (2021). Mapping, structure and modulation of PPI. *Frontiers in Chemistry* 9: 718405.
46 Cao, S., Kang, S., Mao, H. et al. (2022). Defining molecular glues with a dual-nanobody cannabidiol sensor. *Nature Communications* 13 (1): 815.
47 Titani, K. and Walsh, K.A. (1988). Human von Willebrand factor: the molecular glue of platelet plugs. *Trends in Biochemical Sciences* 13 (3): 94–97.
48 Faux, M.C. and Scott, J.D. (1996). Molecular glue: kinase anchoring and scaffold proteins. *Cell* 85 (1): 9–12.
49 Tan, X., Calderon-Villalobos, L.I.A., Sharon, M. et al. (2007). Mechanism of auxin perception by the TIR1 ubiquitin ligase. *Nature* 446 (7136): 640–645.
50 Schreiber, S.L. (2024). Molecular glues and bifunctional compounds: therapeutic modalities based on induced proximity. *Cell Chemical Biology* 31 (6): 1050–1063.
51 Holdgate, G.A., Bardelle, C., Berry, S.K. et al. (2023). Screening for molecular glues – challenges and opportunities. *SLAS Discovery* 29: 100136–6.
52 Chen, Y.N.P., LaMarche, M.J., Chan, H.M. et al. (2016). Allosteric inhibition of SHP2 phosphatase inhibits cancers driven by receptor tyrosine kinases. *Nature* 535 (7610): 148–152.
53 Nishimura, K., Fukagawa, T., Takisawa, H. et al. (2009). An auxin-based degron system for the rapid depletion of proteins in nonplant cells. *Nature Methods* 6 (12): 917–922.

54 Natsume, T., Kiyomitsu, T., Saga, Y., and Kanemaki, M.T. (2016). Rapid protein depletion in human cells by auxin-inducible degron tagging with short homology donors. *Cell Reports* 15 (1): 210–218.

55 Bartlett, J.B., Dredge, K., and Dalgleish, A.G. (2004). The evolution of thalidomide and its IMiD derivatives as anticancer agents. *Nature Reviews Cancer* 4 (4): 314–322.

56 Atra, E. and Sato, E.I. (1993). Treatment of the cutaneous lesions of systemic lupus erythematosus with thalidomide. *Clinical and Experimental Rheumatology* 11 (5): 487–493.

57 Gutiérrez-Rodríguez, O. (1984). Thalidomide a promising new treatment for rheumatoid arthritis. *Arthritis & Rheumatism* 27 (10): 1118–1121.

58 Zhu, Y.X., Kortuem, K.M., and Stewart, A.K. (2012). Molecular mechanism of action of immune-modulatory drugs thalidomide, lenalidomide and pomalidomide in multiple myeloma. *Leukemia & Lymphoma* 54 (4): 683–687.

59 Fischer, E.S., Böhm, K., Lydeard, J.R. et al. (2014). Structure of the DDB1–CRBN E3 ubiquitin ligase in complex with thalidomide. *Nature* 512 (7512): 49–53.

60 Lopez-Girona, A., Mendy, D., Ito, T. et al. (2012). Cereblon is a direct protein target for immunomodulatory and antiproliferative activities of lenalidomide and pomalidomide. *Leukemia* 26 (11): 2326–2335.

61 Ito, T., Ando, H., Suzuki, T. et al. (2010). Identification of a primary target of thalidomide teratogenicity. *Science* 327 (5971): 1345–1350.

62 Petzold, G., Fischer, E.S., and Thomä, N.H. (2016). Structural basis of lenalidomide-induced CK1α degradation by the $CRL4^{CRBN}$ ubiquitin ligase. *Nature* 532 (7597): 127–130.

63 Krönke, J., Udeshi, N.D., Narla, A. et al. (2014). Lenalidomide causes selective degradation of IKZF1 and IKZF3 in multiple myeloma cells. *Science* 343 (6168): 301–305.

64 Sievers, Q.L., Petzold, G., Bunker, R.D. et al. (2018). Defining the human C2H2 zinc finger degrome targeted by thalidomide analogs through CRBN. *Science* 362 (6414): eaat0572.

65 Matyskiela, M.E., Zhang, W.H., Man, H. et al. (2017). A cereblon modulator (CC-220) with improved degradation of Ikaros and Aiolos. *Journal of Medicinal Chemistry* 61 (2): 535–542.

66 Chamberlain, P.P., Lopez-Girona, A., Miller, K. et al. (2014). Structure of the human Cereblon–DDB1–lenalidomide complex reveals basis for responsiveness to thalidomide analogs. *Nature Structural & Molecular Biology* 21 (9): 803–809.

67 Matyskiela, M.E., Zhu, J., Baughman, J.M. et al. (2020). Cereblon modulators target ZBTB16 and its oncogenic fusion partners for degradation via distinct structural degrons. *ACS Chemical Biology* 15 (12): 3149–3158.

68 Watson, E.R., Novick, S.J., Matyskiela, M.E. et al. (2022). Molecular glue CELMoD compounds are regulators of cereblon conformation. *Science* 378 (6619): 549–553.

69 Hansen, J.D., Correa, M., Alexander, M. et al. (2021). CC-90009: a cereblon E3 ligase modulating drug that promotes selective degradation of GSPT1 for the treatment of acute myeloid leukemia. *Journal of Medicinal Chemistry* 64 (4): 1835–1843.

70 Surka, C., Jin, L., Mbong, N. et al. (2021). CC-90009, a novel cereblon E3 ligase modulator, targets acute myeloid leukemia blasts and leukemia stem cells. *Blood* 137 (5): 661–677.

71 Han, T., Goralski, M., Gaskill, N. et al. (2017). Anticancer sulfonamides target splicing by inducing RBM39 degradation via recruitment to DCAF15. *Science* 356 (6336): eaal3755.

72 Konstantinidou, M. and Arkin, M.R. (2024). Molecular glues for protein-protein interactions: progressing toward a new dream. *Cell Chemical Biology* 31 (6): 1064–1088.

73 Yoshino, H., Ueda, N., Niijima, J. et al. (1992). Novel sulfonamides as potential, systemically active antitumor agents. *Journal of Medicinal Chemistry* 35 (13): 2496–2497.

74 Owa, T., Yoshino, H., Okauchi, T. et al. (1999). Discovery of novel antitumor sulfonamides targeting G1 phase of the cell cycle. *Journal of Medicinal Chemistry* 42 (19): 3789–3799.

75 Faust, T.B., Yoon, H., Nowak, R.P. et al. (2020). Structural complementarity facilitates E7820-mediated degradation of RBM39 by DCAF15. *Nature Chemical Biology* 16 (1): 7–14.

76 Słabicki, M., Kozicka, Z., Petzold, G. et al. (2020). The CDK inhibitor CR8 acts as a molecular glue degrader that depletes cyclin K. *Nature* 585 (7824): 293–297.

77 Lu, L., Chen, P.H., Cao, L. et al. (2020). Discovery of a molecular glue promoting CDK12-DDB1 interaction to trigger cyclin K degradation. *Elife* 9 (9): e59994.

78 Mayor-Ruiz, C., Bauer, S., Brand, M. et al. (2020). Rational discovery of molecular glue degraders via scalable chemical profiling. *Nature Chemical Biology* 16 (11): 1199–1207.

79 Dieter, S.M., Siegl, C., Codó, P.L. et al. (2021). Degradation of CCNK/CDK12 is a druggable vulnerability of colorectal cancer. *Cell Reports* 36 (3): 109394–4.

80 Jorda, R., Havlíček, L., Peřina, M. et al. (2022). 3,5,7-Substituted pyrazolo[4,3-*d*]pyrimidine inhibitors of cyclin-dependent kinases and cyclin K degraders. *Journal of Medicinal Chemistry* 65 (13): 8881–8896.

81 Ahmad, F., Murata, T., Shimizu, K. et al. (2014). Cyclic nucleotide phosphodiesterases: important signaling modulators and therapeutic targets. *Oral Diseases* 21 (1): e25–e50.

82 Kim, Y.R., Yi, M., Cho, S. et al. (2020). Identification and functional study of genetic polymorphisms in cyclic nucleotide phosphodiesterase 3A (*PDE3A*). *Annals of Human Genetics* 85 (2): 80–91.

83 Maurice, D.H., Ke, H., Ahmad, F. et al. (2014). Advances in targeting cyclic nucleotide phosphodiesterases. *Nature reviews Drug Discovery* 13 (4): 290–314.

84 Tian, F.-M., Zhong, C.-Y., Wang, X.-N., and Meng, Y. (2017). PDE3A is hypermethylated in cisplatin resistant non-small cell lung cancer cells and is a

modulator of chemotherapy response. *European Review for Medical and Pharmacological Sciences* 21 (11): 2635–2641.
85 Chen, J., Liu, N., Huang, Y. et al. (2021). Structure of PDE3A–SLFN12 complex and structure-based design for a potent apoptosis inducer of tumor cells. *Nature Communications* 12 (1): 6204.
86 de Waal, L., Lewis, T.A., Rees, M.G. et al. (2015). Identification of cancer-cytotoxic modulators of PDE3A by predictive chemogenomics. *Nature Chemical Biology* 12 (2): 102–108.
87 Lewis, T.A., de Waal, L., Wu, X. et al. (2019). Optimization of PDE3A modulators for SLFN12-dependent cancer cell killing. *ACS Medicinal Chemistry Letters* 10 (11): 1537–1542.
88 An, R., Liu, J., He, J. et al. (2019). PDE3A inhibitor anagrelide activates death signaling pathway genes and synergizes with cell death-inducing cytokines to selectively inhibit cancer cell growth. *American Journal of Cancer Research* 9 (9): 1905–1921.
89 Li, K., Ou, J., and Gao, S. (2016). Total synthesis of camptothecin and related natural products by a flexible strategy. *Angewandte Chemie International Edition* 55 (47): 14778–14783.
90 Li, D., Chen, J., Ai, Y. et al. (2019). Estrogen-related hormones induce apoptosis by stabilizing Schlafen-12 protein turnover. *Molecular Cell* 75 (6): 1103–1116.e9.
91 Meanwell, N.A. (2023). Anagrelide: a clinically effective cAMP phosphodiesterase 3A inhibitor with molecular glue properties. *ACS Medicinal Chemistry Letters* 14 (4): 350–361.
92 Ai, Y., He, H., Chen, P. et al. (2020). An alkaloid initiates phosphodiesterase 3A–schlafen 12 dependent apoptosis without affecting the phosphodiesterase activity. *Nature Communications* 11 (1): 3236.
93 Cardenas, M.E., Hemenway, C., Muir, R.S. et al. (1994). Immunophilins interact with calcineurin in the absence of exogenous immunosuppressive ligands. *The EMBO Journal* 13 (24): 5944–5957.
94 Blazar, B.R., Taylor, P.A., Sehgal, S.N., and Vallera, D.A. (1993). Rapamycin prolongs survival of murine recipients of fully allogeneic donor grafts when administered during the graft-versus-host disease process. *Annals of the New York Academy of Sciences* 685 (1 Immunomodulat): 73–85.
95 Benjamin, D., Colombi, M., Moroni, C., and Hall, M.N. (2011). Rapamycin passes the torch: a new generation of mTOR inhibitors. *Nature Reviews Drug Discovery* 10 (11): 868–880.
96 Brown, E.J., Albers, M.W., Shin, T.B. et al. (1994). A mammalian protein targeted by G1-arresting rapamycin-receptor complex. *Nature* 369 (6483): 756–758.
97 Sabatini, D.M., Erdjument-Bromage, H., Lui, M. et al. (1994). RAFT1: a mammalian protein that binds to FKBP12 in a rapamycin-dependent fashion and is homologous to yeast TORs. *Cell* 78 (1): 35–43.
98 Liu, J., Farmer, J.D., Lane, W.S. et al. (1991). Calcineurin is a common target of cyclophilin-cyclosporin A and FKBP-FK506 complexes. *Cell* 66 (4): 807–815.
99 Khan, Z.M., Real, A.M., Marsiglia, W.M. et al. (2020). Structural basis for the action of the drug trametinib at KSR-bound MEK. *Nature* 588 (7838): 509–514.

100 Simanshu, D.K., Nissley, D.V., and McCormick, F. (2017). RAS proteins and their regulators in human disease. *Cell* 170 (1): 17–33.

101 Malumbres, M. and Barbacid, M. (2003). RAS oncogenes: the first 30 years. *Nature Reviews Cancer* 3 (6): 459–465.

102 Pylayeva-Gupta, Y., Grabocka, E., and Bar-Sagi, D. (2011). RAS oncogenes: weaving a tumorigenic web. *Nature Reviews Cancer* 11 (11): 761–774.

103 Schulze, C.J., Seamon, K.J., Zhao, Y. et al. (2023). Chemical remodeling of a cellular chaperone to target the active state of mutant KRAS. *Science* 381 (6659): 794–799.

104 Qin, W., Cho, K.F., Cavanagh, P.E., and Ting, A.Y. (2021). Deciphering molecular interactions by proximity labeling. *Nature Methods* 18 (2): 133–143.

105 King, E.A., Cho, Y., Hsu, N.S. et al. (2023). Chemoproteomics-enabled discovery of a covalent molecular glue degrader targeting NF-κB. *Cell Chemical Biology* 30: 394–402.e9.

106 Sathe, G. and Sapkota, G.P. (2023). Proteomic approaches advancing targeted protein degradation. *Trends in Pharmacological Sciences* 44 (11): 786–801.

107 Messner, C.B., Demichev, V., Wang, Z. et al. (2022). Mass spectrometry-based high-throughput proteomics and its role in biomedical studies and systems biology. *WILEY Analytical Science*, 2200013–3.

108 Roy, M.J., Winkler, S., Hughes, S.J. et al. (2019). SPR-measured dissociation kinetics of PROTAC ternary complexes influence target degradation rate. *ACS Chemical Biology* 14 (3): 361–368.

109 Crew, A.P., Raina, K., Dong, H. et al. (2017). Identification and characterization of Von Hippel-Lindau-recruiting proteolysis targeting chimeras (PROTACs) of TANK-binding kinase 1. *Journal of Medicinal Chemistry* 61 (2): 583–598.

110 Browne, C.M., Jiang, B., Ficarro, S.B. et al. (2018). A chemoproteomic strategy for direct and proteome-wide covalent inhibitor target-site identification. *Journal of the American Chemical Society* 141 (1): 191–203.

111 Mi, D., Li, Y., Gu, H. et al. (2023). Current advances of small molecule E3 ligands for proteolysis-targeting chimeras design. *European Journal of Medicinal Chemistry* 256: 115444–4.

112 Guzmán U.H., Martínez-Val A., Ye Z. et al. (2023). Narrow-window DIA: Ultra-fast quantitative analysis of comprehensive proteomes with high sequencing depth. *bioRxiv* (Cold Spring Harbor Laboratory). 2023 Jun 6.

113 Demichev, V., Messner, C.B., Vernardis, S.I. et al. (2019). DIA-NN: neural networks and interference correction enable deep proteome coverage in high throughput. *Nature Methods* 17 (1): 41–44.

114 Sathe, G., Albert, M., Darrow, J. et al. (2020). Quantitative proteomic analysis of the frontal cortex in Alzheimer's disease. *Journal of Neurochemistry* 156 (6): 988–1002.

115 Skowronek, P., Krohs, F., Lubeck, M. et al. (2023). Synchro-PASEF allows precursor-specific fragment ion extraction and interference removal in data-independent acquisition. *Molecular & Cellular Proteomics* 22 (2): 100489–9.

116 Watanabe, T. (2006). Identification and characterization of two novel classes of small RNAs in the mouse germline: retrotransposon-derived siRNAs in oocytes and germline small RNAs in testes. *Genes & Development* 20 (13): 1732–1743.

117 Girardini, M., Maniaci, C., Hughes, S.J. et al. (2019). Cereblon versus VHL: hijacking E3 ligases against each other using PROTACs. *Bioorganic & Medicinal Chemistry* 27 (12): 2466–2479.

118 Ito, T., Yamaguchi, Y., and Handa, H. (2021). Exploiting ubiquitin ligase cereblon as a target for small-molecule compounds in medicine and chemical biology. *Cell Chemical Biology* 28 (7): 987–999.

119 Wang, E.S., Verano, A.L., Nowak, R.P. et al. (2021). Acute pharmacological degradation of Helios destabilizes regulatory T cells. *Nature Chemical Biology* 17 (6): 711–717.

120 Nieser C. (2024). Proteomics for PROTACs & Molecular Glues I /Deep/Discovery [Internet]. Sapient I Multi-Omics Services to Discover Dynamic Biomarkers.

121 Dörrbaum, A.R., Kochen, L., Langer, J.D., and Schuman, E.M. (2018). Local and global influences on protein turnover in neurons and glia. *Elife* 7: e34202.

122 Kurzawa, N., Leo, I.R., Stahl, M. et al. (2023). Deep thermal profiling for detection of functional proteoform groups. *Nature Chemical Biology* 19 (8): 962–971.

123 Huang, L., Wang, D., and Zhang, C. (2020). Drug affinity responsive target stability (DARTS) assay to detect interaction between a purified and a small. *Methods in Molecular Biology* 2213: 175–182.

124 Smale, S.T. (2010). Beta-galactosidase assay. *Cold Spring Harbor Protocols* 2010 (5): pdb.prot5423.

125 Molzan, M., Kasper, S., Röglin, L. et al. (2013). Stabilization of physical RAF/14-3-3 interaction by cotylenin A as treatment strategy for RAS mutant cancers. *ACS Chemical Biology* 8 (9): 1869–1875.

126 Ivanenkov, Y.A., Zagribelnyy, B.A., and Aladinskiy, V.A. (2019). Are we opening the door to a new era of medicinal chemistry or being collapsed to a chemical singularity? *Journal of Medicinal Chemistry* 62 (22): 10026–10043.

14

Targeting APC–Asef Protein–Protein Interaction for Drug Discovery in Colorectal Cancer Therapy

Jie Zhong and Xiuyan Yang

Shanghai Jiao Tong University, School of Medicine, Department of Pharmaceutical and Artificial-Intelligence Sciences, Room 1008, Building 2, 280 South Chongqing Road, Shanghai 200025, China

14.1 Introduction

The truncated adenomatous polyposis coli (APC) and Rho guanine nucleotide exchange factor 4 (Asef) protein–protein interaction (PPI) represents a highly promising therapeutic target in colorectal cancer (CRC). Truncating mutations in the *APC* gene are responsible for the majority of both sporadic and familial colorectal tumors [1–3]. The truncated APC loses its tumor suppressor function but continues interacting effectively with Asef, thereby persistently stimulating its guanine nucleotide exchange factor (GEF) activity. Activated Asef facilitates the exchange of guanosine diphosphate (GDP) for guanosine triphosphate (GTP) in Cdc42 or Rac1, leading to the upregulation of matrix metalloproteinase 9 (MMP9) expression primarily through the c-Jun N-terminal kinase (JNK) pathway. This cascade of events promotes cellular flattening, membrane ruffling, and lamellipodia formation, ultimately enhancing cancer cell migration [4–7]. The transition of truncated APC from a tumor suppressor to an oncoprotein has shifted research focus toward its therapeutic potential in CRC.

The *APC* gene is located on chromosome 5q21–q22 and comprises 21 exons. Mutations in *APC* are found in over 80% of sporadic colorectal tumors and are associated with familial adenomatous polyposis (FAP) [8, 9]. Consequently, *APC* has been recognized as a tumor suppressor gene in CRC. The *APC* gene encodes a multi-domain protein weighing 312 kDa, which plays critical roles in various cellular processes, including Wnt/β-catenin signaling [10], cell adhesion [11, 12], cell migration [13, 14], mitosis, and chromosomal instability [15], through different binding partners. The human APC protein features several vital domains: an oligomerization domain (OD), an armadillo repeat domain (ARM), a β-catenin binding region, a basic domain, an end-binding protein 1 (EB1), and a DLG binding domain [16, 17]. The ARM domain is the most conserved region among vertebrate and invertebrate homologs and interacts with various cellular proteins, including

Targeting Protein–Protein Interactions for Drug Discovery, First Edition. Edited by Jian Zhang.

Asef, IQGAP1, kinesin-2, APC membrane recruitment 1 (Amer1), Sam68, and keratin-associated protein (KAP-3) [18–26]. The β-catenin binding region of APC has been extensively studied. It recruits β-catenin into a multi-protein complex that includes axin or its closely related proteins, axin 2/conductin/axil, glycogen synthase kinase 3β, and casein kinase 1α. This recruitment induces proteasomal degradation of β-catenin, thereby negatively regulating Wnt signaling [27–29]. Notably, approximately 60% of APC mutations observed in colon cancers occur within this region, known as the mutation cluster region (MCR) [30, 31]. Additionally, the APC basic domain interacts with mDia, stabilizing microtubules downstream of Rho and promoting cell migration. The EB1 and DLG binding domains also contribute to microtubule stabilization [32, 33].

APC mutations typically result in stable C-terminal truncation of APC in patient samples from FAP and CRC. Truncated APC loses its tumor suppressor function, such as the degradation of β-catenin, leading to the accumulation of β-catenin and activation of Wnt signaling—a critical event in the initiation of CRC [34, 35]. Moreover, the truncated APC effectively interacts with Asef via its ARM domain. Asef comprises several domains, including an APC-binding region (ABR), a Src homology 3 (SH3) domain, a Dbl homology (DH) domain, and a pleckstrin homology (PH) domain. Typically, Asef is auto-inhibited through the binding between the SH3 domain and the DH and PH domains. However, upon binding to truncated APC, this auto-inhibition is released, stimulating the GEF activity of Asef. This activation regulates the cytoskeletal network and influences cell adhesion, migration, angiogenesis, and Golgi fragmentation [7, 25, 36, 37]. These findings suggest that truncated APC exerts dominant functions that may significantly contribute to colon tumorigenesis and metastasis.

The upstream and downstream regulation of APC–Asef PPI is involved in various cellular pathways. (i) Hepatocyte growth factor (HGF), basic fibroblast growth factor (bFGF), and epidermal growth factor (EGF) induce the accumulation and co-localization of APC and Asef in membrane ruffles and lamellipodia. They increase the levels of the APC–Asef PPI in CRCs, thereby promoting cancer metastasis [5]. (ii) Phosphoinositide 3-kinase (PI3K) may also function upstream of APC and Asef within the HGF signaling pathway [38]. (iii) The miR-1-NOTCH3 signaling pathway enhances CRC migration and tumor angiogenesis by upregulating Asef expression [39]. (iv) Truncated APC can induce Golgi fragmentation by activating the Asef-ROCK-MLC2 pathway, resulting in the loss of cell polarity and disrupting regular directed cell migration. This process is considered one of the earliest events in the development of CRC [40]. These findings highlight the complex regulatory networks involving APC and Asef in cancer progression.

Given the clinical implications of APC–Asef PPI, further studies were conducted to investigate the knockdown effects of either APC or Asef. Knocking down either APC or Asef significantly reduced the migration of various colorectal tumor cell lines, including HCT116 and SW480. Additionally, Asef deficiency led to a decrease in both the number and size of adenomas in $Apc^{Min/+}$ mice.

In conclusion, truncated mutant APC in colorectal tumor cells constitutively activates Asef, which enhances tumor cell migration and drives progression to invasive

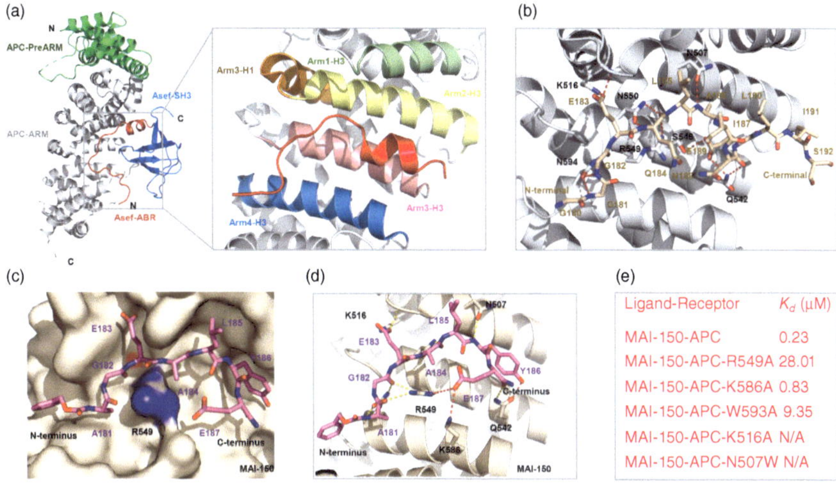

Figure 14.1 (a) APC (PreARM-ARM)/Asef (ABR-SH3) complex structure (PDB: 3NMZ), the PreARM portion of APC is shown in green, and the ARM portions of APC are shown in gray. The ABR domain of Asef is shown in red, and the SH3 domain of Asef is shown in blue. Arm1-H3, Arm2-H3, Arm3-H1, Arm3-H3, and Arm4-H3 are shown in green, yellow, orange, pink, and blue, respectively [41]. (b) Close-up view of interactions between APC and Asef. APC is shown as a gray cartoon, and Asef is shown as a champagne stick. Red dashed lines represent hydrogen bonds. (c) Crystal structure of APC and MAI-150 (PDB: 5IZ6) [42]. APC is shown as a surface representation (gold), and MAI-150 is shown as a stick representation (pink). (d) The interactions of APC (gold cartoon) and MAI-150 (pink stick). Yellow dashed lines represent hydrogen bonds. Red dashed lines represent newly formed hydrogen bonds. (e) The binding affinity of MAI-150 to different APC mutant proteins.

malignancy via multiple pathways. Therefore, the APC–Asef interaction presents a promising therapeutic target for addressing invasive migration in CRC.

14.2 Structural Insights into APC–Asef Interaction

The APC-ARM domain complexed with Asef-ABR was first described by Zhang et al. (PDB: 3NMZ, 42), providing insights into the structural basis of APC–Asef binding mode. APC features a highly conserved surface groove for Asef, formed by the H3 helices of Arm1 through Arm4 and the H1 helix of Arm3 (see Figure 14.1a). The interface between the APC-ARM/Asef-ABR is 1300 $Å^2$ along the exposed surface area. The binding between the APC-ARM/Asef-ABR is tight, with the K_d value of 17.8 nM measured by isothermal titration calorimetry (ITC). An extensive network of hydrogen bonds and van der Waals interactions plays a crucial role in stabilizing this interaction. The hydrogen bond network involves the backbone atoms of residues Gly181, Gly182, Gln184, Ala186, and Asn188, as well as the side chain atoms of residues Glu183 and Glu189 of Asef. Additionally, the aliphatic residues Leu185, Ala186, Ile187, and Leu190 of Asef snugly fit into a hydrophobic pocket on the surface of APC (see Figure 14.1b).

The extensive interface between the APC-ARM and Asef-ABR presents a significant challenge for the design of PPI inhibitors. A critical hot spot, identified as the ORIGIN peptide or MAI-005, comprises the residues ^{181}GGEQLAI^{187} in Asef [42]. Within this hot spot, there are 36 pairs of residue–residue interactions covering an area of 898.6 Å^2, with approximately 60% of the APC-ARM/Asef-ABR interactions concentrated here. Deep mutational scanning indicates that the residues Gly181, Gln184, and Ile187 of MAI-005 are positively influential, potentially enhancing binding affinities by 4–14% through mutations. In contrast, Gly182, Glu183, Leu185, and Ala186 serve as negative residues. These features in the binding hotspot reflect the distinct binding thermodynamic and kinetic profiles and are prerequisites for designing APC–Asef PPI inhibitors.

14.3 Current APC–Asef Inhibitors

14.3.1 Peptides

Targeting PPIs poses a significant challenge in drug discovery due to the large and shallow interface involved. Consequently, initial efforts have focused on developing peptide-based inhibitors, which represent an attractive strategy for PPI modulation. Peptide inhibitors can effectively mimic naturally occurring interactions and are relatively easy to synthesize. Moreover, they have proven valuable in guiding the design of small molecules. As a result, peptide-based approaches are often the first line of strategy in developing PPI modulators. Fortunately, the APC–Asef interface has identified a hot spot among the more extensive interfaces (1300 Å^2), providing an opportunity to discover and optimize origin APC–Asef inhibitors.

14.3.1.1 Minimal Asef Sequence for APC

Through fluorescence polarization (FP) screening of a truncated peptide library, Jiang et al. identified three 7-mer peptides – MAI-102 (^{181}GGEALAW^{187}), MAI-107 (^{181}GGEALAD^{187}), and MAI-108 (^{181}AGEALAD^{187}) as the minimal Asef sequences capable of binding to APC, with K_i values of $3.12 \pm 0.70\,\mu$M, $3.80 \pm 0.72\,\mu$M, and $2.41 \pm 0.88\,\mu$M, respectively [42]. The complexes formed by APC with MAI-102, MAI-107, and MAI-108 demonstrated a consistent and defined binding mode for the MAI peptides. Superimposition of these peptides in the APC structure revealed significant conservation of residues 181–185 while showing variability at the C terminus. This suggests that optimizing the C-terminal region of the MAI peptides could enhance binding affinity. Indeed, the second-generation 7-mer peptide MAI-150 (^{181}AGEALYE^{187}) exhibited a K_i value of $0.12 \pm 0.02\,\mu$M by FP and a K_d value of $0.23\,\mu$M by ITC, which was 20-fold better than that of MAI-108. These results suggest that the 7-mer peptide serves as the minimal sequence for the initial design of APC–Asef PPI inhibitors.

14.3.1.2 MAI-150–Derived Peptidomimetic Inhibitors

The structure of the APC/MAI-150 complex (PDB: 5IZ6) was determined to demonstrate the high binding affinity of MAI-150 to APC (see Figure 14.1c) [42]. In this complex, MAI-150 induces a doughnut-shaped conformational change in APC-Arg549, enhancing its druggable properties and facilitating improved structure-based mapping. The C-terminal residue Glu187 of MAI-150 forms salt-bridge bonds with APC-Arg549 and APC-Lys586 (see Figure 14.1d). Additionally, the hydroxyl group of Tyr186 in MAI-150 establishes a hydrogen bond with APC-Asp539. MAI-150 also preserves critical salt bridges between Glu183 and APC-Lys116, as well as van der Waals interactions between Leu185 and APC-F458 and APC-F510. Furthermore, the N-terminal benzyloxy carbonyl (Z) cap of MAI-150 significantly increases hydrophobic contacts with Trp553 and Trp593 of APC.

Mutations of critical residues in APC provide further confirmation of the structure–activity relationship (SAR) observed. Specifically, the APC mutants N507W and K516A completely abolish the binding of MAI-150, highlighting these residues as essential for binding affinity. The APC R549A mutant reduces the binding affinity of MAI-150 for APC by 121-fold, underscoring the significant impact of the extended conformation of the Arg549 side chain (see Figure 14.1e). In addition, the APC W593A and K586A mutants moderately weaken MAI-150 binding to APC by 40-fold and 3.6-fold, agreeing with their additional interactions in the APC-ARM/MAI-150 structure. These findings suggest that the optimization strategy should focus on the side chains of Leu185 and Tyr186, as well as the capping group.

As for the Tyr186 side chain of MAI-150, attempts to optimize its position and substitutions do not result in enhanced activity. Similarly, the optimization of the N-capping group leads to decreased potency. On the other hand, the exploration of various hydrophobic side chains of Leu185, including cyclopropyl, cyclobutyl, cyclopentyl, cyclohexyl, and phenyl, reveals that cyclopentyl derivatives exhibit the best binding affinity. These optimization efforts culminated in the development of the peptidomimetic compound MAI-203, which demonstrated the highest potency, with a K_i of $0.015 \pm 0.001\,\mu M$ by FP and a K_d of $0.036\,\mu M$ by ITC. This represents an eightfold improvement over MAI-150 and a remarkable 2900-fold enhancement compared to MAI-005. Notably, the binding affinity of MAI-203 is approximately equivalent to that of the APC–Asef interaction, with a K_d of $0.018\,\mu M$.

14.3.1.3 MAI-203 as the First-in-Class Inhibitor of APC–Asef PPI with Anti-CRC Migration Activity

MAI-203, a representative inhibitor of the APC–Asef interaction, was selected to evaluate its *in vivo* activity. The co-immunoprecipitation (co-IP) experiments in HEK293T cells transiently transfected with Flag-tagged APC and HA-tagged Asef demonstrated that MAI-203 efficiently blocked the APC–Asef interaction in cell lysates in a dose-dependent manner. Notably, this inhibition occurred without affecting the expression levels of either APC or Asef (see Figure 14.2a).

However, MAI-203 has inherent limitations in membrane permeability. To enhance its cell penetration, a cell-penetrating peptide, the trans-activating

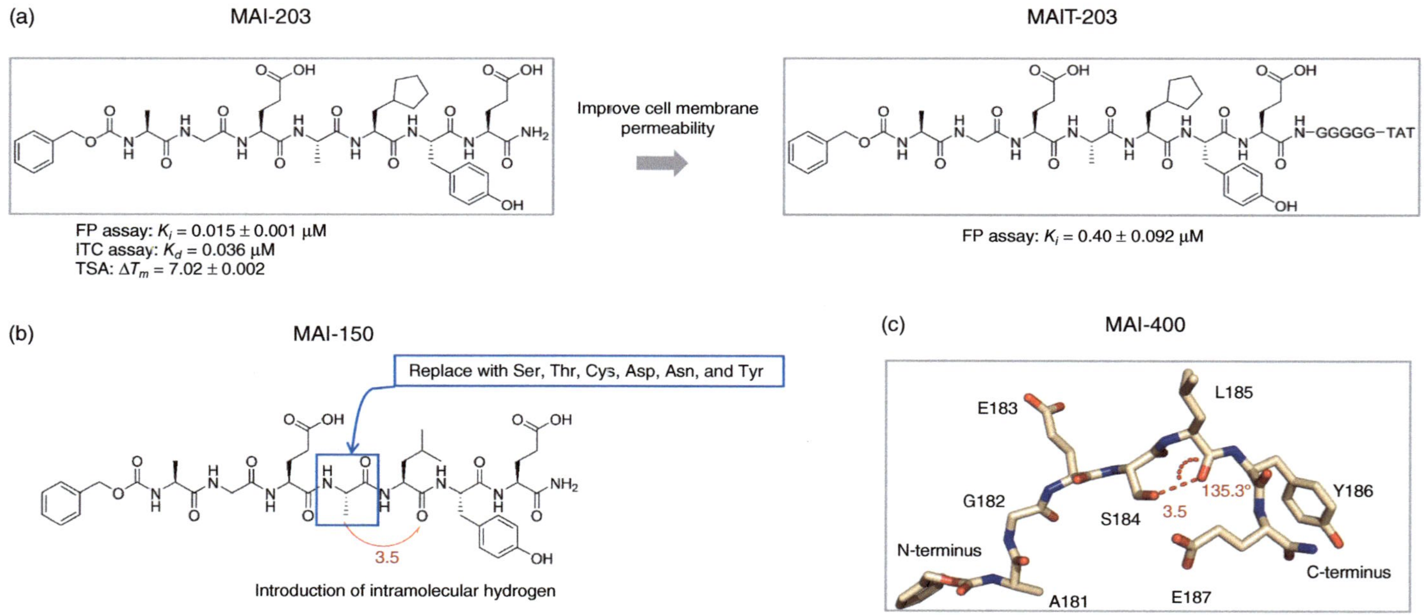

Figure 14.2 (a) The structure of MAI-203 and MAIT-203. (b) Chemical structure of MAI-150 and the design strategy. (c) Perspective view showing the intramolecular hydrogen bond observed in the crystal structure of APC/MAI-400 (PDB: 5Z8H). The red dashed lines represent the distance and the angle of the hydrogen bond. Source: Adapted from Yang et al. [43].

transcriptional activator (TAT) from human immunodeficiency virus, was conjugated to the C-terminus of MAI-203 using an optimized "GGGGG" linker, resulting in the cell-permeable compound MAIT-203. Flow cytometry analysis demonstrated considerable cellular uptake of the MAIT-203 in HEK293T, SW480, and HCT116 cells. Subsequent co-IP assays in HEK293T cells transiently transfected with Flag-tagged APC and HA-tagged Asef revealed that MAIT-203 effectively reduced the Asef–APC interaction in a dose-dependent manner, highlighting its efficacy in blocking this interaction within cells. Additionally, a cellular thermal shift assay (CETSA) conducted with MAIT-203 in SW480 cells confirmed its direct binding to APC. Besides, the specificity of MAIT-203 for the disruption of APC–Asef interaction in HEK293T cells was assessed. MAIT-203 selectively inhibited the APC–Asef PPI without affecting APC–Sam68 or APC–striatin interactions at a concentration of 10 μM. Endogenous co-IP in SW480 cells further validated the selectivity of MAIT-203 in disrupting the APC–Asef interaction.

MAIT-203 effectively binds to APC and specifically inhibits APC–Asef interaction in CRC cells. Next, the xCELLigence real-time cell analysis (RTCA), a transwell assay, and a wound-healing assay were performed to determine whether MAIT-203 could impede APC–Asef-mediated cell migration and invasion. The results revealed that MAIT-203 at 10 μM significantly inhibited colorectal cell migration in SW480 and HCT116 cells by disrupting the APC–Asef interaction. Additionally, the RTCA invasion assay demonstrated that MAIT-203 at the same concentration markedly reduced the invasive capacity of CRC cells. The data above indicated that MAIT-203 inhibits the metastasis of CRC cells by suppressing APC–Asef-mediated migration and invasion.

14.3.1.4 Intramolecular Hydrogen Bond Strategy Leading to the Best-in-Class Inhibitor MAI-400

Encouraging by the cellular results of MAIT-203, further effort was made to optimize the MAI-series peptides. Most linear APC peptide inhibitors exhibit flexible conformations, which can limit their binding affinity. Therefore, increasing the rigidity and stability of the peptide presents a considerable optimization strategy to improve the binding affinities. Yang et al. adopted an intramolecular hydrogen bond-based pseudocyclic peptide strategy for optimization [43].

The structure of the APC/MAI-150 complex suggested that the hydrogen atom of the A184 side chain and the backbone oxygen atom of L185 in MAI-150 were ideally positioned to form an intramolecular hydrogen bond (H···C=O) with an angle of 120.9° and an interaction distance of 3.5 Å (see Figure 14.2b). To introduce an intramolecular hydrogen bond donor, the A184 residue was replaced with six amino acids (Ser, Thr, Cys, Asp, Asn, and Tyr) containing hydrogen-bond donors in their side chains. Among these, only the substitution of Ala with Ser, resulting in MAI-400, improved the binding affinity fourfold ($IC_{50} = 0.25 \pm 0.01$ μM) compared to MAI-150. To confirm the formation of an intramolecular hydrogen bond in MAI-400, the crystal structure of MAI-400 bound to APC was determined. The APC/MAI-400 complex structure with unambiguous electron density showed that MAI-400 contained a robust intramolecular hydrogen bond (distance = 2.7 Å, angle = 135.3°) formed by the hydroxy group of the S184 side chain and the carbonyl oxygen atom of the L185 backbone (see Figure 14.2c). This intramolecular hydrogen bond resulted in a pseudo-nine-atom ring within MAI-400, stabilizing its conformation by restricting its flexibility. Furthermore, the intramolecular hydrogen bond of MAI-400 induced a conformational change in W553 of APC, which could have a positive effect on stabilizing the conformation of APC. The ITC results of MAI-400 and MAI-150 showed that the stable conformation of MAI-400 maintained by an intramolecular hydrogen bond contributed to a favorable increase in enthalpic energy. In addition, compared to MAI-150, MAI-400 exhibited a longer residence time (RT = 15 s) and a superior kinetic K_D value (k_{off}/k_{on}) of 0.22 ± 0.01 μM for binding to APC by surface plasmon resonance (SPR). Collectively, these factors account for the improved binding affinity of MAI-400 to APC.

The SAR analysis of various hydrophobic groups at position 185 of MAI-400 indicated that the APC pocket for binding the side chain of residue 185 was rigid to induce. Encouraged by the intramolecular hydrogen bond strategy, Yang et al. attempted to stabilize this bond to create a nine-atom ring. However, the design of multiple scaffolds failed to obtain positive results. In the future, additional efforts should focus on developing novel scaffolds with the potential rigid ring to enhance the binding affinity of peptide inhibitors.

14.3.1.5 A More Sensitive FP Assay Led to the Discovery of a Highly Efficient and Potent Inhibitor, MAI-516

The new round of optimization included different substituents at the N terminal to enhance the π–π interaction between the benzyl moiety of MAI-400 and the indoyl moiety of APC-W593 and replacement of the residue L185 and E187 of MAI-203 and MAI-400 to improve the hydrophobic effects surrounding APC hydrophobic pocket [44]. Using a more sensitive FP assay, we identified MAI-516 as a more potent inhibitor, demonstrating an improved binding affinity with an IC_{50} of 0.041 ± 0.004 μM. MAI-516 exhibited a K_d value of 4.4 nM by ITC and a K_D value of 18 ± 2 nM by SPR.

The crystal structure of the APC/MAI-516 complex revealed that the N-terminal 4-methoxy-phenyl forms parallel-displaced π–π stacking with the indoyl of APC-W593 with a distance of 3.7–4.6 Å. This configuration allows the methoxy moiety of MAI-516 to establish a hydrogen bond with the NH_2 group of APC-N641. The newly formed π–π stacking interaction and hydrogen bond are crucial for the increased binding affinity of MAI-516. Besides, APC mutants W593A and N641A reduced the binding affinity of MAI-516 to APC by approximately 100-fold and 2-fold, respectively. Overall, these results indicated that the newly established π–π stacking interaction and hydrogen bond were the primary determinants for the improved activity of MAI-516.

MAI-516 attenuated the APC–Asef PPI dose-dependently in cell lysates from HEK293T cells transiently transfected with Flag-tagged APC and HA-tagged Asef, demonstrating greater efficacy than MAI-203. However, MAI-516 struggled to cross the cell membrane and failed to reach intracellular targets. To enhance the permeability of MAI-516, a cell-penetrating peptide (TAT) was conjugated to the

C-terminus of MAI-516, resulting in the cell-permeable inhibitor MAIT-516 with a K_d value of 67 nM. MAIT-516 significantly diminished the APC–Asef PPI in a dose-dependent manner in HEK293T cells transiently transfected with Flag-tagged APC and HA-tagged Asef. Notably, MAIT-516 was more potent at blocking the APC–Asef PPI than MAIT-203 at a concentration of 50 μM. Additionally, MAIT-516 specifically inhibited the truncated APC–Asef PPI, showing minimal impact on the endogenous interaction between wild-type APC and Asef. Furthermore, MAIT-516 significantly inhibited the migration of SW480 cells in wound-healing, transwell, and RTCA migration assays. Importantly, MAIT-516 did not affect cell proliferation, apoptosis, or differentiation in colorectal epithelial cells. The low cytotoxicity and cancer-specific effects of MAIT-516 highlight its potential for future drug development.

14.3.1.6 Computer-Aided Design of Tripeptide Inhibitors Disrupting APC–Asef Interaction

Based on the binding mode of MAI-400 with APC, Wang et al. sought to simplify the molecular structure by truncating MAI-400 to tripeptide [45]. They preserved the core sequence ^{183}ESL^{185}, which fits into the APC surface cavity in a U-shaped conformation. Subsequently, structure-based modifications, combined with our own N-layered integrated molecular orbital and molecular mechanics (ONIOM) model-driven optimization of the ESL N- and C-terminal caps, resulted in compound AT46, which exhibited improved affinity (IC_{50} = 0.11 μM). Additionally, they designed a prodrug, AT47, by enclosing the carboxyl group of AT46 in a tert-butyl protective ester. Both AT46 and AT47 demonstrated significantly enhanced in vivo half-lives compared to MAI-203, indicating that structure-guided optimization effectively improved potency and metabolic stability. Moreover, AT47 displayed anti-migration properties and low cytotoxicity in colorectal cells expressing truncated APC. Notably, AT47 also exhibited pronounced anti-migration effects against CRC *in vivo.*

14.3.2 Small Molecule Inhibitors of APC–Asef PPI

Utilizing the structure of APC-PreARM-ARM/Asef-ABR (PDB: 3NMZ), Yan et al. reported a number of 2-*H* pyrazole derivatives featuring morpholine groups as inhibitors of APC–Asef PPI [46]. Among these, compound 7g exhibited the most favorable IC_{50} value of 0.18 μM. Its binding affinity to APC was further validated with a K_d value of 12.8 μM by ITC. Compound 7g demonstrated significant anti-proliferative activity across six cancer cell lines, particularly against HCT-116 cells. Further studies indicated that 7g elevated reactive oxygen species (ROS) levels, induced apoptosis, and diminished the adhesion and invasion of HCT-116 cells at a concentration of 2 μM. In the mouse xenograft and orthotopic transplantation models, 7g notably suppressed tumor growth.

Through further optimization of 7g, compound Q19 was developed as a next-generation inhibitor of APC–Asef, exhibiting an IC_{50} value of 1 μM [47]. Compound Q19 was found to induce mitochondrial depolarization, trigger apoptosis, and inhibit the proliferation of HCT-116 cells. However, Q19 was not as potent as 7g in the FP and anti-proliferation assays. The binding of 7g to APC still needs further confirmation, such as co-crystal structure, CETSA, etc.

Another research group led by Jadav et al. aimed to identify new small molecular inhibitors of APC–Asef PPI through docking methods and molecular dynamics simulations [48]. They validated the central binding region of APC for the MAI-peptide and carried out the structure-based virtual screening (SBVS) of ZINC13, National cancer institute (NCI), and Maybridge public compound libraries. Their induced fit docking analysis and molecular dynamics simulations yielded 16 ligand hits featuring amide groups among the top-scoring candidates. However, further bioassays are necessary to confirm the actual targets and activities of these compounds [49].

14.4 A More Sensitive FP Method for Identifying Highly Active APC–Asef Inhibitors

The FP assay has proven effective in screening for inhibitors of PPIs. The sensitivity of the FP assay is crucial for successfully identifying PPI inhibitors. Previous studies have utilized an FP competition assay to screen for APC–Asef PPI inhibitors, such as MAI-150, MAI-203, and MAI-400. However, as the binding affinity of APC–Asef inhibitors increases, the existing FP method is hindered by the low binding affinity of the original tracer, which limits its ability to distinguish highly potent compounds. Therefore, there is an urgent need to enhance the binding affinity of tracers used in FP assays to facilitate the discovery of highly active inhibitors [43].

The sensitivity of the FP assay is contingent upon the binding affinities of the FP tracers to their target proteins. The original tracer 1 exhibited a limited binding affinity ($EC_{50} = 0.54 \pm 0.009\ \mu M$). The tracer used in FP assays commonly contains three main subgroups: a peptide ligand, a fluorescent label, and a linker. Typically, the peptide ligand is responsible for binding affinity to the target protein, so ligand optimization was adopted to enhance the tracer binding affinity. Recent studies have also suggested that the FITC subgroup can contribute to the binding affinity for specific targets. Inspired by these findings, Zhong et al. optimized both the ligand and the linker of the tracer to improve the FP assay sensitivity. Through molecular dynamics (MD) simulations, they designed three tracers with different ligands and linkers. Among these, tracer 7 (Z-AGESLYEK[FITC]-NH_2) demonstrated the highest binding affinity ($EC_{50} = 0.016 \pm 0.0003\ \mu M$; $K_d = 0.078\ \mu M$) and signal dynamic range ($\Delta mp = 306 \pm 2$ mp), with a 33-fold increase in EC_{50} and a 25% enhancement in dynamic range compared to the original tracer (see Figure 14.3a). The APC/tracer 7 complex structure revealed that not only does the ligand peptide subgroup of tracer 7 bind to APC tightly but also the FITC subgroup of tracer 7 serves as an additional binding anchor to its favored surface of APC through hydrophobic interaction and a hydrogen bond. In the improved FP assay, tracer 7 showed an excellent resolution for inhibitors with IC_{50} values <1 μM, while for those with IC_{50} values >1 μM, tracer 7 exhibited a good correlation with that of tracer 1 (see Figure 14.3b). Moreover, a higher Z' factor of 0.92 was achieved in the FP assay with tracer 7 (see Figure 14.3c).

In conclusion, the increased binding affinity of the tracer 7 significantly improved the dynamic range, sensitivity, and resolution of the FP assay. Consequently, the affinity-enhanced tracer 7 in FP assays will be employed in screening for highly potent APC–Asef inhibitors.

14.5 Conclusions and Outlook

The APC–Asef PPI represents a promising druggable target for metastatic CRC. Peptide-based inhibitors provide specificity and exhibit improved safety profiles, minimizing off-target effects commonly associated with traditional small-molecule drugs. Numerous APC–Asef PPI peptide inhibitors have been developed and shown significant anti-CRC efficacy in both *in vitro* and *in vivo* models. The successful identification of APC–Asef peptide inhibitors highlights the potential for early clinical development in CRC and serves as proof of principle for broader applications in PPI-targeted drug discovery. This success can inspire the exploration of other challenging PPI targets, ultimately expanding the arsenal of therapeutic options available for various cancers and other diseases characterized by dysregulated protein interactions.

However, despite their promise, optimizing drug-like properties for these peptide inhibitors remains a critical hurdle. Factors such as permeability, stability, and bioavailability are essential for translating these promising candidates into effective clinical therapies. Enhancing these properties will be crucial to ensure that peptide inhibitors can reach therapeutic concentrations at the site of action within the body.

In contrast, developing small molecule inhibitors targeting the APC–Asef interaction presents a unique set of challenges. The inherent characteristics of the APC–Asef interface – flat, featureless, and expansive – complicate the identification of effective small molecule binders. This necessitates a more sophisticated approach to drug discovery, leveraging advanced methodologies such as virtual drug screening, DNA-encoded library (DEL) screening, and high-throughput bioassays. These technologies can significantly enhance the likelihood of identifying small molecules that can effectively disrupt the APC–Asef interaction.

In summary, while the APC–Asef PPI presents both opportunities and challenges in drug development, the ongoing research and innovative approaches in this field hold promise for advancing targeted therapies that could significantly improve patient outcomes in metastatic CRC and beyond. Continued collaboration between academia and industry will be essential to navigate these challenges and translate research findings into viable clinical solutions.

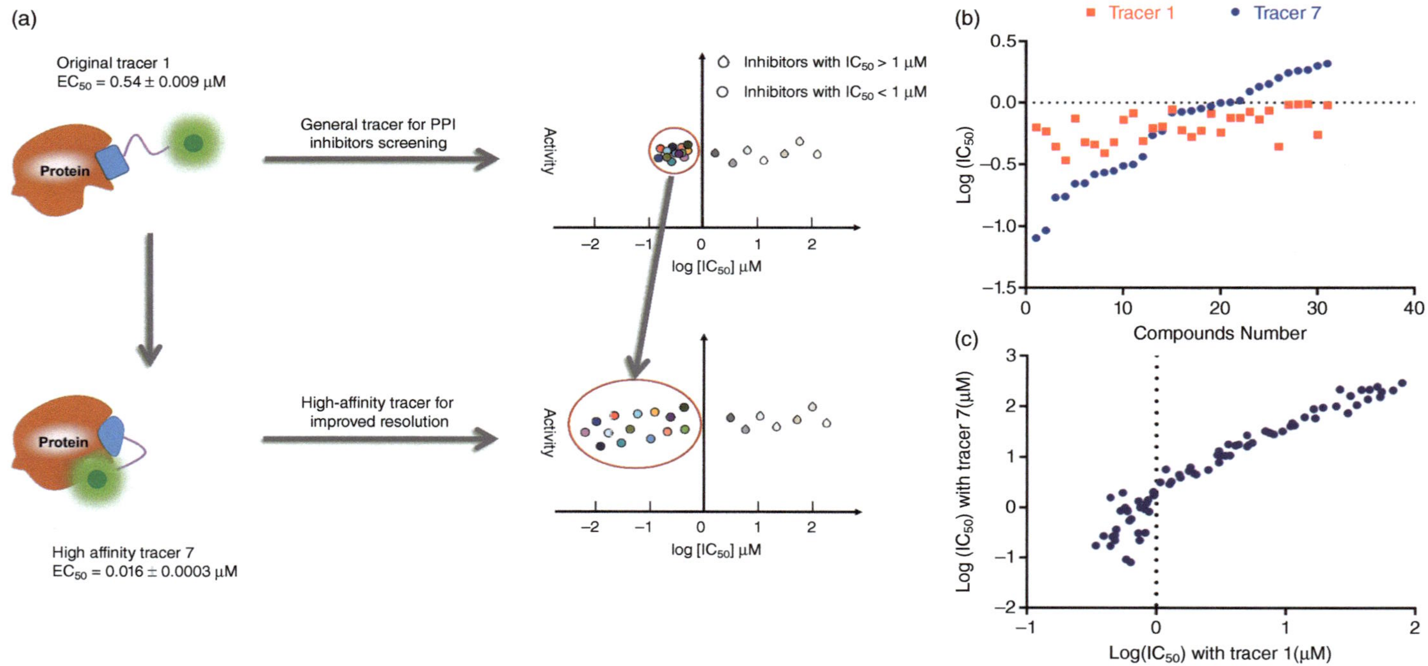

Figure 14.3 Optimization of the FP tracer to improve the resolution of the FP assay. (a) Schematic illustration of the resolution of FP assay for APC–Asef inhibitor screening using tracer 1 and tracer 7. FP assays using tracer 1 were unable to discriminate inhibitors with IC_{50} values <1 μM. FP assays using tracer 7 could clearly distinguish inhibitors with IC_{50} values <1 μM. (b) Log(IC_{50}) of 31 APC–Asef PPI inhibitors was determined by FP assays with tracer 1 (red) and tracer 7 (blue). (c) The correlation of log(IC_{50}) of 80 APC–Asef PPI inhibitors determined by FP assays with tracer 1 and tracer 7. Inhibitors with IC_{50} less than 1 μM under tracer 1 are better distinguished by tracer 7. Source: Zhong et al. /Springer Nature [44].

References

1 Fearnhead, N.S., Britton, M.P., and Bodmer, W.F. (2001). The ABC of APC. *Human Molecular Genetics* 10 (7): 721–733.
2 Groden, J., Thliveris, A., Samowitz, W. et al. (1991). Identification and characterization of the familial adenomatous polyposis coli gene. *Cell* 66 (3): 589–600.
3 Rowan, A.J., Lamlum, H., Ilyas, M. et al. (2000). APC mutations in sporadic colorectal tumors: a mutational "hotspot" and interdependence of the "two hits". *Proceedings of the National Academy of Sciences of the United States of America* 97 (7): 3352–3357.
4 Gotthardt, K. and Ahmadian, M.R. (2007). Asef is a Cdc42-specific guanine nucleotide exchange factor. *Biological Chemistry* 388 (1): 67–71.
5 Itoh, R.E., Kiyokawa, E., Aoki, K. et al. (2008). Phosphorylation and activation of the Rac1 and Cdc42 GEF Asef in A431 cells stimulated by EGF. *Journal of Cell Science* 121 (16): 2635–2642.
6 Kawasaki, Y., Sato, R., and Akiyama, T. (2003). Mutated APC and Asef are involved in the migration of colorectal tumour cells. *Nature Cell Biology* 5 (3): 211–215.
7 Mitin, N., Betts, L., Yohe, M.E. et al. (2007). Release of autoinhibition of ASEF by APC leads to CDC42 activation and tumor suppression. *Nature Structural and Molecular Biology* 14 (9): 814–823.
8 Beroud, C. and Soussi, T. (1996). APC gene: database of germline and somatic mutations in human tumors and cell lines. *Nucleic Acids Research* 24 (1): 121–124.
9 Laurent-Puig, P., Béroud, C., and Soussi, T. (1998). APC gene: database of germline and somatic mutations in human tumors and cell lines. *Nucleic Acids Research* 26 (1): 269–270.
10 MacDonald, B.T., Tamai, K., and He, X. (2009). Wnt/β-catenin signaling: components, mechanisms, and diseases. *Developmental Cell* 17 (1): 9–26.
11 Hamada, F. and Bienz, M. (2002). A drosophila APC tumour suppressor homologue functions in cellular adhesion. *Nature Cell Biology* 4 (3): 208–213.
12 Bienz, M. and Hamada, F. (2004). Adenomatous polyposis coli proteins and cell adhesion. *Current Opinion in Cell Biology* 16 (5): 528–535.
13 Fang, X. and Svitkina, T.M. (2022). Adenomatous polyposis Coli (APC) in cell migration. *European Journal of Cell Biology* 101 (3): 151228.
14 Akiyama, T. and Kawasaki, Y. (2006). Wnt signalling and the actin cytoskeleton. *Oncogene* 25 (57): 7538–7544.
15 Caldwell, C.M. and Kaplan, K.B. (2009). The role of APC in mitosis and in chromosome instability. *Advances in Experimental Medicine and Biology* 656: 51–64.
16 Zhang, L. and Shay, J.W. (2017). Multiple roles of APC and its therapeutic implications in colorectal cancer. *Journal of the National Cancer Institute* 109 (8): djw332.
17 Sieber, O.M., Tomlinson, I.P., and Lamlum, H. (2000). The adenomatous polyposis coli (APC) tumour suppressor--genetics, function and disease. *Molecular Medicine Today* 6 (12): 462–469.

18 Mahajan, V., Das, A.K., Taggar, R.K. et al. (2015). Polymorphism of keratin-associated protein (KAP) 3.2 gene and its association with wool traits in Rambouillet sheep. *The Indian Journal of Animal Sciences* 85 (3): 262–265.

19 Morishita, E.C., Murayama, K., Kato-Murayama, M. et al. (2011). Crystal structures of the armadillo repeat domain of adenomatous polyposis coli and its complex with the tyrosine-rich domain of Sam68. *Structure* 19 (10): 1496–1508.

20 Grohmann, A., Tanneberger, K., Alzner, A. et al. (2007). AMER1 regulates the distribution of the tumor suppressor APC between microtubules and the plasma membrane. *Journal of Cell Science* 120 (21): 3738–3747.

21 Tanneberger, K., Pfister, A.S., Kriz, V. et al. (2011). Structural and functional characterization of the Wnt inhibitor APC membrane recruitment 1 (Amer1). *The Journal of Biological Chemistry* 286 (22): 19204–19214.

22 Zhang, Z.Y., Akyildiz, S., Xiao, Y.F. et al. (2015). Structures of the APC-ARM domain in complexes with discrete Amer1/WTX fragments reveal that it uses a consensus mode to recognize its binding partners. *Cell Discovery* 15016.

23 Weiner, A.T., Lanz, M.C., Goetschius, D.J. et al. (2016). Kinesin-2 and Apc function at dendrite branch points to resolve microtubule collisions. *Cytoskeleton* 73 (1): 35–44.

24 Watanabe, T., Wang, S.J., Noritake, J. et al. (2004). Interaction with IQGAP1 links APC to Rac1, Cdc42, and actin filaments during cell polarization and migration. *Developmental Cell* 7 (6): 871–883.

25 Kawasaki, Y., Senda, T., Ishidate, T. et al. (2000). Asef, a link between the tumor suppressor APC and G-protein signaling. *Science* 289 (5482): 1194–1197.

26 Zhang, Z.Y., Lin, K., Gao, L. et al. (2011). Crystal structure of the armadillo repeat domain of adenomatous polyposis coli which reveals its inherent flexibility. *Biochemical and Biophysical Research Communications* 412 (4): 732–736.

27 Ha, N.C., Tonozuka, T., Stamos, J.L. et al. (2004). Mechanism of phosphorylation-dependent binding of APC to β-catenin and its role in β-catenin degradation. *Molecular Cell* 15 (4): 511–521.

28 Wang, L.L., Liu, X.Y., Gusev, E. et al. (2014). Regulation of the phosphorylation and nuclear import and export of β-catenin by APC and its cancer-related truncated form. *Journal of Cell Science* 127 (8): 1647–1659.

29 Xing, Y., Clements, W.K., Le Trong, I. et al. (2004). Crystal structure of a β-Catenin/APC complex reveals a critical role for APC phosphorylation in APC function. *Molecular Cell* 15 (4): 523–533.

30 Iwama, T. (2001). Somatic mutation rate of the APC gene. *Japanese Journal of Clinical Oncology* 31 (5): 185–187.

31 Su, L.K., Johnson, K.A., Smith, K.J. et al. (1993). Association between wild-type and mutant Apc gene-products. *Cancer Research* 53 (12): 2728–2731.

32 Wen, Y., Eng, C.H., Schmoranzer, J. et al. (2004). EB1 and APC bind to mDia to stabilize microtubules downstream of Rho and promote cell migration. *Nature Cell Biology* 6 (9): 820–830.

33 Kaplan, K.B., Burds, A.A., Swedlow, J.R. et al. (2001). A role for the adenomatous polyposis coli protein in chromosome segregation. *Nature Cell Biology* 3 (4): 429–432.

34 Smith, K.J., Johnson, K.A., Bryan, T.M. et al. (1993). The APC gene product in normal and tumor cells. *Proceedings of the National Academy of Sciences of the United States of America* 90 (7): 2846–2850.

35 Miyoshi, Y., Nagase, H., Ando, H. et al. (1992). Somatic mutations of the APC gene in colorectal tumors: mutation cluster region in the APC gene. *Human Molecular Genetics* 1 (4): 229–233.

36 Murayama, K., Shirouzu, M., Kawasaki, Y. et al. (2007). Crystal structure of the rac activator, Asef, reveals its autoinhibitory mechanism. *The Journal of Biological Chemistry* 282 (7): 4238–4242.

37 Kawasaki, Y., Tsuji, S., Muroya, K. et al. (2009). The adenomatous polyposis coli-associated exchange factors Asef and Asef2 are required for adenoma formation in Apc(Min/+)mice. *EMBO Reports* 10 (12): 1355–1362.

38 Kawasaki, Y., Tsuji, S., Sagara, M. et al. (2009). Adenomatous polyposis coli and Asef function downstream of hepatocyte growth factor and phosphatidylinositol 3-kinase. *The Journal of Biological Chemistry* 284 (33): 22436–22443.

39 Furukawa, S., Kawasaki, Y., Miyamoto, M. et al. (2013). The miR-1-NOTCH3-Asef pathway is important for colorectal tumor cell migration. *Plos One* 8 (11): e80609.

40 Rodríguez-Cruz, F., Torres-Cruz, F.M., Monroy-Ramírez, H.C. et al. (2018). Fragmentation of the golgi apparatus in neuroblastoma cells is associated with tau-induced ring-shaped microtubule bundles. *Journal of Alzheimer's Disease* 65 (4): 1185–1207.

41 Zhang, Z., Chen, L., Gao, L. et al. (2012). Structural basis for the recognition of Asef by adenomatous polyposis coli. *Cell Research* 22 (2): 372–386.

42 Jiang, H.M., Deng, R., Yang, X.Y. et al. (2017). Peptidomimetic inhibitors of APC–Asef interaction block colorectal cancer migration. *Nature Chemical Biology* 13 (9): 994–1001.

43 Yang, X.Y., Zhong, J., Zhang, Q.F. et al. (2018). Rational design and structure validation of a novel peptide inhibitor of the adenomatous-polyposis-coli (APC)-rho-guanine-nucleotide-exchange-factor-4 (Asef) interaction. *Journal of Medicinal Chemistry* 61 (17): 8017–8028.

44 Zhong, J., Guo, Y., Lu, S. et al. (2022). Rational design of a sensitivity-enhanced tracer for discovering efficient APC–Asef inhibitors. *Nature Communications* 13 (1): 4961.

45 Wang, X., Du, Z., Guo, Y. et al. (2024). Computer-aided molecular design and optimization of potent inhibitors disrupting APC–Asef interaction. *Acta Pharmaceutica Sinica B* 14 (6): 2631–2645.

46 Yan, X.Q., Wang, Z.C., Qi, P.F. et al. (2019). Design, synthesis and biological evaluation of 2-H pyrazole derivatives containing morpholine moieties as highly potent small molecule inhibitors of APC–Asef interaction. *European Journal of Medicinal Chemistry* 177: 425–447.

47 Qi, P.F., Fang, L., Li, H. et al. (2020). Discovery of novel pyrazoline derivatives containing methyl-1-indole moiety as potential inhibitors for blocking APC–Asef interactions. *Bioorganic Chemistry* 99: 103838.

48 Jadav, S.S., Macalino, S.J.Y., and Alluri, R. (2020). Structure-based discovery of small molecule APC–Asef interaction inhibitors: in silico approaches and molecular dynamics simulations. *Journal of Molecular Modeling* 26 (8): 207.
49 Yang, X., Zhong, J., Zhang, Q. et al. (2021). Advances and insights of APC–Asef inhibitors for metastatic colorectal cancer therapy. *Frontiers in Molecular Biosciences* 8: 662579.

15

Computational Methods Applied to Drug Discovery of Protein–Protein Interaction Systems

Zhiyong Gu[2,3] *and Xi Cheng*[1,2,3]

[1] *Department of Pharmaceutical and Artificial Intelligence Sciences, School of Medicine, Shanghai Jiao Tong University, 227 Chongqing South Road, 200025, China*
[2] *Institute of Materia Medica, Chinese Academy of Sciences, State Key Laboratory of Drug Research, 555 Zuchongzhi Road, Shanghai 201203, China*
[3] *University of Chinese Academy of Sciences, 19 Yuquan Road, Beijing 10049, China*

15.1 Introduction

Proteins rarely act alone but interact with each other to regulate various biological functions [1, 2]. The interactions between proteins (protein–protein interactions, PPIs) are central for most signaling pathways in cells and are considered promising targets for therapeutic intervention [3]. A comprehensive understanding of the PPIs not only offers insights into the molecular mechanisms of complex biological processes but also facilitates drug target identification and drug discovery [4]. The number of PPIs is enormous. The human interactome may contain over 130,000 PPIs [5, 6]. Despite the remarkable experimental efforts undertaken to date to determine the PPIs, many of them still remain unknown [7]. Meanwhile, PPI prediction through computational methods has been gaining increasing attention, as these methods offer advantages of low cost and high throughput compared to experimental approaches [8, 9]. These computational strategies can accelerate PPI identification by significantly narrowing down the number of alternative interactions to be confirmed in laboratory experiments [10, 11].

Computational methods can help reveal the principles of PPIs at multiple scales, ranging from the proteome level, molecular level, residual level to the atomic level (Figure 15.1). They have been developed to address different issues of PPI systems, including PPI network mapping, protein complex prediction, and PPI modulator discovery. To date, many computational methods for PPI prediction have been developed based on a variety of biological data, including genomics, proteomics, drug response information, binding affinity, and protein structures [12–14]. The following section describes several state-of-the-art computational methods for PPI prediction.

Targeting Protein–Protein Interactions for Drug Discovery, First Edition. Edited by Jian Zhang.

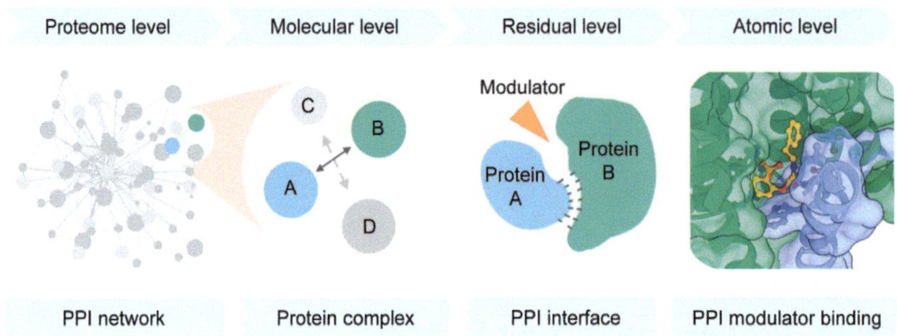

Figure 15.1 The protein–protein interaction systems at different scales. Source: Zhiyong Gu, Xi Cheng.

15.2 Computational Methods for PPI Prediction

15.2.1 PPI Network Mapping

The PPIs can be represented as a network where nodes correspond to proteins and edges represent pairwise interactions [15]. The biological functions of a particular protein can be infered form the global connectivity pattern of the PPI network [15]. A complete understanding of a PPI network within specific cells would represent an important milestone toward a comprehensive description of cellular mechanisms and functions [16]. In order to map the PPI networks of major organisms, extensive experimental efforts have been made using laboratory-based methods, including yeast two-hybrid, tandem affinity purification (TAP)-tagging, protein chips, synthetic lethal analysis, and correlated mRNA expression profiling [17]. Although the quality and volume of experimentally identified PPIs have improved dramatically over the last decade, the PPI networks of many organisms remain incomplete and elusive.

Computational methods play a crucial role in facilitating PPI network mapping and further characterizing signal flow within these networks. As one of the major PPI network mapping approaches, the network-based computational methods use known PPIs in a network representation to predict missing PPIs [18]. These methods intend to score connectivity between protein nodes in the PPI network and aim to determine whether pairwise proteins are interacting or not. Recently, the International Network Medicine Consortium performed an evaluation of 26 representative network-based prediction methods across six different interactomes of four different organisms: *A. thaliana*, *C. elegans*, *S. cerevisiae*, and *H. sapiens* [19]. They found that the advanced similarity-based algorithms, including maximum protein similarity models, exhibit superior performance in both computational and experimental validations. In general, the similarity-based algorithms quantify the likelihood of PPI connectivity based on predefined similarity functions among protein nodes in the PPI network. Advanced similarity-based algorithms leverage specific connectivity properties, such as L_3 principle, to consider interactions of two proteins if they are connected by multiple paths of length three in the network [20]. However, it is worth noting that this evaluation study did not assess the capability of these methods in predicting PPI directions. Compared with undirected PPI networks, the directed ones could provide valuable information on cellular signaling pathways and enhance protein function prediction [21, 22]. To address this, Silverbush and Sharan developed a network-based method, i.e. Diffuse2Direct [23] (Table 15.1), to predict the directed human PPI network using drug response and cancer genomic data. The input of Diffuse2Direct is a PPI topology network along with a collection of experiments, inducing causal proteins and effect proteins. For drug response data, causal proteins are drug targets, and affected ones correspond to genes, whose expression has changed under drug treatments. In the context of cancer genomics, causal proteins correspond to mutated genes and affected proteins correspond to genes that are differentially expressed between the tumor and normal tissues. A diffusion value for each protein is calculated according to its proximity to causal and affected proteins in the network. All diffusion values are combined to score possible PPI directions of the PPI network. Diffuse2Direct significantly outperforms alternative methods on five testing sets involving E3 ubiquitination interactions, kinase–substrate interactions, protein–DNA interactions, signal–transduction interactions in mammals, and interactions from the epidermal growth factor receptor (EGFR) pathway. Collectively, for effective PPI network mapping, the network-based methods need to leverage more inherent properties of the network and incorporate additional related biological data, including drug response data and cancer genomic data.

In recent years, many public databases have been established to make PPI data and related biological information available for academic researchers. The Database of Interacting Proteins (DIP) [24] (Table 15.1) documents experimentally determined PPIs from more than 800 organisms. It evaluates the reliability of PPIs and provides a reliable PPI core subset for each species. The Biological General Repository for Interaction Datasets (BioGRID) [25] (Table 15.1) is another database that archives interaction data from major organism species. It includes 347,966 interactions (170,162 genetic, 177,804 protein) curated from both high-throughput datasets and individual studies [51–53]. The Human Protein Reference Database (HPRD)[26] (Table 15.1) specializes in curated proteomic information, covering over 15,000 human proteins. It documents PPIs derived from in vivo and in vitro experiments, as well as yeast two-hybrid assays. The Molecular Interaction Database (MINT) [30] (Table 15.1) is a public repository for experimentally verified PPIs from peer-reviewed journals. It has mined more than 6,000 scientific publications to collect over 130,000 interactions. Similarly, the IntAct database [28] (Table 15.1) also curates molecular interaction information from the scientific literature and from direct data depositions. It provides more than one million binary interactions, curated by global partners of the International Molecular Exchange consortium. The STRING database [29] (Table 15.1) collects, scores, and integrates public sources of PPIs, supplementing these with computational predictions. It includes physical interactions as well as functional ones. The latest version of STRING

Table 15.1 Currently available computational methods mentioned in this chapter.

Category	Name	Source
PPI network mapping	Diffuse2Direct [23]	https://github.com/danasilv/Diffuse2Direct
PPI databases	Database of interacting proteins (DIP) [24]	http://dip.doe-mbi.ucla.edu/dip/Main.cgi
	Biological general repository for interaction datasets (BioGRID) [25]	https://thebiogrid.org/
	Human protein reference database (HPRD) [26]	http://www.hprd.org
	Molecular interaction database (MINT) [27]	http://mint.bio.uniroma2.it
	IntAct database [28]	https://www.ebi.ac.uk/intact/
	STRING database [29]	https://string-db.org/
Network-based protein complex prediction	Markov cluster (MCL) [30]	https://micans.org/mcl
	Clustering-based on maximal cliques (CMC) [31]	https://www.comp.nus.edu.sg/~wongls/projects/complexprediction/CMC-26may09/
Sequence-based protein complex prediction	SAAMBE-SEQ [32]	http://compbio.clemson.edu/saambe_webserver/indexSEQ.php#started
	AttABseq [33]	https://github.com/ruofanjin/AttABseq/
Protein structure database	Protein data bank (PDB) [34]	https://www.rcsb.org
Protein docking	HawkDock [35]	http://cadd.zju.edu.cn/hawkdock/

Protein structure prediction	AlphaFold2 [36]	https://alphafold.ebi.ac.uk/
	RoseTTAFold [37]	https://github.com/RosettaCommons/RoseTTAFold
Motif-based PPI prediction	PPI-miner [38]	https://github.com/Wang-Lin-boop/PPI-Miner
Molecular dynamics simulation	CHARMM [39, 40]	https://academiccharmm.org/program
	GROMACS [41]	https://manual.gromacs.org/
	Amber [42]	https://ambermd.org/
	NAMD [43]	https://www.ks.uiuc.edu/Research/namd/
Molecular docking	Schrödinger Glide [44]	https://www.schrodinger.com/
	GOLD [45]	https://www.ccdc.cam.ac.uk/solutions/software/gold/
Molecular modelling	Rosetta molecular modeling suite and scripts for protein design [46]	https://www.rosettacommons.org/software/license-and-download
	Molecular operating environment (MOE) [47]	https://www.chemcomp.com/en/Products.htm
Binding energy estimation	FoldX [48]	https://foldxsuite.crg.eu/
	Rosetta FlexDDG [49]	https://github.com/Kortemme-Lab/flex_ddG_tutorial/blob/master/analyze_flex_ddG.py
Antibody design	Generative unconstrained intelligent drug engineering (GUIDE) [50]	https://github.com/LLNL/CRPCA https://github.com/LLNL/improvwf

documents more than 20 billion interactions for more than 59 million proteins from 12,535 organisms. Among these databases, MINT, IntAct, and STRING provide the scores of PPIs from various sources to indicate their reliability.

15.2.2 Protein Complex Prediction

Protein complexes are multimolecular machines composed of proteins and promising targets for therapeutic intervention [3, 54]. They can be predicted from PPI networks, protein sequences, and protein structures using computational methods. The network-based methods detect meaningful clusters of protein nodes from PPI networks as protein complexes. In contrast, the sequence-based methods extract sequence-based features to predict the protein–protein binding for a complex assembly. The structure-based methods use protein structures to model protein complexes. Notably, these three categories of methods are not completely exclusive. In fact, protein sequence information is frequently incorporated into network-based and structure-based methods [38, 55].

The network-based methods identify protein complexes by clustering PPI networks, which are constructed by mapping large-scale PPI data sets [56]. The Markov cluster (MCL) [30] method is one of the most widely used clustering methods for PPI networks. MCL clusters protein nodes based on simulation of stochastic flow in networks and rapidly assigns proteins. It scales to millions of nodes and hundreds of millions of connectivity edges, all while running on a single machine with multiple CPUs. Clustering based on maximal cliques (CMC)[31] is a computational method developed to discover complexes from weighted PPI networks. CMC uses an iterative method to weight PPI networks, generates all the maximal cliques from such networks, and then removes or merges overlapping clusters, to reduce the impact of random noise from the experimental data. In a systematic evaluation of 17 state-of-the-art clustering methods on two benchmark datasets of the DIP dataset [24] and the Collins dataset [57], the CMC demonstrated superior performance in identifying large protein complexes [17].

The sequence-based methods only need protein sequences to predict PPIs at the residual level. They are feasible for large-scale analysis of disease-causing mutations in protein complexes without requiring structural information [58]. For instance, the SAAMBE-SEQ [32] (Table 15.1) is a sequence-based machine learning method for predicting the binding affinity changes ($\Delta\Delta G$) upon a single mutation in protein complexes. Instead of structural information, it utilizes 80 features representing evolutionary information, sequence-based features, and physical properties changes upon mutation. The method achieved a high Pearson correlation coefficient of 0.83 in a benchmarking test against the experimentally determined binding free energy change ($\Delta\Delta G$). The sequence-based prediction methods can also be developed for a specific type of protein complexes, such as antigen–antibody protein complexes. The AttABseq [33] (Table 15.1) is an end-to-end sequence-based deep learning model utilizing antigen–antibody complex sequences for the predictions of the antigen–antibody binding affinity changes associated with antibody mutations. On three benchmark datasets of antigen–antibody protein complexes, AttABseq is significantly more accurate than other sequence-based models in terms of the Pearson correlation coefficient between the predicted and experimental binding affinity changes. Such sequence-based method offers a fast and convenient approach to antibody optimization without requiring structural data.

The structure-based prediction methods construct protein complex models to elucidate the molecular basis of PPI systems. When high-resolution experimentally determined three-dimensional structures of individual proteins are available, the protein docking methods can predict protein complexes in a precise way [59]. The principle of steric complementarity occurring at the protein–protein interface is central to the docking methodology. The protein–protein interfaces are mostly tightly packed, as observed in experimentally co-crystallized complexes in the Protein Data Bank (PDB) [34] (Table 15.1). The steric complementarity has been the major driving force in the development of docking approaches, often supplemented with physicochemical complementarity and statistics-based propensities [60–63]. Over time, protein docking has evolved significantly by incorporating more adequate scoring functions and various types of biological information [55]. For instance, HawkDock [35] (Table 15.1) is a web server that predicts protein complexes based on protein docking. In HawkDock, the protein complex structures are predicted by the ATTRACT [64] docking and the HawkRank [65] re-scoring, followed by the identification of key PPI residues through free energy decomposition. With the recent advancement of AlphaFold2 [36] (Table 15.1), it is possible to use a protein sequence to obtain a three-dimensional protein structure model with near experimental accuracy for many proteins. Intrigued by AlphaFold2, Baek et al. developed a three-track neural network algorithm RoseTTAFold [37] (Table 15.1) to accurately predict protein structures and further construct protein complex models. Remarkably, Humphreys et al. combined AlphaFold2 and RoseTTAFold to systematically identify and build accurate models of core eukaryotic protein complexes within the *Saccharomyces cerevisiae* proteome [66]. They screened 8.3 million pairs of yeast proteins, identified 1505 likely to interact, and built structural models for 106 previously unidentified complexes. AlphaFold-Multimer [67] was developed as an extension of the AlphaFold2 to predict protein complex, showing improved accuracy of predicted multimeric interfaces over AlphaFold2 while maintaining high intra-chain accuracy. In some cases, it is challenging to obtain full-length protein structures (or models) for protein complex prediction [68]. In addition to the full-length structures, the short motifs of proteins can also be utilized to predict PPIs, as most of protein–protein communications are mediated by functional motifs [69]. Wang et al. developed the PPI-miner [38] (Table 15.1), which can identify potential interacting protein pairs and predict their complex structures based on protein motifs. In general, reliable protein complex structures are often central to inform structure-based drug discovery of PPI systems, and these complex structure prediction methods significantly accelerate the discovery of PPI modulators.

15.2.3 PPI Modulator Discovery

As potential drug candidates targeting PPI systems, PPI modulators can inhibit (or stabilize) PPIs in several different ways. They may bind directly to the PPI

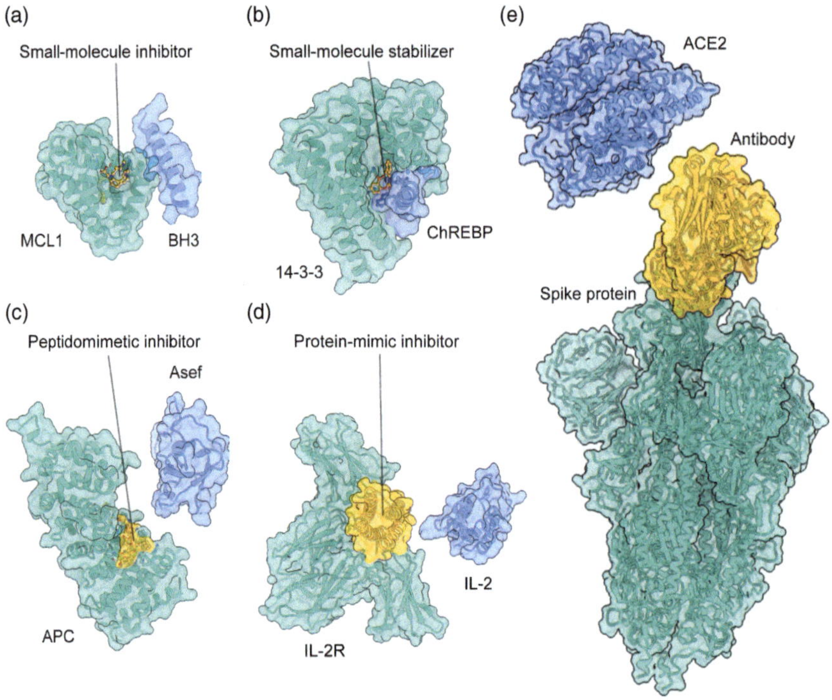

Figure 15.2 Different types of PPI modulators binding to their target proteins. (a) A small molecule inhibiting the interaction between MCL1 (green) and BH3 (blue) (PDB codes: 2NL9 and 4WMR). (b) A small molecule stabilizing the interaction between 14-3-3 protein (green) and ChREBP (blue) (PDB code: 6YGJ). (c) A peptide inhibiting the interaction between APC (green) and Asef (blue) (PDB codes: 3NMZ and 5B6G). (d) A protein mimetic inhibiting the interaction between IL-2R (green) and IL-2 (blue) (PDB codes: 2B5I and 6DG5). (e) An antibody inhibiting the interaction between spike protein (green) and ACE2 (blue) (PDB codes: 7WPA and 7WPD). Small molecules, peptides, and key ligand residues are shown as sticks. Proteins are shown in cartoon and surface depicts. Source: Xi Cheng.

interface to prevent (or facilitate) the formation of PPIs or change the structure of a PPI through allosteric modulation [3]. There are several types of PPI modulators, including small molecules, peptides, antibodies, and related biologics (Figure 15.2). Generally, small molecules can often permeate cells but typically cover only a modest surface area on a protein complex. Peptides are suited for binding to large surfaces, although their metabolic stability can pose challenges. Antibodies are highly potent and specific but do not readily cross cell membranes or the blood–brain barrier [70]. Due to the distinct characteristics of different modulators, certain discovery strategies may be more applicable to a particular modulator type than others [71]. The following section provides examples of successful discovery efforts to illustrate the computational methods that have proved useful for each modulator type.

15.2.3.1 Small Molecules

Due to the relatively flat and flexible surfaces of the protein-binding interfaces, discovering small-molecule PPI modulators presents a significant challenge [72, 73]. Considerable advancement has been made to develop computational methods to identify selective and druggable binding sites of small molecules on proteins. These small-molecule binding sites may lie in the hot spots of a protein. Hot spots are key residues that are involved in PPIs and are typically located at the interfaces [74]. In a study of discovering small-molecule severe acute respiratory syndrome coronavirus 2 (SARS-CoV-2) entry blockers, Wang et al. identified hot spots to develop inhibitors preventing interactions between the virus spike protein and the human receptor protein (angiotensin-converting enzyme 2, ACE2) [75]. They used a PPI prediction method, i.e., Fd-DCA [55], to detect hot spots on the receptor binding domain structure of the spike protein. Targeting these hot spots, five small-molecule natural products were identified and then validated by both native mass spectrometry and surface plasmon resonance. Remarkably, an optimal molecule, i.e., **H69C2** (Table 15.2), showed a strong binding affinity of 0.0947 μM and an anti-virus IC_{50} of 85.75 μM.

The chemical properties of hot spot residues give rise to their dynamical behaviors, which can be modeled by molecular dynamics simulations. The molecular dynamics simulation approaches, such as CHARMM [39, 40], GROMACS [41], Amber [42] and NAMD [43] (Table 15.1), have been widely used to provide an ensemble of probable protein conformations, revealing cryptic modulator-binding pockets. In a study focused on discovering small-molecule inhibitors for treating osteoporosis, Huang et al. hypothesized that protein thermal fluctuation can lead to discovery of the small-molecule binding sites to selectively inhibit the interaction of soluble RANKL and RANK [80]. They performed molecular dynamics simulations using Amber to mimic thermal fluctuation and then identified a selective small-molecule binding site on RANKL. This led to the discovery of a series of small-molecule inhibitors that showed efficacy in antiosteoporosis activity without causing immunosuppression side effects in vivo and in vitro.

Molecular docking is one of the major computational approaches in virtual screening of small-molecule PPI modulators [3]. It is a process through which small molecules are docked into the macromolecular structures to score their complementary values at the binding sites. There are many molecular docking methods feasible for screening small molecules, including Schrödinger Glide [44] and GOLD [45] (Table 15.1). To discover small-molecule modulators for the interaction of a phosphor-binding protein 14-3-3 with the carbohydrate-response element-binding protein (ChREBP), Sijbesma et al. performed molecular docking to virtually screen compounds [76]. The docking-based screening was performed using a crystal structure of the 14-3-3/ChREBP protein complex [81]. Two compounds (compounds **1** and **2**) (Table 15.2) were found to increase 14-3-3/ChREBP binding in a dose-dependent fashion up to 10-fold and 4-fold, with EC50 values of 0.7 and 45 μM, respectively. This study demonstrates the capability of molecular docking in the development of small-molecule PPI stabilizers.

Integration of molecular docking and molecular dynamics simulation approaches is sometimes necessary to guide small-molecule modulator design. In a study of an

Table 15.2 PPI modulators mentioned in this chapter.

Target PPI	Modulator	Structure	Therapeutic area
Spike protein (S protein) & Angiotensin-converting enzyme 2 (ACE2)	H69C2 [75]	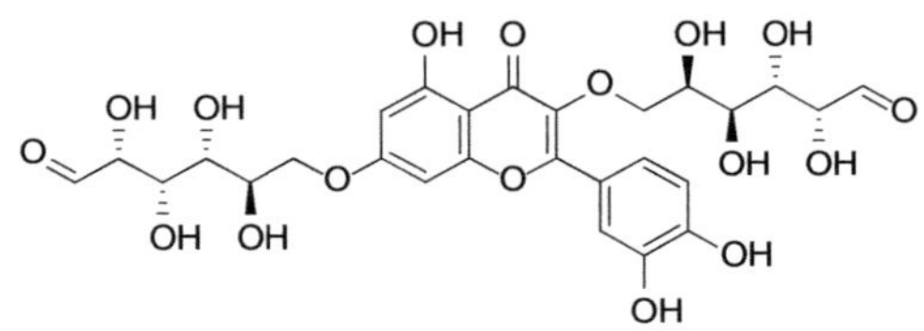	Antivirus
Phosphor-binding protein (14-3-3) and Carbohydrate-response element-binding protein (ChREBP)	Compound **1** [76]	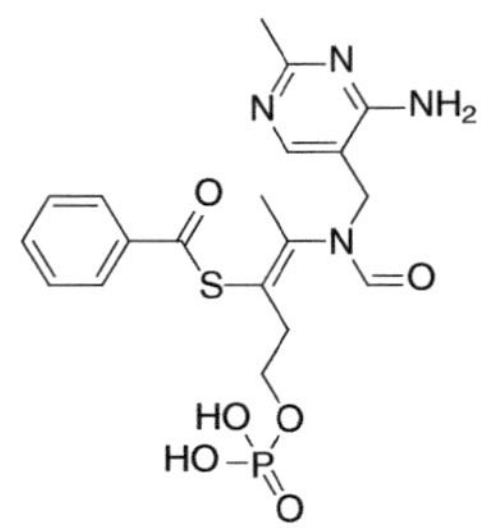	Alzheimer's disease
	Compound **2** [76]	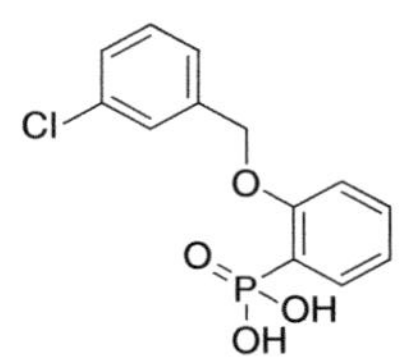	Alzheimer's disease

Anti-apoptotic protein Mcl-1 and Pro-apoptotic proteins	FWJ-D5 [77]	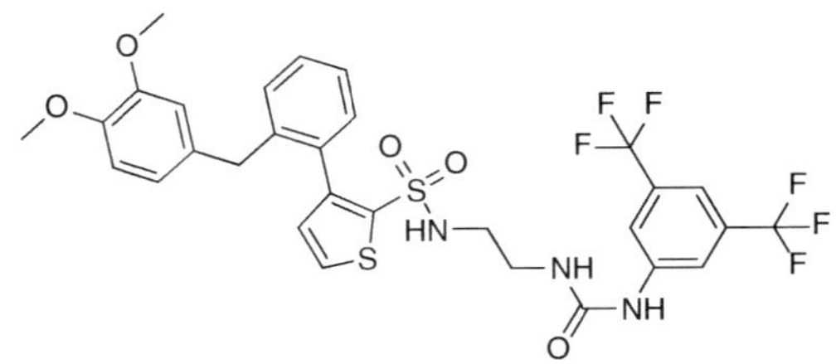	Anticancer
Class III phosphatidylinositol-3-kinase (PI3KC3) complex	i7–01s-31 [78]	Ac-[Tat]-VLFN(R8)LVDVIK(S5)RKV-NH2	Anticancer
Interleukin-2 (IL2) and IL-2 receptor βγc heterodimer	Neoleukin-2/15 [79]	> Neoleukin-2/15 GSHMPKKKIQLHAEHALYDALMILNIVKTNSPP AEEKLEDYAFNFELILEEIARLFESGDQKDEAEKA KRMKEWMKRIKTTASEDEQEEMANAIITILQSWIFS	Anticancer
Spike protein (S protein) and Angiotensin-converting enzyme 2 (ACE2)	2130–1–0114-112 [50]	>Fab LLNL-199 HC EVQLVESGGGLVKPGGSLRLSCAASGFTFRDVWM SWVRQAPGKGLEWVGRIKSKIDGGTTDYAAPVKG RFTISRDDSKNTLYLQMNSLKTEDTAVYYCTTAGS YYYDTVGPELPEGKFDYWGQGTLVTVSS >Fab LLNL-199 LC DIVMTQSPDSLAVSLGERATINCKSSQSVLYAAN NKNYLAWYQQKPGQPPKLLMYWASERESGVP DRFSGSGSGAEFTLTISSLQAEDVAIYYCQQY YSTLTFGGGTKVEIKR	Antivirus

anti-apoptotic protein Mcl-1, Li et al. optimized previously discovered PPI inhibitors using structural analysis and molecular modeling [77]. The binding modes of lead compounds to the BH3-binding groove on Mcl-1 were predicted by molecular docking and molecular dynamics simulation, which then guided structure-based optimization. A total of 85 new compounds were synthesized and tested in binding assays and cytotoxicity assays. As compared to the lead compound, new compound **FWJ-D5** (Table 15.2) achieved higher affinity and excellent selectivity toward Mcl-1 in the binding assay. Treatment with **FWJ-D5** at 50 mg/kg every two days on an RS4; 11 xenograft mouse model for 22 days led to 75% reduction in tumor volume without body weight loss.

15.2.3.2 Peptides and Protein Mimics

In cases where PPI modulation requires full displacement of proteins, small molecules may not be suitable. Instead, peptides or protein mimics offer better solutions [82, 83]. Rational design of peptide-based modulators generally requires reliable information of the PPI, ideally from experimental structures [3]. Such modulators typically derive from one of the protein partners and often mimic typical secondary structures, such as an α-helix, β-sheet or β-turn. Optimized modulators with higher affinity can be obtained after a few rounds of single-point or multiple-point mutation and even incorporation of unnatural amino acids [84–86]. During such a process, molecular modeling tools, such as Rosetta [46] and molecular operating environment (MOE) [47] (Table 15.1), can be employed to construct the possible complexes of designed modulators and the target proteins. Molecular dynamics simulations can be performed to characterize the interaction patterns between the modulators and target proteins. Based on the simulation-resulting conformation ensembles, the MM/PB(GB)SA [87] technique is feasible for calculating the per-residue decomposition energy, indicating the contribution of residues on the binding interface. Such information can be used to guide selection or optimization of modulators. For example, in the development of the Beclin 1-targeting staple peptides, Yang et al. presented the peptide optimization by staple scanning and sequence permutation [78]. An α-helical segment of Beclin 1 was truncated as the template for designing stapled peptides. Molecular models of all designed stapled peptides were constructed using MOE, followed by molecular dynamics simulation and binding energy prediction. Based on the predicted binding energies, they successfully designed optimized peptides with significantly improved Beclin 1-binding affinities. These peptides (Table 15.2) showed potent antiproliferative efficacy in EGFR- and HER2-overexpressing cancer cells.

When the experimental structure of the partner protein (other than the target one) is not available, a de novo protein design method can be used to discover peptide-based (or protein-mimic) modulators for PPI systems. To design high-affinity modulators binding to protein targets, Cao et al. developed a de novo protein design approach of Rosetta [46] that starts with a broad exploration of the vast space of possible binding modes to a selected region of a protein surface and then intensifies the search in the vicinity of the most promising binding modes. Based on this approach, Silva et al. designed mimics (neoleukin-2/15) (Table 15.2) of the central immune cytokine interleukin-2 (IL-2) that are hyper-stable and bind human and mouse IL-2 receptor βγc heterodimer with higher affinity than the natural cytokines [79]. The mimics have superior therapeutic activity to IL-2 in mouse models of melanoma and colon cancer, with reduced toxicity and undetectable immunogenicity.

15.2.3.3 Antibodies

Antibodies serve important roles by counteracting antigens, which are recognized as foreign by the immune system. The antigen-binding fragment of an antibody contains a paratope that recognizes a unique antigen via the antigen's epitope [88]. Because of their high specificity, antibodies are effective PPI modulators targeting large proteins [89]. Computational approaches to developing and designing these molecules are increasingly used to complement traditional lab-based processes [90].

Nowadays, in silico methods fill multiple elements of the antibody discovery stage, such as construction of complex structures, evaluation of antibody binding affinity, estimation of antibody stability, and more. Molecular modeling and molecular dynamics simulation can be used to construct and sample a large number of structural conformations for protein complexes. The structure-based or ensemble-based algorithms, such as Rossetta Flex [49] (Table 15.1) and FoldX [48] (Table 15.1), are feasible to predict antigen binding affinity. The antigen thermal stability can be estimated using the free energy perturbation method [91]. In a recent study on SARS-CoV-2 antibody design, Desautels et al. sought to computationally optimize a clinical antibody (COV2–2130) against Omicron BA.1 and BA.1.1 strains while maintaining efficacy against the dominant Delta variant [50]. They developed a computational approach, i.e. generative unconstrained intelligent drug engineering (GUIDE) (Table 15.1), combining molecular modeling, molecular dynamics simulation and machine learning for optimizing binding affinity to multiple antigen targets, along with other critical attributes such as thermostability. This computational approach operates in a "zero-shot" setting, in which designs are generated without iterative input from wet laboratory experiments on proposed antibody candidates. This zero-shot approach successfully generated an optimized antibody 2130-1-0114-112 (Table 15.2), which increases neutralization potency against Delta and subsequent variants of concern and provides protection in vivo against the strains tested: WA1/2020, BA.1.1 and BA.5.

15.3 Conclusions and Outlook

PPI systems present both challenges and exciting new opportunities in drug discovery. Numerous disease-associated proteins are currently being mapped within PPI systems with the assistance of computational approaches, which hold promise for future drug discovery. For these computational models, the quality of training data is a critical issue determining prediction accuracy of PPI prediction. However, given the high false-positive and false-negative rates observed in PPIs produced by high-throughput technologies, it is necessary to preprocess the PPI data and appropriately evaluate the performances of prediction models. Meanwhile, in addition to the PPI data, other sources of biological information of proteins are considered to compensate for the negative influences resulting from the problematic reliability of PPI data. Hence, how to effectively integrate multiple sources of biological information for PPI prediction is still one of the major challenges for future work.

Biological characterization and pharmacological targeting of specific PPIs are crucial for the identification of more efficient, more selective, and safer drugs. Advancements in computational resources and algorithms, coupled with an atomic-level understanding of protein dynamics, have made in silico approaches successful in PPI drug discovery. Recent advances in protein structure prediction have inspired further computational study of protein complex prediction. Entirely in silico design of peptides, protein mimetics or antibodies is within reach, and when fully achieved, it will usher in a new era in therapy development. Currently, most efforts to develop PPI modulators by preventing or changing the interaction between two proteins. An intriguing future direction is instead to stabilize PPIs [92]. Efforts toward augmenting PPIs with small molecules are emerging [93], and similar strategies are now being explored with antibodies and nanobodies [94]. Another promising avenue for PPI could be ligand development for targeted protein degradation [95, 96]. Degradation of protein partners could offer an alternative and subtler way to interfere with protein complexes. Overall, if the drug discovery community is able to step up to the challenge posed by PPIs, there are tremendous opportunities to develop new approaches to the treatment of disease in this diverse and fascinating target class.

References

1 Vidal, M., Cusick, M.E., and Barabási, A.L. (2011). Interactome networks and human disease. *Cell* 144 (6): 986–998.

2 Rolland, T., Tasan, M., Charloteaux, B. et al. (2014). A proteome-scale map of the human interactome network. *Cell* 159 (5): 1212–1226.

3 Scott, D.E., Bayly, A.R., Abell, C., and Skidmore, J. (2016). Small molecules, big targets: drug discovery faces the protein–protein interaction challenge. *Nature Reviews. Drug Discovery* 15 (8): 533–550.

4 Menche, J., Sharma, A., Kitsak, M. et al. (2015). Uncovering disease-disease relationships through the incomplete interactome. *Science* 347 (6224): 1257601.

5 Stumpf, M.P.H., Thorne, T., de Silva, E. et al. (2008). Estimating the size of the human interactome. *Proceeding of the National Academy of Sciences of the United States of America* 105 (19): 6959–6964.

6 Venkatesan, K., Rual, J.F., Vazquez, A. et al. (2009). An empirical framework for binary interactome mapping. *Nature Methods* 6 (1): 83–90.

7 Luck, K., Kim, D.K., Lambourne, L. et al. (2020). A reference map of the human binary protein interactome. *Nature* 580 (7803): 402–408.

8 Keskin, O., Tuncbag, N., and Gursoy, A. (2016). Predicting protein—protein interactions from the molecular to the proteome level. *Chemical Reviews* 116 (8): 4884–4909.

9 Xiang, H.G., Zhou, M., Li, Y. et al. (2023). Drug discovery by targeting the protein-protein interactions involved in autophagy. *Acta Pharmaceutica Sinica B* 13 (11): 4373–4390.
10 Albert, I. and Albert, R. (2004). Conserved network motifs allow protein-protein interaction prediction. *Bioinformatics* 20 (18): 3346–3352.
11 Wang, X.W., Chen, Y.Z., and Liu, Y.Y. (2020). Link prediction through deep generative model. *Iscience* 23 (10): 101626.
12 Tang, T., Zhang, X.C., Liu, Y.S. et al. (2023). Machine learning on protein–protein interaction prediction: models, challenges and trends. *Briefings in Bioinformatics* 24 (2): bbad07.
13 Hu, X.T., Feng, C., Ling, T.Y., and Chen, M. (2022). Deep learning frameworks for protein-protein interaction prediction. *Computational and Structural Biotecnology Journal* 20: 3223–3233.
14 Soleymani, F., Paquet, E., Viktor, H. et al. (2022). Protein–protein interaction prediction with deep learning: a comprehensive review. *Computational and Structural Biotecnology* 20: 5316–5341.
15 Vazquez, A., Flammini, A., Maritan, A., and Vespignani, A. (2003). Global protein function prediction from protein-protein interaction networks. *Nature Biotechnology* 21 (6): 697–700.
16 Stelzl, U., Worm, U., Lalowski, M. et al. (2005). A human protein-protein interaction network: a resource for annotating the proteome. *Cell* 122 (6): 957–968.
17 Wu, Z.R., Liao, Q., and Liu, B. (2020). A comprehensive review and evaluation of computational methods for identifying protein complexes from protein–protein interaction networks. *Briefings in Bioinformatics* 21 (5): 1531–1548.
18 Hu, L., Wang, X.J., Huang, Y.A. et al. (2021). A survey on computational models for predicting protein–protein interactions. *Briefings in Bioinformatics* 22 (5): bbab036.
19 Wang, X.W., Madeddu, L., Spirohn, K. et al. (2023). Assessment of community efforts to advance network-based prediction of protein–protein interactions. *Nature Communications* 14 (1): 1588.
20 Kovács, I.A., Luck, K., Spirohn, K. et al. (2019). Network-based prediction of protein interactions. *Nature Communications* 10: 1240.
21 Cao, M.F., Pietras, C.M., Feng, X. et al. (2014). New directions for diffusion-based network prediction of protein function: incorporating pathways with confidence. *Bioinformatics* 30 (12): 219–227.
22 Vinayagam, A., Stelzl, U., Foulle, R. et al. (2011). A directed protein interaction network for investigating intracellular signal transduction. *Science Signaling* 4 (189): rs8.
23 Silverbush, D. and Sharan, R. (2019). A systematic approach to orient the human protein–protein interaction network. *Nature Communications* 10, 3015.
24 Xenarios, I., Salwínski, L., Duan, X.Q.J. et al. (2002). DIP, the database of interacting proteins: a research tool for studying cellular networks of protein interactions. *Nucleic Acids Research* 30 (1): 303–305.
25 Stark, C., Breitkreutz, B.J., Chatr-aryamontri, A. et al. (2011). The BioGRID interaction database: 2011 update. *Nucleic Acids Research* 39: D698–D704.

26 Prasad, T.S.K., Goel, R., Kandasamy, K. et al. (2009). Human protein reference d0atabase-2009 update. *Nucleic Acids Research* 37: D767–D772.

27 Licata, L., Briganti, L., Peluso, D. et al. (2012). MINT, the molecular interaction database: 2012 update. *Nucleic Acids Research* 40 (D1): D857–D861.

28 del Toro, N., Shrivastava, A., Ragueneau, E. et al. (2022). The IntAct database: efficient access to fine-grained molecular interaction data. *Nucleic Acids Research* 50 (D1): D648–D653.

29 Szklarczyk, D., Gable, A.L., Lyon, D. et al. (2019). STRING v11: protein-protein association networks with increased coverage, supporting functional discovery in genome-wide experimental datasets. *Nucleic Acids Research* 47 (D1): D607–D613.

30 Enright, A.J., Van Dongen, S., and Ouzounis, C.A. (2002). An efficient algorithm for large-scale detection of protein families. *Nucleic Acids Research* 30 (7): 1575–1584.

31 Liu, G.M., Wong, L., and Chua, H.N. (2009). Complex discovery from weighted PPI networks. *Bioinformatics* 25 (15): 1891–1897.

32 Li, G., Pahari, S., Murthy, A.K. et al. (2021). SAAMBE-SEQ: a sequence-based method for predicting mutation effect on protein-protein binding affinity. *Bioinformatics* 37 (7): 992–999.

33 Jin, R., Ye, Q., Wang, J. et al. (2024). AttABseq: an attention-based deep learning prediction method for antigen-antibody binding affinity changes based on protein sequences. *Briefings in Bioinformatics* (4): 25, bbae304.

34 Berman, H.M., Westbrook, J., Feng, Z. et al. (2000). The protein data bank. *Nucleic Acids Research* 28 (1): 235–242.

35 Weng, G.Q., Wang, E.C., Wang, Z. et al. (2019). HawkDock: a web server to predict and analyze the protein-protein complex based on computational docking and MM/GBSA. *Nucleic Acids Research* 47 (W1): W322–W330.

36 Jumper, J., Evans, R., Pritzel, A. et al. (2021). Highly accurate protein structure prediction with AlphaFold. *Nature* 596 (7873): 583.

37 Baek, M., DiMaio, F., Anishchenko, I. et al. (2021). Accurate prediction of protein structures and interactions using a three-track neural network. *Science* 373 (6557): 871–876.

38 Wang, L., Li, F.L., Ma, X.Y. et al. (2022). PPI-Miner: a structure and sequence motif co-driven protein–protein interaction mining and modeling computational method. *Journal of Chemical Information and Modeling* 62 (23): 6160–6171.

39 Brooks, B.R., Brooks, C., Mackerell, A. et al. (2009). CHARMM: The biomolecular simulation program. *Journal of Computational Chemistry* 30 (10): 1545–1614.

40 Wu, E.L., Cheng, X., Jo, S. et al. (2014). CHARMM-GUI membrane builder toward realistic biological membrane simulations. *Journal of Computational Chemistry* 35 (27): 1997–2004.

41 Páll, S., Zhmurov, A., Bauer, P. et al. (2020). Heterogeneous parallelization and acceleration of molecular dynamics simulations in GROMACS. *The Journal of Chemical Physics* 153 (13): 134110.

42 Case, D.A., Aktulga, H.M., Belfon, K. et al. (2023). The ambertools. *Journal of Chemical Information and Modeling* 63 (20): 6183–6191.

43 Phillips, J.C., Hardy, D.J., Maia, J.D.C. et al. (2020). Scalable molecular dynamics on CPU and GPU architectures with NAMD. *The Journal of Chemical Physics* 153 (4): 044130.

44 Halgren, T.A., Murphy, R.B., Friesner, R.A. et al. (2004). Glide: a new approach for rapid, accurate docking and scoring. 2. Enrichment factors in database screening. *Journal of Medicinal Chemistry* 47 (7): 1750–1759.

45 Jones, G., Willett, P., and Glen, R.C. (1995). Molecular recognition of receptor-sites using a genetic algorithm with a description of desolvation. *Journal of Molecular Biology* 245 (1): 43–53.

46 Cao, L.X., Coventry, B., Goreshnik, I. et al. (2022). Design of protein-binding proteins from the target structure alone. *Nature* 605 (7910): 551–560.

47 Boyd, S. (2005). Molecular operating environment. *Chemistry World-UK* 2 (9): 66–66.

48 Schymkowitz, J., Borg, J., Stricher, F. et al. (2005). The FoldX web server: an online force field. *Nucleic Acids Research* 33: W382–W388.

49 Barlow, K.A., Conchúir, S.O., Thompson, S. et al. (2018). Flex ddG: rosetta ensemble-based estimation of changes in protein-protein binding affinity upon mutation. *The Journal of Physical Chemistry. B* 122 (21): 5389–5399.

50 Desautels, T.A., Arrildt, K.T., Zemla, A.T. et al. (2024). Computationally restoring the potency of a clinical antibody against omicron. *Nature* 629 (8013): 878–885.

51 Gavin, A.C., Bösche, M., Krause, R. et al. (2002). Functional organization of the yeast proteome by systematic analysis of protein complexes. *Nature* 415 (6868): 141–147.

52 Uetz, P., Giot, L., Cagney, G. et al. (2000). A comprehensive analysis of protein—protein interactions in. *Nature* 403 (6770): 623–627.

53 Ho, Y., Gruhler, A., Heilbut, A. et al. (2002). Systematic identification of protein complexes in by mass spectrometry. *Nature* 415 (6868): 180–183.

54 Spirin, V. and Mirny, L.A. (2003). Protein complexes and functional modules in molecular networks. *Proceeding of the National Academy of Sciences of the United States of America* 100 (21): 12123–12128.

55 Bai, F., Morcos, F., Cheng, R.R. et al. (2016). Elucidating the druggable interface of protein–protein interactions using fragment docking and coevolutionary analysis. *Proceeding of the National Academy of Sciences of the United States of America* 113 (50): E8051–E8058.

56 Wang, J.X., Peng, X.Q., Li, M., and Pan, Y. (2013). Construction and application of dynamic protein interaction network based on time course gene expression data. *Proteomics* 13 (2): 301–312.

57 Collins, S.R., Kemmeren, P., Zhao, X.C. et al. (2007). Toward a comprehensive atlas of the physical interactome of of Saccharomyces cerevisiae. *Molecular and Cellular Proteomics* 6 (3): 439–450.

58 Jemimah, S., Sekijima, M., and Gromiha, M.M. (2020). ProAffiMuSeq: sequence-based method to predict the binding free energy change of protein-protein complexes upon mutation using functional classification. *Bioinformatics* 36 (6): 1725–1730.

59 Vakser, I.A. (2014). Protein-protein docking: from interaction to interactome. *Biophysical Journal* 107 (8): 1785–1793.

60 Vakser, I.A. and Aflalo, C. (1994). Hydrophobic docking - a proposed enhancement to molecular recognition echniques. *Proteins* 20 (4): 320–329.

61 Mandell, J.G., Roberts, V.A., Pique, M.E. et al. (2001). Protein docking using continuum electrostatics and geometric fit. *Protein Engineering* 14 (2): 105–113.

62 Kozakov, D., Brenke, R., Comeau, S.R., and Vajda, S. (2006). PIPER: An FFT-based protein docking program with pairwise potentials. *Proteins: Structure, Function, and Bioinformatics* 65 (2): 392–406.

63 Mintseris, J., Pierce, B., Wiehe, K. et al. (2007). Integrating statistical pair potentials into protein complex prediction. *Proteins: Structure, Function, and Bioinformatics* 69 (3): 511–520.

64 Zacharias, M. (2003). Protein-protein docking with a reduced protein model accounting for side-chain flexibility. *Protein Science* 12 (6): 1271–1282.

65 Feng, T., Chen, F., Kang, Y. et al. (2017). HawkRank: a new scoring function for protein-protein docking based on weighted energy terms. *Journal of Cheminformatics* 9: 66.

66 Humphreys, I.R., Pei, J.M., Baek, M. et al. (2021). Computed structures of core eukaryotic protein complexes. *Science* 374 (6573): eabm4805.

67 Evans, R.; O'Neill, M.; Pritzel, A.; Antropova, N.; Senior, A.; Green, T.; Žídek, A.; Bates, R.; Blackwell, S.; Yim, J.; Ronneberger, O.; Bodenstein, S.; Zielinski, M.; Bridgland, A.; Potapenko, A.; Cowie, A.; Tunyasuvunakool, K.; Jain, R.; Clancy, E.; Kohli, P.; Jumper, J. M.; Hassabis, D., Protein complex prediction with AlphaFold-Multimer. *bioRxiv* **2021**.

68 Perrakis, A. and Sixma, T.K. (2021). AI revolutions in biology: the joys and perils of AlphaFold. *EMBO Reports* 22 (11): e54046.

69 Van Roey, K., Uyar, B., Weatheritt, R.J. et al. (2014). Short linear motifs: ubiquitous and functionally diverse protein interaction modules directing cell regulation. *Chemical Reviews* 114 (13): 6733–6778.

70 Jhoti, H., Williams, G., Rees, D.C., and Murray, C.W. (2013). The 'rule of three' for fragment-based drug discovery: where are we now? *Nature Reviews. Drug Discovery* 12 (8): 644.

71 Van Molle, I., Thomann, A., Buckley, D.L. et al. (2012). Dissecting fragment-based lead discovery at the von Hippel-Lindau protein:hypoxia inducible factor 1α protein-protein interface. *Chemistry & Biology* 19 (10): 1300–1312.

72 Rehman, A.U., Khurshid, B., Ali, Y. et al. (2023). Computational approaches for the design of modulators targeting protein-protein interactions. *Expert Opinion on Drug Discovery* 18 (3): 315–333.

73 Eyrisch, S. and Helms, V. (2009). What induces pocket openings on protein surface patches involved in protein–protein interactions? *Journal of Computer-Aided Molecular Design* 23 (2): 73–86.

74 Clackson, T. and Wells, J.A. (1995). A hot-spot of binding-energy in a hormone-receptor interface. *Science* 267 (5196): 383–386.

75 Wang, L., Wu, Y., Yao, S. et al. (2022). Discovery of potential small molecular SARS-CoV-2 entry blockers targeting the spike protein. *Acta Pharmacologica Sinica* 43 (4): 788–796.

76 Sijbesma, E., Visser, E., Plitzko, K. et al. (2020). Structure-based evolution of a promiscuous inhibitor to a selective stabilizer of protein–protein interactions. *Nature Communications* 11 (1): 3954.

77 Li, Y., Fan, W.J., Gong, Q.N. et al. (2021). Structure-based optimization of 3-phenyl-(2-(3-phenylureido)ethyl)thiophene-2-sulfonamide derivatives as selective Mcl-1 inhibitors. *Journal of Medicinal Chemistry* 64 (14): 10260–10285.

78 Deng, G.T., Li, C.R., Chen, L. et al. (2022). BECN2 (beclin 2) negatively regulates inflammasome sensors through ATG9A-dependent but ATG16L1-and LC3-independent non-canonical autophagy. *Autophagy* 18 (2): 340–356.

79 Silva, D.A., Yu, S., Ulge, U.Y. et al. (2019). De novo design of potent and selective mimics of IL-2 and IL-15. *Nature* 565 (7738): 186–191.

80 Huang, D., Zhao, C., Li, R.Y. et al. (2022). Identification of a binding site on soluble RANKL that can be targeted to inhibit soluble RANK-RANKL interactions and treat osteoporosis. *Nature Communications* 13 (1).

81 Sato, S., Jung, H.M., Nakagawa, T. et al. (2016). Metabolite regulation of nuclear localization of carbohydrate-response element-binding protein (ChREBP). *The Journal of Biological Chemistry* 291 (20): 10515–10527.

82 Aarts, M., Liu, Y.T., Liu, L.D. et al. (2002). Treatment of ischemic brain damage by perturbing NMDA receptor-PSD-95 protein interactions. *Science* 298 (5594): 846–850.

83 Bach, A., Clausen, B.H., Moller, M. et al. (2012). A high-affinity, dimeric inhibitor of PSD-95 bivalently interacts with PDZ1-2 and protects against ischemic brain damage. *Proceeding of the National Academy of Sciences of the United States of America* 109 (9): 3317–3322.

84 Rogers, J.M., Passioura, T., and Suga, H. (2018). Nonproteinogenic deep mutational scanning of linear and cyclic peptides. *Proceeding of the National Academy of Sciences of the United States of America* 115 (43): 10959–10964.

85 Chu, Y.Y., Zhang, Y., Wang, Q.K. et al. (2022). A transformer-based model to predict peptide-HLA class I binding and optimize mutated peptides for vaccine design. *Nature Machine Intelligence* 4 (3): 300–311.

86 Liu, T., Ma, X.J., Yu, J.H. et al. (2021). Rational generation of lasso peptides based on biosynthetic gene mutations and site-selective chemical modifications. *Chemical Science* 12 (37): 12353–12364.

87 Gohlke, H. and Case, D.A. (2004). Converging free energy estimates: MM-PB(GB)SA studies on the protein-protein complex Ras-Raf. *Journal of Computational Chemistry* 25 (2): 238–250.

88 Chothia, C., Lesk, A.M., Tramontano, A. et al. (1989). Conformations of immunoglobulin hypervariable regions. *Nature* 342 (6252): 877–883.

89 Wang, M.L., Cang, Z.X., and Wei, G.W. (2020). A topology-based network tree for the prediction of protein-protein binding affinity changes following mutation. *Nature Machine Intelligence* 2 (2): 116–123.

90 Wilman, W., Wróbel, S., Bielska, W. et al. (2022). Machine-designed biotherapeutics: opportunities, feasibility and advantages of deep learning in computational antibody discovery. *Briefings in Bioinformatics* 23 (4): bbac267.

91 Zhu, F.Q., Bourguet, F.A., Bennett, W.F.D. et al. (2022). Large-scale application of free energy perturbation calculations for antibody design. *Science Report-UK* 12 (1): 12489.

92 Thiel, P., Kaiser, M., and Ottmann, C. (2012). Small-molecule stabilization of protein–protein interactions: an underestimated concept in drug discovery? *Angewandte Chemie. (International Ed.)* 51 (9): 2012–2018.

93 Arkin, M.R., Tang, Y.Y., and Wells, J.A. (2014). Small-molecule inhibitors of protein–protein interactions: progressing toward the reality. *Chemistry and Biology* 21 (9): 1102–1114.

94 Manglik, A., Kobilka, B.K., and Steyaert, J. (2017). Nanobodies to study G protein-coupled receptor structure and function. *Annual Review of Pharmacology* 57: 19–37.

95 Burslem, G.M., Smith, B.E., Lai, A.C. et al. (2018). The advantages of targeted protein degradation over inhibition: an RTK case study. *Cell Chemical Biology* 25 (1): 67–77.

96 Bensimon, A., Pizzagalli, M.D., Kartnig, F. et al. (2020). Targeted degradation of SLC transporters reveals amenability of multi-pass transmembrane proteins to ligand-induced proteolysis. *Cell Chemical Biology* 27 (6): 728–739.

Index

Targeting Protein–Protein Interactions for Drug Discovery, First Edition. Edited by Jian Zhang.

d

h

i

q